Ticket to Write

Writing Skills for Success

Annotated Instructor's Edition

SUSAN SOMMERS THURMAN
Henderson Community College

WILLIAM L. GARY, JR.
Henderson Community College

PEARSON

Boston Columbus Indianapolis New York San Francisco Upper Saddle River
Amsterdam Cape Town Dubai London Madrid Milan Munich Paris Montreal Toronto
Delhi Mexico City São Paulo Sydney Hong Kong Seoul Singapore Taipei Tokyo

Executive Editor: Matthew Wright
Senior Development Editor: Gill Cook
Assistant Editor: Amanda Dykstra
Editorial Assistant: Kristen Pechtol
Marketing Manager: Kurt Massey
Senior Media Producer: Stefanie A. Snajder
Digital Project Manager: Janell Lantana
Senior Digital Media Editor: Robert A. St. Laurent
Senior Supplements Editor: Donna Campion
Production Manager: Savoula Amanatidis
Project Coordination, Text Design, and Electronic Page Makeup: Cenveo Publisher Services
Cover Design Manager: John Callahan
Photo Researcher: Bill Smith Group
Senior Manufacturing Buyer: Dennis J. Para
Printer and Binder: Courier Corporation–Kendallville
Cover Printer: Lehigh-Phoenix Color Corporation–Hagerstown

For permission to use copyrighted material, grateful acknowledgment is made to the copyright holders on pp. C.1–C.2, which are hereby made part of this copyright page.

Library of Congress Cataloging-in-Publication Data is on file at the Library of Congress

10 9 8 7 6 5 4 3 2 1—CRK—15 14 13 12

www.pearsonhighered.com

Student Edition
ISBN-10: 0-205-82275-4; ISBN-13: 978-0-205-82275-1

Annotated Instructor's Edition
ISBN-10: 0-205-05993-7; ISBN-13: 978-0-205-05993-5

With love and gratitude,
we dedicate this book to our spouses,
Mike Thurman and Randa Gary

Brief Contents

Detailed Contents

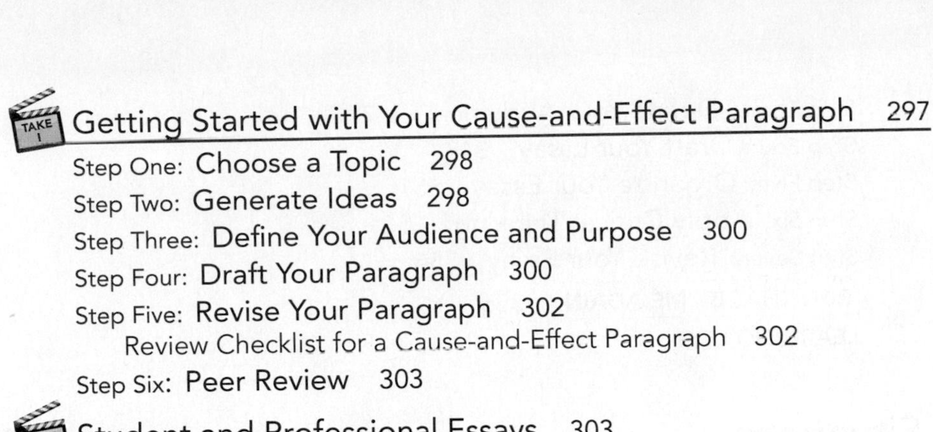

Preface to the Student

Welcome to college and welcome to *Ticket to Write*. Even though you are just beginning your collegiate life, you know that writing will be a big part of this class and most other classes you will take.

You may feel overwhelmed by this idea, but the purpose of *Ticket to Write* is to guide you as you advance from a novice writer to one well prepared for higher-level college classes. The activities and exercises in this text will help you increase and improve your writing skills for your college classes and beyond.

In Part 1 of the text, you learn and practice the five steps in the writing process: *prewriting, discovery drafting, revising, editing and proofreading*, and *publishing*. Part 2 helps in composing nine different types of academic paragraphs and essays. These include *descriptive, narrative, illustration, process, definition, compare-and-contrast, classification, cause-and-effect*, and *persuasive*. Part 3 introduces you to other types of writing you may use in college classes, like resource-based writing, in-class writing, personal and business writing, online writing, and writing in newspapers and journals. Part 4 covers grammar and mechanics, giving you answers to questions you may have as you fine-tune your papers. Part 5 opens with a chapter on reading strategies and provides additional essays that reinforce and enhance the material you study in other parts of the book. Part 6, Study Skills, is provided online through the MyWritingLab/eText version of *Ticket to Write*. As college freshmen, the various study skills it addresses will be particularly helpful to you not only in this class but in your other classes.

We look forward to hearing about your successes with the skills you develop using *Ticket to Write*. If you would like to share any paragraphs or essays you write while using this text, please e-mail them to us at tickettowrite@gmail.com (put your paragraph or essay in the body of the e-mail, not as an attachment). You may also enter Pearson's Writing Rewards Student Essay Contest. Read about this contest, download entry forms and rules, and review past winning essays at http://www.pearsonhighered.com/writingrewards/.

William Gary Susan Shurman

Preface to the Instructor

Ticket to Write is designed to aid novice composition students as they progress from paragraph writing to essay writing. This all-inclusive text features a thorough treatment of the writing process, nine types of academic writing (at both paragraph and essay level), non-academic writing, grammar and mechanics, and study skills, which are provided in the MyWritingLab/eText version of the book. *Ticket to Write* focuses on delivering the skills students need to enter freshmen-level composition classes as prepared, ready-to-write college students. The text presents material in ways that allow writers at various levels the opportunity to apply what they learn, based on their individual writing needs.

A major difference between *Ticket to Write* and other college writing texts is that this text deals directly with the varying levels of writing skills students bring to the classroom. Novice writers may need instruction in composing paragraphs before they move on to an essay. Others who have had more writing practice may need additional instruction on essay development. *Ticket to Write* includes instruction in both paragraph and essay writing in every chapter in Part 1, "The Writing Process," and Part 2, "Types of Paragraphs and Essays." Each chapter uses similar formats in paragraph development and essay writing, with techniques that guide students to a finished, polished piece of writing. This parallel approach allows the instructor to adapt assignments to fit the needs of the class or an individual student.

Ticket to Write reaches all freshmen, including novice and ELL writers, via issues in pop culture and contemporary life, and it also addresses students' increased use of technology. Throughout the text, students are directed to numerous video tutorials, lessons, practice material, and other assessment through online videos, Pearson's MyWritingLab, and additional online sites devoted to writing, grammar, or learning skills.

Purpose

Ticket to Write is intended for students who are not yet ready for the freshman-level composition courses offered by their colleges. The purpose of the text is to teach those students how to build on their fundamental writing skills so they have the proficiency to move into more advanced composition classes.

We have long been involved with freshman students, many of whom feel apprehensive about attending college because of their age, their developmental placement, their ELL concerns, or myriad other reasons. Unfortunately, this apprehension often translates into anxiety about their abilities to succeed. With this in mind, we decided to gather and share material that has been successful in our classrooms. The result is *Ticket to Write*, which is designed to help students

- advance from paragraph to essay writing
- move from personal essays to source-based essays

- move into more analytical and detailed methods of writing
- foster their critical thinking skills
- connect with contemporary topics related to their experience
- investigate topics of a collegiate nature
- engage more deeply in various writing topics
- increase their vocabulary skills
- improve their grammar skills
- improve their study skills
- increase confidence in their abilities

The treatment for *Ticket to Write* is both practical and expansive. The text is written in a basic, easy-to-understand form and employs attention-grabbing graphics and explanatory material to engage and instruct students.

Content Overview

The book is organized into five parts:

PART 1: THE WRITING PROCESS introduces the five stages of the writing process: prewriting, discovery drafting, revising, editing and proofreading, and publishing. Each stage is explained in a way that is easy for beginning writers to comprehend, and each offers a variety of techniques to use within the stages. Paragraph and essay writing are covered concurrently so students can understand the parallels between them and how they relate to each other. Part 1 also presents an examination of various patterns of essay organization, including the five-paragraph essay.

PART 2: TYPES OF PARAGRAPHS AND ESSAYS highlights instruction and practice in writing both paragraphs and essays using nine types of academic writing: *description, narration, illustration, process, definition, compare and contrast, classification, cause and effect*, and *persuasion*. Each chapter focuses on the purpose of one writing type. Easy-to-follow, detailed instruction guides students through practice in creating both paragraphs and essays, allowing flexibility for students who need extra work with paragraphs.

PART 3: WRITING SITUATIONS teaches students how to write source-based essays and read and answer in-class writing assignments. In both their collegiate and personal lives, today's students often face writing situations beyond academic writing, so Part 3 also provides instruction and guidance with writing in other disciplines, writing for journals and periodicals, electronic writing, and personal writing.

PART 4: GRAMMAR AND MECHANICS provides comprehensive coverage to aid all students, including developmental and ELL students who are often apprehensive about their use of spoken and written grammar. The easy-to-use handbook presents conventional concepts of grammar in a student-friendly format through the use of

- clear and simple language
- various options for correcting common problems
- attention-getting graphic design
- contemporary examples

PART 5: READING TIPS AND ADDITIONAL READINGS opens with a chapter on strategies for critical reading that contains information on reading textbooks, skimming and scanning, active reading, and annotation. It also provides a supplement, "Additional Readings," to the many paragraphs, essays, and partial essays embedded in the text. The essays in this section address diverse topics, including medicine, popular culture, and appreciation—or lack of appreciation—of college life. Following each essay are critical thinking questions that ask students to delve into the essay's content and structure, group discussion questions, vocabulary, and ideas for further writing.

PART 6: STUDY SKILLS is provided online through the MyWritingLab/eText version of *Ticket to Write* and offers practical information and strategies students can use to study more effectively. This section addresses topics such as learning styles, time management, organization, note-taking, reading tips, stress management, vocabulary building, and exam strategies. Many students arrive at college with inadequate skills that must be strengthened and developed, and these chapters offer strategies students can use to be successful in their college careers.

MyWritingLab™

The Part 6 study skills chapters are also available through Pearson Custom Library (PCL) and any of the chapters can be added to a print version of the book by contacting your local Pearson sales representative and involving Pearson's Custom Publishing Division.

Features

The following features are designed to support and enhance the effectiveness of the text for both the instructor and the student:

- **Paragraph and Essay Composition Taught in Parallel** In Parts 1 and 2, paragraph and essay development and writing are covered in every chapter, highlighting the similarities between them, showing how the writing process works for each, and explaining how the paragraph serves as the basic unit of the essay and how students can use the paragraph as a springboard to the essay.
- **Use of Contemporary, Pop Culture Topics** The writing assignments, fastwrites, practice activities, and sample student and professional essays reflect topics and issues that are part of the lives of contemporary college students. These are subjects they discuss, issues they face, and beliefs they value. Because students are more comfortable with writing about topics they know and with which they connect, they feel more at ease with these topics and their writing assignments.

- **Integration of Visuals** Today's tech-savvy students are visual learners. Their environment is rich with scrolling news and podcasts, info blurbs and factoids, and texts with graphics that help them comprehend and recall information more completely. In line with this, *Ticket to Write* uses a wide range of graphic elements, including callouts that provide added details or definitions of specific terms, pertinent quotes that give extra insight, and directions to videos and other online tutorials that supplement chapters with audio-visual information.

- **Integration of Technology** *Ticket to Write* integrates technology to supplement information and activities. To enhance chapter-specific skills, students are directed to MyWritingLab (online units featuring additional review materials, practice exercises, and pretests and posttests students can use to sharpen writing skills); sent to online videos (videos reinforcing chapter lessons and topics through unique and clever presentations); and directed to additional, relevant online sites through Techno Tips.

- **Linking of Critical Thinking and Critical Writing** Through a close examination and analysis of writing, each chapter in Part 2 provides students with reflective questions and directions that guide them toward expanding their critical thinking skills and help them transfer their critical thinking into critical writing.

- **Engaging Format** The text has an engaging, reader-friendly format, and it speaks clearly and directly to students. Throughout *Ticket to Write*, assignments, sample essays, and activities center on contemporary life and issues, and they all target traditional and nontraditional students.

- **Fastwrites** Fastwrites, which appear at the beginning of Chapters 6 through 19, are quick, timed writing techniques that help writers—especially novice writers—generate ideas without fear of criticism. With fastwrites, students realize that no answer is wrong, so they enjoy the comfort and freedom fastwriting affords them.

- **Writing at Work: Snapshot of a Writer** In each of the patterns-of-organization chapters (6–14), workplace writing is profiled through interviews with people who use the pattern discussed in the chapter in their daily work lives and explain and illustrate how they do so.

- **Ticket to Write Activities** Throughout the text are numerous "Ticket to Write" practice activities that reinforce chapter-specific skills and build on skills from previous "Tickets." For example, in Part 2, students not only practice writing specific elements of paragraphs and essays, but they also build on activities to complete a final draft of a paragraph or essay.

- **Student and Professional Essays** In each pattern-of-organization chapter, a sample student essay and a professional reading model the pattern being discussed; both are accompanied by questions that encourage students to analyze the structure and techniques the writers use.

- **Techno Tips** Throughout the text, these tips provide students with directions to online sites where they can find useful videos, instruction, and information related to the topic being discussed.
- **Run That by Me Again** This feature, found at the end of each chapter, provides a concise summary of the main instructional points and can be used for review and test preparation.
- **Learning Logs** Functioning as mastery tests, these lists of questions are based on the major topics of each chapter.
- **Comprehensive Grammar Coverage** *Ticket to Write* includes a condensed handbook covering grammar, mechanics, punctuation, and spelling. The handbook addresses problematic writing concerns many developing writers and ELL students face. It offers them straightforward, easy-to-comprehend explanations, examples showing how specific problems occur and how they can be corrected, different methods for discovering specific errors, end-of-chapter reviews, and learning logs. Various chapters that deal with composing paragraphs and essays also highlight pertinent grammar skills.
- **MyWritingLab** Within each chapter, students are referred to Pearson's MyWritingLab site for further review, additional practice, and topic mastery opportunities.

MyWritingLab™

Writing Resources and Supplements

Annotated Instructor's Edition for *Ticket to Write: Writing Skills for Success*

ISBN 0205059937

The Annotated Instructor's Edition (AIE) is an exact replica of the student text but includes all the answers to all the activities and practices. It also includes various teaching tips and helpful advice to instructors.

Instructor's Resource Manual for *Ticket to Write: Writing Skills for Success*

ISBN 0205059872

A detailed instructor's manual, written by Julie Yankonich of Camden Community College, provides teaching suggestions, additional activities, additional exercises, chapter overviews, and sample syllabi that instructors can use with each chapter of the text. In addition, the manual also provides a full and complete test bank that can be used for chapter quizzes or tests.

A La Carte Version of *Ticket to Write: Writing Skills for Success*

ISBN 0321782550

An unbound, three-hole punched version of *Ticket to Write* is available for the bookstore to order at a substantial savings.

PowerPoint Presentation for *Ticket to Write: Writing Skills for Success*

ISBN 0321853318

This PowerPoint presentation set consists of chapter-by-chapter classroom-ready lecture outline slides, lecture tips and classroom activities, and review questions. It is available for download from the Instructor Resource Center.

MyTest Test Bank for *Ticket to Write: Writing Skills for Success*

ISBN 032185330X

Pearson MyTest is a powerful assessment generation program that helps instructors easily create and print quizzes, study guides, and exams. Questions designed to accompany *Ticket to Write* or from other writing test banks are included. You can also create and add your own questions. Save the finished test as a Word document or PDF or export it to WebCT, Blackboard, or other CMS systems. Available at www.pearsonmytest.com.

CourseSmart eText for *Ticket to Write: Writing Skills for Success*

ISBN 0205043178

CourseSmart is one of the world's largest providers of digital course materials.

Additional Resources for Instructors and Students

The Pearson Writing Package Pearson is pleased to offer a variety of support materials to help make teaching writing skills easier for instructors and to help students excel in their coursework. Many of our student supplements are available free or at a greatly reduced price when packaged with *Ticket to Write*. For more information, please visit www.pearsonhighereducation.com, contact your local Pearson sales representative, or review a detailed listing of the full supplements package in the *Instructor's Resource Manual*.

MyWritingLab™ **Where better practice makes better writers!**
www.mywritinglab.com

MyWritingLab, a complete online learning program, provides additional resources and better practice exercises for developing writers. MyWritingLab accelerates learning through layered assessment and a personalized learning path utilizing the Knewton Adaptive Learning Platform™ which customizes standardized educational content to piece together the perfect personalized bundle of content for each individual student. With over 8,000 exercises and immediate feedback to answers, the integrated learning aids of MyWritingLab reinforce learning throughout the semester.

What makes the practice in MyWritingLab better?

- **Diagnostic Testing** MyWritingLab's diagnostic Path Builder test comprehensively assesses students' skills in grammar. Students are provided an individualized learning path based on the diagnostic's results, identifying the areas where they most need help.
- **Progressive Learning** The heart of MyWritingLab is the progressive learning that takes place as students complete the Overview, Animations, Recall, Apply, and Write exercises along with the Posttest within each topic. Students

move from preparation (Overview, Animation) to literal comprehension (Recall) to critical understanding (Apply) to the ability to demonstrate a skill in their own writing (Write) to total mastery (Posttest). This progression of critical thinking, not available in any other online resource, enables students to truly master the skills and concepts they need to become successful writers.

- **Online Gradebook** All student work in MyWritingLab is captured in the Online Gradebook. Instructors can see what and how many topics their students have mastered. They can also view students' individual scores on all assignments throughout MyWritingLab, as well as overviews of student and class performance by module. Students can monitor their progress in new Completed Work pages, which show them their totals, scores, time on task, and the date and time of their work by module. They can also open and review any of their assignments directly from these pages.

- **Where Print and Media Connect** Students can also complete the chapter-opening Fastwrite activities, Writing Assignments, and end-of-chapter Learning Log features of the *Ticket to Write* printed text within the *Ticket to Write* book-specific module in MyWritingLab. These unique activities are clearly identified in the print text by the MyWritingLab logo and/or new icons including and similar to the following example

  Complete these **Fastwrites** at **mywritinglab.com**

  Therefore, for the first time ever, students can complete and submit exercises from the printed text within MyWritingLab.

- **eText** The *Ticket to Write* eText is accessed through MyWritingLab. Students now have the eText at their fingertips while completing the various exercises and activities within MyWritingLab. Students can highlight important material in the eText, tab pages and areas of importance, add notes to any section for reflection and/or further study, and access all of the study skills e-chapters from Part 6 of the book. The eText also includes additional media and links to all sites referenced in the book.

Acknowledgments

The authors especially thank their spouses, Randa Gary and Mike Thurman. Without the support and patience they have given us, this book would still be in its infancy.

Many personal friends (Paula Fowler, in particular) have also helped and encouraged us along the way. In addition, cyberfriends (especially members of the Conference on Basic Writing discussion list) have been generous with ideas about successful teaching methods and also about the diverse writing approaches and requirements at their institutions.

At our institution, Henderson Community College, colleagues have assisted us in myriad ways. We thank them all, especially Dr. Kris Williams, Dr. Patrick Lake, Rebecca Emerson, Sandra Ross, Mike Knecht, Lynda Sinnett, Cheryl McKendree, Sharon Burton, Jon Reidford, Kim Conley, Tracy Sword, Tony Strawn, and Joey Goebel.

In addition to our colleagues, our students have been extremely helpful in the development of this book. Because of the many suggestions they gave us and the insight they provided, we owe them a special debt of gratitude.

We have had the pleasure to work with two editors extraordinaire, and we are deeply indebted to them both. Executive Editor Matt Wright has shepherded us and this project from its inception, and Gillian Cook, Senior Development Editor, has provided innumerable eagle-eyed observations and suggestions. Both editors helped us hone our work into what readers see today. We thank them for their encouragement and professional expertise all the way.

We are also indebted to the reviewers who aided us during the many legs of this journey. Their suggestions have been invaluable, and we thank each of them for the guidance we gleaned from their incisive comments:

Lisa Avendano, Lincoln Land Community College; Craig Barto, Charleston Southern University; David Beighley, Pierpont Community and Technical College; Tom Bellomo, Daytona Beach College; Linda Bernhagen, Highland Community College; Linda Black, St. Johns River Community College; Frederick Brown, Bunker Hill Community College; Shirley Buttram, Northeast Alabama Community College; Martha Campbell, St. Petersburg College; Christopher Commodore, Middlesex Community College; Carol Copenhefer, Central Ohio Technical College; Bill Corby, Berkshire Community College; Marcia Cree, Camino College; Kennette Crockett, Harold Washington College Kate Cross, Phoenix College; Rochelle Dahmer, Jones County Junior College; Shelley DeBlasis, New Mexico State University Carlsbad; Eric Devlin, Tarrant County College; Shari Dinkins, Illinois Central College; Claudia Edwards, Piedmont Technical College; Leslie Fredericks, Camden County College; Angelina Gonzales, Santa Monica College; Carissa Gray, Georgia Perimeter College; Judith Hague, Massasoit Community College; Angela Hebert, Hudson County Community College; Vickie Kelly, Hinds Community College; Julia Lafoon-Jackson, Hopkinsville Community College; Michelle LaFrance, UMass Dartmouth; Kevin Lamkins, Capital Community College; Rita Lammot, Central Florida Community College; Dana Dildine Lopez, Eastern New Mexico University; Deborah Mael, Newbury College; Peter Omar Manuelian, Seattle Central Community College; Denise Marchionda, Middlesex Community College; Lisa Martin, Piedmont Technical College; Marti Miles-Rosenfield, Collin County Community College; Carol Miter, Norco College; Steve Moiles, Southwestern Illinois College; Kimberly Muff, Cloud County Community College; Melissa Nicholas, Florida Keys Community College; Ida Nunley, Eastern Arizona College; Mary Poole, Madisonville Community College; Jessica Rabin, Anne Arundel Community College; Cheryl Reed, San Diego Miramar College; Joan Reeves, Northeast Alabama Community College; Deborah Repasz, San Jacinto College; Minati Roychoudhuri, Capital Community College; Becky Rudd, Citrus College; Danielle Santos, Middlesex Community College; Gail Schilling, NHTI–Concord's Community College; Brandon Shaw, Mohawk Valley Community College; Janice Shelton, New River Community College; Deneen Shepherd, St. Louis Community College; Alison Stachera, Lincoln Land Community College; Brenda Tuberville, Rogers State University; Maria Villar-Smith, Miami Dade College; Sabrina Walters, Miami Dade College; Colleen Weeks, Arapahoe Community College; Elizabeth Wilber, Palm Beach Community College; Jonathon Wild, Cloud County Community College; Lisa Williams, Kirkwood Community College; Lynda Wolverton, Polk State College; DeBorah Zackery, Dunwoody Community College.

A Walk-through of the Features in *Ticket to Write*

Your *Ticket to Write* . . .

Contemporary and cutting-edge, *Ticket to Write* engages novice and ELL writers via issues in pop culture and contemporary life and addresses students' increased use of technology.

LEARNING GOALS Listed at the beginning of each chapter, the learning goals tell students what the chapter is about, what topics are most important, and what skills they will have mastered by the end of the chapter.

LEARNING GOALS

In this chapter, you'll learn and practice how to

1. Revise for purpose
2. Revise for topic, unity, and coherence
3. Use the RAMS method to improve your writing
4. Use revision checklists

GETTING THERE

- After you have completed prewriting and discovery drafting, revising will help you find areas that need to be reworded, strengthened, or even eliminated.
- The RAMS method of replacing, adding, moving, or subtracting is another technique for revising your work.

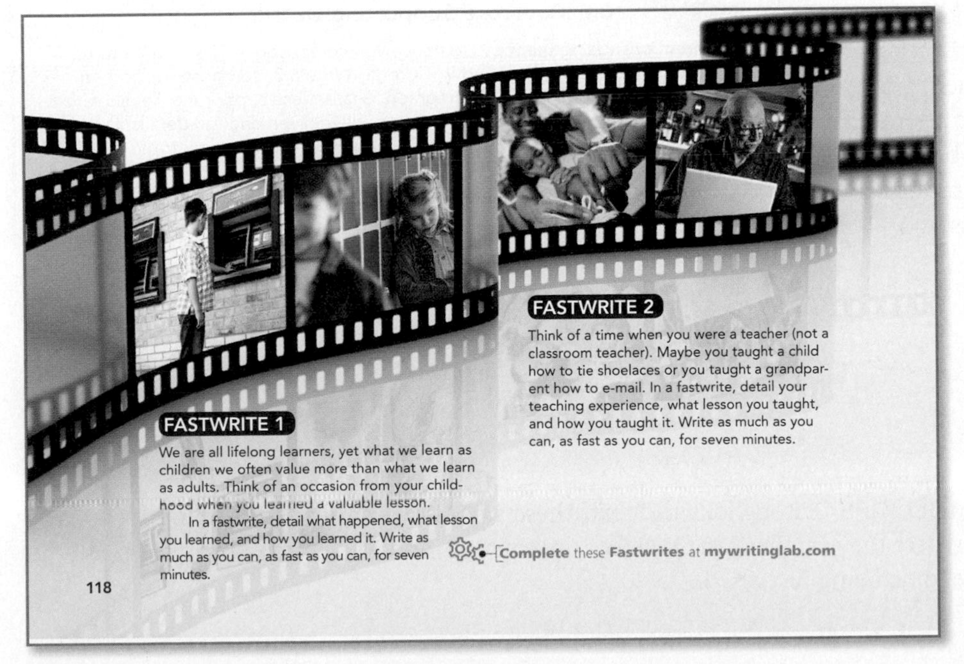

FASTWRITES These quick, timed writing activities appear at the beginning of Chapters 6 through 19 and can be completed in the *Ticket to Write* MyWritingLab course. They can help students get their ideas down on paper without fear of criticism. Since fastwrites have no "right answers," students can enjoy writing.

Types of Paragraphs and Essays

PART 2

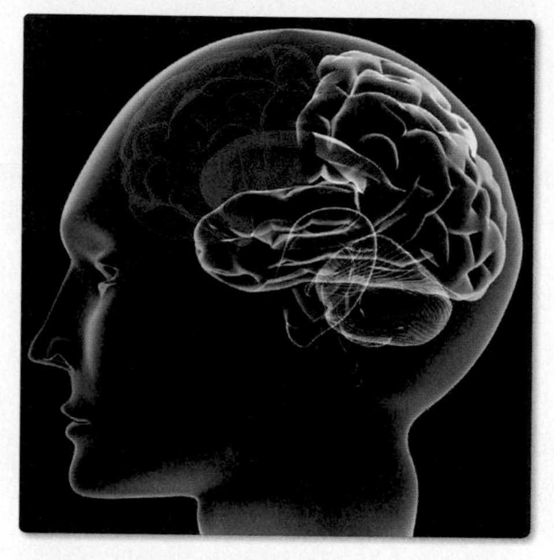

Traditional academic writing is based in nine types of writing: *description, narration, illustration, process, definition, compare and contrast, classification, cause and effect,* and *persuasion*. The type of writing you use depends on the purpose behind your writing. Part 2 offers you instruction and practice in writing paragraphs and essays using each type of academic writing.

SIMULTANEOUS PARAGRAPH AND ESSAY COVERAGE Paragraph and essay writing are covered concurrently so students can see and understand the parallels between them and how they relate to each other. This corresponding approach allows instructors to adapt assignments to fit the varying needs of their class or the individual students whose abilities warrant progressing more rapidly (self-pacing in print).

***TICKET TO WRITE* EXERCISES** The "Ticket to Write" practice activities reinforce chapter-specific skills and build on skills from previous "Tickets." For example, in Part 2, students not only practice writing specific elements of paragraphs and essays, but they also build on activities to complete a final draft of a paragraph or essay.

6.6 Compose Supporting Details

Directions: Using the topic sentence you composed in Ticket to Write 6.5 and the ideas you generated in Ticket to Write 6.3, compose supporting details for your descriptive paragraph. Share these supporting details with your writing group. Ask members if you have given enough details so that your readers appreciate the sentiment you expressed in your topic sentence. Answers will vary.

TECHNO TIP

For additional ideas on proofreading and editing, and a PowerPoint presentation on peer review, visit Purdue OWL (Online Writing Lab) at owl.english.purdue.edu/owl/ then go to **General Writing, The Writing Process, Proofreading.**

TECHNO TIPS Throughout the text, these tips direct students to online sites where they can find useful videos, instruction, and information related to the topic being discussed.

WRITERS AT WORK Chapters 6 through 14 (patterns of organization) include unique interviews with workers who use the pattern discussed in the chapter in their daily work lives and explain and show how they do so. These distinctive snapshots of writers in the workplace give students a context in which to see the importance writing has in their careers.

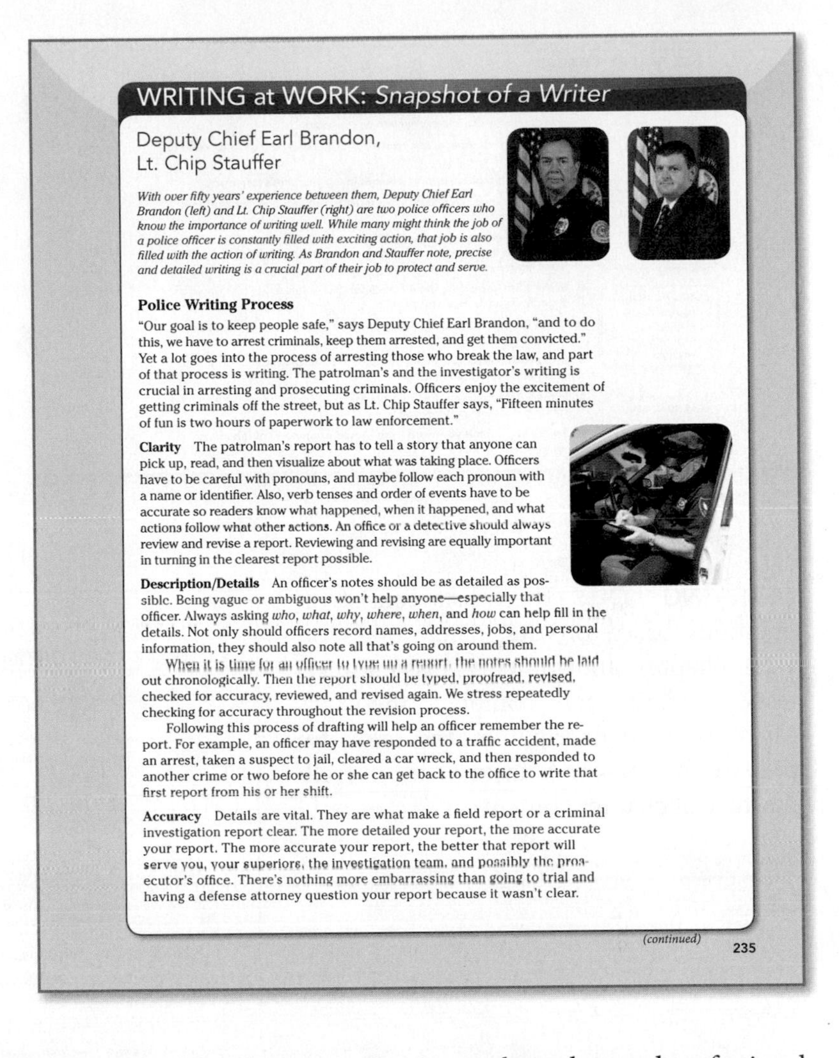

WRITING at WORK: *Snapshot of a Writer*

Deputy Chief Earl Brandon,
Lt. Chip Stauffer

With over fifty years' experience between them, Deputy Chief Earl Brandon (left) and Lt. Chip Stauffer (right) are two police officers who know the importance of writing well. While many might think the job of a police officer is constantly filled with exciting action, that job is also filled with the action of writing. As Brandon and Stauffer note, precise and detailed writing is a crucial part of their job to protect and serve.

Police Writing Process

"Our goal is to keep people safe," says Deputy Chief Earl Brandon, "and to do this, we have to arrest criminals, keep them arrested, and get them convicted." Yet a lot goes into the process of arresting those who break the law, and part of that process is writing. The patrolman's and the investigator's writing is crucial in arresting and prosecuting criminals. Officers enjoy the excitement of getting criminals off the street, but as Lt. Chip Stauffer says, "Fifteen minutes of fun is two hours of paperwork to law enforcement."

Clarity The patrolman's report has to tell a story that anyone can pick up, read, and then visualize about what was taking place. Officers have to be careful with pronouns, and maybe follow each pronoun with a name or identifier. Also, verb tenses and order of events have to be accurate so readers know what happened, when it happened, and what actions follow what other actions. An office or a detective should always review and revise a report. Reviewing and revising are equally important in turning in the clearest report possible.

Description/Details An officer's notes should be as detailed as possible. Being vague or ambiguous won't help anyone—especially that officer. Always asking *who, what, why, where, when,* and *how* can help fill in the details. Not only should officers record names, addresses, jobs, and personal information, they should also note all that's going on around them.

When it is time for an officer to type up a report, the notes should be laid out chronologically. Then the report should be typed, proofread, revised, checked for accuracy, reviewed, and revised again. We stress repeatedly checking for accuracy throughout the revision process.

Following this process of drafting will help an officer remember the report. For example, an officer may have responded to a traffic accident, made an arrest, taken a suspect to jail, cleared a car wreck, and then responded to another crime or two before he or she can get back to the office to write that first report from his or her shift.

Accuracy Details are vital. They are what make a field report or a criminal investigation report clear. The more detailed your report, the more accurate your report. The more accurate your report, the better that report will serve you, your superiors, the investigation team, and possibly the prosecutor's office. There's nothing more embarrassing than going to trial and having a defense attorney question your report because it wasn't clear.

(continued)

235

STUDENT AND PROFESSIONAL ESSAYS In each of the chapters in Part 2, sample student and professional essays illustrate the type of writing students are studying. The "A Closer Look" questions that follow each essay ask students to closely examine the readings for the characteristics and stylistic elements of the specific type of writing.

Exceedingly Extraordinary X Games
by Jeff Blake

1 My father's job as a **photojournalist** takes him to a lot of exciting places and events, and I've been lucky enough to have joined him on a few of his photo shoots. I saw **Kobe** hit fifty points in a single game to lead the Lakers to a win over New Orleans. I saw **Sam "Hardluck" Hornish**'s miracle win at the Indy 500. And I even saw France's leading scorer, **Zindine Zidane**, head butt an Italian player and hand Italy the World Cup in Berlin. But, I'm not really a basketball or a stock car fan, and even though Berlin was an eye-opening experience for me, I'm not a big soccer buff. I am a winter sports fanatic, so I was really excited last January when my dad surprised me with a trip to the twelfth annual **X Games**. Attending the X Games for the first time was my favorite sporting adventure.

2 The lodge's guests were unlike any hotel guests I'd seen before. When we walked into the lobby, leaving the 19° weather outside, I was amazed by the hustle and bustle of people. Skiers and boarders were coming and going, bringing a constant blast of cold air into the lobby. Sitting around large oak coffee tables were a dozen people my age, getting ready for the slopes. They were lacing boots, attaching ski-lift tickets to jackets, smearing on ChapStick,

Willie, My Thirteen-Year-Old Teacher
by Scott Leopold

1 Thirteen. That's how old I was when I learned poverty existed not just in my hometown but also in my own backyard. I was in the eighth grade when classmate Willie Reidford taught me more in a thirty-minute bus ride than had all my eighth-grade teachers combined.

2 I remember that sweltering May 30 in 1993. It was the last day of eighth grade, and I would finally emerge from my three-year-long **pupa stage** as a middle-schooler and embark on the final and most **prestigious** stage in my childhood development: high-schooler. We would be young adults, and so our teachers were sending us off in style.

3 Everyone was relaxed and all were getting along, even Mrs. Brokaw, our homeroom and biology teacher. Then, she gave each of us a brand-new pocket dictionary. "Remember," she said, handing out the small gift-wrapped paperbacks, "that half of knowledge is knowing where to find it." After that, we signed yearbooks, ate pizza, and accepted graciously her teacherly gift. Many of us brought in music, and we listened to an alternating blend of rock and country. We bounced from Everlast, Barenaked Ladies, and Lenny Kravitz to George Strait, Tim McGraw and the Dixie Chicks. The rednecks rolled their eyes at the rock; the **preps** rolled theirs at the country; yet the goths, geeks, skaters, jocks, and motorheads seemed to tolerate these two primary **cliques**. I sat back and reveled in our **camaraderie**. Here

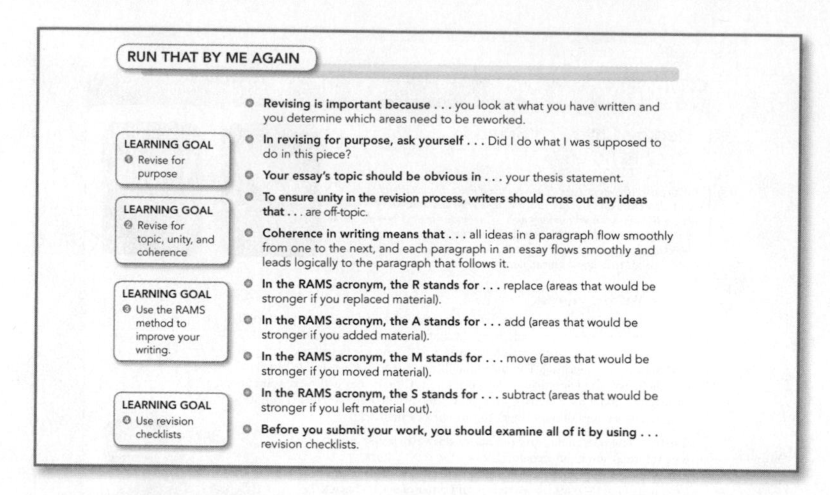

RUN THAT BY ME AGAIN

LEARNING GOAL
1 Revise for purpose

LEARNING GOAL
2 Revise for topic, unity, and coherence

LEARNING GOAL
3 Use the RAMS method to improve your writing.

LEARNING GOAL
4 Use revision checklists

- **Revising is important because . . .** you look at what you have written and you determine which areas need to be reworked.
- **In revising for purpose, ask yourself . . .** Did I do what I was supposed to do in this piece?
- **Your essay's topic should be obvious in . . .** your thesis statement.
- **To ensure unity in the revision process, writers should cross out any ideas that . . .** are off-topic.
- **Coherence in writing means that . . .** all ideas in a paragraph flow smoothly from one to the next, and each paragraph in an essay flows smoothly and leads logically to the paragraph that follows it.
- **In the RAMS acronym, the R stands for . . .** replace (areas that would be stronger if you replaced material).
- **In the RAMS acronym, the A stands for . . .** add (areas that would be stronger if you added material).
- **In the RAMS acronym, the M stands for . . .** move (areas that would be stronger if you moved material).
- **In the RAMS acronym, the S stands for . . .** subtract (areas that would be stronger if you left material out).
- **Before you submit your work, you should examine all of it by using . . .** revision checklists.

RUN THAT BY ME AGAIN This feature, at the end of each chapter, provides a concise summary of the main instructional points and can be used for review and test preparation.

LEARNING LOGS These lists consist of questions based on the major topics of each chapter and function as mastery tests. Students can complete these in MyWritingLab or right in the book to prepare for chapter quizzes or for overall mastery of chapter content.

Compare-and-Contrast Writing Learning Log 261

COMPARE-AND-CONTRAST WRITING LEARNING LOG MyWritingLab™

Answer the questions below to review your mastery of compare-and-contrast writing. Answers will vary.

Complete this Exercise MyWritingLab™

1. Beyond reporting the similarities and differences of two subjects, what might a compare-and-contrast paragraph or essay do?
   It may propose that one subject is superior to the other in specific ways, that both subjects are equal in importance, or that one subject is far less important.
2. What is a Venn diagram?
   A Venn diagram is a drawing of two intersecting circles that note features that are either unique or common to two concepts.
3. When contrasting two subjects, what do you focus on?
   You focus on their different qualities.
4. What could be your purpose in comparing and contrasting two subjects?
   Your purpose could be to convince readers that one subject is better than the other, show readers that two subjects are equal, or inform readers of unique qualities of each subject.
5. How should you create your starter statement?
   Combine the answers to your starter questions into a single sentence.
6. What should you create from your starter statement?
   You should create your working thesis statement.
7. What two elements should be evident in your working thesis statement?
   Your purpose for comparing or contrasting subjects should be evident.
8. After you have determined what you want to write about your two subjects, what is the next step?
   Determining who will benefit from reading what you write about the subjects is the next step.
9. In compare-and-contrast writing, what are two patterns of organization you must decide between?
   You must decide between an alternating pattern and a divided pattern.
10. In applying critical thinking to your essay, what questions help you in examining your purpose?
    Will my readers understand why my comparison or contrast of these two subjects is important? Have I clearly stated or clearly implied my purpose?

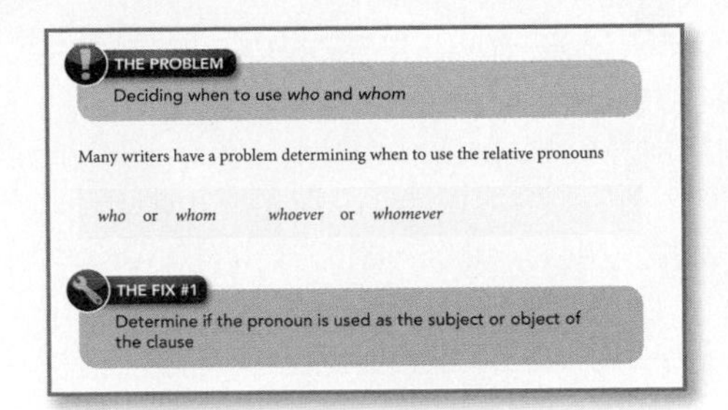

GRAMMAR COVERAGE: THE PROBLEM AND THE FIX This unique presentation offers students straightforward, easy-to-comprehend explanations; examples showing how specific problems occur and how they can be corrected; and different methods for discovering specific errors and then fixing them before finalizing their paper.

MYWRITINGLAB, THE ETEXT, AND ADDITIONAL MEDIA The highly successful MyWritingLab is completely integrated into each chapter of *Ticket to Write* and provides students an opportunity to master the various writing topics within this powerful assessment-based technology. Furthermore, students have access to the *Ticket to Write* eText via MyWritingLab and can access various media and links discussed within the print text. Students can also complete the end-of-chapter Learning Log features and the chapter-opening Fastwrite activities of *Ticket to Write* within the book-specific module in MyWritingLab. These unique activities are clearly identified in the print text by this new icon: Therefore, for the first time students can complete and submit exercises in the printed text within MyWritingLab.

MyWritingLab™ Visit *MyWritingLab.com* and complete the exercises and activities in the **Paragraph Development-Narrating** and **Essay Development-Narrating** topic areas.

The Writing Process

Whatever you're writing—an e-mail to your college advisor, or an assigned paragraph or essay for class—you use the five steps in the writing process: *prewriting* (brainstorming to get ideas), *discovery drafting* (putting ideas on paper), *revising* (adjusting, adding, or deleting ideas), *editing and proofreading* (looking at the fine points), and *publishing* (presenting your work in its final form). Part 1 helps guide you through each separate stage of the writing process.

"The scariest moment is always right before you start."
– Stephen King, *On Writing*

CHAPTER 1

The Writing Process and Prewriting

LEARNING GOALS

In this chapter, you'll learn

1. How the writing process works
2. How to use prewriting techniques (listing, clustering, fastwriting, reporter's questions, and journaling) to discover ideas for paragraphs and essays

GETTING THERE

- The **writing process** is a series of steps that make up the activity of writing.
- **Prewriting** is the first step writers use to gather ideas for their paragraphs or essays.
- Writers use several prewriting techniques to help generate topic ideas and support to develop those ideas.

Writing as a Winding Path

LEARNING GOAL

1. How the writing process works

Many people consider writing a difficult and aggravating task that seems impossible to "get right." Often, frustrated writers are put off because they are trying to write a perfect piece all at once, and yet they stop and stumble along the way, running into spelling and punctuation errors and even brick walls commonly referred to as "writer's block."

Over the decades, professional writers and writing teachers have looked at writing as an activity that proceeds on a straight path, moving from beginning to end in a straight line:

Recently, these writers and teachers have considered writing more of a winding path, crossing over itself time and time again, but still with a beginning and an end.

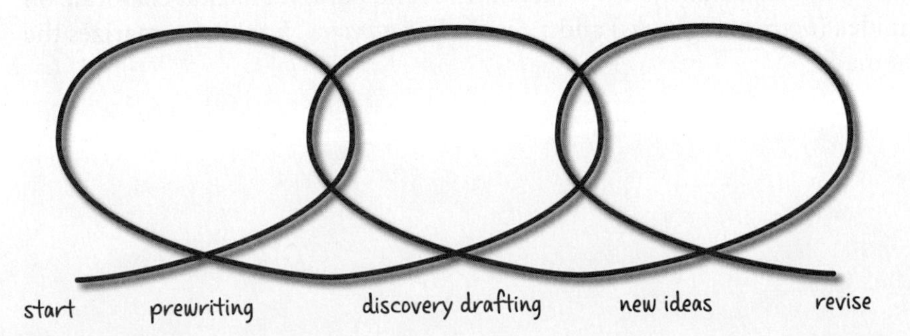

You may find this view more helpful because it allows you to go back and forth from Point A (your beginning idea) as many times as necessary on your path to Point Z (your conclusion).

The Writing Process

For many of you, the term *writing process* is nothing new. Since you were in elementary school, you've heard teachers use this phrase. For others, this is the first time you've seen this phrase, and you may be intimidated by it.

Simply put, the *writing process* is the name for the five steps everyone goes through in order to get the desired finished product.

These are the steps:

- *prewriting* (brainstorming to get ideas)
- *discovery drafting* (putting ideas on paper)
- *revising* (adjusting, adding, or deleting ideas)
- *editing and proofreading* (looking at the fine points)
- *publishing* (presenting your work in its final form)

While everyone goes through the five steps of the writing process, different writers find success with different techniques throughout the process. So that you will be able to discover which writing techniques help you create your best writing, Part 1 of *Ticket to Write* offers instruction and practice in a number of methods for each step in the writing process. By the end of this part, you will have the tools you need to write a basic paragraph or essay.

The Building Blocks of Paragraphs and Essays

Academic writing involves clear and concise composition of both paragraphs and essays. A **paragraph** focuses on one subject, theme, or idea that you state in an introductory sentence (a *topic sentence*). Next come several sentences that elaborate on what you stated in your topic sentence (*supporting sentences*) and a *concluding sentence* that summarizes the points you made.

You could describe what your cell phone looks like in a paragraph, but you'd need an essay (several paragraphs) to describe the features of it you frequently use.

In some ways, an **essay** is a paragraph that you expand because your subject is too complex to cover in a single paragraph. An essay, like a paragraph, focuses on one subject, theme, or idea, but it needs several paragraphs to make its point. The first paragraph, the introductory paragraph, contains a sentence with the main idea of your essay (the *thesis statement*). Next come several paragraphs that elaborate on your main idea (*body paragraphs*) and a *concluding paragraph* that summarizes the points you made.

Paragraph	Essay
Introduction in **one sentence**, a *topic sentence* that gives the main idea	Introduction in **one paragraph** that includes a *thesis statement* that gives the main idea
Sentences that elaborate on the main idea	**Paragraphs** that elaborate on the main idea
Concluding **sentence** summarizing the points about the main idea	Concluding **paragraph** summarizing the points about the main idea

In Part 1 of *Ticket to Write*, you will learn the steps—the writing process—in composing a basic paragraph and essay. Part 2 introduces you to nine types of academic writing, and you can hone your paragraph and essay skills with each type.

Prewriting Techniques

LEARNING GOAL

❷ How to use prewriting techniques to discover ideas for paragraphs and essays

Everyone needs a little push to get going. The scariest sight for many novice writers is a blank page. If getting started writing makes you nervous, then do a bit of writing *before* you start writing.

A marathoner will jog before a race; a cyclist will ride before an event. Just as these athletes run before they *run* and ride before they *ride*, writers also need to warm up. That's where prewriting techniques come in. **Prewriting techniques** are unique methods of brainstorming that help you discover ideas, find support for those ideas, and fill in that empty screen or blank page.

"The best way to get a good idea is to get lots of ideas."
—Linus Pauling

Listing

In ***listing***, you let your mind go free about a particular subject, idea, or question. Jot down random thoughts as they occur, and don't worry about their relevance to each other. Use this technique for developing subjects of entire essays, for determining topics of paragraphs, or for finding small details that support your subject.

If you have the freedom to choose what you write about, then answering a few general questions can start your brain down the path of subject discovery:

- What interests do I have?
- What special knowledge do I have?
- What subjects interest me?
- What issues do I care about?

In Alma's first writing class, her instructor directed the students to list, for five minutes, all their thoughts on any issue they cared about. Here is Alma's list:

Recycling

make world better for future generations
city underwrites some recycling
find place to send used batteries
why don't others recycle—too lazy?
don't care about future?
wish trucks picked up stuff every week
second-hand stores forms of recycling
wish college had more places for recycling
homeless man who collects cans is recycler

In college, you'll often be given topics on which you *have* to write. Expanding the same listing questions can also help you discover your thoughts and knowledge on topics assigned to you:

- What interest do I have in this topic?
- What special knowledge do I have about this topic?
- What related subjects interest me?
- What issues do I care about concerning this topic?

Once you begin to answer one of these questions, you might find that you have more than one answer, so just keep listing your answers. You can go back later and put similar answers together.

For another specific assignment, Alma's instructor asked the students to list ideas on the topic of *fast-food restaurants*. Here is Alma's list for that topic:

convenient when I'm in a hurry
worked at McDonald's 6 months
not good for date night
too many fried things
my cholesterol always high
not many cola alternatives
more fast-food places than sit-down
wait in line not always "fast"
don't usually make nutritional choices
I love cheeseburgers

1.1 Listing

Directions: Use listing to discover ideas about one of the following topics:

sleep habits
study habits
job market
Facebook
career planning
cell phones
role models
military life
American symbols

Clustering

Clustering is writing a key word or phrase and then jotting down other ideas that spring from it, tracking this path of ideas with connecting lines. Your key word or phrase may generate a number of ideas that radiate from it; each of these, in turn, may produce additional ideas.

In this example, Isabelle took the topic of *mountain biking* and used it as her key idea, which she put in a large circle in the middle of her paper. From this idea, Isabelle branched out to other ideas as they occurred to her. With each new idea, she drew another circle and filled it in. Every time she could go no further, Isabelle returned to the nucleus *mountain biking* and began a new train of thought.

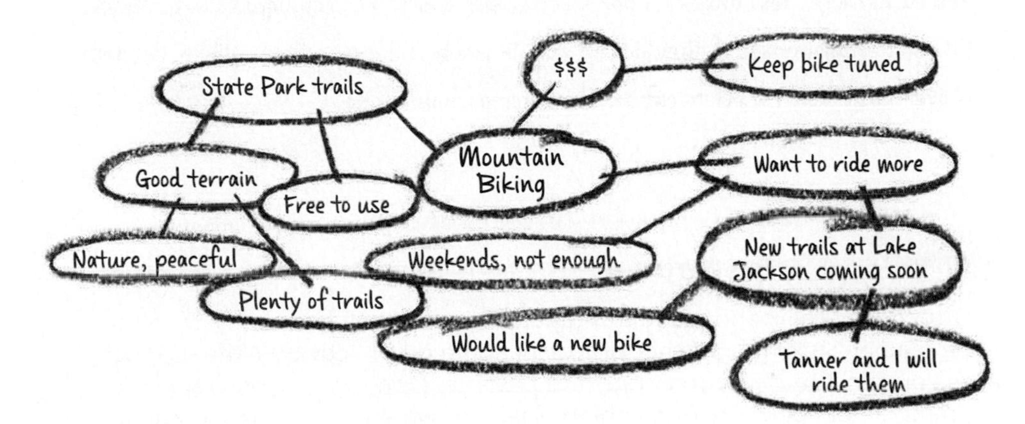

1.2 Clustering

Directions: Use clustering to discover ideas about one of these topics:

tattoos	procrastination	job interviews	anxiety	fake IDs
computer privacy	advertising	gambling	smart phones	caffeine

Fastwriting

Fastwriting is exactly what it sounds like, *writing fast*. It's also known as *freewriting* because the goal of this technique is to write freely, without worrying about correct grammar, punctuation, and spelling. Fastwriting is your license to write fast . . . and make mistakes without consequences.

"Creativity consists of coming up with many ideas, not just that one great idea." —Charles Thompson

Jamal's instructor asked his class to freewrite for five minutes about *test anxiety*. Here is what Jamal wrote:

This happens when I'm unprepared. If I feel like I haven't studied enough, I get nervous. Even if I have studied enough. I don't sweat or anything. I just feel uneasy. I can't eat anything for a couple of hours before a test. And I worry about my teacher knowing that I'm not prepared. I know that's kind of ??? of me, but I don't like to look up from a test and see the teacher looking at the class. I know the teacher has to do that, but I think she's looking at me wondering if I'm trying to cheet. Andrea once told me that some different breathing techniques helped her with test anxiety. I don't remember what the techniques were. Maybe I'll ask her again—see if they'll help me. Or maybe I'll google them. When the test is over I still don't want to eat anything for an hour or so.

Heads Up!
When Jamal couldn't think of the right word, he just inserted some question marks, which told him to come back to that part later.

Heads Up!
Jamal didn't worry about grammar or spelling in his freewriting. Also, he noted several details that probably won't end up in his final paper.

TICKET to WRITE

1.3 Fastwriting

Directions: Think about a relaxing pastime you enjoy frequently. It could be reading, texting, gaming, biking, or some other activity. Write as much as you can about why you enjoy this pastime. Don't worry about spelling, grammar, punctuation, or anything else. Just write as much as you can as fast as you can for five minutes.

Reporter's Questions

Reporter's questions are the six questions (*Who*? *What*? *Why*? *Where*? *When*? and *How*?) that journalists use to guide readers through a news article. These work well in the writing process to help you discover specific ideas and details about a particular topic. Isabelle developed these questions and answers regarding mountain biking:

Topic: *Mountain biking*

Who usually rides with me?	Tanner, my best friend
What is special about mountain biking?	helps me unwind
Why do I enjoy mountain biking?	get away from city and work; relax
Where do I usually ride?	trails at state park
When do I usually ride?	weekends . . . time to unwind
How do I prepare for a day of biking?	tune up bike, pack daypack

Reporter's questions work even better when they are reused, taking answers to the first set of questions as starting points for new questions:

Where do I usually ride?	trails at state park
Why do I usually ride there?	trails are long, a lot of features, peaceful
How long are the trails?	six trails = 56 miles
What features do the trails have?	flat pine forests, breezy, quiet, secluded . . . picnic areas

TICKET to WRITE

1.4 Using Reporter's Questions

Directions: Use the reporter's questions to discover ideas about one of these topics:

Twitter
buying a vehicle
celebrations
bank accounts
student expectations
instructor expectations
job expectations
my music
my television list
blue jeans

Journaling

Journaling is writing about personal experiences or reflections. You may think journaling sounds a bit like keeping a diary, and it is. Like writing in a diary, journaling can help you examine and reexamine certain events, topics, or ideas.

Below is what college freshman Jeong wrote when asked to create a journal entry about the differences between high school and college.

College is nothing like high school was. I don't have to attend class because not all my profs take roll. If I miss class I have to get the notes from the class website or another student. Seems like everyone was pushing me to do well—Ms. Connie, the counselor, Mr. Russell, senior homeroom. He was the best. Mr. R. always made sure we had our homework assignments before we left for the day. If I missed it in some class he'd give me a hall pass and send me to get the homework I needed. All my teachers were available afterschool for help—some stayed longer than others but all would stay if asked. Mr. R. brought in snacks if some-one was dragging. My college professors are not always in their offices when I go

(continued)

looking for them. Some part-time faculty don't even have office hrs. No one checks up on me every day to see if I'm missing an assignment or falling behind in a class. There's free tutoring for a lot of classes (but not all) and I have to go to Academic Services to sign up and find a tutor. It's across campus and that's a hassle. Tutors are usu. students. The writing ctr has student tutors and faculty tutors but it's not open all week. Online tutoring's available 24/7 and doesn't cost anything. Bulletin boards all over campus advertize study groups anyone can join. I study in the library a lot because I can bring in a soda.

TICKET to WRITE

1.5 Journaling

Directions: Write a journal entry about one of these topics:

my reaction to a local news item
my reaction to a national news item
impressions of college
I wonder why . . .
Never again will I . . .
If I could change something at work . . .
In my culture . . .
When I was lost . . .
If I could do something over, I'd . . .
a national event I'll never forget

TECHNO TIP

For additional ideas about prewriting, search the Internet for this video:
The Writing Process: Prewriting Strategies Video

MyWritingLab™ Visit *MyWritingLab.com* and complete the exercises and activities in the **Prewriting** and **Writing Process** topic areas.

RUN THAT BY ME AGAIN

LEARNING GOAL
1 How the writing process works

- **You should think of writing not as a straight line but as . . .** a winding path.
- **The five steps of the writing process are . . .** (1) prewriting, (2) discovery drafting, (3) revising, (4) editing and proofreading, and (5) publishing.
- **A paragraph focuses on . . .** one subject, theme, or idea called a topic.
- **A topic sentence states . . .** the topic of a paragraph.
- **Supporting sentences elaborate on . . .** the idea stated in the topic sentence.
- **A concluding sentence summarizes . . .** the points made in a paragraph.
- **An essay is different from a paragraph because . . .** its subject is too complex to cover in a single paragraph.

- **The first paragraph of an essay is called . . .** the introductory paragraph.
- **The sentence that contains the main point of an essay is called . . .** the thesis statement.
- **The paragraphs that elaborate on an essay's main idea are called . . .** body paragraphs.
- **The last paragraph that summarizes the essay's points is called . . .** the conclusion.
- **Prewriting techniques are unique methods that can help you . . .** discover ideas and find support for those ideas.
- **Five prewriting techniques are . . .** (1) listing, (2) clustering, (3) fastwriting, (4) reporter's questions, and (5) journaling.
- **When you list, you . . .** let your mind go free about a particular subject or idea and list all your thoughts as they occur.
- **When you fastwrite, you . . .** write freely all that comes to mind, without worrying about correct grammar, punctuation, and spelling.
- **When you cluster, you . . .** write a key word or phrase, and then jot down other ideas that spring from it, tracking this path of ideas with connecting lines.
- **When you answer reporter's questions, you . . .** answer the questions *Who? What? Why? Where? When?* and *How?* to help you discover specific ideas and details about a particular topic.
- **When you journal, you . . .** write about personal experiences or reflections.

LEARNING GOAL

❷ How to use prewriting techniques to discover ideas for paragraphs and essays

THE WRITING PROCESS AND PREWRITING LEARNING LOG MyWritingLab™

1. **What is the writing process?**
   The writing process is the name given to five specific steps that writers go through to get their desired finished product.
2. **What are the five steps of the writing process?**
   The five steps of the writing process are (1) prewriting, (2) discovery drafting, (3) revising, (4) editing and proofreading, and (5) publishing.
3. **What sentence states the main point or topic of a paragraph?**
   The topic sentence states the main point or topic of a paragraph.
4. **What sentence states the main point of an essay?**
   The thesis statement states the main point of an essay.
5. **How is an essay's topic different from a paragraph's topic?**
   An essay's topic is too complex to cover in a single paragraph.
6. **What are body paragraphs?**
   Body paragraphs are the paragraphs that elaborate on an essay's main idea.
7. **How can prewriting techniques help in the writing process?**
   Prewriting techniques can help writers discover ideas and find support for those ideas.
8. **What are five prewriting techniques?**
   Five prewriting techniques are (1) listing, (2) clustering, (3) fastwriting, (4) reporter's questions, and (5) journaling.
9. **What is listing?**
   Listing is letting your mind go free about a particular subject or idea and listing all your thoughts as they occur.
10. **What is fastwriting?**
    Fastwriting is writing freely all that comes to mind, without worrying about correct grammar, punctuation, or spelling.
11. **What is clustering?**
    Clustering is writing a key word or phrase, and then jotting down other ideas that spring from it, tracking this path of ideas with connecting lines.

12. **What are the reporter's questions, and how can answering them help you in prewriting?**

    The reporter's questions are *Who? What? Why? Where? When?* and *How?* and answering them can help you discover specific ideas and details about a particular topic.

13. **What is journaling?**

    Journaling is writing about personal experiences or reflections.

CHAPTER 2

Discovery Drafting

LEARNING GOALS

In this chapter, you'll learn and practice how to

1. Draft a paragraph by developing a topic sentence and discovering and organizing support
2. Draft an essay by developing a thesis statement and organizing support
3. Ensure unity and coherence
4. Write an introduction, conclusion, and title for an essay

GETTING THERE

- One essential part of writing is narrowing your topic and clearly stating it in your topic sentence or thesis statement.
- Another fundamental step in the writing process is developing ideas that support your topic sentence or thesis statement.
- Important to your essay's structure are the opening and closing paragraphs, which establish and reemphasize your main point.

Drafting a Paragraph

LEARNING GOAL

1. Draft a paragraph by developing a topic sentence and discovering and organizing support

Once you have completed your prewriting, the moment has come to start drafting. Drafting in a paragraph centers on developing your main idea. You state this main idea in your **topic sentence**. Your topic sentence is the essence of your paragraph; everything else in your paragraph depends on it and expands on it.

A **topic sentence** should do the following:

- introduce your readers to the general subject you're writing about
- establish the position or point you're taking about your subject
- state a single idea in one straightforward, declarative sentence
- set the tone for the rest of your paragraph

Now that you have learned different prewriting techniques, you can develop ideas and refine them into working topics, topic sentences, and complete paragraphs. As you write, continue thinking about your assignment as a process, as writing in motion, and not a piece of work completed in a single draft.

Narrowing Your Topic

At this stage of the game, you need to narrow your topic so it becomes a manageable assignment you can cover in a single paragraph. First, take stock of the ideas you generated through prewriting. Doing this will help you focus your topic. You might find something in the **reporter's questions** or a specific notion you ended with from **clustering** that holds some importance for you.

"Words are a lens to focus one's mind."
—Ayn Rand

For instance, if your general subject is *computers*, you'll need many paragraphs—many books, even—to cover that subject. But in looking at your prewriting about computers, you might discover that your main focus is about trouble you've had learning PowerPoint. That's a far more manageable topic, one you can probably cover in a single paragraph.

Isabelle's class read and discussed an article about *leisure activities*. Following that, Isabelle's instructor asked the class to freewrite on the same topic, with the goal of creating a paragraph. This is Isabelle's freewriting:

Like to do lots of things, but wouldn't call myself athletic. I play some board games and online games. probably spend too much time with online games. Last night played solitaire to pass time instead of doing homework. Finally made myself stop and get math work done. like to watch football and basketball; play some softball in the summer if there's a pick-up game. have played tennis, but don't like to be on the concrete too much. too hot . . . look forward to biking. like to be outside . . . helps me unwind from work . . . like the nature, like the time I spend with Tanner, when he can go. guess that's my favorite leisure activity.

In reading over her freewriting, Isabelle discovered a great deal of her material focused on mountain biking. She decided she could narrow the wide topic of *leisure activities* and create a paragraph that focused only on *mountain biking*.

2.1 Prewrite for a Subject

Directions: Using any prewriting technique (listing, clustering, fastwriting, reporter's questions, or journaling), prewrite on three of the subjects below. Use a different prewriting technique for each subject you select. This will also help you discover ideas about the subject and support for these ideas.

driving age	worry	commercials
pollution	birthdays	body piercings

Discovering and Focusing Your Topic Sentence

A **topic sentence** has two parts:

- a *general subject*
- the *position* you're taking or *point* you're making about that subject.

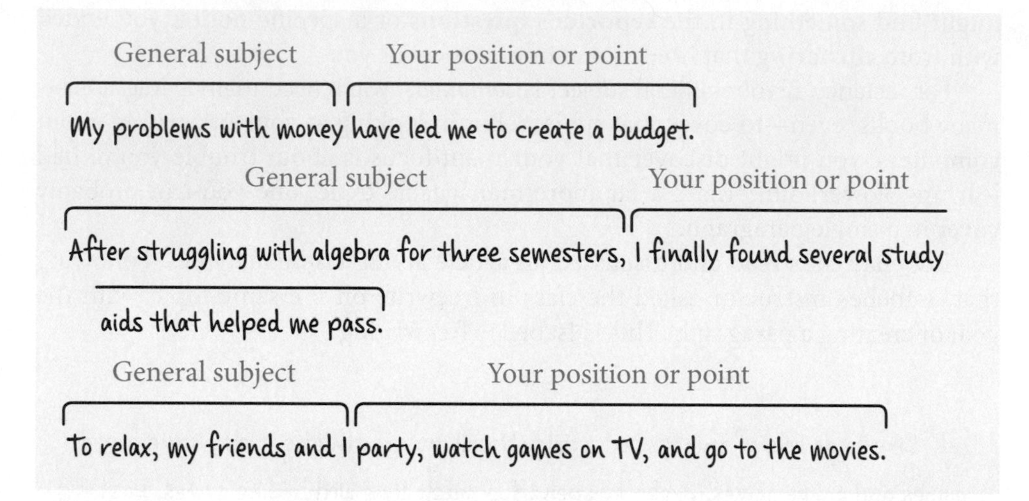

If a specific point doesn't jump out at you, you may need to do additional freewriting.

To discover and focus your topic sentence, follow these steps:

- ***Review*** everything you created during prewriting. Look for similar or related ideas, beliefs, or points that keep surfacing in your prewriting. These will most likely lead you to the subject you want to write about.
- ***Highlight*** these recurring ideas, beliefs, or points (sentences, fragments, phrases, or words) everywhere you see them.
- ***Examine*** all the elements you've highlighted and identify a particular point they have in common, an idea that links them together.
- ***Write*** a single sentence that expresses your general subject and that specific idea—your position or point. You now have your topic sentence.

Isabelle, who had decided to write about the general subject *mountain biking*, developed and answered reporter's questions to determine and narrow her topic sentence. Then she studied her responses and highlighted an idea that recurred: how mountain biking helped her *unwind*.

Topic: *Mountain biking*

Who usually rides with me?	Tanner, my best friend
What is special about mountain biking?	helps me unwind
Why do I enjoy mountain biking?	get away from city and work . . . relax
Where do I usually ride?	trails at state park
When do I usually ride?	weekends . . . time to unwind
How do I prepare for a day of mountain biking?	tune up bike, pack daypack
Why do I usually ride there?	trails are long, a lot of features, peaceful

How long are the trails? six trails = 56 miles

What features do the trails have? flat pine forests, breezy, quiet, secluded . . . picnic areas

Isabelle used the specific idea of *unwinding* as the basis of her topic sentence, which she composed this way:

General subject	Position or point
I enjoy mountain biking	because it helps me unwind.

2.2 Discover and Focus Topic Sentences

Directions: Below is a list of subjects and some background information on each. Working alone or in groups, choose three of the subjects and generate a topic sentence for each. You may use the background information given, or you might generate other ideas through additional freewriting.

Subject	Background information
Problems with air travel	have to get to the airport so early, extra charges with luggage, long wait after getting off plane, worries about security
"Staycation"	limited days off, too expensive to travel, sports commitments
Eco-friendly options	more online textbooks, fewer hard copy assignments, additional recycling bins
Collectables	baseball cards, angel figurines, *Star Wars* memorabilia
Classic movies	*Back to the Future, Rear Window, A Fistful of Dollars*

2.3 Discover and Focus Your Topic Sentences

Directions: Return to the subjects you wrote about in Ticket to Write 2.1. On a separate sheet of paper, use the steps in the bulleted list on page 16 (review, highlight, examine, write) to discover and focus topic sentences for these three subjects.

Supporting Your Topic Sentence

The largest part of your paragraph is made up of **supporting sentences**. These elaborate on, develop, or explain the position or point you state in your topic sentence.

To generate supporting sentences for her paragraph, Isabelle reviewed her prewriting a second time and put a check beside details that supported the idea she expressed in her topic sentence—all the details that elaborated on or explained how mountain biking helped her to unwind. Then her reporter's questions looked like this:

Topic: *Mountain biking*

Who usually rides with me?	Tanner, my best friend
What is special about mountain biking?	✓ helps me unwind
Why do I enjoy mountain biking?	✓ get away from city and work . . . relax
Where do I usually ride?	✓ trails at state park
When do I usually ride?	✓ weekends . . . time to unwind
How do I prepare for a day of mountain biking?	tune up bike, pack daypack
Why do I usually ride there?	trails are long, a lot of features, peaceful
How long are the trails?	✓ six trails = 56 miles
What features do the trails have?	✓ flat pine forests, breezy, quiet, secluded . . . picnic areas

Supporting sentences are the details that prove or illustrate the point of your topic sentence. They may be examples, explanations, facts, reasons, steps in a process, or any combination of these types of support. In the following examples, topic sentences are boldfaced.

Examples as supporting details

You have several ways to focus on crime prevention in your workplace. First, keep your purse or wallet with you or locked at all times. This also goes for any other valuables that you have. If you have personal items that can't be locked, like a coffee pot or radio, mark those things with your name or initials. Second, keep your personal life to yourself. Don't talk openly about your party or vacation plans. It's okay to share that information with close friends but not with those you don't know well. Last, let someone else know where you'll be, especially if you're working late or at odd hours. A little precaution can keep you much safer at work.

Explanations as supporting details

Because of good business practices, Greenville saw an explosion in population in the second part of the twentieth century. Beginning in the 1960s, civic leaders created and funded an economic development council that had the purpose of recruiting new businesses and industries. This led to the city eventually authorizing lucrative tax incentives for businesses that were deciding where to start and for businesses that needed money to expand. Because of these reasons, new businesses came, and existing ones expanded and hired more workers.

Facts as supporting details

Movies, a popular form of entertainment, have an interesting timeline. The first motion-picture camera was called a Kinetograph and was built in 1889. Six years after that camera was invented, the first Kinetoscope parlor opened in New York, and people paid a quarter to see a film there. *The Great Train Robbery*, a famous silent picture and the first Western, was shown in 1903. "Talkies" became possible in 1910, but they weren't shown widely until the 1920s. Winsor McCay introduced the first animated cartoon in 1914, and Walt Disney created his first cartoon ten years after that. Feature-length talkies debuted in 1927 with Al Jolson's movie *The Jazz Singer.* This is the type of movie that has remained the most popular and is what is usually shown in today's movie theaters.

Reasons as supporting details

Darla has vowed to never go to the Thrifty Palace again. On her last trip, she saw that the price that was marked for the microwaveable dinners she wanted was almost double what she would pay at a grocery closer to her home. Also, when she asked a clerk where she could find paper plates, the clerk said he didn't know. His tone told Darla that he thought she was really rude to ask, too. The third reason that Darla won't return to Thrifty Palace is that she noticed when she got home that the clerk had overcharged her for three items she had bought. Because of this terrible shopping experience, Thrifty Palace will never see Darla again.

Steps in a process as supporting details

Everyone should know how to wash clothes, and learning how is easy. First, sort your clothes into three piles—one for whites, one for bright colors, one for dark colors. As you're doing this, check your pockets for any extra change or other things you might have left in them. Next, look for any stains you have and use a stain remover on them. Then look at the detergent bottle or box and measure whatever amount the manufacturer tells you. Pour that amount into the washer or the dispenser in the washer. On the washer, choose the water temperature you need for the kind of clothes you're washing. Also adjust the water level depending on the size of the load. Finally, put your clothes in the washer and close the lid. The washing cycle usually takes about forty-five minutes.

2.4 Generate Your Support

Directions: Return to the three topics you wrote about in Ticket to Write 2.1 and 2.3. Choose the topic that interests you the most and freewrite on this topic to develop supporting details. Your supporting details could be *examples, explanations, facts, reasons, steps in a process*, or a combination of these.

Organizing Your Supporting Sentences

For your supporting sentences to be effective, arrange them in a logical way. Depending on your subject, you might choose to arrange them by

Time order is also called *chronological order.*

- **time** (usually earliest to latest)
- **space** (from front to back, top to bottom, side to side, or any other series of direction)
- **importance** (most to least important and vice versa)

Heads Up!
Isabelle didn't use all the details she listed in her reporter's questions; she eliminated ones that didn't support the idea of unwinding.

Isabelle decided to arrange her supporting details in **time** (**chronological**) order. Using her topic sentence and supporting details, she developed this paragraph:

Topic sentence

I enjoy mountain biking because it helps me unwind. On the trails at our state park, I can get away from the stress of work and enjoy nature. Every Sunday I meet my friend Tanner at the trail head, and we leave behind all the sights and sounds of the city. Once we hit one of the long trails, I let go of the tension from work and enjoy the company of a good friend on a peaceful ride.

In the flat pine forest we enjoy the shade and cool breezes. This is where we always stop for rest, water, and a quick snack at one of the picnic tables. At the end of the ride, I'm physically tired but my mind is relaxed and I can face another week at work.

2.5 Generate Supporting Details

Directions: The topic sentences below introduce paragraphs with supporting sentences arranged in one of the three ways listed above. One supporting sentence is provided for each topic sentence. Compose two other possible supporting sentences for each topic sentence. Answers will vary.

1. **Paragraph with supporting sentences arranged in time order**

   Topic Sentence: I can make it to my favorite weekend getaway in just under two hours.

   1. First, I leave my house and take the south ramp to get on the highway.

2. **Paragraph with supporting sentences arranged in order of importance**

   Topic Sentence: A recent survey at work rated the qualities of a good boss.

   1. Being open to suggestions was the third most popular response.

3. **Paragraph with supporting sentences arranged in space order**

   Topic Sentence: I'm really proud of my first sports car.

   1. On the front I put a new vanity plate that says "Smokin' Hot."

4. **Paragraph with supporting sentences arranged in time order**

   Topic Sentence: Downloading a free ringtone from a computer is an easy process.

   1. First, find a site that offers the song you want.

5. **Paragraph with supporting sentences arranged in importance order**

   Topic Sentence: If you're hungry, getting to the closest off-campus restaurant is easy.

   1. Exit campus through the east gate and turn left.

2.6 Organize Your Support

Directions: Return to the topic and supporting details you created in Ticket to Write 2.4. Organize your supporting sentences by time, space, or importance.

Identifying Irrelevant Sentences

Sometimes writers get carried away and include ideas that aren't directly linked to the position or point they take in their topic sentence. Read this paragraph and find the sentence that doesn't support either the writer's subject or main point.

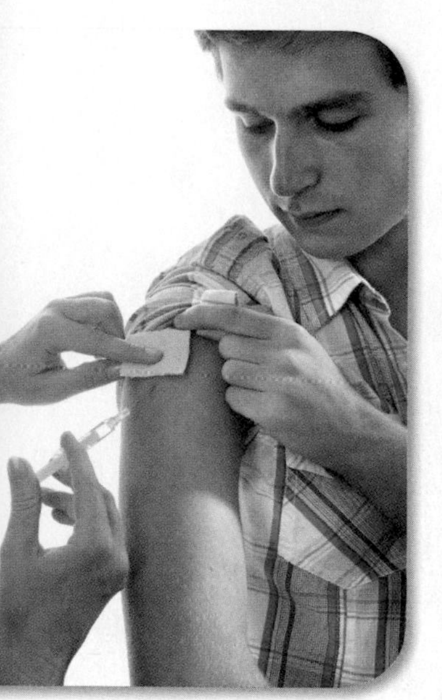

Position or point — General subject

You can take several preventative measures to avoid getting the flu. When flu shots become available, be sure to get one. Also, wash your hands after you've been in a crowd. If you're like me, you don't like to shake hands at all. After blowing your nose, throw the tissue away and then wash your hands. If you don't have a tissue, sneeze into your shoulder, not your hands. You will lessen your chances of getting the flu by following as many of these procedures as you can.

Here the writer presents several *preventative measures* to support her topic sentence. All four of these ideas illustrate ways *to avoid getting the flu.* However, the sentence

If you're like me, you don't like to shake hands at all.

doesn't belong because it isn't directly linked to the position or point set up in the topic sentence. The writer's dislike of shaking hands doesn't elaborate on measures to prevent getting the flu.

2.7 Identify Irrelevant Details

Directions: Read the paragraphs below and cross through the sentence or sentences that should be deleted because they contain irrelevant details.

1. I can get to my favorite weekend getaway in just under two hours. ~~It takes my friend Marty only a half hour, but he doesn't have to face as much traffic as I do~~. First, I leave my house and take the south ramp to get on the highway. Then I travel east on Route 54 for 65 miles. ~~Route 54 is designated a Scenic Highway, and I enjoy driving it.~~ Route 54 comes to an end, and I take a left on US 37. ~~Five miles after that, I usually stop to get some groceries at Lucy's Little Store~~. After that, I continue on another ten miles to the turn-off that takes me up the hill to the cabin I rent on Indian Lake.
2. I'm really proud of my first sports car. On the front I put a new vanity plate that says "Smokin' Hot." It's a mirror-backed plate

with metallic blue lettering, which matches the metallic blue of the car. I had it pinstriped with a thin line of smoke curling from the front quarter panels all the way down the sides to the rear quarter panels. ~~My cousin Jimmy does awesome detail work and gave me a good deal on the paint job~~. On the front quarter panels, I installed chrome alloy side vents that swoop upward toward the hood. Above the side vents, on the outer edge of the hood, I put two chrome hood scoops with five vent slots. ~~Once again, Jimmy came through for me and didn't charge me anything over cost for the eye candy~~. Behind the quarter panel vents, the doors were trimmed out in chrome that matched the vents and my new 22-inch spoked chrome wheels. ~~My dad says I wasted a lot of money, but he just doesn't understand the importance of a smoking hot car.~~

3. A recent survey at work rated the qualities of a good boss. ~~I work at Talbot Productions, and three people in management recently left for other jobs~~. Being open to suggestions was the third most popular response. Next came being willing to support the people who work under the bosses. This included being helpful and giving constructive criticism. ~~My boss, Anthony Hardinger, has never helped me, even though I've asked him to a bunch~~ of times. The quality that was rated highest was recognition of good workers. ~~Mr. Hardinger has hardly ever spoken to me, much less recognized that I've done a good job.~~

4. Downloading a free ringtone from a computer is an easy process. ~~I recently got Rihana's "Rude Boy," and I love it~~. First, find a site that is trustworthy and offers the types of ringtones you want. Next, preview the ringtones and listen to the sound bytes to determine if the sound quality is adequate. After you find the ringtone you want, enter the name of your carrier and your cell phone number. Then, before you send the ringtone to your phone, make sure you haven't agreed to receive any unwanted e-mail subscription or push messages. ~~I already get a lot of pushes I don't want from my carrier, advertising texting and calling deals~~. Finally, have the ringtone sent to your phone.

Drafting an Essay

Paragraph writing centers on a topic sentence. Essay writing centers on a **thesis statement**. Just as your topic sentence is the main idea of your paragraph, your **thesis statement** is the main idea of your essay. All details or other support in your essay must relate to and support your thesis statement. Put another way, your thesis statement is the key thought you want your readers to understand fully after having read your essay.

LEARNING GOAL

❷ Draft an essay by developing a thesis statement, organizing support, and ensuring unity and coherence

An **essay** is a nonfiction composition addressing a single subject.

A **thesis statement** should be

- your opinion, observation, or idea about your topic
- a straightforward, declarative sentence
- a reflection of your knowledge, experience, or beliefs
- only one idea

A declarative sentence issues a statement, *not* a question, command, or exclamation.

Here are four common mistakes new writers sometimes make when developing thesis statements:

- **Make an announcement**

Unacceptable In this paper, I will discuss four reasons that the college should lengthen spring break to two weeks.

Acceptable Dellwood College should lengthen spring break to two weeks.

Unacceptable The thesis of this essay is that George Washington's first week in office shaped the course of US history.

Acceptable George Washington's first week in office shaped the course of US history.

- **State a fact**

Unacceptable The Ohio River is formed when the Allegheny and Monongahela rivers meet in Pittsburgh. *(Since this is a fact, the writer has nothing to create an essay about.)*

Unacceptable The first area code in New York City was 212. *(Since this is a fact, the writer has nothing to create an essay about.)*

- **Make a statement that is too broad**

Unacceptable America's health-care industry will face many changes in the coming years.

Acceptable The fastest-growing segment of the health-care delivery system is the home health business.

Unacceptable Rules have changed in baseball since it began in the 1800s.

Acceptable Major League Baseball's allowing limited instant replay has improved the game.

- **Make a statement that is too narrow**

Unacceptable The cost of a sirloin steak at Ricky's Surf and Sirloin is shocking.

Acceptable At several local restaurants, prices of food and drinks are shocking.

Unacceptable Wednesday is quiz day in my history class.

Acceptable Because we have a twenty-question quiz every Wednesday in history class, students are overworked.

Discovering and Focusing Your Thesis Statement

To discover your thesis statement, take the same approach as you did to discover your topic sentence. Begin by prewriting about your topic, using any technique you prefer (see Chapter 1). Next, highlight all the statements that reflect a recurring main idea or point. Then, review what you highlighted and answer these **focus questions:**

- **What *idea* dominates in my prewriting?** (What is the main *subject* of my prewriting?)
- **What is the *main point* I want to make about my subject?**

Now write a single sentence that expresses the specific idea you want to share. Use this as your thesis statement. During discovery drafting, a thesis statement is referred to as a **working thesis statement** because writers often tweak the wording of their working thesis statement as they develop support for it.

Jeong's assignment was to freewrite about his college experience and then highlight recurring points. Here is what Jeong composed and the points he highlighted:

For more about thesis statements and focusing, search the Internet for these videos:

Thesis: How to Write a Thesis Statement for Your English Essay

Thesis Statements, Focus (Video).

"When genuine passion moves you, say what you've got to say, and say it hot."
—D. H. Lawrence

College is nothing like high school. don't have to attend class . . . not all my profs take roll. If I miss class gotta get notes from class website or student. Seems like everyone pushed me to do well—Ms. Connie—counselor, Mr. Russell, homeroom teacher. He was the best. Learned the hard way not to rely on others for notes. Brian's notes too messed up to copy. Mr. R. always made sure we had our homework and stuff before we left. If you missed it in some class he'd give you a hall pass to go get it. All teachers stayed after school—some longer than others. Mr. R. brought in snacks if someone was dragging. college profs aren't always in their offices when I go looking for them. Part-time faculty don't have offices hrs. surprised to find I get a daily grade for participation. No one checks up on me every day to see if I'm missing an assignment or falling behind. free tutoring for lots of classes—not all—have to go to Academic Services to sign up and find a tutor. It's across campus . . . hassle. Tutors are usu. students. writing ctr has student tutors and faculty tutors but it's not open all week. Online tutoring 24/7 and doesn't cost anything. Bulletin boards advertize study groups anybody can join. I study in the library . . . can bring in a soda. Can get 5 extra points on final grade if I don't miss any classes.

Reviewing his highlighting, Jeong noted he kept emphasizing the idea of *personal responsibility*, so he answered the two focus questions this way:

- **What *idea* dominates in my prewriting?**

  what others did for me in high school, I have to do for myself in college

In **discovery drafting,** you explore the directions your writing might take. A *discovery draft* is also called a *rough draft, working draft,* or *first draft.*

- **What *main point* do I want to make about this idea?**

  college success is up to me

Here is Jeong's working thesis:

My success in college is up to me and does not rely on others.

TICKET to WRITE

2.8 Develop a Working Thesis Statement

Directions: Below is a list of subjects and three topic ideas about each. First, work alone or in groups to generate a fourth topic idea for each subject. Then, compose a working thesis statement for each subject. Your working thesis statement will reflect the main idea (the subject) and will focus on the point you want to make about it.

Subject	***Topic ideas***
Class reunion	old friends, childhood rivals, former teachers, ____________
Fallen heroes	drug use, marital problems, alcohol abuse, ____________
Spring break	work extra hours, catch up on class work, get out of town, ____________
Weight loss	exercise, calorie reduction, appetite suppressants, ____________
Academic dishonesty	failing grade in class, withhold degree, expulsion, ____________

2.9 Develop Your Working Thesis Statement

Directions: Choose one of the subjects below or one that your instructor gives you and freewrite about that subject. When you have completed your freewriting, review it and highlight statements that reflect a recurring idea. Then reexamine your highlighting, looking for a dominant idea and the main point that you make about this idea. Use these to compose a working thesis statement for an essay. Share your working thesis statement with your writing group.

unusual occupations
personal happiness
youth organizations
favorite charities
eating disorders
student government
leisure activities
campus diversity
study habits
family dinners

Supporting Your Working Thesis Statement

You used one color to highlight all your statements that reflected a recurring main idea or point. That led you to develop your working thesis statement. Now use a different color to highlight each sentence, fragment, phrase, or word that illustrates or explains your working thesis. These are the ideas that could make up the topics of supporting paragraphs. Jeong reviewed his freewriting to discover those supporting ideas. He wrote his working thesis at the top of his freewriting. With this statement at the top of his page, Jeong could keep focused on the idea he wanted to support.

My Working Thesis: My success in college is up to me and does not rely on others.

College is nothing like high school. don't have to attend class . . . . not all my profs take roll. . If I miss class gotta get notes from class website or student. Seems like everyone pushed me to do well—Ms. Connie—counselor, Mr. Russell, homeroom teacher. He was the best. Learned the hard way not to rely on others for notes. Brian's notes too messed up to copy. Mr. R. always made sure we had our homework and stuff before we left. If you missed it in some class he'd give you a hall pass to go get it. All teachers stayed after school—some longer than others. Mr. R. brought in snacks if someone was dragging. college profs aren't always in their offices when I go looking for them. . Part-time faculty don't have office hrs. surprised to find out I get a daily grade for participation. No one checks up on me every day to see if I'm missing an assignment or falling behind. free tutoring for lots of classes—not all—. have to go to Academic Services to sign up and find a tutor. . It's across campus . . . . hassle. have to get notes firsthand . . . supposed to take part in discussions . . . not used to doing that . . . if I'm in class, I hear the lecture and not my friend's version of it . . . office hours for profs not always good for me . . . can get help for classes here if I go to it . . . . Tutors are usu. students. writing ctr has student tutors and faculty tutors but it's not open all week. Online tutoring 24/7 and doesn't cost anything. Bulletin boards advertize study groups anybody can join. I study in the library . . . can bring in a soda . . . Can get 5 extra points on final grade if I don't miss any classes.

Organizing Your Ideas

Jeong noticed his supporting ideas fell into two main groups: *attending class* and *finding outside study help*. He used these two ideas to begin organizing and expanding on support for his working thesis statement.

Attending class	Outside study help
no one checks up on me	college professors not always in their office
get notes firsthand	Is up to me to find
take part in class discussion	office hours (not always convenient for me)
hear the real lecture, not student version	free tutoring @ Academic Serv. (across campus)
no absences = extra 5 pts	writing center not open all week study groups

Next, Jeong wrote out topic sentences for the two ideas that supported his working thesis. These two topic sentences will form the basis of two paragraphs to support his working thesis statement. Paragraphs supporting the thesis are called **body paragraphs**.

- **Working thesis statement**
  My success in college is up to me and does not rely on others.
- **Topic sentence for first body paragraph**
  To be successful, I have to attend class and keep up with class work.
- **Topic sentence for second body paragraph**
  Now that I'm in college, I need to find and take advantage of study help.

With these key sentences, Jeong was ready to begin drafting his body paragraphs by consulting his prewriting. Here is his discovery draft of these two body paragraphs:

To be successful, I have to attend class and keep up with my class work. When I miss class, I have to depend on others to give me the notes, but I can never be sure that another student will take notes the way I would or that the notes are as reliable as mine would be. Also, if I'm in class, then I can participate in class discussion. For example, I've noticed that when I take part in the discussion in Mr. Reid's Early American History class, I understand the material better and the causes behind certain events. Hearing the lecture and taking part in the discussion firsthand gives me Mr. Reid's perspective without the bias of another student reporting it to me. And if those aren't reasons enough to attend class some of my teachers give extra credit points for attending class.

Now that I'm in college, I need to find and take advantage of study help. Unlike high school, when your grades start falling, counselors don't call you up or pull you out of class. Instead, I have to find ways to raise my grades on my own. I learned the hard way not to depend on catching some profs in their office because they sometimes don't keep office hours or their hours don't match up with mine.

Organizing by Outlining

Some writers find that creating an outline during their prewriting and discovery drafting helps them organize their thoughts, note relationships between ideas, and discover gaps that need additional attention.

Writing Tip: Type your thesis statement at the top of your outline to remind yourself of your focus and to ensure you don't stray from your main point.

In a formal outline, you

- use **Roman numerals** for your main points (I, II, III, etc.)
- indent and label each subordinate point using **capital letters** (A, B, C, etc.)
- further indent and label each subordinate point about the topic in capital letters using **Arabic numbers** (1, 2, 3, etc.)
- indent again and label each subordinate point about the topic in Arabic numbers using **lowercase letters** (a, b, c)
- ensure that you have **at least two entries** for each category (Roman numerals, capital letters, Arabic numbers, lowercase letters)

Amanda wrote about shopping online and shopping in a mall. After she examined her prewriting, she organized her thoughts into an outline, which looked like this:

I. Online
    A. Lack of restrictions
        1. More comfortable
            a. Don't have to be dressed
            b. Don't have to walk around
                (1) Can sit in favorite chair
                (2) Can lie on floor
        2. Can shop anytime
            a. Can stop for phone or bathroom breaks
            b. Stores don't close at 9 p.m.
        3. No controls about food/drink
            a. Can eat or drink what I want
            b. Save money on food and drink

(continued)

B. More choices of stores

1. Some national stores not located locally

2. Some international stores not located locally

II. Mall

A. Can try on to check for fit

1. Clothes

2. Shoes

B. Can feel texture of items for sale

C. Can get questions answered sooner

2.10 Developing an Outline

Directions: When writing the outline above, Amanda noticed she had fewer details about shopping in a mall than she had about shopping online. Working alone or in groups, brainstorm additional items Amanda could add about shopping in a mall. Copy the second major section (11.Mall) and its outlined ideas. Then, add the new items to Amanda's formal outline.

2.11 Find Support for a Thesis

Directions: Below are five working thesis statements and topic sentences for their body paragraphs. Working alone or in groups, compose additional supporting sentences for each body paragraph. Answers will vary.

1. ***Working thesis statement:*** The American Red Cross says that three steps for disaster preparedness are assembling a basic supply kit, making a plan, and being informed.

   Body paragraph 1 topic sentence: A basic supply kit should have food supplies, medical supplies, and documents.

   Supporting details sentences: ______________________________

   Body paragraph 2 topic sentence: Everyone in the house should be involved in planning.

   Supporting details sentences: ______________________________

   Body paragraph 3 topic sentence: All family members should also be informed.

   Supporting details sentences: ______________________________

2. **Working thesis statement:** Among the activities that are available this weekend are going to the movies, watching television, and attending sporting events.

   Body paragraph 1 topic sentence: A number of movies are showing this weekend.

   Supporting details sentences: ______________________

   Body paragraph 2 topic sentence: Television always has a lot to offer on the weekends.

   Supporting details sentences: ______________________

   Body paragraph 3 topic sentence: Many different sports are played in our area.

   Supporting details sentences: ______________________

3. **Working thesis statement:** My best friend recently moved across the country, and I miss many things about her.

   Body paragraph 1 topic sentence: I miss her sense of humor.

   Supporting details sentences: ______________________

   Body paragraph 2 topic sentence: I also miss her impulsiveness.

   Supporting details sentences: ______________________

   Body paragraph 3 topic sentence: Most of all, I miss the way we depended on each other.

   Supporting details sentences: ______________________

4. **Working thesis statement:** Josh went through several stages when he came down with the flu.

   Body paragraph 1 topic sentence: Josh first had both chills and a fever.

   Supporting details sentences: ______________________

   Body paragraph 2 topic sentence: Although he stayed home from work and school to rest, Josh couldn't get much sleep.

   Supporting details sentences: ______________________

   Body paragraph 3 topic sentence: Josh finally began feeling better after about four days of misery.

   Supporting details sentences: ______________________

5. **Working thesis statement:** Today I have three reasons to smile.

   Body paragraph 1 topic sentence: The first reason I'm smiling has to do with my work.

   Supporting details sentences: ______________________

   Body paragraph 2 topic sentence: I'm also smiling because of something that happened at home.

   Supporting details sentences: ______________________

   Body paragraph 3 topic sentence: My biggest smile comes because of college.

   Supporting details sentences: ______________________

Writing Tip: Remember that you might not use every detail you listed in your freewriting, and you might find additional ideas as you compose your topic sentences and supporting details.

2.12 Develop Supporting Details for Your Essay

Directions: Return to your freewriting, working thesis statement, and organizing. Reexamine your work and look for at least two different ideas that support your working thesis statement. Compose topic sentences for each of those separate ideas. Then, organize supporting details for each of your topic sentences and create your body paragraphs.

Ensure Unity and Coherence in Paragraphs and Essays

LEARNING GOAL
3 Ensure Unity and Coherence

Unity

Unity centers on **relevance**—a direct connection with your subject. Everything you write in a **paragraph** must be relevant to your **topic sentence**; everything you write in an **essay** must be relevant to your **thesis statement**. Unity is lost when you include

- ideas that are out of place
- ideas that don't logically relate to each other
- ideas that don't further explain what you state in your topic sentence or thesis statement
- ideas that are not relevant to the topic of your paragraph

2.13 Eliminate Irrelevant Details

Directions: The following body paragraphs come from an essay about changes a student recommends to improve college football. Find and underline the topic sentence of each body paragraph. Then, ask yourself which sentences are not relevant to what the writer expresses in the topic sentences (which sentences don't contribute to the paragraph's unity) and cross them out.

Thesis statement: To decide the national champion, the NCAA should replace the current system and use playoffs games.

Currently, the college football championship is decided by polls from coaches and the media. The people who vote in these polls might not be objective, so rankings that come from their votes might not be objective either. ~~President Obama said that he is favor of a playoff system.~~

Also, a team can lose early in the season and then be out of the running to be champion, even though it might improve the rest of the season. ~~Here at Oak Grove State, we're currently 4-1, and fans hope we're not going to lose any other games.~~ Look at last year's records. Burton University lost the first game and then won all the rest of them. McCarty College lost its second game when its star quarterback was out with mono, and then it won all the rest by at least twenty points.

A third reason to have playoffs is that sometimes schools schedule games against weaker opponents just so they can have a good record at the end of the season. If a school with a great team plays a school with a lousy team, everybody knows the great team will almost always win. ~~I think that's what Bell A&T has done for years~~. It's not fair to schedule games against weaker teams just so a team can have a good record.

Coherence

Coherence in writing comes when all the parts are clearly connected and logically organized.

Coherence in a Paragraph

In a **paragraph**, coherence means that ideas in individual sentences all relate to the topic sentence and all flow logically. Also, each sentence should relate to the sentence before it and the sentence after it. For example, a sentence may do the following:

- **Illustrate an idea**

  Topic sentence: The national economy is experiencing an upswing.

  Illustration of idea in topic sentence: According to *Time*, US auto sales are up 11.5% in the last month.

- **Offer an explanation**

  Topic sentence: Hockey star David Lemonde is out for the season.

  Explanation of idea in topic sentence: He reinjured his knee in last night's game and will have surgery later this week.

- **Relate steps or stages**

  Topic sentence: Weddings aren't supposed to be peculiar, but the one I was in last week certainly was.

  Explanation of stages in topic sentence: First, the rehearsal dinner took place on a miniature golf course.

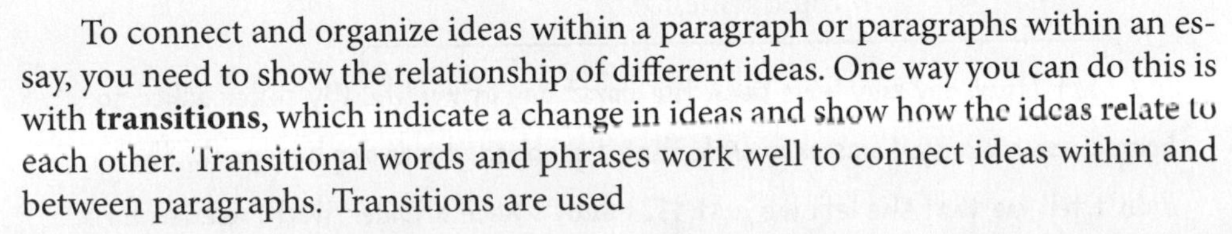

To connect and organize ideas within a paragraph or paragraphs within an essay, you need to show the relationship of different ideas. One way you can do this is with **transitions**, which indicate a change in ideas and show how the ideas relate to each other. Transitional words and phrases work well to connect ideas within and between paragraphs. Transitions are used

- to show how individual ideas relate to topic sentences or thesis statements
- to show the relation between ideas
- to help the writing read smoothly and logically

The following chart lists some common types of transitions and defines their purpose.

Writing Tip: Some transitions can be used in more than one way.

Common Transitions	
Addition	again, along with, also, and, further, in addition, likewise, thus
Chronology	after, another, before, during, finally, first (second, third), immediately, in the meantime, last, next
Concession	although, at least, (even) though, in spite of, of course, still
Comparison or **Contrast**	at the same time, comparatively, conversely, however, in spite of, instead, likewise, moreover, nevertheless, on one hand, on the contrary, rather, similarly, yet
Detail	including, in particular, namely, specifically, to list
Emphasis or **Clarification**	above all, again, certainly, especially, furthermore, in addition, particularly, truly
Example	chiefly, for example (instance), in other words, in particular, mainly
Space	above, along, amid, atop, behind, below, near(by), next (to), opposite, surrounding, under
Similarity or **Difference**	along with, but, despite, however, in contrast, likewise, regardless too, yet

In the paragraph below, Cameron made no mistakes in grammar, punctuation, or spelling. The problem with his paragraph is that readers can't see how his statements relate to each other.

Topic sentence

Last Thursday may have been the worst day of my life. My sister asked to borrow some money. I was okay with that. I grabbed the wrong backpack. She didn't tell me that she left me just $5. I didn't have a folder that I needed for a meeting. My boss was fired up at me for all the trouble I'd caused. My $10 pledge to the flower fund at work was due that day. My boss told me to make a quick trip home to get the folder. I couldn't find the cell phone anywhere. I flagged down a friend who happened to be passing by. I got a ride home, got the folder and got back to work. My truck died. I was still late for the meeting.

When you read Cameron's topic sentence, you knew what the rest of his paragraph should have told you—the details about why the day had been so terrible for him. However, because the paragraph lacked coherence (the sentences weren't clearly connected or logically organized), you probably found it confusing.

For the paragraph to be coherent, Cameron needed to rearrange some of the sentences and also to add a few details and transitions. In his rewritten example below, Cameron organizes his details chronologically, and the transitions he includes to ensure coherence are highlighted.

Topic sentence

Last Monday may have been the worst day of my life. First, my sister asked to borrow some money. While I was okay with that, she didn't tell me that she left me just $5. My $10 pledge to the flower fund at work was due that day, so I couldn't pay it. When I got to work, I realized that I'd grabbed the wrong backpack. Because of that, I didn't have a folder that I needed for a meeting. My boss told me to make a quick trip home to get the folder, but on the way home my truck died. I searched for my cell phone to call for help, but I couldn't find it anywhere. Finally, I flagged down a friend who happened to be passing by. With her help I got a ride home, got the folder and got back to work. But I was still late for the meeting, and my boss was fired up at me for all the trouble I'd caused.

Cameron's revised paragraph reads more smoothly not only because he puts the causes for his bad day in **time** (chronological) order, but also because he uses transitional words and phrases to connect the details of his day.

Cameron arranged his body paragraphs in time order because that fit his topic of relating events as they occurred on a particular day. Other types of order, however, might accent the point of your thesis statement.

Common Types of Order

Order	Definition	Example
Time	from first to last or latest	*first* days of your college life
Importance	from least to most important	*qualities to look for* in an apartment
Space	any spatial order (top to bottom, left to right, or any other directional arrangement)	*looking down* to the bottom of the Grand Canyon

TICKET to WRITE

2.14 Order of Time, Space, and Importance

Directions: Read the following thesis statements. Working alone or in groups, determine if the body paragraphs that would follow them should be arranged according to time, space, or importance. In the spaces provided, write *T* for time, *S* for space, or *I* for importance.

__S__ 1. Looking around the classroom, I see several pieces of equipment that are unfamiliar.

__I__ 2. Sportscasters are citing three reasons the Mustangs will win the Super Bowl this year.

__T__ 3. Brandon's outlook on education was influenced by his first day in kindergarten, his seventh grade teacher, and an encounter with his high school principal.

__S__ 4. Surveying the crime scene, Sheriff Abbott noted the ceiling sprayed with white paint, the walls with blue paint, and the floor with green paint.

__T__ 5. In the early 1900s, women were told that their names should appear in the paper only when they were born, when they married, and when they died.

__I__ 6. After I stopped smoking, my clothes didn't smell anymore, I had more spending money, and my health improved dramatically.

__T__ 7. After much deliberation the previous night, Tyler decided that the best use of his time was to finish his homework, then study for his test, then go to the party.

__S__ 8. After the student center is completed, looking left from there you'll see Moss Library, Sommers dorm, and then Busby Student Center.

__I__ 9. Wearing a bicycle helmet is inexpensive, is sometimes mandatory, and can save lives.

__T__ 10. I decided my party affiliation, registered to vote, and then proudly cast my first ballot.

2.15 Order of Body Paragraphs in Your Essay

Directions: Look back at your working thesis statement, your topic sentences, and your supporting details. Decide which order would be best for arranging your body paragraphs and write them in that order. If you have word processed your paragraphs, you can use the copy and paste features to arrange your paragraphs. Then share your ordered body paragraphs with members of your writing group and see if they agree with the order you have chosen.

Coherence in an Essay

In an **essay**, the relationship of ideas and their logical flow is important

- within each paragraph
- in how individual paragraphs relate to each other
- in how individual paragraphs relate to the thesis statement

Ta'Nia was drafting an essay about the advantages of Phi Omega Alpha membership. Below are her working thesis statement and two body paragraphs. The problem with her paragraphs is that readers can't see how the first body paragraph relates to the second one.

Working thesis statement

Becoming a member of Phi Omega Alpha has a number of advantages.

They are allowed to park in the faculty and staff lots. POA members can check out library materials for an extra week. They are allowed to register a week early for classes.

Businesses offer discount coupons to POA members. Some companies offer hiring preferences to students who are members of Phi Omega Alpha.

Ta'Nia's two body paragraphs are choppy. She realized transitions and additional details would improve the flow of these paragraphs. Below is her first revision (transitions are highlighted and new details are in blue):

Working thesis statement

Becoming a member of Phi Omega Alpha has a number of advantages.

First, members are allowed to park in the faculty and staff lots, including the coveted lots that are next to classroom buildings. POA members can also check out library materials for an extra week, even during finals week. Plus, they are allowed to register a week early for classes.

In addition to on-campus advantages, POA members also receive perks from area businesses. Businesses, especially restaurants, offer discount coupons to POA members. Above all, some companies offer hiring preferences to students who are members of Phi Omega Alpha.

2.16 Add Transitions

Directions: Add the appropriate transitional words and phrases to the paragraph below.

along with	before	but	finally
first	however	next	then

To copyright a DVD you create, first log on to the Web site for the US Copyright Office. Next, fill out the form labeled CO (copyright office). Then submit your form electronically, which will begin the process. You will, however, need to send a hard copy via the mail, along with your DVD. Finally, pay the registration fee. Your work is registered, but you may have to wait up to six months before you receive your official certificate.

2.17 Add Transitions and Details

Directions: The paragraph below lacks both the details and transitions needed to make it read well. Working alone or in groups, rewrite this paragraph, adding appropriate details and transitions.

Watching a movie at home is different than watching one in a theater. At home, you can wear what you want. You can eat or drink what you want. You can watch the movie when you want. A theater has a larger screen than you have at home. A theater has a better sound system. In a theater, you get better popcorn.

2.18 Adding Transitions and Details to Your Essay

Directions: Look at your working thesis statement and the body paragraphs you have composed. Circle or underline the transitional words or phrases you used. Then, trade drafts with a member of your writing group. As you read each other's draft, look for

- places where additional transitions are needed
- places where transitions need revising

Introductions, Conclusions, and Titles

A strong introduction that engages your reader and clearly states your point and a conclusion that sums up your main point without introducing new ideas will make your paragraph or essay more effective.

LEARNING GOAL

❹ Write an introduction, conclusion, and title for an essay

Introductions

One of the most common challenges for all writers—not just student writers—is getting started. In your introduction, you have three obligations to your readers. You must (1) hook them, (2) alert them to your topic, and (3) make your point of view clear to them.

If you're writing a **paragraph**, your introduction comes in your topic sentence alone, so you must be succinct. Your instructor may prefer your topic sentence to come first so that your readers will immediately recognize your topic.

If you're writing an **essay**, your instructor may prefer your thesis statement be near or at the end of your introductory paragraph. In that case, the first sentences of your introductory paragraph will give information—background, explanatory, chronological, or some other type—about your thesis statement.

Types of introductions for essays

- **General subject**

  This is probably the most common way to introduce an essay. With this, you first write something about the general subject of your essay and then narrow the focus of your introduction until it ends with your thesis statement.

> Every generation has its own catch phrases and popular slang words. Some of these now popular phrases used to have dirty meanings. **However, today's younger generation has given new, less offensive meaning to some slang.**

- **Background detail**
  In this introduction, you offer your readers points that explain the circumstances or events that lead to or explain the position or point in your thesis statement.

> In my family, my father was the disciplinarian. Although he was my biggest fan in every sport I played, he was also my biggest critic. When I was growing up, I vowed I'd never copy my Dad's bad habits. **Now that I have children of my own, I realize I have become my father in many ways.**

- **Anecdote**
  Providing an anecdote (a short account of an incident) is a good way to illustrate either your subject or the point you are making about the subject. Your anecdote acts as a lead-in to or explanation of your thesis statement.

> Aunt Noelle was a second mother to me. When her husband, Uncle Larry, called to say he had taken her to the hospital, I went there as fast as I could. Noelle hadn't been herself for quite some time, but no one had said she was ill. Uncle Larry met me in the hall and said Aunt Noelle had been diagnosed with Alzheimer's. When I went into Noelle's room, I said hi to her, but she didn't answer me. Then she almost screamed at me, "Who are you and what are you doing here?" **That's when I realized that Alzheimer's is one cruel disease.**

- **Question or questions**
  One method of piquing your readers' interest is opening with a thought-provoking question or series of questions that your essay will answer.

> We all have people we are close to, people we call good friends. Sometimes these are people we've known for a long time, and sometimes they're people we've just recently bonded with. **What makes someone become a good friend?**

- **Surprising information**
  Giving your readers some unexpected information can immediately hook them and stir their interest in your topic.

During World War II, many British airmen were held in German war camps. As prisoners, these men were entitled to receive care packages from charity groups at home. What the Germans didn't know, though, was that sometimes the care packages contained a compass, metal tools, and maps of the area. All of these were intended to help the POWs escape from the camps. British groups sneaked these items past the Germans through, of all things, the board game Monopoly.

- **Pertinent quotation**
  A quotation from a respected or well-recognized source can give credibility to your position and alert your readers to the significance of your topic.

Late-night comedian David Letterman said, "This isn't brain surgery; it's just television." I've become fed up with the media attention so many TV stars get for doing nothing more than being on television. **Many seem to think just because somebody's on TV for one reason or another, that celebrity's every word is important, but it's not.**

In your introduction . . .

- **Do not directly announce your topic and point of view**

  Unacceptable In this essay, I will classify three types of foods that are beneficial to everyone.

  Acceptable Three types of foods are beneficial to everyone.

- **Do not give a dictionary definition**

  Unacceptable According to *Webster's* dictionary, "beneficial" means "producing or promoting a favorable result; advantageous."

  Acceptable Beneficial foods promote healthy living.

- **Do not indicate that a topic is a personal opinion**

  Unacceptable I think that people should eat more foods that are beneficial.

  Acceptable People should eat more foods that are beneficial.

TECHNO TIP

For more about introductions to essays, search the Internet for these videos:

How to Write an Introduction Paragraph for Your Essay

English Shorts Writing an Introduction

TICKET to WRITE

2.19 Identifying Types of Introduction

Directions: Review the paragraphs below and determine what type of introduction each is. Write *GS* for general subject, *BD* for background detail, *A* for anecdote, or *SI* for surprising information.

SI 1. As identity theft increases, many people are using their cell phones instead of land lines. They feel the information land lines often provide (caller ID numbers, name and address information listed in telephone books) makes their lives too open. What they don't realize, though, is that certain cell phone services provide access to information not only about callers' names and addresses, but also about the callers' neighbors, address history, and a number of public records that relate to them.

GS 2. All over the country, many colleges are facing budget cuts. Our college is no different. One way our administration can address this is by capping faculty and staff salaries. Another way is to increase tuition fees. The best way, however, is to save utility fees by shutting down dorms, offices, and the library during the times when classes are not in session, like from December 23–January 5.

A 3. John Standers really wanted a job at Critman Industrials, and his uncle was able to pull some strings to get him an interview. Needless to say, John didn't get the job after he arrived fifteen minutes late and came dressed in a shirt that had a coffee stain on the front. Six months later, John's uncle arranged for another interview, but this time John was late again. Sometimes, people who get a second chance don't realize their good fortune and don't learn from their mistakes.

BD 4. In the two years since Speedy Fuel and Food opened at the corner of Drury Lane and Highway 61, a number of problems have arisen. The site, often scattered with garbage, is unsightly. While the food it offers comes at a low price, the unsanitary conditions of the kitchen have been cited six times by the Health Department. Maybe the worst offense is the prices that Speedy Fuel advertises on its signs are often as much as five cents lower than the prices the pumps actually charge. Because of all of these problems, Speedy Fuel and Food needs to be closed.

TICKET to WRITE

2.20 Adding an Introductory Paragraph to Your Essay

Some instructors require that the thesis statement be the last sentence of the introductory paragraph.

Directions: Look at your working thesis statement and the body paragraphs you have composed. Now compose an introductory paragraph that leads readers into your first body paragraph. Remember to include your working thesis statement in your introductory paragraph.

Conclusions

The **conclusion** is the last impression you leave with your readers. This is your final opportunity to stress the main point you want your readers to know. Many essay writers think of their work as a conversation with their readers. When that conversation is over, writers want to end thoughtfully and without overstressing or repeating themselves.

Your conclusion should

- reference the main points of your essay
- be relevant to what you stated in your introduction and body paragraphs
- be free of new ideas or opinions

"Don't mistake a good setup for a satisfying conclusion."
—Stanley Schmidt

Types of conclusions in essays

- **End with a thesis reminder**

This is probably the most common way to conclude an essay because it is effective and can work for almost any method of writing.

Heads Up!
Be sure to reword your thesis in your conclusion. Make it unique.

By following these few simple steps, you can create a smart password that is hard to crack. Hackers will have a harder time discovering your online password. You'll protect yourself from identify theft, and you'll feel safer when you're online.

- **Provocative question**

The question should reflect the thought-provoking ideas in the topics you elaborate on in your body paragraphs. Your readers should be able to answer the question through their understanding of your thesis.

All community service projects will help your fellow man or your fellow woman. They can also save money for your city or state. Surprisingly, performing community service will make you feel good, too. Can you think of any good reason to refuse to participate in community service?

- **Proposal or suggestion**

A proposal or suggestion in the concluding paragraph logically wraps up the points you have made and gives your readers the opportunity to make your thesis into a practical application. If you have presented your thesis and support well, your readers will accept your proposal or suggestion as valid and may even act upon it.

For more ideas about conclusions, search the Internet for this video: **Academic Writing Tips: How to Write the Conclusion of an Essay.**

The tutoring service here at MHCC is a treasure. It is free, it's convenient, and it's staffed by people who help you bring your grade up. Everyone who needs help should make an appointment at MHCC's Student Success Center.

TICKET to WRITE

2.21 Identifying Types of Conclusions

Directions: Review the paragraphs below and determine what type of conclusion each is. Write *TR* for thesis reminder, *PQ* for provocative question, or *P/S* for proposal or solution.

__TR__ 1. My life away from campus is somewhat different from that of many of my younger classmates. We share the experience of taking classes, worrying about finding time to study, keeping up with class lectures and completing assignments on time. Yet, as an older student, I have work and family obligations most traditional students don't have.

__P/S__ 2. The people of Johnstown have made clear that they do not want additional revenue to come from increased taxes. Still, all realize the city is in desperate need of a new high school and of a solution to ongoing flooding problems. The only solution to these problems is to allow riverboat gambling, which would generate revenue without increasing taxes.

__PQ__ 3. For the last ten years, population numbers in our state have been declining, but not here in Webster City. The low cost of living, the nearby access to outdoor activities, and the growing number of industries are key to drawing new residents and keeping native citizens. With these advantages, who wouldn't want to live in Webster City?

__TR__ 4. That's when I realized DJ, the man whose company I'd always relished, wasn't who I'd thought he was. Like other people, he was human and had failings. Maybe it was that insight or maybe it was my age, but from that point on, I began to look at adults with more skepticism and, I think, with far more wisdom—the wisdom to recognize that nobody's perfect.

TICKET to WRITE

2.22 Adding a Conclusion Paragraph to Your Essay

In your concluding paragraph, remind your readers of your thesis statement. Be sure to use different wording than you used in your introductory paragraph.

Directions: Look at your working thesis statement, introductory paragraph, and body paragraphs. Now compose a concluding paragraph. Be sure to keep it focused on your essay's main point. Share your completed discovery draft with one of your writing group members. As you read each other's draft, check that the introduction and conclusion paragraphs frame the body paragraphs by opening and closing the draft with a focus on the main idea.

Titles

The title of your essay is your readers' first impression of your work. Because of this, readers should get an idea of what to expect in your essay from the title alone.

Usually your title reveals your general subject or alludes to your thesis. In other words, your title may include the topic you're addressing or the direction you're taking on that topic. Here are some tips for writing a title:

- **Write the title last:** Most writers have an easier time composing their title after they have written their essay.
- **Keep it simple:** Titles of academic essays usually use language that is straightforward rather than quirky, clever, or full of jargon.
- **Use your essay as a guide:** Key words or phrases you've used in your text—or synonyms for them—may help you find the right title.

2.23 Adding a Title to Your Essay

Directions: Look at your completed essay and compose an appropriate title for your work. Ask group members who have read your discovery draft if your title fits your thesis and main points.

MyWritingLab™ Visit ***MyWritingLab.com*** and complete the various exercises and activities in the **The Topic Sentence, Thesis Statement,** and **Essay Introductions, Conclusions, and Titles** topic areas.

RUN THAT BY ME AGAIN

- **A paragraph's main idea is stated in . . .** the topic sentence.
- **A topic sentence should . . .** (1) introduce the general subject; (2) establish the position or point you're taking; (3) state a single idea in a straightforward sentence; (4) set the tone for the rest of the essay.
- **After prewriting to discover a subject, you should . . .** begin narrowing your focus to discover a topic.
- **To discover your topic sentence, you should . . .** use a variety of prewriting techniques.
- **Supporting sentences are . . .** the details that prove or illustrate the point of your topic sentence.
- **Different types of supporting details include . . .** examples, explanations, facts, reasons, and steps in a process.

LEARNING GOAL

1. Draft a paragraph by developing a topic sentence and discovering and organizing support

- **Three logical methods of organization include . . .** time, space, and importance.

LEARNING GOAL
❷ Draft an essay by developing a thesis statement, organizing support, and ensuring unity and coherence

- **The main idea of an essay is . . .** the thesis statement.
- **During discovery drafting, a thesis statement is referred to as . . .** a working thesis statement.
- **A thesis statement should be . . .** (1) your opinion; (2) a straightforward, declarative sentence; (3) a reflection of your knowledge, experience, or beliefs; (4) only one idea.
- **Two focus questions that can help you discover your topic after prewriting are . . .** (1) What idea dominates in my prewriting? (2) What is the main point I want to make about my subject?
- **Creating an outline during discovery drafting can help you . . .** (1) organize your thoughts; (2) note relationships between ideas; (3) discover gaps in your writing that need attention.
- **Prewriting techniques are unique methods that can help you . . .** discover ideas and find support for those ideas.

LEARNING GOAL
❸ Ensure Unity and Coherence

- **Everything you write in an essay must be relevant to . . .** the thesis statement.
- **Coherence in writing comes when all parts are . . .** clearly connected and logically organized.

LEARNING GOAL
❹ Write an introduction, conclusion, and title for an essay

- **The three obligations you have in your introduction are to . . .** (1) hook the readers' interest; (2) alert them to your topic; (3) make your point of view clear to them.
- **Your conclusion should . . .** (1) reference the main points of the essay; (2) be relevant to what you stated in your introduction and body paragraphs; (3) be free of new ideas or opinions.
- **The first impression readers get of your writing is . . .** your title.

DISCOVERY DRAFTING LEARNING LOG MyWritingLab™

1. What is the sentence that contains the main idea of a paragraph?
   The main idea of a paragraph is the topic sentence.
2. What is the sentence that contains the main idea of an essay?
   The main idea of an essay is the thesis statement.
3. What can help you narrow your focus and discover your topic?
   Prewriting techniques can help you narrow your focus and discover a topic.
4. What are three common and logical methods of organization?
   Three logical methods of organization are time, space, and importance.
5. What are four criteria for a thesis statement?
   A thesis should be (1) your opinion; (2) a straightforward, declarative sentence; (3) a reflection of your knowledge, experience, or beliefs; (4) only one idea.
6. What is *unity* in writing?
   In writing, unity centers on relevance. Everything you write in a paragraph should be relevant to the topic sentence; everything you write in an essay should be relevant to the thesis statement.
7. What is *coherence* in writing?
   Coherence comes in writing when all points in a piece of writing are clearly connected and logically organized.
8. What are three obligations you have to your readers?
   Your three obligations are to (1) hook the readers' interest; (2) alert them to your topic; (3) make your point of view clear to them.
9. Where are these obligations usually fulfilled?
   These obligations are usually fulfilled in the introduction.
10. What should you accomplish in your conclusion paragraph?
    In your conclusion paragraph, you should (1) reference the main points of the essay; (2) stay relevant to what you wrote in your introduction and body paragraphs; (3) not introduce new ideas or opinions

CHAPTER 3

Revising

LEARNING GOALS

In this chapter, you'll learn and practice how to

1. Revise for purpose
2. Revise for topic, unity, and coherence
3. Use the RAMS method to improve your writing
4. Use revision checklists

GETTING THERE

- After you have completed prewriting and discovery drafting, revising will help you find areas that need to be reworded, strengthened, or even eliminated.
- The RAMS method of replacing, adding, moving, or subtracting is another technique for revising your work.

Why You Should Revise

"Write your first draft with your heart. Rewrite with your head."
—from the movie *Finding Forrester*

Prewriting and discovery drafting are the first two steps in the writing process. The third step is **revising**. In some ways, this step is the most important because it is here that you look at what you have written, and you determine which areas need to be reworked.

Polished writers usually make key changes in their work after their discovery draft—even after their second or third drafts. These changes can improve organization and often help develop additional ideas or help clarify points the writers are trying to make. The best way to approach revision is to look for one kind of improvement at a time.

Purpose

LEARNING GOAL
1. Revise for purpose

Start with the big picture (your **purpose**) and ask yourself this question: Did I do what I was supposed to do in this piece? Think about the type of piece you are writing. If you wrote a narrative, did you, in fact, relate a story? If you wrote a classification essay, did you narrow a general topic into simpler or more helpful categories? Look at the definition and explanation of the approach you were assigned, and make sure your piece meets those requirements. If you have not yet become familiar with various types of essays, just review your essay for what you intended to accomplish.

3.1 Revising for Purpose

Directions: Return to the essay you wrote in Chapter 2 or to another one you wrote recently. Reread this essay and review your purpose. At the top of your essay, complete this sentence about your purpose:

My purpose in this essay is to persuade / describe / narrate / show (underline one) my audience that ______________________________.

In your essay, underline particular sentences that focus on your purpose. Then share your essay and your essay's purpose with your writing group and see if members agree that you accomplished your intended purpose.

Topic, Unity, and Coherence

Three key areas for all writers are *topic*, *unity*, and *coherence*. Each has its own role in good discovery drafting and good revising. Until you develop your own revision process, focus on making improvements in these three areas. The following chart shows what to concentrate on when revising for these three elements.

LEARNING GOAL

❷ Revise for topic, unity, and coherence

Revising Stage	The purpose is to make sure . . .
Topic	**Paragraph**—the topic of each paragraph is clear and stated in a topic sentence at the beginning of the paragraph. **Essay**—the thesis statement clearly states the topic and purpose of the essay, and each paragraph focuses on the thesis statement.
Unity	**Paragraph**—all sentences elaborate on the paragraph's topic. **Essay**—each body paragraph elaborates on only one idea that supports the thesis statement.
Coherence	**Paragraph**—each sentence fits with the one before and after it. **Essay**—each paragraph flows logically and smoothly from one point to the next.

Topic

Writers sometimes have so many good ideas to say that they lose sight of their focus. In a *paragraph*, your topic sentence establishes your subject and position—the

"Writing and rewriting are a constant search for what it is one is saying."
—John Updike

"I have never thought of myself as a good writer. . . . But I'm one of the world's great rewriters."
—James Michener

focused idea of your paragraph. Everything else should branch out from or support the subject and position you state in your topic sentence.

The same is true for an *essay*. Your essay's topic—your subject and position on the subject—should be obvious in your thesis statement. The rest of the essay should branch out from or support what your thesis statement asserts.

Below is Martina's discovery draft on brightening the lives of soldiers who are stationed overseas. Martina's next step in the writing process is to revise this draft, and she focuses on *topic* first.

3.2 Revising for Topic

Directions: Working alone or in groups, complete the following tasks:

First, double underline Martina's thesis statement. Then underline the topic sentence in each body paragraph. Write *MTS* where you think she may have a Missing Topic Sentence. Compose a topic sentence for each paragraph you have marked *MTS*. Finally, cross out any sentences you think stray from her main point (her thesis statement).

Staying in Touch

When my cousin Serena was deployed, I began to think about the military on a personal level. I admit, prior to this I had never kept up with current events, but now that I have family involved, I see the importance of supporting our soldiers and helping them cope with being away. Friends and relatives can brighten the lives of overseas soldiers in a couple of simple ways.

The Internet is a great way to stay in touch with family and friends who have been deployed. For Serena, access to the Internet means access to e-mail. E-mail keeps her connected to the latest local news and family gossip. Because of e-mail's speed, Serena found out about my engagement the same day my boyfriend proposed. Last June, Serena sent us an e-mail saying that she was feeling really down because she would miss our town's annual week-long Jazz and Jambalaya Festival. She had not missed a festival for ten years. Even though she was thousands of miles away, we took videos of some of the performers, posted them to YouTube, and e-mailed the links to Serena.

MTS Serena says her unit commander enjoys getting letters from his seven-year-old daughter so much that he shares and shows off each letter he gets from her. The post office is also the way to send soldiers little luxuries that we often take for granted. While Serena and the other members of her squad appreciate the packages, Serena likes the pictures and handwritten notes we send. Those are what keep her most connected, she says.

Sending postal mail involves more effort and a slower response time, but it's easy and offers a unique opportunity to soldiers. Our family regularly sends care packages not only to Serena but also to her entire unit. ~~Once we found out that Serena was sharing her packages, Aunt Eva decided we needed to learn what made the best care package items. She went to the armory and attended a Family Services class that detailed what to send and how to pack it. Who would've known that baby wipes and canned food were so popular?~~ Now every package we send includes deodorant, chewing gum, snack cakes, power bars, baby wipes, canned foods, lots of candy, and macaroni & cheese.

Imagine being able to lift up a soldier's spirits for the cost of a stamp or the length of time it takes to type an e-mail. Sending a note, a letter, an e-mail, a video, or care package is a small gesture to make for those who have made such large sacrifices.

Martina's instructor asked the class to make notes in the margins of their drafts as they revised for topic and focus. She asked them to record their thoughts on the left and explain their revisions on the right. Read Martina's revisions and her notes to see how she revised and strengthened her essay's topic and focus. Martina used Track Changes to make her revisions.

Staying in Touch

When my cousin Serena was deployed, I began to think about the military on a personal level. I've got to admit, prior to this I had never kept up with current events, but now that I have family involved, I see the importance of supporting our soldiers and helping them cope with being away. Friends and relatives can brighten the lives of overseas soldiers in a couple of simple ways.

Underlined thesis—can find when revising.

Thesis at end of intro so readers can't miss my point.

The Internet is a great way to stay in touch with family and friends who have been deployed. For Serena, access to the Internet means access to e-mail. E-mail keeps her connected to the latest local news and family gossip. Because of e-mail's speed,

Serena was able to find out about my engagement the same day my boyfriend proposed. Last June, Serena sent us an e-mail saying that she was feeling really down because she would miss our town's annual week-long Jazz and Jambalaya Festival. She had not missed a festival for ten years. Even though she was thousands of miles away, we were able to take videos of some of the performers, posted them to YouTube, and e-mailed the links to Serena.

2nd body para focus— snail mail.

Sending postal mail involves more effort and a slower response time, but it's easy and offers a unique opportunity to soldiers overseas. Serena says her unit commander enjoys getting letters from his seven-year-old daughter so much that he shares and shows off each letter he receives from her. The post office is also the way to send soldiers little luxuries that we often take for granted. While Serena and the other members of her squad appreciate the packages, Serena likes the pictures and handwritten notes that we send. Those are what keep her most connected, she says.

2nd body para needed topic sent, so added this from the next para.

Know what I mean . . . haven't found words yet.

~~Sending postal mail involves more effort and a slower response time, but it's easy and offers a unique opportunity to soldiers overseas.~~ Our family regularly sends care packages not only to Serena but also to her entire unit. ~~Once we found out that Serena was sharing her packages, Aunt Eva decided we needed to learn exactly what made the best care package items. She went to the armory downtown and attended a Family Services class that detailed what to send and how to pack care package items. She learned a lot and passed her knowledge on to us. Who would've known that baby wipes and canned food were so popular?~~ Now every package we send includes deodorant, chewing gum, snack cakes, power bars, baby wipes, canned foods, lots of candy, and macaroni & cheese.

off-topic??

This didn't support either topic . . . deleted it all.

This para needs more support . . . join two paras??

Details focus on stuff sent snail mail . . . moved to that para.

Imagine being able to lift up a soldier's spirits for the cost of a stamp or the length of time it takes to type an e-mail. Sending a note, a letter, an e-mail, a video, or care package is a small gesture to make for those who have made such large sacrifices.

3.3 Revising Your Writing for Topic

Directions: In Ticket to Write 3.1, you revised an essay for *purpose*. Make two copies of that essay and now revise it for *topic*. On the first copy, double underline your thesis statement, underline the topic sentence in each body paragraph, and write *MTS* beside paragraphs that are missing a topic sentence. Give the second copy of your essay to a member of your writing group and ask him or her to do the same.

After the reader has completed this exercise, review what he or she underlined and wrote. Do you and the reader agree on which sentence is the thesis statement, which are the topic sentences, and which paragraphs (if any) are missing a topic sentence? If not, consider revising your essay to make your thesis statement and topic sentences stronger and more apparent.

Unity

Sometimes, even when the topic sentence is clearly stated and the topic is well established, a few sentences wander off the range and need rounding back up. While these ideas or insights may be valid, in the revision process you should cross out any ideas that are off-topic. These ideas may be fine for other paragraphs or essays, but you should delete them if they don't fit the flow and continuity of that particular paragraph or essay.

"First drafts are learning what your novel or story is about. Revision is working with that knowledge to enlarge or enhance an idea, or reform it."
—Thomas Wolfe

For an essay, Kristin decided to expand on her interest in urban legends. She discovered a number of interesting facts about them, and after her prewriting she decided to focus her essay on two themes often found in urban legends: *caution* and *morality*. Using her prewriting, she wrote her discovery draft, consulting a number of the facts she had noted in her prewriting.

After making sure her overall purpose was clear and her essay focused on her topic, Kristin's next step was to revise for unity, checking her work to be sure she hadn't wandered off-topic.

TICKET to WRITE

3.4 Revising for Unity

Directions: Below are Kristin's introductory paragraph and two body paragraphs. Working alone or in groups, underline Kristin's thesis statement. Then cross out any sentences that stray from the main point of her essay or the main point of a paragraph.

People hear them all the time. Well-meaning friends send e-mails, text messages, or phone calls with the news. They're spreading stories that have little or no basis in truth—stories we know as "urban legends." These tales can include

(continued)

any number of subjects, but modern urban legends often center on themes of caution or morality.

The basis of many urban legends is a warning that something awful has happened to others and is on the verge of happening to those who are hearing the warning. One legend like this relates the story of a customer who was grossed out to find a rodent in his order at a particular fast-food restaurant. ~~A popular television show features bizarre foods, and one segment highlighted fried rats that are popping up on dinner plates in certain Asian countries. Restaurants billed these as alternative sources of protein.~~ Another cautionary tale that has made the rounds talks about temporary tattoos that are popular with children. Parents are warned that those tattoos are financed by drug dealers, who spike the tattoos with LSD. ~~LSD is an acronym for lysergic acid diethylamide, which was first synthesized in 1938 from a grain fungus.~~ Children who use these tattoos supposedly absorb the drug through their skin and then become drug addicts. The kids become customers of those who are behind the scheme.

Another popular theme of urban legends is morality. For decades, adolescents have heard the tale of "Hook Man," the story of a dating couple who are parked on a remote country road. Odd scraping sounds interrupt their making-out, so they abruptly leave and drive into town, where they stop at a gas station. ~~When they stop, the radio is playing "their song," a ballad by Lady Gaga. They like this song because of the way she sings it and because it has several lines that they think fit their relationship.~~ As they get out of the car, they're horrified to find a bloody hook hanging from the handle of the car's door. The moral of the story is that teenagers shouldn't "go parking" in cars. A second urban legend dealing with morality tells how a certain wife got even with her cheating husband when she bought space on a billboard and advertised his infidelities. ~~Surprisingly, outdoor advertising brings in almost $7 billion each year. The wife was heavily concerned with ecology, and she insisted the billboard be printed on recyclable material.~~ On the billboard, she said that she'd caught her husband's extramarital activities on tape, and she bragged that she had paid for the bulletin board with money from their joint checking account.

3.5 Revising Your Writing for Unity

Directions: In Tickets to Write 3.1 and 3.3, you revised an essay for *purpose* and *topic*. Now revise it for *unity*. Trade it with a member of your writing group and ask the reader to circle any sentence (1) that strays from your thesis statement—write *NTR* (Not Thesis Related) in the circle; and (2) that strays from a topic sentence in a body paragraph—write *NTSR* (Not Topic Sentence Related) in the circle.

After you have both completed this exercise, review what the other has circled. If your reader noted any problems with unity, consider revising those sentences.

Coherence

Writing can be full of good thoughts, details, and examples, and even though all are on the same topic, the reader may not be able to see how some of these ideas are related. Literally meaning "to stick together," **coherence** in writing means that all the ideas in a *paragraph* flow smoothly from one to the next, and each paragraph in an **essay** flows smoothly and leads logically to the paragraph that follows it.

"Revision is just as important as any other part of writing and must be done *con amore*."
—Evelyn Waugh

For a class essay, Radley was assigned to write about his cultural identity. Through class discussions and prewriting, he discovered his culture is defined in ways he had not considered.

After Radley composed his discovery draft, he began his revision process. Now he is revising his draft for coherence.

3.6 Revising for Coherence

Directions: Below are paragraphs from Radley's essay on defining his culture. Working alone or in groups, underline Radley's thesis statement. Then revise the paragraphs by providing any missing transitions, repeated key words, or additional linking details that would improve the *coherence* of these paragraphs. Answers may vary.

TECHNO TIP

For more about transitions and key words in revisions, search the Internet for this video:
Revising Your Essay: Part 4 of 4

Defining My Culture

At one time, I assumed I really had no specific culture outside of being an American. Looking deeper into how my family and I spend our time, however, I realized my culture was more than just being an American. My culture is a collection of my behaviors or deeds, beliefs, and values.

A common characteristic of culture is the behavior or deeds we often associate with a particular group of people. Behavior involves more than just the way people act.

(continued)

It includes the way they communicate and even spend their time. For example, from my parents I learned to volunteer and participate in my community. As far back as I can remember, my family has volunteered at the Salvation Army on Thanksgiving Day. Dad always worked in the kitchen cooking and cleaning, Mom served food, and my sister and I went around refilling drinks and clearing tables. Another example of my family's volunteering was my father's participation in scouts. My father was one of my scout leaders from the time I entered cub scouts into my venture scouting years. I am not a scout leader, but I work two days a week in an after-school program and help elementary school kids with their reading skills. Volunteering is part of my culture. It is a part of who I am.

My culture is also partly defined by my belief in respecting nature. This is something stressed to me in scouting. As a scout, I was taught to respect the outdoors and to leave a natural area better than I had found it. Because of this, I picked up trash at campsites, replaced grass and dirt after driving in tent stakes, and never started a fire outside an existing fire ring. I came to learn that when a part of nature isn't respected, it can be lost forever. The farm I grew up on was surrounded by so much natural beauty that I took a lot of it for granted—even what I thought were weeds. One yellow flowering plant seemed to crop up every summer. The yellow flowers were everywhere. I didn't notice when I mowed down all of them that grew next to our house. The older I got, the fewer of these yellow flowers popped up on the farm. By the time I was sixteen, they were all gone. That was when I found out the "weeds" were Short's goldenrod—something that's now endangered. On our farm it is extinct.

TICKET to WRITE

3.7 Revising Your Writing for Coherence

Directions: Return to the essay you have been revising; now revise it for *coherence*. After trading your essay with a member of your writing group, complete the following. Circle any transitional words or phrases the writer has used. Place a check mark inside the circle if the writer used the transitional word or phrase correctly. Place a question mark inside the circle if the writer used the transitional word or phrase incorrectly. Circle an area that would benefit from a transitional word or phrase, and write an appropriate word or phrase inside the circle.

After you have both completed this exercise, review what the other has noted and consider revising those areas.

RAMS: Replace, Add, Move, Subtract

You have looked at your discovery draft and examined how to strengthen your essay through revising for *purpose, topic, unity*, and *coherence*. Now look for additional ways to improve your work by replacing, adding, moving, or subtracting material. As you revise, use the acronym **RAMS** (Replace, Add, Move, Subtract) to remind yourself of the following revision options:

LEARNING GOAL
❸ Use the RAMS method to improve your writing.

Revise Using RAMS

Replace	Look for areas that would be stronger if you **replaced** material. What would be improved if you used more vivid or more precise words?
Add	Look for areas that would be stronger if you **added** material. What would be stronger with more explanation, more description, or additional transitions?
Move	Look for areas that would be stronger if you **moved** material. What would be more logical or convincing if you put it in a different place?
Subtract	Look for areas that would be stronger if you **subtracted** material. What is confusing or seems off your topic?

Replace

Using vivid and precise words helps writers present a more accurate picture, a more compelling argument, or a more clear-cut point. For instance, when describing a dinner, if a writer says that "the food was good," readers get only a vague idea of the meal. If the writer changes the wording to "Within ten minutes, the baked ham, sweet potatoes, broccoli casserole, and chocolate pie were all gone from our table," then readers understand what food was presented and how good it must have been. If the writer says that "students are going from the Administration Building to the Student Center," were they walking, ambling, strolling, meandering, scurrying, dashing, creeping, roaming, wandering, crawling, rushing, darting, running—or some other vivid word? Replacing everyday words with vivid and precise words gives readers more of an appreciation of the exact picture you want to convey.

"The difference between the right word and the almost right word is the difference between lightning and a lightning bug."
—Mark Twain

3.8 Revising with Precise and Vivid Language

Directions: Working alone or in groups, revise the sentences below so the words in italics are more vivid and precise. Use a separate sheet of paper. Answers will vary.

1. The *sad* fact is that we paid $8 each to see a movie that we *didn't like*.
2. Seeing the oncoming traffic, three adults *went after* the little children.
3. Bobby *said* to Bobbi, "What a *good* friend you are."
4. When we talked with her via Skype, Leslie reported that she had had a *great* day.
5. Increasing levels of mercury are a *big* concern to the fishing industry.

"I can't write five words but that I change seven."
—Dorothy Parker

Add

When drafting, writers are always looking for topics or ideas that need developing. They want to make sure they've covered their main points well enough so their readers will easily understand them; therefore, they look for places that might need additional details, definitions, or examples. For instance, an essay detailing the steps to take in order to save water might include "Take a shower instead of a bath." That step would have more impact with additional details such as "A bath typically uses about seventy gallons of water, and a five-minute shower uses only about ten to twenty-five gallons." Writers also check for areas in both sentences and paragraphs where they can strengthen the continuity of their writing by adding transitional words and phrases.

3.9 Revising with Additional Details

Directions: Below are several suggestions a writer gave for an essay that dealt with ways to save money. Working alone or in groups, add details to the sentences below or compose additional sentences so readers will have a clearer understanding and appreciation of the topic. Answers will vary.

1. Saving gas money is easy.
2. Food costs can be cut in several ways.
3. Food costs account for a large part of everyone's budget.
4. The telephone is another big money-guzzler.
5. Cutting down on utility bills will save money.

"I can't understand how anyone can write without rewriting everything over and over again."
—Leo Tolstoy

Move

As they are drafting, writers often think of additional supporting ideas that will strengthen the points they already have on paper. Writers usually jot these details down as they think of them, not taking time to consider their placement. In revising, however, writers often see that these details would be more effective if they were moved because they better support another part of the paragraph.

3.10 Revising by Moving Details

Directions: Below is a body paragraph James composed for an essay about online classes at his college. Working alone or in groups, revise the paragraph by determining which sentence(s) James should move for his paragraph to be more logical and convincing. Then, using the numbers at the beginning of the sentences, write out the correct order in which the sentences should appear. Correct order: 5, 2, 1, 4, 3, 7, 6

(1) Semester-based classes have the same schedule as the campus. (2) Both types let students resolve scheduling conflicts that may prevent them from taking classes on campus. (3) Fall semester begins in August and winter semester begins in January. (4) Students in these classes must complete homework and tests by specific dates. (5) Bakerville Community College offers online classes that are either semester-based or self-paced. (6) The completion dates vary with the courses, but most must be finished in a year. (7) Self-paced courses offer more flexibility in their completion date.

Subtract

When working on their discovery draft, many writers include every detail they noted in their prewriting. In the revising stage, however, they often see that some details are excessive or even off-topic and should be deleted from the next version of their drafts.

"Writing is 1% inspiration, and 99% elimination."
—Louise Brooks

3.11 Revising by Subtracting Unnecessary Material

Directions: Below is a body paragraph Marcus composed for an essay about a previous job. Working alone or in groups, revise the paragraph by crossing out material that is confusing or seems off-topic.

Still, I miss my job at the warehouse because of the feeling of accomplishment it offered. Each morning when I got to work, I would count the trucks lined up at the receiving docks. On some days, we would have as many as ten trucks to unload. ~~My brother Joe worked with me for awhile, and sometimes we rode to work together.~~ One semi-trailer might have 140 pairs of washers and dryers and a dozen dishwashers that had to be unloaded with a hand truck, while another

(continued)

might have a few hundred water heaters of assorted sizes that could be removed with a forklift. No matter the product or the method of off-loading, seeing a full warehouse and an empty parking lot at the end of the day was rewarding. ~~We saw semis of all kinds from all across the US and even a few from Canada.~~

Checking the Parts

LEARNING GOAL
4 Use revision checklists

Revision is an ongoing process, and most writers continue to find areas for improvement as they check their work. The following checklists can help as you revise the content and organization of any paragraph or essay. If you find any problems, use the information in this chapter to make the necessary changes.

"The pleasure is the rewriting. . . . The completion of any work automatically necessitates its revisioning."
—Joyce Carol Oates

REVIEW CHECKLIST for a Paragraph

Focus

Read your topic sentence; then skip to your last sentence. After reading the two, ask yourself if you've said essentially the same thing. You shouldn't have the same wording, but you should have the same focus. If your topic sentence and last sentence don't focus on the same key idea, you need to revise. Ask these questions:

- ☐ Does each sentence say what you intend it to say?
- ☐ Can you make your point more clearly?

Topic Sentence

- ☐ Does your topic sentence make clear the position or point of your paragraph?

Body

- ☐ Does each sentence in your paragraph relate to your topic sentence?
- ☐ Do you use appropriate transitional words and phrases so your work reads smoothly and guides your readers from one thought to the next?
- ☐ Have you given your readers enough examples, facts, or other details to support or explain your thoughts and ideas? Have you presented your thoughts in a logical order?
- ☐ Is any vocabulary inappropriate or does any need defining?

RAMS Checklist What areas would be improved if you . . .

- ☐ *replaced* vocabulary with more vivid or more precise words?
- ☐ *added* explanations, descriptions, or transitions?
- ☐ *moved* ideas to a different place?
- ☐ *subtracted* confusing or off-topic material?

☑ REVISION CHECKLIST for an Essay

Focus

Read the opening paragraph; then skip to the conclusion. After reading the two, ask yourself if they have essentially the same focus. If your introduction and conclusion don't center on the same key idea, you need to revise. Ask these questions:

- ☐ Does each sentence say what you intend it to say?
- ☐ Can you make your point more clearly?

Introduction

- ☐ Is your introduction clear?
- ☐ Does your introduction contain enough details so readers will know what to expect?

Thesis Statement

- ☐ Does your introduction contain your thesis statement?
- ☐ Is your thesis statement clear?
- ☐ Does your thesis statement show readers the main point of your essay?

Body of a Paragraph

- ☐ Does each sentence in your paragraph relate to your topic sentence?
- ☐ Do you use appropriate transitional words and phrases so your work reads smoothly and guides your readers from one thought to the next?
- ☐ Have you given your readers enough examples, facts, or other details to support or explain your thoughts and ideas?
- ☐ Have you presented your thoughts in a logical order?
- ☐ Is any vocabulary inappropriate or in need of additional definition?

Body Paragraphs in an Essay

- ☐ Does each of your body paragraphs relate to your thesis statement?
- ☐ Does each of your body paragraphs focus on a topic sentence?
- ☐ Do you use appropriate transitional words and phrases so your work reads smoothly and guides readers from one idea, sentence, or paragraph to the next?
- ☐ Does each sentence in each body paragraph relate to the topic of that paragraph?
- ☐ Have you given your readers enough examples, facts, or other details to support or explain your thoughts and ideas?
- ☐ Have you presented your body paragraphs in a logical order? (Think about arranging them in chronological or emphatic order.)
- ☐ Is any vocabulary inappropriate or in need of additional definition?

Conclusion

- ☐ Have you summarized the main points you presented in your paper?
- ☐ Is your conclusion effective?

(continued)

"Grab a pen and put down some words—your name even—and a title: something to see, to revise, to carve, to do over in the opposite way."
—Jacques Barzun

- [ ] Is your conclusion relevant?
- [ ] Have you made sure you didn't include any new ideas or opinions in your conclusion?

RAMS Checklist What areas would be improved if you . . .

- [ ] *replaced* vocabulary with more vivid or more precise words?
- [ ] *added* explanations, descriptions, or transitions?
- [ ] *moved* ideas to a different place?
- [ ] *subtracted* confusing or off-topic material?

MyWritingLab™ Visit ***MyWritingLab.com*** and complete the exercises and activities in the **Revising** topic area.

RUN THAT BY ME AGAIN

- **Revising is important because . . .** you look at what you have written and you determine which areas need to be reworked.

LEARNING GOAL
❶ Revise for purpose

- **In revising for purpose, ask yourself . . .** Did I do what I was supposed to do in this piece?

LEARNING GOAL
❷ Revise for topic, unity, and coherence

- **Your essay's topic should be obvious in . . .** your thesis statement.
- **To ensure unity in the revision process, writers should cross out any ideas that . . .** are off-topic.
- **Coherence in writing means that . . .** all ideas in a paragraph flow smoothly from one to the next, and each paragraph in an essay flows smoothly and leads logically to the paragraph that follows it.

LEARNING GOAL
❸ Use the RAMS method to improve your writing.

- **In the RAMS acronym, the R stands for . . .** replace (areas that would be stronger if you replaced material).
- **In the RAMS acronym, the A stands for . . .** add (areas that would be stronger if you added material).
- **In the RAMS acronym, the M stands for . . .** move (areas that would be stronger if you moved material).
- **In the RAMS acronym, the S stands for . . .** subtract (areas that would be stronger if you left material out).

LEARNING GOAL
❹ Use revision checklists

- **Before you submit your work, you should examine all of it by using . . .** revision checklists.

REVISING LEARNING LOG MyWritingLab™

MyWritingLab™

1. How often do polished writers usually make key changes in their work?
   They usually make key changes after their discovery draft and even after their second or third drafts.
2. What is the best approach to revision?
   The best way to approach revision is to look for one kind of improvement at a time.
3. When looking at your purpose, what question should you ask yourself?
   Did I do what I was supposed to do in this piece?
4. In revision for topic in an essay, of what should you be sure?
   Be sure your thesis statement clearly states the topic and purpose of the essay and that each paragraph focuses on your thesis statement.
5. In revision for unity in an essay, of what should you be sure?
   Be sure each body paragraph elaborates on only one idea that supports the thesis.
6. In revision for coherence in an essay, of what should you be sure?
   Be sure each paragraph flows logically and smoothly from one point to the next.
7. When revising using the RAMS method, what should you ask yourself about material that might be replaced?
   What would be improved if you used more vivid or more precise words?
8. When revising using the RAMS method, what should you ask yourself about material that might be added?
   What would be stronger with more explanation, more description, or additional transitions?
9. When revising using the RAMS method, what should you ask yourself about material that might be moved?
   What would be more logical or convincing if you put it in a different place?
10. When revising using the RAMS method, what should you ask yourself about material that might be subtracted?
    What is confusing or seems off your topic?

CHAPTER 4 Editing and Proofreading

LEARNING GOALS

In this chapter, you'll learn and practice how to

1. Fine-tune individual parts of your manuscript
2. Benefit from peer reviewers

GETTING THERE

- Editing and proofreading your work helps you learn how to polish it to its best form.
- Peer reviewing can be a valuable tool in refining your manuscript.

Check . . . 1, 2, 3 . . . Check . . . Readjust

LEARNING GOAL

1. Fine-tune individual parts of your manuscript

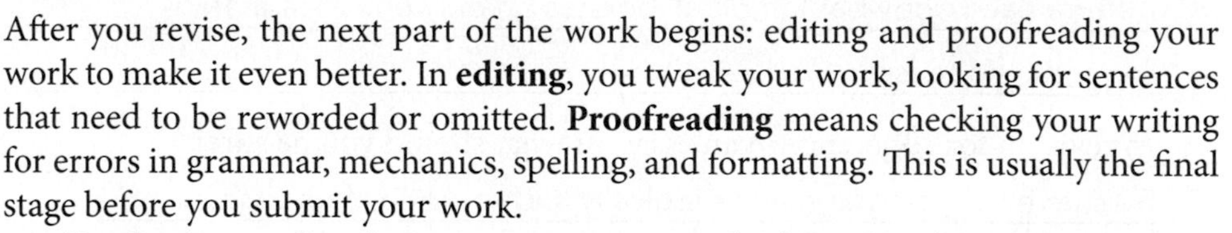

After you revise, the next part of the work begins: editing and proofreading your work to make it even better. In **editing**, you tweak your work, looking for sentences that need to be reworded or omitted. **Proofreading** means checking your writing for errors in grammar, mechanics, spelling, and formatting. This is usually the final stage before you submit your work.

You begin to edit when you start to revise, looking for places to strengthen your topic, unity, and coherence. As you move into editing, continue to ask yourself about these areas. Put yourself in your reader's place. If you were reading your paper for the first time, would you understand the big picture and all the smaller points in the paper?

An important area to address is sentence structure (the way your sentences are worded). Look through your writing for sentences that could be improved just by changing their structure. If, for instance, you're writing about the reasons why a family should adopt a collie, you may have several sentences that begin in the same way, like these:

> A collie is an elegant dog.
> A collie is ideal as a family dog.
> A collie is both loyal and affectionate.

While none of these sentences is incorrect, reading them is boring. You can easily improve them by combining some or all of them.

> A collie is ideal as a family dog because it is elegant, loyal, and affectionate.

You could also improve your sentence structure by rewording one or more of the sentences.

Heads Up!
For more on sentence variety, see Chapter 25.

Loyal and affectionate, a collie is an ideal family dog.

Here is a detailed editing and proofreading checklist you can use as you complete the final steps in writing your paragraph or essay. Additional instruction and practice for many of the items is provided in Part 4, Grammar and Mechanics.

REVIEW CHECKLIST for Editing and Proofreading

Editing

- ☐ **Have you varied your sentence structure?** If many sentences begin with the same word or phrase, rearrange some of the sentences so your style isn't repetitive.
- ☐ **Are you using first, second, or third person?** Did your instructor require that your paper be written in first, second, or third person? If so, use your word processing software's *Find* function and check that you have used personal pronouns correctly. Search for the pronouns you are not to use. For example, if you are not supposed to use first person, then search for all first person pronouns: *I, me, my, mine, we, us, our,* and *ours.*
- ☐ **Have you used any unnecessary slang, jargon, or clichés?**
- ☐ **Have you given your piece an appropriate title?**
- ☐ **Do you have any spelling errors?** Use a separate read-through to search for errors in spelling, punctuation, and pronouns.
- ☐ **Have you used *Spell Check* or *Grammar Check*?** Use your computer's *Spell Check* and *Grammar Check* functions, but do not rely solely on either of these.

Eye ewes spell check and grammar check on this sentence and the computer found their where no miss steaks.

Grammar

- ☐ **Have you repaired any sentence fragments?** Check that each group of words has a subject, a verb, and expresses a complete thought.
- ☐ **Have you repaired any sentence run-ons?** Two or more complete thoughts should be separated with some form of punctuation. Check to see if you have used the correct punctuation.
- ☐ **Is your verb tense consistent?** Change any unnecessary shifts in verb tense.
- ☐ **Do you have any problems with subject-verb agreement?** Check that you used plural verbs with plural subjects, and singular verbs with singular subjects.
- ☐ **Do you have parallelism in sentence construction?** Check for words, phrases, and clauses joined with *and* or *or*. Make sure you have used these words, phrases, and clauses in the same pattern.

(continued)

Heads Up!
Review pronouns in Chapter 23.

- ☐ **Have you repaired any misplaced modifiers?** Check for words, phrases, and clauses that describe or elaborate on something else in your sentence. Make sure these are not improperly separated from the word being described or elaborated on.
- ☐ **Are your pronoun references clear?** Because pronouns refer to a particular noun, make sure your readers know to which noun you're referring.
- ☐ **Have you used pronoun gender correctly?** If the noun your pronoun refers to can be either male or female, have you used the correct pronoun?
- ☐ **Do your pronouns agree with the nouns they refer to in number?** Your pronoun must be singular if the noun it refers to is singular. Your pronoun must be plural if the noun it refers to is plural.
- ☐ **Do your pronouns agree with the nouns they refer to in person?** Check to be sure you steer clear of any shifts between *I*, *you*, *we*, *he*, *she*, *they*, and *one*.
- ☐ **Have you used the objective and subjective case correctly with your pronouns?** Check for subjective pronouns (*I*, *we*, *you*, *he*, *she*, *it*, *they*) if your pronoun is the subject or predicate nominative in the sentence. Check for objective pronouns (*me*, *us*, *you*, *him*, *her*, *it*, *them*) if your pronoun is a direct or indirect object or an object of a preposition in the sentence. Check for subjective pronouns (*I*, *we*, *you*, *he*, *she*, *it*, *they*) if your pronoun is the subject or predicate nominative in the sentence. Check for objective pronouns (*me*, *us*, *you*, *him*, *her*, *it*, *them*) if your pronoun is a direct or indirect object or object of a preposition in the sentence.

Proofreading

"Proofread carefully to see if you any words out."
—author unknown

- ☐ **Read your paper out loud.** Doing this helps slow down your brain, and you can often spot errors that might otherwise slip by. Also, this helps you hear the rhythm of your words, and you could find places where you should reword to keep from being monotonous.
- ☐ **Record your reading of your paper and play it back later.** You may find additional mistakes or hear sentences that don't flow smoothly. This will also help you find problems with sentences that have similar lengths or patterns.
- ☐ **As you read, pretend you're back in elementary school and touch each word with your pencil or your finger.** Or use a straightedge (a ruler or blank piece of paper) and cover the lines below what you're reading. Either trick will help you focus on only the words in front of you so you won't skip ahead and overlook possible mistakes.
- ☐ **Start at the end of the paragraph or essay and read backward, one sentence at a time.** This unfamiliarity can increase your focus and help you find fragments, misspellings, omitted words, and other problems.

Keep this checklist handy for future assignments, and add to it as you see other frequent mistakes.

- ☐ **Make a checklist of errors that often crop up in your work.** Do you have problems with run-on sentences? Are you prone to overusing commas? If so, read through your paper again, concentrating on only those problems. If you often confuse *they're* and *their*, for example, use the *Find* function in your word processing software to isolate those words. Then you can make doubly sure you used them correctly.

- ☐ **Print your paper and read it.** Most people find far more errors when they proofread a hard copy rather than one on-screen.
- ☐ **If you must proofread your manuscript on-screen, change its look by altering the font style or size.** This may help you get a different perspective. After you've finished, be sure to change the document back to the font size and style required by your instructor.
- ☐ **Let your paper get cold.** Leave it alone for several hours. Then go back and reread it.

Formatting

- ☐ **Formatting requirements.** Did you adhere to the formatting your instructor required (e.g., margin and font size, font style, point size, spacing requirements)?
- ☐ **Identifying information.** Are your name, class heading, instructor's name, and date in their appropriate place?
- ☐ **Page numbering.** Are your pages numbered correctly?

"Sleep on your writing; take a walk over it; scrutinize it of a morning; review it of an afternoon; digest it after a meal; let it sleep in your drawer a twelvemonth . . ."
—A. Bronson Alcott

TECHNO TIP

For additional ideas on proofreading, search the Internet for these videos:

Writing Capstone - Proofreading

Husson University Proofreading

Husson Proofreading Planning

TICKET to WRITE

4.1 Editing and Proofreading Your Essay

Directions: In Chapter 3, you revised your essay specifically for *purpose*, *topic*, *unity*, and *coherence*. Take that essay or another one you have developed through the revising stage and edit and proofread it. Use the checklist above to make sure you have your essay in the best form possible.

Getting Feedback: Peer Review

LEARNING GOAL

❷ Benefit from peer reviewers

You might think you've found all your errors and made all the improvements you can, but having other people read your paragraph or essay and offer suggestions will almost always improve your writing. These are your **peer reviewers**. In peer reviews, you work with one or more students to find ways to improve your paragraph or essay. You might or might not agree with the suggestions you receive, but peer reviewers often see places for improvement or find errors that have slipped by you. Listen to or read closely your peer reviewers' suggestions. Keep an open mind, and don't be shy about asking questions to clarify their feedback. In the end, only you will decide whether or not you should make changes in your work.

Heads Up!
Peer reviews are sometimes called writing workshops.

Peer Review Options

Peer reviewers often find that reading each other's papers aloud aids in uncovering errors.

Your instructor can approach peer reviewing in several ways, including one-on-one peer review, group peer review, or online peer review.

- In a **one-on-one peer review**, you meet with only one other student and read each other's papers, providing written or verbal feedback on each other's work.
- In a **group peer review**, you are a member of a small group, usually three to five people. You may be in the same small group for every review session, or your group may change often. For your review, you trade a copy of your current paragraph or essay with members of your group and you in turn receive a copy of other members' work. These members read your writing and comment on it, and you do the same for the pieces you receive. Your instructor will tell you whether your group will discuss these comments verbally or in a written format.

  Your instructor may ask your group to review only particular features (like topic sentences or use of transitions) or all aspects of the paper. In this case, you might use a checklist like the Checklist for Editing and Proofreading on page 65 or the Checklist for Peer Reviewing on page 69 in this chapter.
- In an **online peer review**, you post your paragraph or essay to a class blog, discussion board, or some other online platform. Your peers review your work, responding through the same blog, discussion board, or e-mail with specific suggestions about how you can improve your paragraph or essay. Again, your instructor will let you know how many peers will review your work and how many you will review.

How to Be a Helpful Peer Reviewer

For your peer review, your instructor may provide a scoring rubric or a peer review checklist that asks you to address specific writing concerns. At other times, you may review for all the elements of a paragraph or essay (organization, topic, thesis, unity, coherence, mechanics, grammar, spelling) using an all-purpose revision checklist (see page 69). To be a helpful peer reviewer, keep in mind these tips:

Tips for Effective Peer Review

- point out specific concerns with the piece
- identify why the concerns are problems
- suggest what the writer might do to address the problems
- offer insightful comments
- avoid vague comments like "Good job" or "That was interesting"
- offer solid recommendations for how the piece can be improved

To read other ideas about peer reviewing, search the Internet for this site: **CSU - An Introduction to Peer Review.**

When peer reviewing, writing group members should focus on providing **constructive criticism**, pointing out both strengths and weaknesses of a piece. Constructive criticism is often a difficult concept for novice writers because they don't know how to offer it. Here are some suggestions about ways you might offer constructive criticism.

- **Begin by pointing out at least one positive quality about the piece.** "I like the examples in your second body paragraph" or "I can see the place you describe here; you used language well."
- **Be tactful and specific with your feedback.** You might word your comments about the areas you think need revision in one of these ways: "I think this paragraph is a little vague. How can you elaborate on your (topic, details)?" or "I don't get your point here. What's another way you could say this?"

For additional ideas about peer review, search the Internet for these videos: **Writing & Education: What Is Peer Review?**

Otis College: Peer Writing Review Process.

REVIEW CHECKLIST for Peer Reviewing

Title

- ☐ Is the title appropriate for the piece?
- ☐ Is the title capitalized correctly?

Introductory Paragraph

- ☐ Does this paragraph identify the subject that will be addressed?
- ☐ Does this paragraph provide the necessary background information?
- ☐ Does this paragraph include the thesis statement?
- ☐ Does this paragraph create interest?

Thesis Statement

- ☐ What is the thesis statement?
- ☐ Is the thesis statement clear?
- ☐ Does the thesis statement reflect the purpose of the paper? (By reading the thesis statement alone, can you identify what the rest of the paper will address?)

Body Paragraphs

- ☐ Can you identify the topic sentences of all the body paragraphs?
- ☐ Does each topic sentence support the thesis statement?
- ☐ Does any sentence in a body paragraph stray from the paragraph's topic?
- ☐ Does any body paragraph need additional details, examples, or explanations to support its topic?
- ☐ Has the author used transitional words and phrases appropriately?
- ☐ Is the information the author presents logical? Is it thorough?

Conclusion

- ☐ Does the conclusion reflect what the author expressed in the thesis statement?
- ☐ Does the conclusion contain any ideas, thoughts, or opinions not presented in the body paragraphs?

"I try to leave out the parts that people skip."
—Elmore Leonard

(continued)

Transitions

- ☐ Does the essay use transitions in appropriate places?
- ☐ Are transitions used correctly?
- ☐ Does the essay have enough transitions?

Grammar, Spelling, Capitalization, and Other Problems

- ☐ Do you see any problems with grammar? Spelling? Capitalization? Word choice? Sentence fragments or run-on sentences? Punctuation?
- ☐ Do you see any places where words are used incorrectly or words should be defined?
- ☐ Do you see any sentences that read in an awkward manner? (As you read the piece, were you confused by any sentences or passages?)
- ☐ Does the author adhere to all the formatting requirements (spacing, font and font size, margins, headings, page numbering) for the essay?

4.2 Editing and Proofreading

Directions: Below is an essay Kyle presented to her peer review group. After reading the essay, use the Review Checklist for Peer Reviewing to find areas that need improvement. Answers will vary.

Amstead 1

Kyle Amstead

Professor Burton

English 099

9 October 2011

Changing my Reading Habit's

I've always been what my friends call a "readaholic." I used to know everybody at the library, and I used to go to all the used book sales. Books and Magazines were stacked all over my apartment. I live just two blocks from the library, so checking out books was convenient for me. But every since my birthday, things have changed. My sisters pooled their money and bought me a great present. It's an e-reader, and

Amstead 2

it's changed my reading habit's. Using an e-reader has many advantages over reading printed material.

One reason I like my e-reader is that I have alot of reading material right at hand. This one machine can hold hundreds of books, this is an advantage for me because I usually have several different books going at the same time. I like to read mysteries, biographies, and books about travel to foreign lands like India and Brazil. I used to keep library books out for so long that I'd have several dollars in overdue fines. Also, I have subscribed to three newspapers on my e-reader, including one from Germany. The newspapers are delivered to me every morning, and I don't even have to go outside to pick them up. I just start drinking my morning coffee and turn on my e-reader.

Now all the books Im reading are in one place and I don't have to track down or carry several different ones. I put my e-reader in my backpack and have anything I want close at hand. My e reader weighs a little over ten ounces and it's thinner than most paperback books, so I can take it anywhere without any trouble. Since it's portable, I can download material anywhere Im sitting. The other day I saw a man on television talking about his new book, and he said the first three chapters could be downloaded free. The book was about different ways to save money, and Im in a little financial bind at the moment due to some problems with my credit card. I took my e-reader out of my backpack and had the chapters in my delivered in less than a minute. Also, because it's so portable, I can read it while Im curled up on the couch, working out on the treadmill, or sitting on the bus. My bus ride to campus usually takes about twenty minutes.

The third and most important reason I'm thrilled with this present is that it's versatile. One feature I like is that I can choose from six different font sizes. My vision is 20/20 with my contacts in. Sometimes, though, I want to read after I take them out, and with my e-reader I just push a button and make the font big enough to see. The dictionary in my e-reader is also handy. When Im stumped by a word. I can look it up without tracking down my printed dictionary.

Amstead 3

While my friends haven't stopped calling me a "readaholic," some other things have changed. I haven't been to the library since my birthday, and I've saved money by not buying anything at used book sales. Im proud my e-reader helps the environment, too, since trees aren't cut down for e-books like they are for printed books. I read more then ever, but that's because of the portability and versatility of my e-reader and how much reading material it stores.

4.3 Peer Reviewing for Specific Points

Directions: In Ticket to Write 4.1, you put the finishing touches on your essay. Now you're ready to get feedback from your peer reviewers. In peer review groups of three or four, read and respond to each of your group members' essays. After reading an essay, respond to the questions below. Once you are finished, trade essays with another group member and answer the questions again.

Reviewer's Name: ____________________

Writer's Name: ____________________

1. In your words, what is the main point of this essay?
2. In your words, what is the main point of each body paragraph?
3. In your opinion, what part of the essay is the most successful?
4. What makes this part successful?

4.4 Peer Reviewing with the Review Checklist for Peer Reviewing

Directions: In the same peer review groups, read and respond to each of your group members' essays, using the Review Checklist for Peer Reviewing. After reading an essay, record your responses.

Once you are finished, trade essays with another group member and answer the questions again. When all in the group are finished, discuss each essay, your feedback, and your suggestions.

Collect the responses your peers wrote about your essay. Review the responses and make any suggested changes you believe will improve your essay.

TECHNO TIP

For additional ideas on proofreading and editing, and a PowerPoint presentation on peer review, visit Purdue OWL (Online Writing Lab) at owl.english.purdue.edu/owl/ then go to **General Writing, The Writing Process, Proofreading.**

MyWritingLab™ Visit ***MyWritingLab.com*** and complete the exercises and activities in the **Editing Your Paragraph** and **Editing Your Essay** topic areas.

RUN THAT BY ME AGAIN

LEARNING GOAL
1. Fine-tune individual parts of your manuscript

- **After the revising stage of the editing process comes . . .** the editing and proofreading stage.
- **When editing in the writing process, you . . .** tweak your work, looking for sentences that need to be reworded or omitted.
- **When proofreading in the writing process, you . . .** check your writing for errors in grammar, mechanics, spelling, and formatting.
- **Seven techniques of proofreading include . . .** (1) reading your paper out loud; (2) recording your paper and playing it back later; (3) touching each word with your pencil or your finger; (4) reading your paper backward, one sentence at a time; (5) making a checklist of errors that often crop up in your work; (6) reading a printed copy of your paper; (7) changing the look of your paper on-screen by altering the font style or size.
- **When proofreading for formatting, you should check that you have . . .** (1) adhered to the required assignment format; (2) included your name, class heading, instructor's name, and the date appropriately; (3) numbered pages correctly.

LEARNING GOAL
2. Benefit from peer reviewers

- **Peer reviewers are . . .** students who read your writing and offer suggestions to improve your writing.
- **Three peer review options are . . .** (1) one-on-one, (2) group, (3) online.
- **When peer reviewing, writing group members should focus on providing . . .** constructive criticism.
- **Constructive criticism is . . .** pointing out both strengths and weaknesses of a piece of writing.

EDITING AND PROOFREADING LEARNING LOG MyWritingLab™

MyWritingLab™

1. What is editing?
Editing is part of the fourth stage of the writing process; in it, you tweak your work, looking for sentences that need to be reworded or omitted.

2. What is proofreading?
Proofreading is part of the fourth stage of the writing process; in it, you check your writing for errors in grammar, mechanics, spelling, and formatting.

3. When proofreading, for what three specific errors should you check?
Answers may include: You should check for sentence fragments, run-ons, verb tense errors, subject-verb agreement errors, nonparallel constructions, misplaced modifiers, and pronoun errors.

4. When proofreading, how can reading your paper out loud help you?
Reading your paper out loud can help you reword mistakes that might otherwise slip by you.

5. When proofreading, with what can a voice recording of your paper help you?
Recording your paper and playing it back later can help you find sentences that don't flow, are too short, or are too long.

6. When proofreading, how can touching each word help?
Touching each word can help you focus on each word without skipping over any.

7. When proofreading, what can reading your paper backwards, a sentence at a time, help you locate?
Reading your paper backwards, one sentence at a time, can help you locate fragments, misspellings, omitted words, and other errors.

8. What kind of checklist should you make for proofreading?
Make a checklist of errors that often occur in your writing.

9. Why should you proofread from a printed copy of your paper?
You find far more errors when proofreading a hard copy than an on-screen view of your paper.

10. If you proofread on-screen, what should you do to gain a different perspective of your paper?
If proofreading on-screen, alter the paper's look by changing the font size and style.

11. **When proofreading, what are three formatting elements you should review?**
When proofreading, check that you have (1) adhered to any required formatting instructions, (2) included the necessary identifying information (name, class, instructor's name, date), and (3) numbered your pages correctly.

12. **What are three methods of peer reviewing?**
Three methods of peer reviewing include (1) one-on-one, (2) group, and (3) online.

13. **What is constructive criticism?**
Constructive criticism is pointing out both the strengths and weaknesses of a piece of writing.

CHAPTER 5

Publishing and Academic Writing

LEARNING GOALS

In this chapter, you'll learn how to

1. Publish and format a paper for class
2. Publish your writing in a personal, online, or print format based on your purpose
3. Choose an organizational structure for an essay based on your purpose

GETTING THERE

- Publishing your work—in one of several formats—is the final step in the writing process.
- Publishing may be personal, for an online or print forum, or for a classroom assignment.
- Each type of academic writing accomplishes a different purpose for a specific writing situation and can be organized in various ways.

Publishing

In the writing process, **publishing** is the final step, the feel-good presentation of any material you've written for any audience. Your audience may be small; if you're writing in a diary, for instance, your audience is just yourself. On the other hand, your audience may be the rest of the world if you post something on the Internet. In addition, the material you publish may be informal (a note to a friend) or distinguished (your great American novel). No matter who your audience is or what your material includes, all publishing—classroom, personal, online, and print—centers on your **purpose**.

Classroom Publishing: Formatting Your Paper

LEARNING GOAL

1. Publish and format a paper for class

Eventually, you turn your paper in to your instructor. In a classroom, that's what *publishing* means—sending your finished product from your hands to your instructor's.

You have one final step before you turn your paper in—being certain you followed any formatting requirements your instructor gave. These include specific directions about how to format the heading, title, and margins; spacing; font type and size; and page numbering.

Usually in classes devoted to the humanities (like writing, literature, languages, philosophy, religion, and public speaking) instructors require you to follow MLA (Modern Language Association) style. For an essay in a humanities class, you will follow these MLA requirements for formatting:

Heads Up!
Different instructors have different requirements for formatting. Follow the specific directions your instructor has given.

MLA Formatting Requirements

- Use standard, white 8½-by-11-inch paper.
- Use a common font (like Times New Roman) set at 12 points.
- Use 1-inch margins on all sides (top, bottom, left, right) of your paper.
- Double-space your material.

For the first page:

- Do not type a title page for your essay.
- Number your pages with a header in the upper right-hand corner of the paper, one-half inch from the top of the paper and flush with the right margin. Format your header with your last name, followed by the page number in Arabic numbers (1, 2, 3, etc.). Do not use p. or # before the page number.
- In the upper left-hand corner, type this information, each on a separate line with double-spacing between: (1) your first name, then your last name; (2) your instructor's name; (3) the course name; (4) the date.
- Double-space under the date and type the title of your essay. Center this title. Do not use underlining, italics, or quotation marks on your title.
- Double-space between your title and the first line of your essay.

For your text:

- Double-space your text.
- Indent the first line of all paragraphs one-half inch from the left margin.
- Use only one space after end punctuation (periods, question marks, exclamation marks).
- Do not right justify your text.

Heads Up!
For help with MLA style essay formatting using Microsoft Word, search the Internet for this video: **MLA Style Essay Formatting Tutorial for MS Word.**

Cooper wrote an essay about becoming more environmentally involved. His instructor directed that the essay be published in MLA style. The first page of Cooper's essay follows on the next page.

Bartlett 1

Correct header (every page) Last name Page number

Correct heading (first page only) Your name Instructor's name Class name Current date All double-spaced

Cooper Bartlett

Professor Reagan Abbott

English 099

7 March 2012

Correct title Centered. Main words capitalized. No bolding, underlining, or italicizing. No other punctuation.

Now I'm Proud of Being Green

Correct indent One-half inch indent for every paragraph.

Correct font choice 12 point, Times New Roman

Twice when I was growing up people used the word "green" to describe me, and not in a good way. When I was twelve and had the flu, my mom said I looked "green around the gills." In my senior year, I messed up three times in the same basketball game, and Coach Lindenberg told me I was playing like a "green freshman." Those were both negative definitions of "green." Now I've become interested in the environment, and being green has a positive definition. At home, in my truck, and on campus I'm now proud of being "green."

Correct margins ↑1" ←1" 1"→ 1"↓ One-inch margins all around

Correct spacing All material double-spaced.

I use some energy-saving measures in my apartment. I run cold water instead of warm when I'm washing clothes. That reduces carbon dioxide from my machine, and I also save water and electricity. Another way I save is with power strips. I plug my several electronics into those and turn the power strip off overnight. That reduces electrical use for those electronic devices with just one flip of a switch. I also recycle glass, paper, and plastics.

In my English class here at Fowler College, I read an article about conserving energy in vehicles and I put these to work in my truck. During the winter, I'd been letting my truck idle to get it warmed up, but I learned warming it up more than thirty seconds not only wastes fuel but hurts the engine, and is bad for the environment.

Formatting for Other Disciplines

In classes outside the humanities, you may be asked to use a formatting style other than MLA. Popular styles include the following:

- **APA (American Psychological Association),** usually used in psychology, education, and other social sciences

- **AMA (American Medical Association),** usually used in medicine, health, and biological sciences
- **Turabian,** often used for term papers, theses, and dissertations in all subjects
- **CMS (*Chicago Manual of Style*),** often used by books, magazines, and non-scholarly publications

You will study classroom publishing in more detail in Chapter 16, "In-Class Writing."

Publishing and Purpose: Personal, Online, and Print

As a college student, your publishing opportunities go beyond the paragraph and essay, and they include personal, online, and print publishing outside the classroom. Each of these types of publishing centers on your *purpose*.

LEARNING GOAL
❷ Publish your writing in a personal, online, or print format based on your purpose

Personal Publishing

Personal publishing may take many forms, including the following:

- **Journal or diary entries** Your *purpose* might be to reflect on a movie, to record your reaction to a meeting, or to note your feelings about a troublesome situation.
- **Memoirs** Your *purpose* might be to document your travel, to record a part of your family history, or to examine your personal struggles.
- **E-mail** Your *purpose* might be to post a question about a meeting, to answer an invitation, or to alert others to a class assignment.
- **Letters** Your *purpose* might be to complain about a purchase, to recommend a friend for a job, or to thank someone who was kind to you.

"The age of technology has both revived the use of writing and provided ever more reasons for its spiritual solace. E-mails are letters, after all, more lasting than phone calls, even if many of them r 2 cursory 4 u."
—Anna Quindlen

Your purpose in personal publishing also dictates whether your writing is formal (maybe a letter to your congressional representative) or informal (maybe an e-mail saying hello to a friend). You'll study personal publishing in more detail in Chapter 17, "Personal and Business Writing."

Online Publishing

Online publishing takes a number of forms. If you send an e-mail to a friend, you publish online. If you use your company's Web site to advertise a new product, you publish online. If you post a question to an online discussion group, you publish online.

Heads Up!
Increasingly, published material is offered in several formats: print, online, and other digital media.

Other forms of online publishing are mushrooming in popularity and include personal blogs, social networks (like Facebook, Twitter, Myspace), and digital media.

You will study online publishing in more detail in Chapter 18, "Electronic Writing and New Technologies."

Print Publishing

Print publishing is still the favored approach for most writers. You're familiar with commercial books and magazines, but your college or university may also have other material, including

student literary magazines

student showcases

student newspapers and 'zines

professional journals available in the library

You will study print publishing in more detail in Chapter 19, "Writing Newspaper Articles and Examining Journal Articles."

5.1 Connecting What You Have Learned with Published Essays

Directions: Go to www.thisibelieve.org. Click **Explore**, and then click **Featured Essays, 1950s Essays**, or **Most Viewed Essays**. From the lists, read an essay that piques your curiosity or one that your instructor assigns. Use the questions below to evaluate this essay. Compare answers with your writing group and discuss what you find effective about the essay. Answers will vary.

1. What is the essay's thesis statement?
2. Does the thesis statement reflect the purpose of the essay? (By reading the thesis statement alone, can you identify what the rest of the essay will address?) If not, what is the purpose of the essay?
3. What topics does the writer develop to support the thesis statement?
4. What do you like best about this essay and why?
5. Do you agree or disagree with this writer's belief? Support your position.

At www.thisibelieve.org/educators/, you'll find an educational curriculum with additional study and composition ideas for this type of essay, as well as guidelines for submitting your own essay to the *This I Believe* site. Click on **Download the College Curriculum** in the middle of the page. The PDF is intended for instructors, but students can also benefit from the material.

5.2 Publishing in National Magazines

Directions: The world of publishing is just as particular in its requirements as are college instructors. Go to the **Freelance Writing** Web site where you'll find links to the writer's guidelines for almost a thousand magazines. On the home page, the magazines are divided into categories. Open a category that interests you, and then examine the guidelines of three different magazines checking for the information listed below (you may need to click on **Detail**). You'll see that just as no two college instructors have identical requirements for content and formatting, no two magazines do either. Answers will vary.

1. name of magazine
2. audience for magazine
3. requirements to submit articles
4. type of material found in magazine
5. frequency of publication (how often the magazine is published)

5.3 Writing for a National Magazine, Part 1

Directions: Return to the **Freelance Writing** Web site and choose a category that interests you, and then pick one of the magazines in that category. Check to make sure that magazine accepts finished articles (that is, it does not require that you query first with an idea). Fill in all the information below that is available about that magazine (click on **Go to Writer's Guidelines**). Answers will vary.

1. Name of magazine
2. Audience for magazine
3. Type of material found in magazine (If a magazine accepts several types of articles, choose one type.)
4. Word length of material that may be submitted
5. Method for submission (electronic, hard copy, either)
6. Payment for published article (payment may be monetary or in copies of magazine)
7. Additional information in writer's guidelines about what the magazine requires to submit an article in that category

Using the information that you have gathered above, choose and narrow a topic that will appeal to the readers and publishers of this magazine.

8. Topic for article for this magazine:

___

___

Let your instructor review this Ticket to Write to make sure you have narrowed your topic enough and you have an idea for an article that might appeal to the particular magazine.

TECHNO TIP

Publishing Opportunities: To investigate online sites that publish your work for free, search the Internet for these titles: Booksie, Daln, Klatcher, Scribd, and Story Of My Life.

5.4 Writing for a National Magazine, Part 2

Directions: After your instructor has approved your magazine topic, write an article for this magazine. When you have completed your article (after revising, editing, and proofreading), submit your article to the magazine. Pay attention to the particular items in the magazine's writer's guidelines.

Organizational Structures for Academic Writing

LEARNING GOAL
❸ Choose an organizational structure for an essay based on your purpose

In this and the preceding chapters, you learned about the steps in the writing process. In Part 2, you will use what you have learned and apply it to academic writing. Academic writing is divided into nine different types. All of these types are discussed in detail in Part 2, where you will learn how to write both paragraphs and essays using the writing process.

Every writing type has a particular purpose that can help prepare you for writing in other college classes and in the work world.

Heads Up!
You may see **type** referred to as *mode*, *technique*, or *domain*.

Type	In the paragraph or essay, you . . .	Your purpose is to . . .
Descriptive	depict an event, location, person or object	*portray an event, location, person, or object so well readers appreciate its value*
Narrative	relate a story in chronological order	*share an experience so readers understand its significance*
Illustration (Example)	provide examples that prove, show, or explain a particular point	*create interest in an idea, clarify a position or idea, or inform an audience of something new*
Process (How-To)	describe steps or stages necessary to reach some end	*direct how to achieve a goal, analyze steps leading to an outcome, or explain the importance of a procedure*
Definition	explain the meaning of a certain word or term	*support a unique view and new understanding of a subject through connotative meaning*
Compare and Contrast	note similarities and differences between two people, places, or things	*present a deeper understanding of two subjects*
Classification (Classification and Division)	separate a general subject into less complicated or more useful categories	*present a way to more easily understand a general subject*

Cause and Effect	examine the relationship between at least two events, actions, or influences	*illustrate how something came to be, show why an event occurred, or identify reasons something changed*
Persuasion	seek a reaction from readers	*change how readers feel about an issue, or move them to a certain action*

In academic writing, all essays have the same pattern of organization: an introduction, body paragraphs, and a conclusion. Some instructors may ask that you follow an even more specific essay structure—one you may have used before—the five-paragraph essay.

The Five-Paragraph Essay

The **five-paragraph essay** is a highly structured form that can be useful to you as a beginning writer. It can help you discover how the parts of an essay work together. Learning how to compose a five-paragraph essay will give you a standard pattern of organization to use. Some instructors may require that you use only this form, while others may direct you to a less systematic approach.

For additional ideas about the five-paragraph essay, search the Internet for these videos: **Five-Paragraph Essay: Overview and Basic** **Five-Paragraph Essay Format.**

The five-paragraph structure is uncomplicated. The first paragraph is the **introduction**, which ends with the **thesis statement**. The second, third, and fourth paragraphs are the **body paragraphs**. Each of these paragraphs starts with a **topic sentence**. The fifth paragraph is the **conclusion**, which starts with a reminder of the thesis.

This pattern can help you in many writing situations because it pushes you to stay focused. Almost all teachers (not just writing teachers) will likely recognize this pattern since it can be useful in many courses.

Some college writing instructors criticize the five-paragraph model. They feel it constricts writers to compose three similarly fashioned body paragraphs and similarly shaped introduction and conclusion paragraphs. However, many other college writing instructors find this model useful because they feel it provides a flexible foundation for writing lengthier and more complex essays.

Read the five-paragraph essay below and note its introductory paragraph and thesis statement, its body paragraphs and their topic sentences, and its concluding paragraph.

O'Callaghan 1

Aislinn O'Callaghan

Prof. Conrad Andrews

ENG 091

26 Oct. 2011

Good-bye, Facebook

Introduction ending with thesis statement

Social networks can be a lot of fun. During my first year of college, I enjoyed using Facebook to reconnect with old friends I had left back home and to create friendships with new people at college. As the year went on, though, I decided I had to give up FB because it was taking up too much of my time, I had to be friends with people I did not really like, and I had problems with identity theft.

Body paragraph #1 topic sentence

One reason I no longer use Facebook is that it was taking too much of my time. I am not sure how other people are able to stay connected and get their work done, but it was slowly taking over my life and nothing seemed to be getting done. I used to go to the school computer lab three times a week to review notes, check homework, and even do a little writing on my English papers. Each day I would spend about two hours reviewing my class work. Once I got a Facebook account, I would first log on every morning. Before I would even think about studying, I would check for updates on friends and then start looking at their photo albums. Before long, an hour or two would pass and no school work would be done.

O'Callaghan 2

Body paragraph #2 topic sentence

Another reason I canceled my Facebook account is that I felt obligated to maintain online "friendships" with people I would not be friends with in the real world. I could not keep up with the different levels of FB friends I had. In the beginning there were my "friends" who were mostly people from back home and people I graduated from high school with. I also had a few college friends and even some work friends. But in just a few weeks, a handful of friends grew to be over four hundred friends! If one of them commented on my status, I felt I had to comment back; if someone commented on a photo, I returned the kindness and offered some comment in return. One day I realized I was answering questions and carrying on conversations I really did not want to be in with almost total strangers.

Body paragraph #3 topic sentence

The main cause that prompted me to cancel my FB account was when a friend of a friend decided to "borrow" a photo from one of my FB albums and use it as her profile picture. Then, she went to a fan site for the Uptown Café, a local restaurant, and posted hateful comments about the "rotten food." With my picture next to her name, she said that the food was "not fit for a starving buzzard." It did not take long for my Uncle Mike, the cook at Uptown, to hear that I had posted the comment. It took several phone calls and a couple of e-mails before I could repair the damage this FB "friend" had done.

Conclusion with thesis reminder

Because of the time it takes, the fake friendships, and the danger of identity theft, I no longer use Facebook. Now I keep up with friends back home and here mostly through e-mail and texting. I miss seeing my friends' pictures, but the upside is I have much more time for studying.

TICKET to WRITE

5.5 Supporting Details of a Five-Paragraph Essay

Directions: Working alone or in groups, complete the following outline of Aislinn's five-paragraph essay. Fill in the details that support each of her three topic sentences. The topic sentences and some details are provided.

THESIS I decided I had to give up FB because it was taking up too much of my time, I had to be friends with people I didn't really like, and I had problems with identity theft.

(continued)

TOPIC SENTENCE 1 One reason I no longer use Facebook is that it was taking too much of my time.

Detail A go to the school computer lab three mornings a week to review notes, check homework, and work on English papers

Detail B log on every morning before studying, check for updates and look at photos

Detail C an hour or two would pass and no school work would be done

TOPIC SENTENCE 2 Another reason I cancelled my Facebook account is that I felt obligated to maintain online "friendships" with people I wouldn't be friends with in the real world.

Detail A I couldn't keep up with the different levels of FB friends I had.

Detail B If one of them commented on my status, I felt I had to comment back.

Detail C answering questions and carrying on conversations I didn't want to be in with strangers

TOPIC SENTENCE 3 The main cause that prompted me to cancel my FB account was when a friend of a friend decided to "borrow" a photo from one of my FB photo albums and use it as her profile picture.

Detail A Then, she went to a fan site for Uptown Café . . . posted hateful comments about the "rotten food"

Detail B didn't take long for my Uncle Mike, the cook at Uptown, to hear that I'd posted the comment

Detail C It took several phone calls and a couple of e-mails before I could repair the damage this FB "friend" had done.

THESIS REMINDER Because of the time it takes, the fake friendships, and the danger of identity theft, I no longer use Facebook.

In Chapter 2, you learned about various types of introductions for essays. The five-paragraph essay typically uses the *general subject* introduction, in which you write something about the general subject of your essay and then narrow the focus until it ends with your thesis statement.

5.6 Introductory Paragraph of a Five-Paragraph Essay

Directions: Below are thesis statements students have decided to use in their five-paragraph essays. Working alone or in groups, compose sentences that could lead to these thesis statements in an introductory paragraph. Answers will vary.

1. Three guidelines for a good marriage include sustaining open communication, keeping a sense of humor, and dividing chores equally.

2. Because it helps with scheduling my school life, my family life, and my work life, I began keeping a time chart.
3. After all that had happened, I decided to switch doctors.

TECHNO TIP

For more information on the five-paragraph essay, search the Internet for these sites:

The Five-Paragraph Essay - Commnet

Outline of the Five Paragraph Essay - Maricopa

The Five Paragraph Essay - Study Guides and Strategies

JSCC - The Five-Paragraph Essay

Digging Deeper: Beyond the Five-Paragraph Essay

While the five-paragraph essay is a helpful form, it doesn't fit every writing situation. As you progress in college, topics you explore will often have more than three points of development.

For example, suppose you're taking a sociology class and you're assigned to write an essay about causes of stress for college students. You discover that you need to discuss six main causes: *school*, *family*, *work*, *finances*, *health*, and *other personal concerns*. In order to fully explain each of these causes, you need to develop each one in its own body paragraph. Some topics, in fact, may need more than one body paragraph.

TICKET to WRITE

5.7 Body Paragraphs of the Five-Paragraph Essay

Directions: Below are the topic sentences for the sociology essay describing six main causes of stress for college students. Working alone or in groups, brainstorm supporting details for these topic sentences. Then use these details to compose the body paragraphs. Answers will vary.

TOPIC SENTENCE 1 Not surprisingly, a primary cause of stress for college students is college itself.

TOPIC SENTENCE 2 Whether students are away at school or attending college in their hometown, their family life also adds stress.

TOPIC SENTENCE 3 Work issues can affect time that should be devoted to class and studies.

TOPIC SENTENCE 4 Finance problems plague many college students.

TOPIC SENTENCE 5 Unhealthy lifestyles also contribute to stress.

TOPIC SENTENCE 6 Various personal concerns add to college students' anxiety.

TECHNO TIP

Examine this online site for building on the five-paragraph essay:

Beyond the Five-Paragraph Essay (Swarthmore College)

MyWritingLab™ Visit ***MyWritingLab.com*** and complete the exercises and activities in the **Recognizing the Paragraph, Recognizing the Essay, Developing and Organizing a Paragraph,** and **Essay Organization** topic areas.

RUN THAT BY ME AGAIN

LEARNING GOAL
❶ Publish and format a paper for class

- **All publishing centers on . . .** your purpose.
- **In a classroom, *publishing* means . . .** sending your finished product from your hands to your instructor's.
- **Most classes devoted to humanities require you to follow . . .** MLA style.
- **Most classes in social sciences require you to follow . . .** APA style.

LEARNING GOAL
❷ Publish your writing in a personal, online, or print format based on your purpose

- **Personal publishing may take many forms, like . . .** journal or diary entries, memoirs, e-mail, and letters.
- **Examples of online publishing include . . .** e-mailing, advertising on a Web site, posting a to a discussion group, writing a personal blog, and participating in social networks.
- **Your college or university may also have published material, including . . .** student literary magazines, newspapers, 'zines, student showcases, and professional journals.

LEARNING GOAL
❸ Choose an organizational structure for an essay based on your purpose

- **The purpose of descriptive writing is to . . .** portray an event, location, person, or object so well readers appreciate its value.
- **The purpose of narrative writing is to . . .** share an experience so readers understand its significance.
- **The purpose of illustration writing is to . . .** create interest in an idea, clarify a position or idea, or inform an audience of something new.
- **The purpose of process writing is to . . .** direct how to achieve a goal, analyze steps leading to an outcome, or explain the importance of a procedure.
- **The purpose of definition writing is to . . .** support a unique view and understanding of a subject through connotative meaning.
- **The purpose of compare-and-contrast writing is to . . .** present a deeper understanding of two subjects.
- **The purpose of classification (and division) writing is to . . .** present a way to more easily understand a general subject.
- **The purpose of cause-and-effect writing is to . . .** illustrate how something came to be, show why an event occurred, or identify reasons something changed.
- **The purpose of persuasion writing is to . . .** change how readers feel about an issue or move them to a certain action.
- **The parts of a five-paragraph essay are . . .** the introduction, the body paragraphs, and the conclusion.

PUBLISHING AND ACADEMIC WRITING LEARNING LOG MyWritingLab™

MyWritingLab™

1. On what does all publishing center?
   All publishing centers on your purpose.

2. In classroom publishing, the final step before you turn your paper in is to do what?
   The final step is to be certain you followed any formatting requirements your instructor gave.

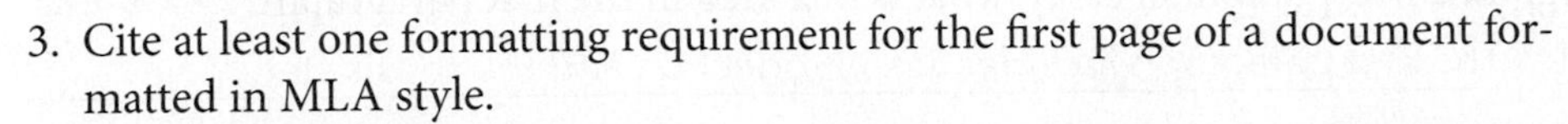

3. Cite at least one formatting requirement for the first page of a document formatted in MLA style.
   Answers will vary. Possible answers include: Do not type a title page; number the pages with a header; double-space under the date and type the title of your essay; double-space between your title and the first line of your essay.

4. Cite at least one formatting requirement for the text of a document formatted in MLA style.
   Answers will vary. Possible answers include: Double-space your text; indent the first line of all paragraphs one-half inch from the left margin; use only one space after end punctuation; do not right justify your text.

5. In personal writing, what dictates whether you use formal or informal writing?
   Your purpose in personal publishing dictates whether your writing is formal or informal.

6. What do you do in descriptive writing?
   You depict an event, location, person, or object.

7. What do you do in narrative writing?
   You relate a story in chronological order.

8. What do you do in illustrative writing?
   You provide examples that prove, show, or explain a particular point.

9. What do you do in process writing?
   You describe steps or stages necessary to reach some end.

10. What do you do in definition writing?
    You explain the meaning of a certain word or term.

11. What do you do in compare-and-contrast writing?
    You note similarities and differences between two people, places, or things.

12. **What do you do in classification (and division) writing?**
    You separate a general subject into less complicated or more useful categories.
13. **What do you do in cause-and-effect writing?**
    You examine the relationship between at least two events, actions, or influences.
14. **What do you do in persuasion writing?**
    You seek a reaction from readers.
15. **In a five-paragraph essay, what is included in the first paragraph?**
    The first paragraph includes the introduction and the thesis statement.

PART

2

Types of Paragraphs and Essays

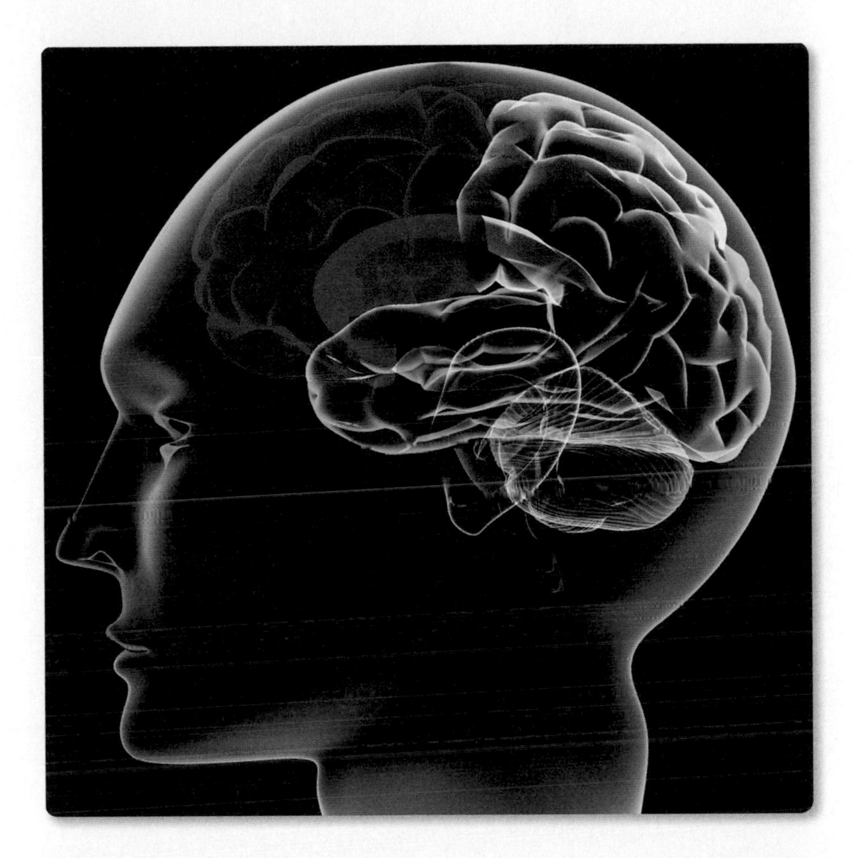

Traditional academic writing is based in nine types of writing: *description, narration, illustration, process, definition, compare and contrast, classification, cause and effect*, and *persuasion*. The type of writing you use depends on the purpose behind your writing. Part 2 offers you instruction and practice in writing paragraphs and essays using each type of academic writing.

"Your purpose is to make your audience see what you saw, hear what you heard, feel what you felt. Relevant detail, couched in concrete, colorful language, is the best way to recreate the incident as it happened and to picture it for the audience."

—Dale Carnegie

CHAPTER 6 Descriptive Writing

LEARNING GOALS

In this chapter, you'll learn and practice how to

1. Get started with descriptive writing
2. Write a descriptive paragraph
3. Read and examine student and professional essays
4. Write a descriptive essay

GETTING THERE

- Effective descriptive writing allows your readers to experience your topic.
- Sensory details—those that describe touch, sound, sight, smell, and taste—make your writing more focused and interesting.

FASTWRITE 1

Look at the pictures above and describe one of the scenes. Write as much as you can, as fast as you can, for three minutes.

FASTWRITE 2

Consider a song that means a good deal to you, perhaps a song that you listen to more than any other. Write a description of that song and how it makes you feel. Write as much as you can, as fast as you can, for five minutes.

Complete these **Fastwrites** at **mywritinglab.com**

Getting Started in Descriptive Writing

LEARNING GOAL
1. Get started with descriptive writing

When you write a **descriptive paragraph or essay**, you re-create a person, place, or thing so your readers can feel either

- they had a part in your experience

or

- they came to know the person, place, or thing you describe.

Descriptive essays are subjective—you're relating your impressions about a topic.

Through your words, readers appreciate how someone or something looked, smelled, felt, sounded, or tasted—or a combination of those.

For instance, if you're describing a bonfire you attended before your football team left for the Cotton Bowl, one sentence might read this way:

Your impressions are different from someone else's, and that's fine. In fact, writers often discover interesting differences when they write about the same topic.

> The constant crackle of burning leaves synchronized with the cheers that erupted when our bowl-bound team paraded down the street.

In this sentence, you **smell** the leaves and **hear** them crackle, you **hear** the cheers, and you **see** both the leaves and the team.

You might describe a place that holds special significance for you and include a sentence like this:

> On spring break at the beach, I watched the sun spread its orange reflection across the gulf, and I listened to the crashing waves and the warm salt breeze that plucked at the porch screens.

In this sentence, you **see** the sun, **hear** the waves and the breeze, **feel** the warmth of the breeze, and **smell** (maybe even **taste**) the salt air.

6.1 Fastwriting a Description

Directions: For three to five minutes, fastwrite a description of a place you have visited or would like to visit. Use as many sensory details as possible. When you're finished, share your description with those in your writing group.

WRITING at WORK *Snapshot of a Writer*

Penny Pennington, Real Estate Broker

Penny Pennington has been selling real estate for twenty-five years. "I had quit my job as a microbiologist to spend more time with my children," said Penny. "Once the kids were older, I was ready to get back in the workforce, but I wanted something new. When I was in my forties, I went back to school at Tallahassee Community College and earned my real estate license. The business has its ups and downs, but I thoroughly enjoy what I do."

Penny's Writing Process

Remember Your Audience I depend a lot on description. I have to be able to write property descriptions that highlight the attention-getting features of a home, attracting potential buyers and satisfying sellers. I need to know what the home buyer is looking for, so I've got to know my audience.

Put Pen to Paper You can't be afraid to just put something to paper. I've got to start somewhere, so once I've talked with the sellers and looked over the property, I make lists, lists of the good and of the bad.

Give It Shape Next I draft a short description that features the unique selling points of that particular property. "Curb appeal" is what we call that head-turning quality passersby latch on to. I've got to sell my readers on a property's location, so whenever I can I lead with the foremost characteristic potential buyers want: location, location, location.

Revise Once I'm satisfied with what I've written, I check again the available ad space I have to fill, the photo I've got to run with the ad, and my notes from my interview with the seller. I make any necessary changes and then pass the ad along to a couple of coworkers for their feedback. They might see something in the home that I've overlooked in my description.

Publish And that's it! Next it's off to the paper, real estate magazine, MLS (Multiple Listing Service) publication, and my company's Web site.

Enjoy the quiet country life! This spacious 3-bedroom, 2½-bath cabin home is situated on 4 acres at the end of a secluded cul-de-sac. Access your wrap-around porch by the front door, the dining room's French doors, or the master bedroom's French doors. To see this property and more, visit Penny's Realty Listings online or call 850/555-5555 and talk to Penny today.

TRY IT!

Write a paragraph describing a dwelling you know well. Assume the role of realtor, and write a brief description of the dwelling for a realty magazine. When you're finished, share your description with those in your writing group.

Getting Started with Your Descriptive Paragraph

LEARNING GOAL
❷ Write a descriptive paragraph

As a writer, you may feel more comfortable writing a descriptive paragraph before you write a descriptive essay. If that is the case, follow steps one through six below. If not, go right on to Take 2 on page 100.

Step One: Choose a Topic

Remember that **topic** and **topic sentence** are not the same. See Chapter 1 to review the difference.

When you're writing a descriptive paragraph, you'll often be assigned a particular topic to develop. If you have the freedom to choose a topic, however, keep in mind that your purpose in descriptive writing is to describe a person, place, or thing that has a special meaning (either good or bad) for you so well that your readers understand your unique connection to your topic.

To get a jumpstart on choosing a topic, complete Ticket to Write 6.2 below.

6.2 Choose a Topic

Directions: Complete the following prompts to start the process of choosing a topic. Answers will vary.

1. List at least three restaurants you like. ______________________________
2. List at least three movies you're fond of. ______________________________
3. List at least three songs you enjoy. ______________________________
4. List at least three toys you remember from childhood. ______________

Step Two: Generate Ideas

Once you've determined your topic, the next step is to begin creating ideas that support your topic. Return to Chapter 1 to review several prewriting methods that help generate ideas. Complete Ticket to Write 6.3 below to begin generating ideas for one of those topics.

6.3 Generate Ideas

Directions: From each list in Ticket to Write 6.2, circle one topic that attracts your attention. Then, freewrite answers to the reporter's questions (*Who? What? Why? Where? When?* and *How?*) for the topics you have chosen from each category. Form your own reporter's questions or use the following: Answers will vary.

1. **Restaurant** . . .
   a. To **whom** does the restaurant cater?
   b. **What** do I like to eat there?
   c. **Why** is this a favorite restaurant of mine?
   d. **Where** is the restaurant located?
   e. **When** do I like it?
   f. **How** is this restaurant unique?
2. **Movie** . . .
   a. **Who** tends to be attracted to this type of movie?
   b. **What** type of movie is this?
   c. **Why** is this one of my favorite movies?
   d. **Where** is the movie set?
   e. **When** is the movie set?
   f. **How** is this movie unique?
3. **Song** . . .
   a. **Who** tends to be attracted to this type of song?
   b. **What** type of song is this?
   c. **Why** is this one of my favorite songs?
   d. **Where** do I usually hear this song?
   e. **When** was this song significant to me?
   f. **How** is this song unique?
4. **Toy** . . .
   a. **Who** tends to be attracted to this type of toy?
   b. **What** type of toy is this?
   c. **Why** is this toy memorable to me?
   d. **Where** did I play with this toy?
   e. **When** did I play with this toy?
   f. **How** is this toy unique?

Step Three: Define Your Audience

For each topic, define an audience that would be interested in reading a description of it. Who, for example, might want to know about a favorite toy from your childhood and why it was your favorite? It might be Aunt Polly, who gave you your favorite Hot Wheels racing set, your first pair of Heelys, or your Malibu Barbie. Or perhaps you want to sell this toy on eBay, emphasizing its remarkable (and marketable) qualities. Or maybe you have another audience in mind.

Think about the movie or song you chose. Who might be interested in reading your description of this movie or song? Is the audience reading your personal blog, a review in your school newspaper, a remembrance posted on your class reunion Web site—or something else? Use Ticket to Write 6.4 to determine your audience and purpose for describing each topic.

6.4 Fastwrite a Description

Directions: Answer these two questions as completely as possible for each of the four topics. Answers will vary.

Who is my audience?
What is my purpose for describing this?

Step Four: Draft Your Paragraph

Because paragraphs center on a topic sentence, review your purpose in order to focus your topic sentence. You should include the focal point of the paragraph in your topic sentence. Now, complete Ticket to Write 6.5 below.

6.5 Create a Topic Sentence

Directions: Choose one of the four topics you have been working with, and write your purpose, audience, and topic sentence. Then share your work with members of your writing group to see if they agree that your topic sentence explains your purpose and is appropriate for your audience. Answers will vary.

After writing your topic sentence, use your responses to the reporter's questions to compose the supporting details of your paragraph. As you write, keep your audience and purpose in mind. Adjust and revise the answers to your reporter's questions as needed. You may not use them all, and you may find that you want to expand further on certain responses. Then complete Ticket to Write 6.6.

6.6 Compose Supporting Details

Directions: Using the topic sentence you composed in Ticket to Write 6.5 and the ideas you generated in Ticket to Write 6.3, compose supporting details for your descriptive paragraph. Share these supporting details with your writing group. Ask members if you have given enough details so that your readers appreciate the sentiment you expressed in your topic sentence. Answers will vary.

Step Five: Revise Your Paragraph

After you've finished your paragraph, save it, print a hard copy, and read it out loud, looking and listening for errors in grammar and content. Refer to the Review Checklist for a Paragraph in Chapter 3, page 60, to make sure you revise as completely as possible.

See Chapter 4 to review proofreading techniques.

REVIEW CHECKLIST for a Descriptive Paragraph

- ☐ Is my topic sentence clear?
- ☐ Do I include sensory details?
- ☐ Do I hear any fragments or run-ons when reading out loud?
- ☐ Do all complete thoughts have appropriate end punctuation?
- ☐ Is my paragraph appropriate for my audience?
- ☐ Does my purpose remain constant?
- ☐ Do any sentences or ideas seem off-topic?

Heads up!
See Chapter 20 to review fragments and run-ons.

Once you've made any necessary changes, save your work and leave it alone for a while. Then come back later, print out a new copy, and look and listen for errors again. Refer again to the Review Checklist for a Paragraph in Chapter 3.

Step Six: Peer Review

You might think you've found all of your errors and made all of the improvements you can, but having someone else read your paragraph and offer suggestions will almost always improve your writing. This person is called your **peer reviewer**.

Your peer reviewer may use the Review Checklist for a Descriptive Paragraph above or the Review Checklist for a Paragraph in Chapter 3, or your instructor may provide a different checklist. You might or might not agree with the suggestions you

Heads up!
See Chapter 4 for more on peer reviewing.

receive, but peer reviewers often find places for improvement or errors that slipped by you. Listen to or read closely your peer reviewers' suggestions, and don't be shy about asking questions to clarify their ideas and suggestions.

6.7 Peer Review

Directions: Once you have completed revising your paragraph, share it with your peer reviewers. Ask them to review your paragraph using the Review Checklist for a Descriptive Paragraph above and the Review Checklist for a Paragraph from Chapter 3 (page 60). Then ask them to record their suggestions for revision. Answers will vary.

Student and Professional Essays

LEARNING GOAL
❸ Read and examine student and professional essays

Below are two descriptive essays. The first was written by a student and the second by a professional. Read these two essays and answer the questions that follow them.

Student Essay

Exceedingly Extraordinary X Games
by Jeff Blake

1 My father's job as a **photojournalist** takes him to a lot of exciting places and events, and I've been lucky enough to have joined him on a few of his photo shoots. I saw **Kobe** hit fifty points in a single game to lead the Lakers to a win over New Orleans. I saw **Sam "Hardluck" Hornish**'s miracle win at the Indy 500. And I even saw France's leading scorer, **Zindine Zidane**, head butt an Italian player and hand Italy the World Cup in Berlin. But, I'm not really a basketball or a stock car fan, and even though Berlin was an eye-opening experience for me, I'm not a big soccer buff. I am a winter sports fanatic, so I was really excited last January when my dad surprised me with a trip to the twelfth annual **X Games**. Attending the X Games for the first time was my favorite sporting adventure.

2 The lodge's guests were unlike any hotel guests I'd seen before. When we walked into the lobby, leaving the 19° weather outside, I was amazed by the hustle and bustle of people. Skiers and boarders were coming and going, bringing a constant blast of cold air into the lobby. Sitting around large oak coffee tables were a dozen people my age, getting ready for the slopes. They were lacing boots, attaching ski-lift tickets to jackets, smearing on ChapStick,

photojournalist a person who presents a news story primarily through photographs

Kobe a reference to Kobe Bryant, a professional basketball player

Sam "Hardluck" Hornish a professional race car driver

Zindine Zidane a French soccer star

X Games an annual event that centers on extreme action sports

and zipping up pockets they'd just stuffed with cell phones, wallets, and room keycards. I'll always remember our room number was 364 and I still have my room keycard. The teens were smiling, laughing, and eager to hit the slopes. It was an ocean of nylon jackets and bib pants, crazy-colored wool caps, and even a few fuzzy sweaters on those sitting in front of two fireplaces large enough to walk into.

3 After taking our luggage up to our room, we headed out to catch the press courtesy shuttle to our first event, the men's snowboard **slopestyle**. Our short ride in the snowcat, an eight-passenger vehicle with tank-like tracks, made me feel like a VIP because almost everybody else had to walk to the event staging area. The vehicle maneuvered up a snowy path, weaving between groups of people making their way to the slopestyle competition. The ride was smoother than I had thought it would be, but it was loud and cold because the driver had both the front side windows open. The snowcat sounded just like my neighbor's Volkswagen Beetle that has no muffler. My dad talked with the four other press members as I stared out the window. I was like a little kid experiencing a theme park for the first time. I took in everything from the crisp winter air and the snow-tipped trees to the smiling spectators and X Game workers trudging up the hill. People looked at us as if we were athletes, event judges, or some other major players.

4 Watching the first events of the day—the morning runs—was a memorable experience. When I stepped out of the snowcat, the wind stung at my face, bringing with it the pure smell of pines and the clean Colorado winter air. Dad's press passes got us a sweet spot for watching the men's freestyle snowboard event. We stood in a small crowd of photojournalists and watched as one of my favorite **airdogs**, **Shaun White**, spun and twisted his way to a bronze medal. Nicknamed "the Flying Tomato" because of his long red hair, Shaun came out throwing tricks from the start, gaining some big air. He twisted two **540s**, spinning one-and-a-half times first to his left and then immediately to the right. He maneuvered in the air as if suspended by invisible wires, hanging over the earth surrounded by blowing snow. The cheers of the crowd echoed off the tree line behind us as soon as Shaun started his first run. I could see his curly red hair flying behind him, and I could hear his board slap the powder hard after each bit of air—so hard that his board broke during his first run. He looked pretty disappointed afterward, but the crowd still cheered.

5 That afternoon, we were lucky enough to see Shaun take the gold in the men's **superpipe.** The snow had started to fall again and dropped heavy, wet flakes on everything and everybody. I took off my gloves and held the cups of hot chocolate Dad had gotten for us. The warmth felt good on my hands. Again, I stood in the press gallery behind Dad. He squatted below the course tape for the low-angle shot of what he hoped to be Shaun's biggest trick, a **1260**. Even though triples were pretty cool and a number of boarders could turn them, no one had completed three-and-a-half turns . . . yet. Everyone hoped he could do it; even those spectators waving Norwegian and Canadian flags hoped he could. All of us wanted to see a 1260, and we weren't

slopestyle a form of snowboarding and skiing competition in which the skier performs tricks while going through a series of jumps and other obstacles

airdog a slang term meaning "a snowboarder who jumps most of the time and is primarily interested in aerial tricks"

Shaun White a professional snowboarder and skateboarder

540s rotations of 540° (one-and-a-half turns)

superpipe a snow structure built for freestyle snowboarding

1260 a rotation of 1260° (three-and-a-half complete turns)

disappointed. As soon as he completed his 1260, cheers, screams, and ringing cowbells echoed across the mountain.

6 January of 2009 was a month I will never forget. Dad ended up with an awesome photo capturing history in the making that made the sports pages of dozens of newspapers, and I got to see it happen—Shaun White's 1260° spin to victory. I can't wait for next year's Winter X Games.

A CLOSER LOOK

Answer these questions about the essay:

1. Where is the thesis found?

   The thesis is in the last sentence of paragraph 1.

2. What is the topic sentence of paragraph 2?

   The topic sentence of paragraph 2 is the first sentence.

3. In paragraph 2, which sentence does not support the topic and could be taken out for a better paragraph focus?

   I'll always remember our room number was 364 and I still have my room keycard.

4. What is the topic sentence of paragraph 3?

   After taking our luggage up to our room, we headed out to catch the press courtesy shuttle to our first event, the men's snowboard slopestyle.

5. What is the topic sentence of paragraph 5?

   That afternoon, we were lucky enough to see Shaun take the gold in the men's superpipe.

6. List three sensory details in paragraph 3.

   The vehicle maneuvered up a snowy path, weaving between groups of people.

   The snowcat sounded just like my neighbor's Volkswagen Beetle that has no muffler.

   The ride was loud and cold because the driver had both windows open.

7. List three sensory details in paragraph 5.

   The snow had started to fall again and dropped heavy, wet flakes.

   I took off my gloves and held the cups of hot chocolate. The warmth felt good on my hands.

   . . . cheers, screams, and ringing cowbells echoed across the mountain.

8. In paragraph 6, what details make the X Games a memorable event for the writer?

   Dad's awesome photo; seeing Shaun White's victory in person

9. List three slang terms the writer uses and define each without using slang or clichés. Answers may vary. Possible answers:

   Term: gaining big air **Definition:** making high jumps

   Term: throwing tricks **Definition:** performing stunts successfully

   Term: pretty cool **Definition:** fairly impressive

10. What sentence in the conclusion recaps the thesis statement of the essay?

    The first sentence in the last paragraph recaps the thesis statement.

Professional Essay

The professional essay below describes a rock concert a young woman attended with her mother.

Mick, Mom, and Me

by Mary Beth Anderson

1 I was sitting at the kitchen table, nursing another cup of lukewarm coffee and feeling **brain dead** after working a twelve-hour shift. Then Mom sat down beside me, singing the Rolling Stones' "You Can't Always Get What You Want." She slid an envelope across the table to me, and said softly, "Or maybe you *can* get what you want." Curious, I slit the envelope open and out tumbled two tickets to the Rolling Stones' "A Bigger Bang Tour." I shrieked loud enough to wake the neighbors because I knew my mom and I would bond even closer when we rocked out at the concert together.

2 Mom and I often talked about the "**intergenerational bonding**" we felt in our mutual love for the Stones, and when we were alone in the car we sometimes belted out duets of what we called "our **repertoire**"—Stones classics like "Brown Sugar" and "She Saw Me Coming." Now we were going to see the Stones in person.

3 Three long months later, the night of the concert arrived. We had spent what seemed like hours in traffic jams and lines outside the open-air arena. Once we were finally inside the gates, Mom and I made a beeline to the concession stand and laughed when we both grabbed the same soft black t-shirt with the classic Stones logo—a huge red mouth with a tongue sticking out. After choosing shirts, mugs, and key chains, we stuffed our treasures into plastic bags that also sported Stones logos (those bags won't be going into the recycling bin) and made our way to our highly prized seats in third row center. As the sun set and the night sky turned hazy, I began to take notice of the rest of the crowd. We were surrounded by roughly equal parts of baby boomers,

brain dead an idiom meaning "too tired to think"

intergenerational bonding a connection between people of different age groups

repertoire a group of songs

young adults, and teenaged and younger kids. This truly was the intergenerational bonding that Mom had described.

4 Smashing Pumpkins, the opening act, took the stage around 7:30. In my book, the Pumpkins are ticket-worthy themselves, but Mom's not really a fan. I did notice, though, that she tapped her left foot a little when they sang "Perfect." I kept my eyes on Billy Corgan and concentrated on his distinctive, nasal singing. But I still craved the Stones.

5 When they finally came on stage, at almost 9 p.m., the crowd was **abuzz.** Fireworks began blasting in all directions, leaving thick smoke and burnt-sulfur smells below and showering the night sky above. In the rear of the seven-story stage, a huge screen, flanked by two smaller screens, blasted other computer-generated graphics explosions—symbolic, of course, of the **Big Bang**. All of a sudden, the Stones appeared in person and jumped right into their first song, "Start Me Up." Its lyrics "Give it all you got/ You got to never, never, never stop" seemed **apropos** of the band itself.

6 All night long, Mick, dressed in a silver lamé jacket over a black t-shirt—just like the one I'd bought—pranced, preened, skipped, and strutted around the stage. Keith, with the ever-present cigarette dangling from the side of his mouth, traded musical **riffs** with Mick throughout the night. Ron and Charlie generally stayed in the background, occasionally piping in with **droll repartee.**

7 For most of the concert, we stood up, swaying to the blasting beats, our index fingers pointing to the band in **staccato** movements that followed the rhythm of the songs. In the aisles, people were dancing alone or in groups, and as the band played "Jumpin' Jack Flash," we all jumped in unison through the entire song.

8 When the concert finally ended with "Satisfaction," I felt both satisfied and wanting more. As we sauntered out into the still-warm night, I realized this was the first of many Stones concerts that Mom and I would share.

abuzz filled with excitement
Big Bang according to some scientists, the cosmic explosion that began the universe
apropos appropriate, pertinent
riffs short series of notes that form a distinctive part of the accompaniment
droll amusing, funny
repartee clever conversation
staccato rapid, brief, and clipped

A CLOSER LOOK

Answer these questions about the essay:

1. What is the thesis statement of this essay?

   I shrieked loud enough to wake the neighbors because I knew my mom and I would bond even closer when we rocked out at the concert together.

2. Give an example of the imagery of sight in this essay.

   Answers may vary. Possible answers: Anderson sitting at the kitchen table with her coffee, Anderson's mother sliding the envelope across the table, and Mick's onstage antics.

3. Give an example of the imagery of sound in this essay.

   Answers may vary. Possible answers: Anderson shrieking when she discovered the tickets, Anderson and her mother singing duets, and the musical riffs of Mick and Keith.

4. Give an example of the imagery of taste in this essay.

   Answers may vary. Possible answer: lukewarm coffee Anderson was drinking.

5. Give an example of the imagery of touch in this essay.

   Answers may vary. Possible answer: "nursing another cup of lukewarm coffee."

6. Give an example of the imagery of smell in this essay.

   Answers may vary. Possible answer: the burnt-sulfur smell of the fireworks.

7. In paragraph 3, what phrase signals a transition in time?

   Three long months later

8. What does the introduction to this essay detail?

   Answers may vary. Possible answers: The introduction details the author's change from boredom to joy.

9. Which words in paragraph 6 help give a clear description of Mick Jagger?

   dressed in a silver lamé jacket over a black t-shirt, pranced, preened, skipped, strutted

10. Cite at least three vivid verbs in this essay.

    Vivid verbs include *slid, slit, shrieked, tumbled, belted (out), grabbed, stuffed, tapped, craved, pranced, preened, skipped, strutted, swaying, jumped.*

Writing Your Descriptive Essay

Essay writing is much like paragraph writing: you start with a topic and expand on it with details. Remember that just as a paragraph supports a topic sentence, an essay supports a thesis statement.

LEARNING GOAL

4 Write a descriptive essay

Step One: Choose a Topic and Develop a Working Thesis Statement

See Chapter 2 to review thesis statements.

To begin your descriptive essay, ask yourself these **starter questions**:

1. **Whom** or **what** do I want to describe? (Answering this is the easy part.)
2. **What** qualities do I want to concentrate on or make readers most aware of?
3. **Why** do I want to describe this, or **what** is my purpose? ("Because it's an assignment" isn't a good enough answer.)

For instance, in a descriptive piece about attending a live television show, you might answer starter questions this way:

1. **Whom** or **what** do I want to describe?

   attending a live taping of *American Idol*

2. **What** qualities do I want to concentrate on or make readers most aware of?

   what I saw and heard from famous judges and hopeful contestants

3. **Why** do I want to describe this?

   it was a once-in-a-lifetime experience

Answering Starter Questions
Jeff Blake and Mary Beth Anderson

In answering the starter questions, student writer Jeff Blake, the author of "Exceedingly Extraordinary X Games" wrote:

1. I want to describe what it was like attending the twelfth annual Winter X Games.
2. I want to concentrate on the sights, sounds, and wonderment of attending this event for the first time.
3. I want to describe this because I want readers to appreciate my first-time experience.

Mary Beth Anderson, author of "Mick, Mom, and Me" might have addressed the starter questions this way:

1. I want to describe a concert I attended with my mother.
2. I want to make readers aware of the songs, the atmosphere, and the diversity of the crowd.
3. I want to describe this because I want to relate how important this bonding experience was for me.

The writer who wants to describe the *American Idol* taping might fill in the blanks like this:

I will write about a taping of *American Idol* because I want to show readers the sights and sounds of this memorable experience.

Look at how two other writers chose to fill in the blanks for other topics:

I will write about the night I ate at a restaurant owned by a famous TV chef because I want to show readers that spending a great deal of money might not translate into having a good meal.

I will write about the style of clothing that I wear because I want to show readers that my clothing choices reflect my individuality.

TICKET to WRITE

6.8 Answer Starter Questions for a Descriptive Essay

Directions: All essays have thesis statements. To help develop a thesis statement for your descriptive essay, begin with a starter statement. Fill in the blanks below for your starter statement. Answers will vary.

I will write about ____________________ (name a person, place, or thing) because I want to show readers ____________________.

Once you've filled in the starter statement, you've begun to develop the focus of your essay. Return to your starter statement and consider your topic and your purpose. From this starter statement, create your **working thesis statement**. These are working thesis statements created from the starter statements above:

Attending a taping of *American Idol* was a memorable experience because of the emotional highs and lows of the singers and judges.

Chef Edith Quinn may have a hit show on the Let's Eat channel, but spending over $150 in her restaurant does not mean the meal will be worth the price.

Retro clothes may not be for everyone, but I like them because they reflect my fascination with the 70s culture.

6.9 Develop a Working Thesis Statement

Directions: From the starter statement you composed in Ticket to Write 6.8, consider your *topic* and your *purpose*. Now create your *working thesis statement*. Then share it with members of your writing group and ask them if they understand what you will be describing in your essay and why it is important to you.

Step Two: Generate Ideas

Did you know smell is the sense that evokes the most memory?

When you think about whom or what you're describing, ask yourself what sights, sounds, smells, tastes, and textures are dominant, most important, most memorable, most significant, or even most surprising. The idea is to give readers the details they need to share and appreciate your experience with your topic and your understanding of it.

Remember that your working thesis statement needs to relate a specific idea about your topic. With your working thesis statement in mind, **brainstorm** answers to these questions about your topic:

See Chapter 1 to review activities and techniques for brainstorming and freewriting.

- **What did I *see*?** Think of colors, shapes, textures, lights, and facial expressions or gestures.
- **What did I *hear*?** Consider the obvious sounds, but also think about any minor or background sounds.
- **What did I *taste*?** Food has distinctive tastes, but also think of tastes from other sources—like sweat that rolls onto your lips.
- **What did I *feel*?** Think about both things that you touched and things that touched you, like maybe a waterfall, a chilling breeze, the abrasive concrete sidewalk, the heat from a bonfire.
- **What did I *smell*?** Obvious smells, like cologne, contribute to the effect you want to convey, as can the spicy smell of clove cigarettes in a coffeehouse, or the butterscotch smell of ponderosa pine in a Colorado forest.

6.10 Generate Ideas

Directions: Write your working thesis statement and any ideas you generated *in* brainstorming. Then share this with your writing group and ask them to note which ideas seem the most compelling or which elaborate the most on the information in your working thesis. Answers will vary.

Step Three: Define Your Audience

Your next step is to identify the audience for your essay. Defining your audience will help you further focus on your purpose and decide what type of language you will use. Suppose you're describing your first clambake. If you're writing this description for your best friend, your language will be much less formal than if you're describing it for a class assignment. You may relate the same details in both descriptions, but the way you relate them—your language—will change because of your audience.

6.11 Define Your Audience

Directions: Write your working thesis statement and identify your intended audience. Share this with members of your writing group and ask if they agree that your working thesis will be satisfactory for your audience.

Step Four: Draft Your Essay

A cardinal rule of writing is "Show, don't tell."

"Don't tell me the moon is shining; show me the glint of light on broken glass."
—Anton Chekhov

Refer to Chapter 2 to review suggestions about discovery drafting.

In a descriptive essay, **imagery** (writing that forms mental pictures of people, places, or things) helps your readers understand or appreciate your topic. Look at these sentences:

> The day was really hot.
>
> With sweat pouring down my face, I picked up a stray piece of paper, crinkled it accordion-style, and began to fan myself so I could feel just a little cooler.

The first sentence *tells* readers only that the day was hot; the second *shows* readers (that is, it conveys a mental picture of) how hot the day really was.

Keep this use of imagery in mind as you begin the discovery draft of your descriptive essay. You have your topic, your working thesis statement and ideas that support it, and you have determined your audience. Now get your ideas in some kind of order by writing your discovery draft. Remember that, in later revisions, you'll probably change your mind about the order of your ideas. Whether you're writing on paper or computer, the point is to get your ideas down. Your ideas are likely to change—that's the progress of your writing process.

TICKET to WRITE

6.12 Write Your Discovery Draft

Directions: From the material you compiled in Ticket to Write 6.8 through 6.11, you have a starter question, a working thesis statement, ideas, and a definite audience for your descriptive essay. Use this material to write your discovery draft. Refer to Chapter 2 to review discovery drafting.

When you have completed your discovery draft, share your work with members of your writing group. Ask them to note (1) areas that are the most descriptive and (2) areas that may need more development. Answers will vary.

Step Five: Organize Your Essay

Your essay flows much more smoothly when you use transitional words and phrases. These give your readers hints of the direction your thoughts are taking and guide them through your essay.

In a descriptive essay, you might use words and phrases showing **spatial relation**, such as the following:

above	beneath	near
amid	beside	next to
among	between	on top (bottom)
atop	beyond	under
behind	in front (back) of	
below	inside	

Step Six: Apply Critical Thinking

Critical thinking means thinking and rethinking about a topic, with the goal of discovering all the implications you can about it. In thinking critically about a descriptive essay, you often . . .

- examine why a particular person, place, or thing is memorable to you
- analyze the details that contribute to your memory
- reflect on the significance of a particular person, place, or thing
- provide precise details
- reflect on your actions
- discover a deeper appreciation for your topic

The following critical thinking questions are important when you write any type of essay. After you write each of your drafts, ask yourself these questions and then apply your answers to your writing.

1. **Purpose:** Will my readers understand why my topic is important or memorable to me? (This importance or memory may have had a positive or negative impact.) Have I clearly stated or clearly implied my purpose?
2. **Information:** Have I given my readers enough details to replicate what I experienced? Do I see any place to add details that will enhance or further explain my topic?
3. **Reasoning:** Are all my facts accurate? Are they clear? Are they relevant to my topic? Is my reasoning logical?
4. **Assumptions:** Have I made any assumptions that need to be explained or justified? Have I used any vocabulary my audience may not understand?

Critically thinking about specific details means analyzing details for their significance in relation to your topic and for their sensory qualities.

6.13 Apply Critical Thinking

Directions: After revising for organization, ask members of your writing group to read your latest draft and answer the critical thinking questions below. Answers will vary.

(continued)

Purpose

1. After reading my draft, why do you think this topic is important or memorable to me?
2. Did I clearly state or clearly imply my purpose? If not, where should I be clearer?

Information

3. Do you feel you read enough details to replicate what I experienced?
4. Do you see places where adding details would enhance or further explain my topic? If so, where?

Reasoning

5. After reading my draft, do you feel all my facts are accurate? If not, where should I provide more accurate details?
6. Are all my facts clear? If not, where should I change my wording to be clearer?
7. Are all my facts relevant to my topic? If not, where did I stray from my topic?
8. Is my reasoning logical? If not, where do I seem to be illogical?

Assumptions

9. After reading my draft, do you feel I made any assumptions that need to be explained or justified? If so, what are these assumptions?
10. Did I use any vocabulary you don't understand? If so, what words or phrases need definitions?

Step Seven: Revise Your Essay

Heads Up!
Keep hard copies of all your drafts and number them so you'll know which is your latest draft.

As you know, getting your discovery draft on paper is just your first step. In subsequent drafts, you will revise and rewrite, shaping your work into a finished product of which you can be proud. Now, consult the Review Checklist for an Essay in Chapter 3, page 61, and look for places in your essay you need to polish. Think about revisions specific to descriptive essays: *details, topics,* and *language choice*.

Details

"Detail makes the difference between boring and terrific writing. It's the difference between a pencil sketch and a lush oil painting. As a writer, words are your paint. Use all the colors."
—Rhys Alexander

When you revise, look at your details.

- Are your details vivid or specific enough that readers will identify with your impressions?
- Have you given too many details?
- Does each paragraph develop one part of your description?
- Have you written anything that flows away from the paragraph's focus?
- Are your paragraphs arranged in the most logical or effective way?

Topics

Focus each body paragraph on one topic. The topics of your individual paragraphs are narrower than the topic of your essay. Use other sentences in a body paragraph to elaborate on (give more information about) that paragraph's topic.

Language Choice

Also consider the language you use and the word choices you make. First note what seems to be working and what seems to need a bit more work. Then take what is good and make it better. One way you can spice up your writing is by using figurative language.

Figurative Language

Language is divided into two broad categories: *literal* and *figurative*. **Literal language** uses words or phrases in their ordinary meanings, just stating the facts. **Figurative language**, on the other hand, uses words or phrases in ways intended to achieve a particular effect or bring to mind a particular idea or image. Figurative language often provides an unusual or fresh way of expressing a thought or feeling. Here are examples of three types of figurative language:

> **Heads Up!**
> Be careful not to go overboard when you use figurative language.

- **similes** (comparisons that use either *like* or *as*) such as

The wet clothes smelled as sour as fermented pickles.

- **metaphors** (comparisons that don't use *like* or *as*) such as

The air conditioner in the computer lab must have been set to Arctic cold.

- **personification** (giving qualities of a person to something that's not human) such as

The sky cried on the day of the rock star's death.

6.14 Identify Figurative Language

Directions: Figurative language (simile, metaphor, or personification) is underlined in each of the following sentences. Identify the type of figurative language and then explain why the words function as that particular type of figurative language.

(continued)

Example: Around here, August is usually as dry as a bone.

Type of figurative language: simile

Explanation: August is compared to a bone, using the word "as."

Explanation of what phrase means: August is compared to a bone, which is waterless, very dry

1. Since my little boy turned two, he's become as nimble as a monkey.
2. Time heals all wounds.
3. Be sure to wear a coat to room 111; it's an igloo today.
4. Poet Carl Sandburg said, "Life is like an onion: You peel it off one layer at a time, and sometimes you weep."
5. I realized my passing grade would be toast if I didn't turn in my assignment.

MyWritingLab™

Writing Assignment

Consider the topics below or one that your instructor gives you. Choose one topic and expand it into a descriptive essay.

1. A foreign place you visited for the first time
2. A once-in-a-lifetime event
3. An important sporting event as if you had attended it
4. A memorable YouTube or other Internet video
5. An unforgettable sports event you participated in or watched
6. A famous or infamous person
7. A movie, concert, musical, play, or other performance you attended
8. An over-the-top event you witnessed
9. An activity you shared with someone else from an older generation
10. A time you had to battle the elements

Other video sites include **JibJab**, **Metacafe**, and **Google Video.**

TECHNO TIP

For more on descriptive writing, search the Internet for these videos:

Grace Lee on Descriptive Writing

Descriptive Paragraph Slide Show

MyWritingLab™ Visit ***MyWritingLab.com*** and complete the exercises and activities in the **Paragraph Development-Describing** and **Essay Development-Describing** topic areas.

RUN THAT BY ME AGAIN

- **A descriptive paragraph or essay describes . . .** a person, place, or thing that has a special meaning (either good or bad) for you so well that your readers understand your unique connection to your topic.
- **A descriptive paragraph or essay can . . .** re-create a person, place, or thing so your readers can feel either they had a part in your experience or they came to know the person, place, or thing you describe.
- **A descriptive paragraph's focal point should be included in . . .** your topic sentence.
- **Starter questions for a descriptive essay include . . .** Whom or what do I want to describe? What qualities do I want to concentrate on or make my readers most aware of? Why do I want to describe this or what is my purpose?
- **Completing the starter statement leads to . . .** your working thesis statement.
- **In a descriptive paragraph or essay, defining your audience helps . . .** focus your purpose and helps you decide the language to use.
- **In a descriptive paragraph or essay, imagery helps . . .** your readers understand or appreciate your topic by helping them form mental pictures of people, things, or events.
- **Consulting your thesis and purpose will help you determine the order to present . . .** your paragraphs and the support within your paragraphs.
- **Transition words and phrases that are often helpful with descriptive paragraphs or essays are . . .** spatial transitions such as *above, amid, among, atop, behind, below, beneath, beside, between, beyond, in front (back) of, inside, near, next to, on top of,* and *under.*
- **In thinking critically about a descriptive essay, you often . . .** examine why a particular place, event, or person is memorable to you; analyze the details that contribute to your memory; reflect on the significance of a particular person, place, object, or event; provide precise details; discover a deeper appreciation for your topic; and reflect on your actions.
- **When revising a descriptive paragraph or essay, look especially at . . .** details, topics, and language choice.

LEARNING GOAL
❶ Get started with descriptive writing

LEARNING GOAL
❷ Write a descriptive paragraph

LEARNING GOAL
❹ Write a descriptive essay

DESCRIPTIVE WRITING LEARNING LOG MyWritingLab™

MyWritingLab™

Answer the questions below to review your mastery of descriptive writing. Answers will vary.

1. What might a descriptive paragraph or essay re-create for its readers?
   It might re-create a person, place, or thing.
2. What should you review in order to focus your topic sentence?
   You should review your purpose to focus your topic sentence.
3. What is your purpose in writing a descriptive paragraph or essay?
   Your purpose is to describe a person, place, or thing that has a special meaning for you so well that your readers understand your unique connection to your topic.
4. What can completing starter questions help with in a descriptive essay?
   Completing starter questions can help with discovering topics to develop into body paragraphs.
5. What will defining your audience help you do in a descriptive paragraph or essay?
   Defining your audience will help focus your purpose and help you decide what type of language to use.
6. What will completing the starter statement lead to for a descriptive essay?
   Completing the starter statement will lead to your working thesis statement.
7. How can imagery help your readers appreciate your topic?
   Imagery can help readers appreciate or understand your topic by helping them form mental pictures of people, things, or events.
8. In applying critical thinking to your essay, what questions help in examining your purpose?
   Will my readers understand why my topic is important or memorable to me? Have I clearly stated or clearly implied my purpose?
9. In applying critical thinking to your essay, what questions help in examining your information?
   Have I given my readers enough details to replicate what I experienced? Do I see any place to add details that will enhance or further explain my topic?

10. **In using critical thinking in your essay, what questions help in examining your reasoning?**

    Are all my facts accurate? Are they clear? Are they relevant to my topic? Is my reasoning logical?

11. **In using critical thinking in your essay, what questions help in examining your assumptions?**

    Have I made any assumptions that need to be explained or justified? Have I used any vocabulary my audience may not understand?

12. **When you revise for details, what questions should you ask yourself?**

    Are my details vivid or specific enough that readers will identify with my impressions? Have I given too many details? Does each paragraph develop one part of my description? Have I written anything that flows away from the paragraph's focus? Are my paragraphs arranged in the most logical or effective way?

13. **Identify the type of transition words and phrases that are often useful in descriptive writing and name at least five.**

    Transition words or phrases that show spatial relationships include *above, amid, among, atop, behind, below, beneath, beside, between, beyond, in front (back) of, inside, near, next to, on top (bottom) of,* and *under.*

Narrative Writing

LEARNING GOALS

In this chapter, you'll learn and practice how to

1. Get started with narrative writing
2. Write a narrative paragraph
3. Read and examine student and professional essays
4. Write a narrative essay

GETTING THERE

- Effective narrative writing allows your readers to learn from your experience.
- Colorful and precise wording helps reproduce your experience.

FASTWRITE 1

We are all lifelong learners, yet what we learn as children we often value more than what we learn as adults. Think of an occasion from your childhood when you learned a valuable lesson.

In a fastwrite, detail what happened, what lesson you learned, and how you learned it. Write as much as you can, as fast as you can, for seven minutes.

FASTWRITE 2

Think of a time when you were a teacher (not a classroom teacher). Maybe you taught a child how to tie shoelaces or you taught a grandparent how to e-mail. In a fastwrite, detail your teaching experience, what lesson you taught, and how you taught it. Write as much as you can, as fast as you can, for seven minutes.

Complete these **Fastwrites** at **mywritinglab.com**

Getting Started in Narrative Writing

LEARNING GOAL

1. Get started with narrative writing

When you write a **narrative**, you *tell a story*, relating the details of events as they occurred. As a narrative writer, you tell a tale to illustrate a *specific point or lesson* by recounting an event in your life. This might be a lesson you learned, a belief you adopted, or some other realization or awareness.

A narrative is a condensed short story, so it has a setting, a plot, characters, a climax, and an ending. Like most short stories, your narrative should start at the beginning and progress chronologically through the events of your story. Ultimately, your narrative has two concerns:

- sharing an interesting story
- illustrating a point

"Writing well means never having to say, 'I guess you had to be there.'"
—Jef Mallett

Read these examples of topic sentences or thesis statements:

- When I was six years old, I was astounded to learn that I could not fly.
- Not until I was nineteen years old, in school, and working full time did I realize the meaning of "time management."
- After my sister's army unit was deployed overseas, I realized the importance of following current events.

In each of these examples, you see that the author has a story to tell and a point to make.

7.1 Narrative Fastwriting

Directions: For three to five minutes, fastwrite about a time you were frustrated by technology. Use as many details as possible to illustrate what happened. When you're finished, share your narrative with your writing group.

WRITING at WORK *Snapshot of a Writer*

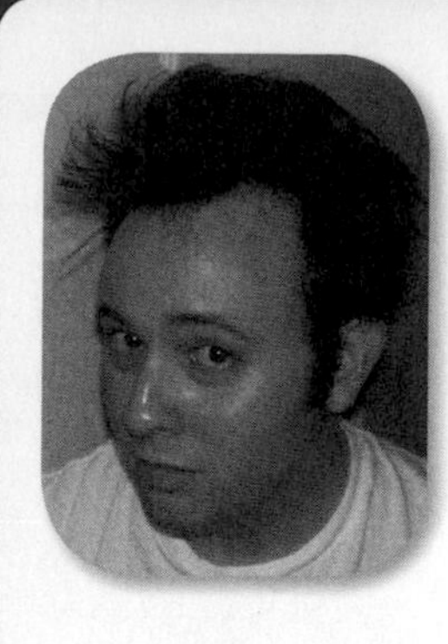

Joey Goebel, Novelist

Joey Goebel's first novel, The Anomalies, *was published in 2003. Since then, his works* Torture the Artist *and* Commonwealth *have been published to high acclaim. In addition to his novels, Joey has written five screenplays. Before his writing success, Joey sang and played guitar in the punk bands The Mullets and Novembrists.*

Joey's Writing Process

Idea Collecting I won't call this part of my process "brainstorming" because when I hear that word, I visualize a writer sitting down with a blank notebook page and generating ideas on the spot. Instead, I "collect" ideas over long periods of time so that I will have them at my disposal when it's time to start drafting. I keep little notepads around the house, and when I get an idea (a concept, a catchy sentence, an unusual combination of words, etc.), I jot it down.

I do this because I've found that if I rely solely on the muses to bring me ideas as I sit to write a draft, the muses always leave me hanging. The cool part about this approach is that my readers won't know the difference. For all they know, my prose all comes out in one glorious sitting, when in reality, they may be reading a line that originated weeks/months/years before.

Notebook Drafting After I have accumulated some material for the narrative (the idea for the narrative itself or dozens of little "safety nets" for lines I can use when the muses let me down), I then sit down with a notebook. I like writing the first (or rough) draft in a notebook because it feels less like work compared to staring at a computer screen.

Computer Drafting Next, I transcribe what's in the notebook into the computer. But I use this transcription as an opportunity to improve upon the notebook draft. As I type out each sentence from the notebook, I'm fine-tuning every sentence as I go along. Sometimes, what ends up on the computer screen is dramatically different from what was in the notebook.

Revising Until I Don't Know What Else to Revise I print out what I've transcribed and get out a red or green pen. Then I take the printout elsewhere (again, anytime I'm away from the computer, it feels more like art and less like work). Then I make changes with the pen. I go back to the computer and transcribe the changes, but again, as I do this, I allow myself to make additional changes on the spot.

I print it out again. I get my pen out again. I repeat these steps until I'm satisfied. Revising can be annoying because the material just gets staler the more times you read it. But I've learned in recent years that revising is just as important—if not more important—than any other step in the creative process.

Putting It Away for Awhile I don't always have time for this, but sometimes after I've finished revising I don't touch the piece for a month or longer. Then I can come back to it with a fresh set of eyes. It's the only way I know of to achieve objectivity with my own writing, because if I allow myself enough time to forget about the piece it's kind of like reading someone else.

Of course, with writing assignments and their due dates, you don't have the luxury of stepping away from your writing for long periods of time. But no matter what your circumstances might be, there are always ways you can systematically approach the creative process. My way seems to work for me, but it's ultimately up to the individual to find out what steps make the writing process feel less torturous.

The Anomalies

Five Misfits Come Together . . . Joey Goebel was twenty-two years old when his debut novel, *The Anomalies*, was published to critical acclaim. The title refers to the name of a small-town rock band that unites five free spirits whose lives are intertwined through their music. These bandmates include rock-star wannabe Luster, octogenarian Opal, Gulf War veteran Ray, eight-year-old Ember, and teenaged Aurora. Following the success of *The Anomalies*, Goebel has written two additional novels that have received national and international praise: *Torture the Artist* and *Commonwealth*.

TRY IT!

Joey's characters are all unusual, to say the least. Write a story about a person you know (maybe someone others would call a misfit) who taught you an unusual lesson. When you're finished, share your narrative with those in your writing group.

Getting Started with Your Narrative Paragraph

As a writer, you may feel more comfortable writing a narrative paragraph before you write a narrative essay. If that's the case, follow steps one through six below. If not, go on to Take 2 on page 126.

LEARNING GOAL

❷ Write a narrative paragraph

A paragraph has both a topic and topic sentence, and students sometimes confuse the two. The **topic** is the subject matter of the paragraph and is often stated in a single word or short phrase. The **topic sentence** (a complete sentence) narrows your topic and often expresses your position or opinion about it.

To compose your narrative paragraph, complete the following steps.

Step One: Choose a Topic

Remember that topic and topic sentence are not the same. See Chapter 1 to review the difference.

Instructors usually allow students to choose the event they write about in a narrative writing assignment. Often instructors also give some kind of prewriting assignment or an umbrella topic to get students started narrowing a topic.

See Chapter 1 for more prewriting techniques, such as clustering, fastwriting, reporter's questions, and journaling.

TICKET to WRITE

7.2 Choose a Topic

Directions: List at least three occasions for each of the following prompts. When finished, share your lists with your writing group and discuss which occasions would make an interesting narrative topic.

1. Times you were frightened
2. Times you were surprised
3. Times you were braver or more cowardly than you thought you would be
4. Events during which you felt yourself changing from youth to adult

Step Two: Generate Ideas

Now that you have some possible topics, consider which would work well as the focus of a narrative paragraph. Remember that you are looking for a topic through which you can (1) share an interesting story and (2) illustrate a point. The point you illustrate should be one that relates to your audience in some way.

Heads Up! Shape your question to fit the topic. Sometimes you may find that one of the reporter's questions won't fit your topic, and that's okay.

An effective method for developing possible topics is to create descriptive details for the topic. Reporter's questions can help you generate descriptive and explanatory details about your topics.

7.3 Generate Ideas

Directions: From each list of topics you generated in Ticket to Write 7.2, circle one topic that attracts your attention. Freewrite on one of the topics you circled, discovering what you know about that topic. Then freewrite answers to the reporter's questions (*Who? What? Why? Where? When?* and *How?*) based on one topic from each category. Form your own reporter's questions or use the following:

1. **Frightened**
   a. Of **whom** or **what** were you frightened?
   b. **Why** were you frightened?
   c. **Where** were you?
   d. **When** did this happen?
   e. **How** did the situation end?
   f. **What** did you learn?
2. **Surprised**
   a. **Who** or **what** surprised you?
   b. **Why** were you surprised?
   c. **Where** did this happen?
   d. **When** did this happen?
   e. **How** did this surprise affect you?
   f. **What** did you learn?
3. **Brave or cowardly**
   a. **Whom** or **what** did you face?
   b. **Why** were you brave or cowardly?
   c. **Where** did this happen?
   d. **When** did this happen?
   e. **How** did you show your bravery or cowardice?
   f. **What** did you learn?
4. **Youth to adult**
   a. **Who** or **what else** was involved?
   b. **Why** did this event move you to adulthood?
   c. **Where** did this happen?
   d. **When** did this happen?
   e. **How** did you react to the change?
   f. **What** did you learn?

When finished, share your freewriting with your writing group.

Step Three: Define Your Audience

For each topic, define an audience that would be interested in reading a narrative about it. Who, for example, might want to know about a time when you were brave and stood up against a bully?

Defining your audience leads to defining your purpose. Why do you want to relate this experience? For instance, is it to show how you were affected or changed by something or someone? Is it to help readers understand something? Identifying your audience and purpose will help keep your narrative focused.

TICKET to WRITE

7.4 Discover Your Audience and Purpose

Directions: To discover your audience and purpose, list your four possible topics from Ticket to Write 7.3. Under each topic, fastwrite a response to each of the following prompts:

1. What is the point I want to illustrate with this topic?
2. What is my purpose for sharing this point?
3. Who would be an appropriate audience for learning about this point?

When finished, share your topics, your audience, and your purpose with your writing group. Use your group's feedback to help you select one topic to develop into a paragraph.

Step Four: Draft Your Paragraph

Now choose one of the four topics to expand into a finished paragraph. Use your prewriting and drafting techniques (see Chapters 1 and 2) to develop your topic sentence. Then use this topic sentence as the paragraph's opening sentence. Remember, this is the sentence that states the purpose of your narrative.

7.5 Create a Topic Sentence

Directions: Choose one of the four topics to expand into a finished paragraph. Write your topic, audience, and purpose. Using these, compose your topic sentence and write it under your purpose. Then share your work with members of your writing group to see if they agree that your topic sentence explains your purpose and is appropriate for your audience.

After writing your topic sentence, use your responses to reporter's questions to tell your story. Adjust and revise the answers to your reporter's questions as needed to help your story unfold. Focus on the details that engage your readers in the story and lead to your point.

7.6 Compose Supporting Details

Directions: Use the topic sentence you created in Ticket to Write 7.5 and the supporting ideas you generated in Ticket to Write 7.3 and 7.4 to compose a paragraph that supports your topic sentence. When finished, share your paragraph with your writing group and ask members if you have given enough details to fulfill your purpose and support your topic sentence.

Step Five: Revise Your Paragraph

After you complete your paragraph, save it, print a hard copy, and read it out loud, looking and listening for errors in content and grammar.

See Chapter 4 to review proofreading techniques.

☑ REVIEW CHECKLIST **for a Narrative Paragraph**

- ☐ Is my topic sentence clear?
- ☐ Do I include vivid details that support the point I want to illustrate?
- ☐ Will this paragraph hold my audience's attention?
- ☐ Does my purpose remain constant?
- ☐ Do any sentences or ideas seem off-topic?
- ☐ Do I hear any fragments when reading out loud?
- ☐ Do all complete thoughts have appropriate end punctuation?

Heads up!
See Chapter 20 to review fragments. See Chapter 30 to review punctuation.

Once you've made any necessary changes, save your work and leave it alone awhile. Then come back later, print out a new copy, and look and listen for errors again.

Step Six: Peer Review

You might think you've found all your errors and made all the improvements you can, but having someone else read your paragraph and offer suggestions will almost always improve your writing: This person is called your **peer reviewer**.

Two heads are better than one. You are the ultimate editor of your own work, but even the most famous writers have editors who suggest revisions and changes.

Your peer reviewer may use the Review Checklist for a Narrative Paragraph in Step Five, or your instructor may provide a different checklist. You might or might not agree with the suggestions you receive, but peer reviewers often find places for improvement or errors that have slipped by you.

Listen to or read closely your peer reviewers' suggestions, and don't be shy about asking questions if you are unclear about their feedback.

TICKET to WRITE

7.7 Peer Review

Directions: Once you have completed revising your paragraph, share it with your peer reviewers. Ask them to read your paragraph using the Review Checklist for a Narrative Paragraph above and the general Review Checklist for a Paragraph from Chapter 3. Then ask them to record their suggestions for revision.

TAKE 2 Student and Professional Essays

LEARNING GOAL

❸ Read and examine student and professional essays

Below are two narrative essays. The first was written by a student and the second by a professional. Read them and answer the questions that follow.

Student Essay

Eyes Opened

by Jill Bryson

1 One Saturday night when I was fourteen, I came home from the movies. My parents and six of their friends were playing poker. That night had been ordinary for me, but then something unusual happened. An adult I admired made me learn that nobody's perfect.

2 When I was growing up, these same people usually came to our house and played poker on Saturday nights with my parents. They didn't play for **big bucks**. They played for what they called "the traveling trophy." It wasn't a trophy at all. It was just a **cheesy** mug with a **royal flush** on one side. Judge Fred Hogan usually won the trophy.

3 I liked all those grown-ups, but I thought Judge Hogan hung the moon. I liked him so much I even had a nickname for him. It was DJ (from Da Judge in the saying "here comes da judge"). DJ always told good stories, and he usually brought me a surprise, like a new paper clip or a pack of mints. All these contributed to me thinking DJ could be my second father. That's how

big bucks large amounts of money

cheesy tacky, tasteless

royal flush the best possible poker hand, consisting of ten through ace of the same suit

much I admired him. Fortunately, my dad is alive and well, and I'm blessed that he enjoys such good health.

4 My feelings about DJ changed that night. When I walked into the kitchen, I saw an ace of diamonds in the cuff of DJ's pant leg.

5 "DJ, be careful," I said. "A card got caught on your pants."

6 "Well, for heaven's sake," DJ said. He grabbed the card and put it on the table. "How'd that get there?"

7 Then everybody at the table got real quiet. Finally, my dad said, "DJ, looks like you've got six cards. Why don't we just **throw this hand in** and deal again?"

8 Next I went to the family room to catch a little TV, and I didn't hear the rest of the conversation around the table. Soon the group left, and I could hear my parents talking. I didn't catch most of it, but I do remember my mom saying, "I know we're not playing for money, but if you cheat your friends . . ." I didn't hear the rest of what she said.

9 That's when I realized that DJ wasn't the saint I'd thought. He was human and had failings. From then on, I looked at adults with the wisdom that nobody's perfect.

throw this hand in an idiom meaning "not count the cards being played"

A CLOSER LOOK

Answer these questions about the essay:

1. Where is the thesis statement found?

   The thesis statement is found in the last sentence of paragraph 1.

2. Cite details from paragraph 2 that help the reader form a mental picture of the traveling trophy.

   The trophy was a mug, it was described as "cheesy," and it had a royal flush on its side.

3. In paragraph 3, what sentence does not support the topic and should be taken out for a better paragraph focus?

   Fortunately, my father is still alive and well, and I'm blessed that he enjoys such good health.

4. What key image, introduced in paragraph 4, supports the author's thesis statement?

   The author saw a card in DJ's pant leg.

5. From whose point of view is "Eyes Opened" narrated?

   "Eyes Opened" is narrated from the point of view of the author as an adult.

6. List the characters who are directly quoted in the story.

   the author, DJ, the author's father, the author's mother

7. Quotation marks are used in what ways in this story?

   (1) to show the title for the mug ("the traveling trophy"), (2) to show direct quotations, (3) to quote the saying "here comes da judge"

8. Which paragraph gives the most background information about the author's relation with DJ?

   Paragraph 3 gives the most background information about the author's relation with DJ.

9. What does the author's mother imply when she says, "I know we're not playing for money, but if you cheat your friends . . ."?

   Answers will vary. Possible answer: The author implies that DJ might cheat others, since he has no bad feelings about cheating his friends.

10. In which paragraphs does the author state the lesson that she learned?

    The author states the lesson she learned in paragraphs 1 and 11.

Professional Essay

The professional essay below tells the story of a young teen who learns a valuable lesson before moving on to high school. Read it and then answer the questions that follow.

Willie, My Thirteen-Year-Old Teacher

by Scott Leopold

(1) Thirteen. That's how old I was when I learned poverty existed not just in my hometown but also in my own backyard. I was in the eighth grade when classmate Willie Reidford taught me more in a thirty-minute bus ride than had all my eighth-grade teachers combined.

(2) I remember that sweltering May 30 in 1993. It was the last day of eighth grade, and I would finally emerge from my three-year-long **pupa stage** as a middle-schooler and embark on the final and most **prestigious** stage in my childhood development: high-schooler. We would be young adults, and so our teachers were sending us off in style.

(3) Everyone was relaxed and all were getting along, even Mrs. Brokaw, our homeroom and biology teacher. Then, she gave each of us a brand-new pocket dictionary. "Remember," she said, handing out the small gift-wrapped paperbacks, "that half of knowledge is knowing where to find it." After that, we signed yearbooks, ate pizza, and accepted graciously her teacherly gift. Many of us brought in music, and we listened to an alternating blend of rock and country. We bounced from Everlast, Barenaked Ladies, and Lenny Kravitz to George Strait, Tim McGraw and the Dixie Chicks. The rednecks rolled their eyes at the rock; the **preps** rolled theirs at the country; yet the goths, geeks, skaters, jocks, and motorheads seemed to tolerate these two primary **cliques**. I sat back and reveled in our **camaraderie**. Here

pupa stage an inactive period in the life of some insects undergoing transformation

prestigious important, impressive

preps those who behave and dress in a traditional manner

cliques groups, factions

camaraderie friendship, companionship, company

we were, thirty-two different kids together merely by alphabetical selection. We were formerly prey and **predator**, different species, now finally come together at a temporary pond in the **savannah**.

4 Jimmy Cadgett, my best friend throughout middle school, signed my yearbook. "Hang loose, dude! Maybe I'll see you chillin' and thrillin' at the beach this summer," he wrote. I couldn't wait. His parents had a beach house on St. Teresa Beach, just a half mile down from my aunt and uncle's place. I was always jealous of Jimmy and my cousins because they spent their entire summer at the beach, and I only had a weeklong visit. Still, I was lucky.

5 I looked around at my **eclectic** classmates. We were indeed the luckiest lot on earth, I thought. We passed around yearbooks, listened to music and laughed, reminiscing days gone by. With only about ten minutes left before we had to pack up the party and say good-bye for real, I remembered my camera.

6 "Hey, Mrs. Brokaw, how 'bout a picture of you and me next to Rattles?" Rattles was the skeleton that hung in the corner of the room by the door.

7 "Sure thing, Scott," she said. I knew I would miss her biology class, and I think she knew this. She knew of my love for biology. She, like all the good teachers, made her subject fun. Next, I got her to take a picture of Jimmy and me and then one of the whole class. They all agreed, though a few were reluctant.

8 "Hey, Linda! Can I get your picture with Cheryl?" Linda and Cheryl were the drama queens of the bunch. Cheryl was a class-A flirt with a fondness for dirty jokes, and Linda was her brainy best friend.

9 "Okay, but only if you invite us to your uncle's beach house," Cheryl said with a wink.

10 "Absolutely!" I said, caught up in the moment, **euphoric** over the summer prospects both real and imagined.

11 An hour later, with our lockers and desks cleaned out and backpacks stuffed, we tittered with laughter and wore ear-to-ear grins as we settled down for our last middle school afternoon announcements. When the bell rang, I was sure the cheering roar of the eighth grade wing could be heard all across campus and across the street at the Publix grocery store. Once out the door, I made my way through a sea of hugs and high-fives and found my seat on the bus right next to Jimmy. As usual, he'd gotten there first and had been holding the seat for me.

12 "I thought you were going to miss your last ride on old 121," Jimmy hollered over the din of bus riders.

13 "Thanks for the seat, dude," I said, throwing my book bag under the seat in front of us. It hit Willie Reidford's foot, and he turned around and looked at me. "Sorry, man. Accident." Willie just stared for a moment, and then he gave me that upward nod, raising his chin quickly and then lowering it smoothly, slowly.

14 "No prob," he said. Willie lived only a few miles from me, but I'd never seen his neighborhood. His bus stop was the last stop, and mine was five

predator hunter, attacker
savannah open grassland
eclectic diverse, assorted, different
euphoric overjoyed, elated, excited

stops before his. The bus was noisy and alive with conversations about summer vacations and high school rivalries.

15 "I'm just glad we're both going to County High," I told Jimmy.

16 "Yeah, I feel sorry for all those dudes going to Lincoln. We'll kick their butts in everything—basketball, football, baseball, even tennis," he laughed. Our conversation fell to the beach again. Jimmy talked of his new carbon composite fishing rod and boogie board he couldn't wait to test out, and I hung on every word in anticipation of my family's trip.

17 As we came to Jimmy's stop, I told him I'd call him that night and that we should talk to our parents and figure out what weekends we'd both be at the beach. He yelled "awesome" and disappeared down the bus steps.

18 I tapped Willie on the shoulder and asked him if he had any special plans for the summer. "Not really," he said wiping the sweat from his brow. "How 'bout you?"

19 I immediately thought of Uncle Rick's beach house, but something told me not to mention the beach anymore and that I'd probably already sounded like a show-off. It wasn't even my parents' house. "I think I'm just going to relax, grab a cold Coke, kick off my shoes, and play some Sega. I want to get out of this dang heat."

20 Willie turned and looked at me with a smile. "You've got a Sega?"

21 "Yep. **Genesis** version. Going to go inside and cool off. How about you, Willie?"

22 And that's when it happened. Willie's smile disappeared slowly. "Well," he said, "I'll bet air conditioning sure is nice. We don't have a Sega game either, but I'll get a Coke," he grinned, "'cause I saved 50 cents out of my lunch money today."

23 A **pregnant pause** filled the air, and I struggled to find any words. Right then the brakes began to squeal, and the bus **lurched** to a stop in front of my house. I gathered my backpack, yearbook, and gym shoes. I **mustered** up a few words and a faint but sincere smile: "Have a good summer, Willie."

24 He smiled back at me, kindly. I stepped off the bus and walked a few feet. I hollered, "Hey, Willie." He turned and dangled his left arm out the window, resting his chin on his right arm. He smiled and waved as I took his picture.

25 Two of my most treasured photos are from that day: One is of Jimmy and me. But the one more dear to me is of Willie Reidford, age thirteen, hanging out the window on Bus 121.

Genesis the name of the initial release of the Sega game system

pregnant pause a silence giving the impression it will be followed by something significant

lurched pitched or staggered suddenly

mustered gathered, collected

A CLOSER LOOK

Answer these questions about the essay:

1. What is the thesis statement of this essay?

   That's how old I was when I learned poverty existed not just in my hometown but also in my own backyard.

2. List the major events of the story.

   paragraphs 3–10: end-of-year party; paragraphs 12–26: the bus ride

3. From whose point of view is "Willie, My Thirteen-Year-Old Teacher" narrated?

   The story is told from the point of view of the author as an adult.

4. Choose one of the story's major events and list all the details the author uses to show that event.

   Answers may vary. Possible answers include: paragraphs 3–10: end-of-year party—listening to music, signing yearbooks, and taking pictures; paragraphs 12–26: the bus ride—talking with best friend Jimmy, talking with classmate Willie, taking Willie's picture.

5. Why does the author state his picture of Willie "is more dear" than his picture of his best friend?

   Answers will vary. Possible answers include: His picture of Willie is more important because it represents his realization of the comforts in life that he took for granted and others didn't have.

6. Why might Scott have struggled "to find any words" in paragraph 24?

   He may not have been sure of what to say because he was embarrassed that he was bragging about the comforts he had in life that Willie didn't have.

7. Identify and explain the meaning of the metaphor in paragraph 2:

   Three-year-long pupa stage is a metaphor for Leopold's three years as a middle-school student.

8. Identify and explain the meaning of one of the metaphors in the last sentence of paragraph 3:

   Scott compares the kids in the class to prey and predators. The temporary watering hole is compared to the end-of-the-year party.

9. In paragraph 3, what phrases signal a transition in time?

   Then, After that

10. Why does the author use slang in the story?

    That was how he spoke in the eighth grade.

Writing Your Narrative Essay

LEARNING GOAL

❹ Write a narrative essay

See Chapter 2 to review thesis statements.

Essay writing is much like paragraph writing: you start with a topic and expand with details. Remember that just as your paragraph supports your topic sentence, your essay supports your **thesis statement**.

To develop a thesis statement for your narrative essay, ask yourself these **starter questions**:

- **What** incident in my life had a strong impact on me?
- **How** and why did this change me or give me insight?
- **How** will my readers benefit, or **what** is my purpose?

For instance, in a narrative piece about a time you misjudged someone, you might answer these questions this way:

1. **What** incident in my life had a strong impact on me?

   The day I was deceived by someone who was very good-looking.

2. **How** or **why** did this change me or give me insight?

   I realized how shallow I'd been.

3. **How** will my readers benefit, or **what** is my purpose?

   My readers might learn not to judge a person by his or her looks alone.

Answering Starter Questions:
Jill Bryson and Scott Leopold

In answering the starter questions above, Jill Bryson, author of "Eyes Opened," wrote:

1. The incident that had a strong impact on me was a time I saw a trusted adult cheat.
2. This changed me because I hadn't realized that someone I liked so well could have a fault.

3. My story will benefit my readers in showing that moving into adulthood can bring insight.

Scott Leopold, author of "Willie, My Thirteen-Year-Old Teacher," might have addressed the starter questions this way:

1. The incident that had a strong impact on me was when a schoolmate made me aware of his poverty.
2. This changed me because I realized that I had taken things for granted.
3. My story will benefit my readers who may identify with my youthful lack of awareness.

Step One: Choose a Topic and Develop a Working Thesis Statement

When you think about the incident you intend to recount, pick something that's interesting and important. You can tell about your trip to the grocery store in chronological order, but if all you do is recap what you bought in each aisle—well, that's really boring. If, though, you found a stash of money hidden behind the cans of green beans you picked up, then you have a story.

"Tell the readers a story! Without a story, you are merely using words to prove you can string them together in logical sentences."
—Anne McCaffrey

7.8 Answer Starter Questions for a Narrative Essay

Directions: To help develop a working thesis statement for your narrative essay, answer each of the following starter questions.

1. *What* incident in my life had a strong impact on me?
2. *How* or *why* did this change me or give me insight?
3. *What* is my purpose?
4. *How* will my readers benefit?

Review your responses to the starter questions.

Use answers to the starter questions to help develop your working thesis statement. Your working thesis statement should incorporate the focus of your essay.

See Chapter 1 to review activities and techniques for brainstorming and freewriting.

Write your working thesis statement and then begin to brainstorm or freewrite about your topic. Ask yourself

- What important incidents contributed to the event I'm narrating?
- In what order did these incidents occur?
- What details would help my readers understand or appreciate each of the incidents?
- What is the climax of my story?
- What details would help my readers understand or appreciate the climax?

The climax is the moment in the story that brings all events together, revealing a theme or lesson.

When you write about an incident, ask yourself which sensory details would add to your readers' understanding of what happened. If you elaborated on how something felt, smelled, tasted, looked, or sounded, would your readers better appreciate your point?

7.9 Develop a Working Thesis Statement

Directions: Review your answers to the starter questions in Ticket to Write 7.8. Then compose your working thesis statement and share it with members of your writing group. Ask them if they believe you state clearly your strong position.

Step Two: Generate Ideas

You won't necessarily use all the material that you wrote when you were prewriting.

In a narrative essay, chronology, details, and main idea are important. In keeping chronology straight, some writers look at the notes they took in their prewriting and simply number the notes, with number 1 being the first event that happened, number 2 the second, and so on.

As you have learned, a narrative essay does more than just relate a series of events. The events must have a point or lesson that you identify in your working thesis statement. Your story should be a series of details showing how this point was made or this lesson was learned. The last part of your essay should wrap up the story so readers will see the conclusion as a natural progression of the details.

Another way to approach a narrative essay is to think of it as a short story. While you're writing, keep these elements in mind:

- **Setting** Give readers enough details about where and when the story takes place so that your audience can feel as if they are there.
- **Plot** In a narrative essay, this usually involves some sort of conflict, crisis, or lesson.
- **Characters** Help readers understand the characters through dialogue and/or descriptive details.

See Chapter 30 to review punctuation rules for writing dialogue.

- **Climax** Be sure to lead readers to a high point in the action of your story.
- **Ending** This may be a summary or reiteration of your thesis statement; it also might include details that occurred after the climax.

7.10 Generate Ideas

Directions: Look at the working thesis statement you created in Ticket 7.9. Then, to support your position, fastwrite about the setting, plot, characters, and climax of your story. Share these ideas with your writing group and ask them to note which seem the most compelling or which elaborate the most on the information in your working thesis statement. Answers will vary.

Step Three: Define Your Audience

Your next step is to identify the audience for your narrative essay. To decide who your audience is, review your purpose and consider who would be interested in reading your narrative. Because you are sharing an interesting story and illustrating a point about that story, the point you make should benefit your audience.

7.11 Define Your Audience

Directions: Write the following:

1. Your working thesis statement
2. A sentence that states your purpose or point you want to illustrate
3. A sentence that states how your audience could benefit from your narrative
4. A sentence that describes a particular audience (or audiences) for this narrative

When finished, share this with members of your writing group and ask if they agree that your working thesis statement will be satisfactory for the audience you describe.

Step Four: Draft Your Essay

To begin discovery drafting your narrative essay, review the ideas you generated from prewriting and starter questions. As you draft, focus on making your narrative flow like a story.

7.12 Write Your Discovery Draft

Directions: Use the material you generated in Tickets to Write 7.8 through 7.11 to write your discovery draft. Refer to Chapter 2 to review discovery drafting.

When you have completed your discovery draft, share your work with members of your writing group. Ask them to note examples that work well to make your point and areas that may need more development.

Step Five: Organize Your Essay

Review your discovery draft and consider how the body paragraphs relate to one another and to the thesis statement. To determine the type of order (time, importance, space) you will use for them, think about how your paragraphs relate to one another and to the thesis statement.

To make connections between sentences, paragraphs, and ideas, use transitional words and phrases. These let your readers see the relationships between topics, and they can keep your writing from being choppy.

See Chapter 2 for lists of other types of transitions, such as emphatic or spatial.

In a narrative, you often use words and phrases that show **chronological order**, such as the following:

afterwards (after)	first (second, third)	now
at first (last)	following that	previously
at the same time	immediately	prior to
before	initially	simultaneously
beginning (at)	in the beginning (end)	soon
during	in the meantime	then
earlier	last	tomorrow
ending (at)	later	when
eventually	meanwhile	while
finally	next	

7.13 Identifying Transitions

Directions: Look at the student and professional essays on pages 126–130 and identify three transitions that show chronological order from each.

"Eyes Opened," Jill Bryson

when, then, finally, next, soon

"Willie, My Thirteen-Year-Old Teacher," Scott Leopold

finally, later, then, after that

Step Six: Apply Critical Thinking

The word **critical** means something different in casual conversation than it means in college. When you're talking with someone, you might say, "My boss has been really critical of me." What you mean is that your boss has looked only at negative aspects of what you've done.

In college, however, *critical* has a different meaning. Critical thinking and critical writing are concerned with careful evaluation and judgment of your work—characteristics that you use to make your work better.

Critical thinking means thinking and then rethinking about a topic, with the goal of discovering all the implications you can about that topic. Applying critical thought to your narrative essay involves examining an experience you had, one that might have caused you to change in some way.

Critical thinking is important when you write any type of essay. After you write each of your drafts, ask yourself these questions and then apply your answers to your writing.

These questions apply to most types of essays, so you'll see them in other chapters.

1. **Purpose:** Will my readers understand why my topic is important or memorable to me? (Your topic could have had a positive or negative impact on you.) Have I clearly stated or clearly implied my purpose?
2. **Information:** Have I given my readers enough details to replicate what I experienced? Would additional details enhance or further explain my topic?
3. **Reasoning:** Are all my facts accurate? Are they clear? Are they relevant to my topic? Is my reasoning logical?
4. **Assumptions:** Have I made any assumptions that need to be explained or justified? Have I used any vocabulary that my audience may not understand?

A key component of critical thinking is examining your thoughts and actions.

In thinking critically about a narrative essay, you often

- scrutinize your suppositions (the ideas that you had taken for granted)
- examine your point of view
- study a problem or issue
- use reflective thinking about a situation
- analyze the circumstances that led to how you solved a problem

7.14 Apply Critical Thinking

Directions: After revising for organization and transitions, ask members of your writing group to read your latest draft and answer the critical thinking questions below.

Purpose

1. After reading my draft, why do you think this topic is important to me?
2. Did I clearly state or clearly imply my purpose? If not, where should I be clearer?

Information

3. After reading my draft, do you feel you read enough examples to understand my point?
4. Do you see any places where adding details would enhance or further explain my topic?

Reasoning

5. After reading my draft, do you feel all my facts are accurate? If not, where should I provide more accurate details?
6. Are all my facts clear? If not, where should I change my wording to be clearer?
7. Are all my facts relevant to my topic? If not, where did I stray from my topic?
8. Is my reasoning logical? If not, where do I seem to be illogical?

Assumptions

9. From reading my draft, do you feel I made any assumptions that need to be explained or justified? If so, what are these assumptions?
10. Did I use any vocabulary you don't understand? If so, what words or phrases need definitions?

Step Seven: Revise Your Essay

"The beautiful part of writing is that you don't have to get it right the first time, unlike, say, a brain surgeon."
—Robert Cormier

Becoming a good writer means applying a critical eye to your writing and asking yourself how and where it could be better. You take another look (that's what the word *revision* literally means). In your discovery draft, you have words on paper, and you've told a story. But those words are a little like a car you've washed and not waxed. For your car to stand out in the crowd, you need to take the time to polish it. In the same way, to have your essay stand out, you need to polish it through revision.

Creating your discovery draft is just your first step. When you revise and rewrite your subsequent drafts, you shape your discovery draft into a finished product of which you can be proud. First, consult the general Review Checklist for an Essay in Chapter 3, page 61, and look for places in your essay you need to polish. Then think about revisions that are specific to illustration essays: **examples**, **details**, and **topics**.

Heads Up!
Keep hard copies of all your drafts and number them so you'll know which is your latest draft.

Examples

When you revise, consider each example you use to support your thesis statement.

- Does each body paragraph develop at least one example?
- Does each example illustrate your thesis statement?
- Do you have enough examples to support your thesis statement?
- Are your examples logically ordered?

Details

When you revise, look at your details.

- Are your details vivid or specific enough that readers will understand the point you are making?
- Have you given too many details?
- Does each paragraph develop one example or multiple examples that illustrate your thesis statement?
- Have you written anything that flows away from a paragraph's focus?

Topics

When you revise, check for topic ideas.

- Does each topic idea directly support your thesis statement?
- Do any topics flow away from your essay's focus (your thesis statement)?
- Do supporting sentences in your body paragraphs connect to the paragraphs' topics?

Language

Precise, strong language and dialogue are important elements of narrative writing.

Precise Language

Using **precise language** helps your readers understand and appreciate your experience. Your use of **sensory details** can turn your plain story into one that is much livelier and more realistic to your readers. As you write, ask yourself

- **What did I *see*?** Use vivid language so your readers can picture, for instance, colors, movements, shapes, appearances, and sizes.

"Don't tell me the moon is shining; show me the glint of light on broken glass."
—Anton Chekhov

Plain: the cat running across the room

Lively: the almond-and-ivory cat bolting across the cluttered kitchen

- **What did I *hear*?** Think of particular sounds that contributed to what occurred in your story, and use words that are as specific as possible.

Plain: a loud sound

Lively: an earsplitting scream that seemed to last for hours

- **What did I *taste*?** Consider details that elaborate on bitterness, saltiness, sourness, sweetness, or other qualities of taste.

Plain: a hot dog and potato chips

Lively: a fresh-off-the-grill hot dog and salt-and-vinegar potato chips

- **What did I *smell*?** Was a taste sweet, fruity, spicy, or even putrid? Use smell-related words with which your audience can identify.

Plain: Aunt Sally's perfume

Lively: my Aunt Sally, with the heavy, musty perfume that always accompanied her

- **What did I *feel*?** Be specific about objects you touch.

Plain: new blue jeans

Lively: unlaundered, brittle new blue jeans

The details you give readers often make or break the quality of your essay. As you revise, ask yourself these questions:

- Are my details clear enough to convey the scene or person I'm portraying?
- Do all my details contribute to the climax of my story?
- Have I given too many details?
- Do I see any details that don't contribute to the impression I want to create?
- Are my paragraphs arranged in the most logical and effective way?
- Is the story I'm telling arranged chronologically?
- If I have used dialogue, have I checked my punctuation for correct use of quotation marks and end marks? Have I begun a new paragraph each time I changed speakers?

Language Choice

Use strong adjectives, adverbs, and verbs.

- Weak adjectives (like *nice* or *great*) don't give your readers a definite idea of what you're describing, so avoid them.
- The same is true of weak adverbs, like *very* and *quite*.
- Your verbs should be as specific as possible. Readers get a far clearer picture if you use specific verbs, like *whispered* instead of *said*, or *bolted* instead of *walked*.

Avoid These Vague Words and Phrases

a bit	kind of	quite
a little	mainly	really
a lot	mostly	some
bad	nice	sort of
generally	ordinarily	usually
good	pretty (much)	very
great	probably	

TECHNO TIP

Using Microsoft Word, you can find a synonym by right-clicking on the word and then clicking on *Synonyms* on the pull-down menu. Alternately, highlight the word and then click *Alt* + *Shift* + *F7*.

If you're stuck, use a thesaurus or dictionary of synonyms and antonyms to find words to more accurately relate what you're describing. If you don't have these at home, you can find them in the reference section of your school's library or by using one of many online dictionaries, like **www.merriam-webster.com/**.

7.15 Revising for Stronger Language

Directions: Rewrite the italicized words or phrases in the sentences below so they are more specific. You may reword the sentences as needed. Answers will vary.

Example: I have *a lot* of homework tonight.

I have three chapters to read and sixteen math problems to work tonight.

1. I bought *a little* gas at the station this morning.
2. I thought $4.00 a gallon was *sort of* high.
3. While filling my tank, I became *quite* annoyed as I watched the numbers climb.
4. Buying gas at a far lower price would be *great*.
5. At these prices, *pretty much* any decrease would help.

Dialogue

Heads Up!
For detailed explanations of quotation marks, see Chapter 30.

Almost all narrative essays involve dialogue. Punctuating dialogue can be a bit tricky, so keep these guidelines in mind as you revise and edit your work.

- Use quotation marks only if you note a speaker's exact words, spoken in the exact order he or she said them.

Kaylon asked, "Does anyone have a pen?" (This is a **direct quote.**)

- If you mix the order of a speaker's words, don't use quotation marks.

Kaylon asked if anyone had a pen. (This is an **indirect quote.**)

- Put your periods and commas *inside* closing quotation marks; put colons and semicolons *outside* closing quotation marks.

"Here you go," I said, digging a pen out of my purse.
When he handed back the pen, Kaylon said, "Thanks for the pen"; I told him he could keep it.

- Every time you change speakers, start a new paragraph. This helps the reader keep up with who's speaking.
- If you quote more than one paragraph from the same speaker, put beginning quotes at the start of each paragraph but put closing quotation marks at the end of the final paragraph only. This lets your reader know the speaker didn't change from one paragraph to the next.
- In dialogue, question marks go inside closing quotations if what is being quoted is a question.

The same rule applies to exclamation marks.

Kaylon asked, "Did you intend to give me this expensive pen?"

- In dialogue, question marks go outside closing quotations if the sentence as a whole is a question.

The same rule applies to exclamation marks.

Did Kaylon say, "Cassandra gave me an expensive pen"?

- To show a quote within a quote, use single quotation marks if you quote the exact words a person says, in the exact order he or she said them.

"Kaylon said, 'Thanks for the pen,' and then he offered to give me his autograph," Cassandra laughed.

7.16 Organize and Develop Your Essay

Directions: Read your draft and answer the following questions. When you have completed these questions, make any necessary revisions.

1. Write your working thesis statement.
2. Write your first topic sentence.
3. What idea does this topic sentence introduce?
4. How does this topic sentence relate to the thesis statement?
5. Write your second topic sentence.
6. What idea does this topic sentence introduce?
7. How does this topic sentence relate to the thesis statement?
8. Write your third topic sentence.
9. What idea does this topic sentence introduce?
10. How does this topic sentence relate to the thesis statement?

Publishing Tip!
Read narratives others have written and share your own at **www.storyofmylife.com**, a free online site that seeks to gather autobiographies.

Writing Assignment

Consider the topics below or one that your instructor gives you. Choose one topic and expand it into a narrative essay.

1. A failure that turned into a success
2. A routine chore that led to an unexpected turn of events
3. A time you reacted to a situation in an unusual way
4. A lesson learned from someone you didn't expect to learn from
5. A horrible day that suddenly changed (or vice versa)
6. A time you won when you should have lost (or vice versa)
7. An idea or belief that was destroyed
8. A time you kicked a bad habit
9. A time you picked up a bad habit

MyWritingLab™

TECHNO TIP

For more on using quotation marks, search the Internet for these videos:

English Grammar & Punctuation: How to Use Quotation Marks in Dialog
Grammar & Punctuation: When to Use Single Quotation Marks
Quotation Mark Song

TECHNO TIP

For more on narrative writing, search the Internet for this video:
How to Write a Narrative Essay

MyWritingLab™ Visit *MyWritingLab.com* and complete the exercises and activities in the **Paragraph Development-Narrating** and **Essay Development-Narrating** topic areas.

RUN THAT BY ME AGAIN

LEARNING GOAL
❶ Get started with narrative writing

- **A narrative paragraph or essay** . . . tells a story, relating the details of events as they occurred.
- **The two concerns of a narrative paragraph or essay are** . . . sharing an interesting story and illustrating a point.

LEARNING GOAL
❷ Write a narrative paragraph

- **The topic of a narrative paragraph refers to . . .** the paragraph's subject matter.
- **The topic sentence of a narrative paragraph expresses . . .** your position or opinion about the topic.
- **A narrative paragraph or essay is like a short story because** . . . it has the elements of setting, plot, characters, climax, and ending.

LEARNING GOAL
❹ Write a narrative essay

- **Starter questions for a narrative essay may include** . . . What incident in my life had a strong impact on me? How and why did this change me or give me insight? How will my readers benefit from reading this?
- **Answering starter questions can** . . . lead to your thesis statement by focusing on the impact an event had on you.
- **In a narrative essay, defining your audience can** . . . help you define your purpose and help keep your narrative focused.
- **Transition words and phrases that are often helpful with narrative essays are** . . . chronological transitions such as *after*, *before*, *during*, *earlier*, *eventually*, *finally*, *first*, *second*, *third*, *in the meantime*, *last*, *later*, *next*, *now*, *then*, *when*, and *while*.
- **In thinking critically about a narrative essay, you often** . . . scrutinize your suppositions, examine your point of view, study a problem or issue, use reflective thinking about a situation, and analyze the circumstances that led to how you solved a problem.
- **When revising a narrative essay, look especially at** . . . the details, language, and imagery you use to make your point.

NARRATIVE WRITING LEARNING LOG MyWritingLab™

Answer the questions below to review your mastery of narrative writing. Answers will vary.

Complete this Exercise

MyWritingLab™

1. What is your purpose in writing a narrative paragraph or essay?
   Your purpose in writing a narrative paragraph or essay is to tell a story in order to make a specific point.
2. How is a narrative paragraph or essay like a short story?
   It is like a short story because it has the elements of setting, plot, characters, climax, and ending.
3. What are three starter questions that can help you discover a topic for a narrative essay?
   What incident in my life had a strong impact on me? How and why did this change me or give me insight? How will my readers benefit from reading this?
4. How can answering these starter questions help you in a narrative essay?
   Completing starter questions can help you discover why an event was memorable to you or why an event changed you in some significant way.
5. What will completing the starter statement lead to for a narrative essay?
   Completing the starter statement will lead to your working thesis statement.
6. What will defining your audience help you do in a narrative essay?
   Defining your audience will help you focus your purpose and help you decide the language you will use in telling your story and illustrating your point.
7. In applying critical thinking to your essay, what two questions help you examine your purpose?
   Will my readers understand why my topic is important or memorable to me? Have I clearly stated or clearly implied my purpose?
8. In applying critical thinking to your essay, what questions help you in examining your information?
   Have I given readers enough details to replicate what I experienced? Do I see any place to add details that will enhance or further explain my topic?
9. In applying critical thinking to your essay, what questions help you in examining your reasoning?
   Are all my facts accurate? Are they clear? Are they relevant to my topic? Is my reasoning logical?

10. In applying critical thinking to your essay, what questions help you in examining your assumptions?

    Have I made any assumptions that need to be explained or justified? Have I used any vocabulary my audience may not understand?

11. Identify three questions you should ask yourself when revising for details.

    Answers include: Are my details clear enough to convey the scene or person I'm portraying? Do all my details contribute to the climax of my story? Have I given too many details? Do I see any details that don't contribute to the impression I want to create? Are my paragraphs arranged in the most logical and effective way? Is the story I'm telling arranged chronologically? If I have used dialogue, have I checked my punctuation for correct use of quotation marks and end marks? If I have used dialogue, have I begun a new paragraph each time I changed speakers?

12. Identify transition words and phrases often useful in narrative writing, and name at least five.

    Transition words or phrases that show chronological relationships are useful in narrative writing. These include *after, before, during, earlier, eventually, finally, first, second, third, in the meantime, last, later, next, now, then, when*, and *while*.

CHAPTER 8 Illustration Writing

LEARNING GOALS

In this chapter, you'll learn and practice how to

❶ Get started with illustration writing

❷ Write an illustration paragraph

❸ Read and examine student and professional essays

❹ Write an illustration essay

GETTING THERE

- Illustration writing shows a general topic and supports it with detailed examples.
- The general topic may be supported through examining several specific examples or one extended example.

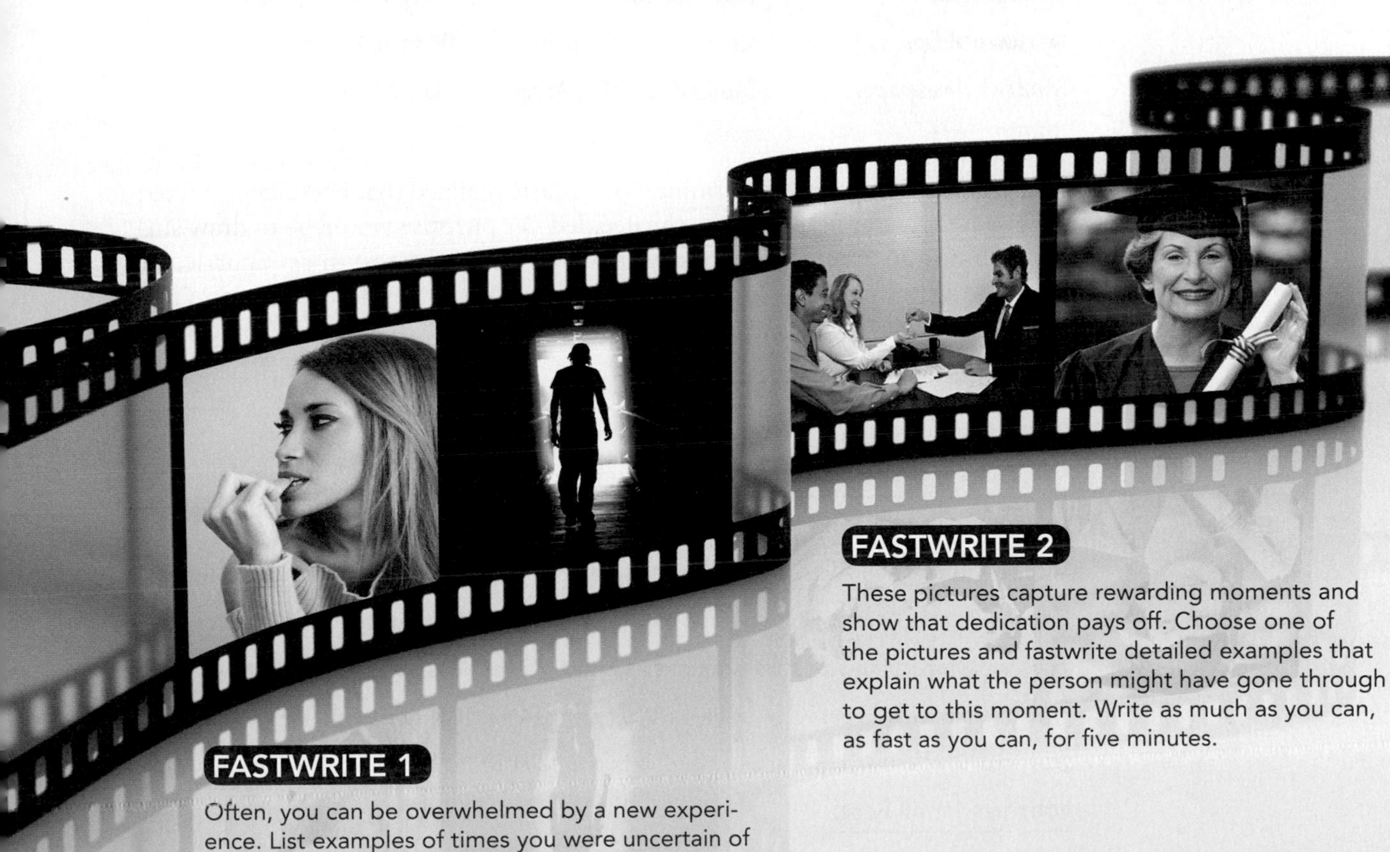

FASTWRITE 1

Often, you can be overwhelmed by a new experience. List examples of times you were uncertain of or nervous about on a *first day*. This could be your first day in high school, in college, on a new job, as a parent, or at another time. Write as much as you can, as fast as you can, for five minutes.

FASTWRITE 2

These pictures capture rewarding moments and show that dedication pays off. Choose one of the pictures and fastwrite detailed examples that explain what the person might have gone through to get to this moment. Write as much as you can, as fast as you can, for five minutes.

Complete these **Fastwrites** at **mywritinglab.com**

Getting Started in Illustration Writing

LEARNING GOAL

1. Get started with illustration writing

"There is nothing as annoying as a good example."
—Mark Twain

Illustration writing, sometimes called **exemplification writing**, shows or proves a specific point through detailed and expressive examples. To determine an illustration topic, consider a specific situation and examples that show the significance of it. Your purpose with illustration could be to

- create interest in that situation
- clarify a position or idea about that situation
- inform your audience of something about that situation

Kim-Marie's assignment was to write about the *extracurricular activities at her school* (the situation). During brainstorming, she developed this list:

Academic Team	Literary Magazine	Film Arts Society
Student Government	Phi Beta Kappa	Community Theatre
Spanish Club	History Club	Singers Ensemble
Intramural Sports	Renewable Energy Org	Bits & Bytes
Student Newspaper	Student Nursing Assoc	Social Council

In looking over her brainstorming, Kim-Marie realized that her college offered many extracurricular activities. So, she decided her purpose would be to draw students' attention to various activities available (to create interest in extracurricular activities). Next, Kim-Marie freewrote to discover ideas to develop her topic. While reviewing her freewriting, she underlined ideas that best expressed her topic. Here is Kim-Marie's freewriting:

<u>There's a lot to do here.</u> <u>There's something for everyone.</u> jocks. actors. singers. history buffs. farmers. Everyone . . . even my sister! I like intramural soccer and softball. Working on the newspaper is great. It's busy but fun . . . like interviewing as sports reporter for paper, like going to the games. Good software programs for laying out stories and newspaper. Sister likes Latin club. I always had trouble with French, so Latin was out of the question for me. Roommate is multicultural committee member and gets a lot out of it. <u>A lot of extracurricular activities for all here.</u>

Kim-Marie realized that the ideas she underlined all emphasized the same point, which led her to this topic sentence:

Our college has a variety of extracurricular activities and organizations for students.

Because Kim-Marie's goal is to inform her readers, she should provide detailed examples that will leave a lasting impression on her readers, allowing them to understand the significance of the main idea she wrote in her topic sentence.

Examples in illustration writing fall into three categories:

- **Personal** (examples of individual experiences)
- **Shared** (examples of common experiences)
- **Hypothetical** (examples of made-up experiences)

"You can't do a fine thing without having seen fine examples."
—William Morris Hunt

To determine what type of illustration she might use, Kim-Marie made a list of what she saw in her freewriting. She noted she could include both **personal** and **shared** examples to support her topic.

Personal	Shared
intramural soccer	sis—Latin club
intramural softball	roommate—multicultural committee
sports reporter	

Next she began writing a discovery draft of her paragraph by expanding on some of the ideas from her fastwriting:

Our college has a variety of extracurricular activities and organizations for students. As a writer for the school newspaper, I enjoy producing sports articles and helping publish our weekly paper. I also participate in intramural softball and soccer. My sister is an active member of Rebus Gestis, the college Latin Club, and seems always to be busy with some club event. Right now she's selling raffle tickets for a new Vespa to raise money for their South American trip. My roommate is a member of the Multicultural Activities Committee and is also busy year-round helping organize special lectures, and music and dance recitals. From the Ag Club to the Future Zoologists of America, our college has something for everyone.

In this paragraph, Kim-Marie illustrates a few different activities available to students at her school. She offers details through *personal* and *shared* examples, and these give readers an idea about each activity or organization.

Kim-Marie's roommate Donja also wrote about extracurricular activities. In her illustration paragraph, Donja focused only on *personal* examples.

Adams College offers me many opportunities outside the classroom. As a member of the Multicultural Activities Committee, I work with more than twenty campus organizations and clubs to create events that celebrate diversity. The committee meets weekly and talks about applications from various campus groups that need help funding, advertising, or finding space for their events. I really like helping plan events. Just last week we finalized plans with the Thai Student Association for the college's first Songkran Festival celebrating the traditional New Year's Day in Thailand. Now I am working with the Green Students Alliance in advertising its upcoming events celebrating Earth Day.

Donja fully develops her paragraph based on her committee-related activities.

8.1 Fastwriting an Illustration

Directions: Brainstorm a list of extracurricular activities, pastimes, hobbies, or any other leisure pursuits you enjoy. Then, for three to five minutes, fastwrite a paragraph that illustrates one of these activities. Focus on illustrating not just the activity but also the reasons you enjoy the activity. When finished, share your illustration with those in your writing group.

WRITING at WORK *Snapshot of a Writer*

Erik Peterson, Urban Planner

After twenty years of working as an urban planner, Erik Peterson knows that his work is essential to many people. "Our department provides research and background knowledge to individual communities within the county, organizations, private citizens, and elected officials." Because Peterson works with so many people and so many groups, efficiency drives much of what he does. "It is key to our planning, and so our writing must be efficient as well."

Erik's Writing Process

I work as a county planner in central Florida, and my job is as much writing as it is anything else. Most of my workplace writing revolves around the

"Rational Planning Model." Whether I am writing a plan detailing the future use of thousands of acres in my county or responding to a commissioner looking for information regarding a problem identified by a citizen, these eight steps help me illustrate plans, problems, and solutions:

Gather Information Having background information is crucial, so I always research the issue I'm writing about so that I understand completely as much as I can about the issue. For example, if there is a storm water runoff ditch that regularly overflows during heavy rains, my office may be asked to investigate the problem and then devise a way to address it. To do this, I have to learn all I can about the issue.

Develop a Goal, Vision, or Direction Once I understand all aspects of the issue, I then develop a goal, vision, or direction share this with my department and the commission. Whether I am dealing with a problem like the storm water runoff issue, or an erosion problem, or a need for an easement, I assess the information I have gathered and then develop a direction to address the problem. Illustrating this vision is essential.

Gather More Information The more we study an issue, the more we discover a need for more information.

Identify Objectives or Indicators My readers like to see objectives. These are like mini goals that show how we are progressing toward our goal in addressing the issue.

Gather Information It's an ongoing need.

Illustrate Policies and Strategies Officials like to see established methods or existing policies for dealing with certain problems. Illustrating these or any strategies for addressing a problem is key to showing my readers exactly how we need to reach our goal.

Gather Information Even this far into a document, I still look for more research. Often I can find past analyses and plans that reference policies or describe strategies that might help me address new issues.

Evaluate In the end, I evaluate the problem and show its causes, its effects, and alternatives for addressing it.

It seems like an arduous process, but after twenty years in the field this process has never failed me. The strength of it is in the information I gather and the way in which I present the information. Often, I am defending a position, and when I can show an issue in vivid detail, I can make my point clearly. I use this method with almost everything I write, from short memos to lengthy analytical documents.

TRY IT!

Write a paragraph illustrating a problem in your neighborhood, on your campus, or elsewhere in your community. Illustrate the problem and a way to fix the problem. When you're finished, share your illustration with those in your writing group.

Getting Started with Your Illustration Paragraph

LEARNING GOAL
❷ Write an illustration paragraph

As a writer, you may feel more comfortable writing an illustration paragraph before you write an illustration essay. If that is the case, follow steps one through six below. If not, go right on to Take 2 on page 156.

Step One: Choose a Topic

Remember that topic and topic sentence are not the same. See Chapter 1 to review the difference.

When writing an illustration paragraph, you'll often be assigned a particular topic to develop. If you have the freedom to choose a topic, however, remember your purpose in illustration writing. You may create interest, clarify a position or idea, or inform your audience. The detailed examples you provide should make clear the point you state in your topic sentence.

To begin brainstorming possible illustration topics, complete Ticket to Write 8.2 below.

TICKET to WRITE

8.2 Choose a Topic

Directions: Complete the following prompts to start the process of choosing a topic. If necessary, freewrite for a few minutes on the subject and then fill in the prompt. Answers will vary.

1. Today's youth are ______________________________.
2. Learning to cook can be ______________________________.
3. Healthy eating requires ______________________________.
4. Stereotyping exists in ______________________________.
5. Too many are dependent upon ______________________________.
6. Being ______________________ has its benefits (or drawbacks).
7. College graduates need ______________________________.
8. Incoming college freshmen need ______________________________.
9. TV advertising is ______________________________.
10. Buying online is ______________________________.

Step Two: Generate Ideas

Once you've decided your topic, the next step is to begin discovering examples that illustrate your topic. Return to Chapter 1 to review several prewriting methods that will help generate ideas about your topic. Complete Ticket to Write 8.3 below to begin discovering examples for one of your topics.

8.3 Generate Ideas

Directions: In Ticket to Write 8.2, you completed prompts on ten topics. Choose three prompts and then prewrite to discover examples that illustrate each of the prompts. Prewrite as much as you can for each prompt.

Step Three: Define Your Audience and Purpose

Once you determine your topic, you need to define an audience who would be interested in reading examples that illustrate your point. For example, you may need to consider who specifically would be interested in the needs of college graduates, the lives of first-time college freshmen, or the politics of TV advertising.

Next, think of your purpose in sharing each topic with the audience you define. Consider why each audience would be interested in a particular topic or what you want to illustrate for this particular audience. Use Ticket to Write 8.4 to determine your audience and purpose for describing the three topics you chose in Ticket to Write 8.3.

8.4 Discovering Your Audience and Purpose

Directions: For each of the three topics you chose in Ticket to Write 8.3, write your audience, purpose, and what you want to illustrate. Answers will vary.

Step Four: Draft Your Paragraph

Because paragraphs center on a topic sentence, review

- who your readers are
- what you want to illustrate to them
- why you want to illustrate this to them

These three points will help you focus your purpose and create a topic sentence your audience will understand.

Now create a topic sentence in Ticket to Write 8.5 below.

8.5 Create a Topic Sentence

Directions: Choose one of the three topics you have been working with, and create a topic sentence for it. Write your purpose, audience, and topic sentence, and then share your work with members of your writing group. See if they agree that your topic sentence explains your purpose and is appropriate for your audience. Answers will vary.

After writing your topic sentence, build your examples from the fastwriting to compose supporting details of your paragraph. As you write, keep your audience and purpose in mind. Work with the prewriting and freewriting you created as part of Ticket to Write 8.2, 8.3, 8.4, and 8.5. Then select the examples that work best to prove your point through illustration.

You may find that some shared examples (others' experiences) are easily replaced with personal examples (individual experiences). Remember, you probably won't use all the ideas from your prewriting. Instead, you will choose the examples that help you make your point the best way. Next, complete Ticket to Write 8.6 below.

8.6 Draft Supporting Details

Directions: Using the topic sentence you composed in Ticket to Write 8.5 and the ideas you generated in Ticket to Write 8.3, draft supporting examples for your illustration paragraph. Then write your topic sentence and your supporting examples and share them with your writing group. Ask members to check that each topic sentence has enough details so your readers fully understand it.

Step Five: Revise Your Paragraph

See Chapter 4 to review proofreading techniques.

After you've finished your paragraph, save it, print a hard copy, and read it out loud, looking and listening for errors in grammar and content. Refer to the general Review Checklist for a Paragraph in Chapter 3, on page 60, to make sure you revise as completely as possible. Then look at the Review Checklist for an Illustration Paragraph for specific areas that may need revision.

REVIEW CHECKLIST for an Illustration Paragraph

- ☐ Is my topic sentence clear?
- ☐ Are my examples detailed?
- ☐ Do I hear any fragments or run-ons when reading out loud?
- ☐ Do all my complete thoughts have appropriate end punctuation?
- ☐ Is my paragraph appropriate for my audience?
- ☐ Do my examples illustrate my position or point?
- ☐ Do any sentences or ideas seem off-topic?

Heads up!
See Chapter 20 to review fragments and run-ons.

Once you've made any necessary changes, save your work and leave it alone for a while. Then come back later, print out a new copy, and look and listen for errors again. Refer again to the general Review Checklist for a Paragraph in Chapter 3.

Step Six: Peer Review

You might think you've found all of your errors and made all the improvements you can, but having someone else read your paragraph and offer suggestions will almost always improve your writing. This person is called your **peer reviewer**.

Peer reviewers bring a fresh perspective. Consider them valuable resources as audience members and fellow writers.

Your peer reviewer may use the review checklist above, the general Review Checklist for a Paragraph in Chapter 3, or a different checklist your instructor provides. You might or might not agree with the suggestions you receive, but peer reviewers often find places for improvement or errors that slipped by you. Listen to or read closely your peer reviewers' suggestions, and don't be shy about asking questions that clarify their ideas and suggestions.

Heads Up!
See Chapter 4 for more on peer reviewing.

8.7 Peer Review

Directions: Once you have revised your paragraph, share it with your peer reviewers. Ask them to review your paragraph, using the Review Checklist for an Illustration Paragraph above and the general Review Checklist for a Paragraph from Chapter 3. Then ask them to record their suggestions for revision. Examine their suggestions and make any revisions you feel are necessary.

TAKE 2 Student and Professional Essays

LEARNING GOAL

3 Read and examine student and professional essays

Below are two illustration essays. The first was written by a student and the second by a professional. Read these two essays and answer the questions that follow them.

Student Essay

I Love My Dog, but I'm Not "In Love" with Her

by Kya Maalouf

1 It seems my family always had a pet when I was a kid. We had a total of four dogs, two cats, one hamster, and too many fish to count. We loved our pets, and no one loved their company more than my sister and I did. Still, and even though we took care of them and enjoyed having them around and playing with them, they were only pets. Some pet owners today seem to overly appreciate the animals in their lives.

2 My Aunt Tina has always had an excessive devotion to her pets. Last May, her cat died. Everybody was upset because we all knew how much the cat meant to her. She was **obsessed** with Mr. Kibbles. As a matter of fact, Kibbles was also the name of my best friend's cat. Aunt Tina had had the cat for seven years and always said he was not a replacement for Miss Kitty who'd died of kidney failure at the age of fourteen. Still, Mr. Kibbles was with Aunt Tina half as long and had a prepaid burial plan and plot that cost $400. His medical insurance, though, was a little cheaper, only $12 per month versus Miss Kitty's $19 per month. Obsession or love, we're not sure, but Miss Kitty's burial plot cost $625.

3 My neighbors, the Robinsons, have a four-year-old Siberian Husky named Fitch. He lives in a large yard enclosed by a tall wooden fence. For a year after my neighbors put up the fence, replacing an old chain link fence, Fitch barked at people walking down the sidewalk and cars passing by. I suppose the Robinsons felt sorry for Fitch not being able to see the people and traffic, so they put in four doggie **peepholes**. These bubble windows stick out from the fence like observation windows on a mini submarine. Now the dog can

obsessed preoccupied with, passionate about

peepholes small openings to look through

see up and down the sidewalk in four different locations, letting him once again bark and **gnash** his teeth at anyone. Mr. Robinson said the windows "only cost $35 each, and $150 for installation." If my math is right, they spent $290 on these windows. I can think of many other things they could have spent that money on, like helping their seventeen-year-old son with a down payment on a car, or replacing their sagging front porch roof, or sending Fitch to obedience school.

4 But it's people like Paris Hilton, Britney Spears, and Lindsay Lohan who have totally lost their perspective. These **fanatics** spend thousands on pet bling and clothes. They treat animals as if they were accessories like a **Louis Vuitton** purse or pair of diamond earrings. Paris even collects these living accessories. She was praised for building a $350,000 Doggy Mansion in Beverly Hills for her thirteen dogs. Why not donate that money to an animal shelter? Also, who can offer quality time to thirteen dogs? According to an article I read, talk show **mogul** Oprah Winfrey employs a pet nanny. I'm all for a dog sitter while I'm out of town, or having a friend come over and let the dog out while I'm at work, but to hire someone to feed, bathe, groom, play with, and basically meet all of your pet's needs is ridiculous. A pet nanny, in my opinion, **defeats the purpose** of pet ownership.

5 Pets are not people and don't need to be treated better than people. Pets are also not **inanimate** and shouldn't be treated like furniture or **Ferragamo pumps**. Still, they should be given food, water, shelter, and their owner's love and care, but not too much. Let's face it: pets and humans aren't the same.

gnash to grind together or bite
fanatics people with an intense devotion to something or someone
Louis Vuitton a maker of luxury leather items
mogul a rich or powerful person; a tycoon
defeats the purpose an idiom meaning "prevents the success of"
inanimate nonliving
Ferragamo an Italian designer of high-priced shoes
pumps women's shoes with medium or high heels

A CLOSER LOOK

Answer these questions about the essay:

1. What is Kya's thesis statement?

   Some pet owners today seem to overly appreciate the animals in their lives.

2. How does Kya illustrate her thesis statement? List these illustrations.

   her Aunt Tina's devotion to her pets; her neighbors, the Robinsons' commitment to their dog Fitch; celebrities' lack of perspective when it comes to their pets

3. What is the topic sentence of Kya's first body paragraph?

   My Aunt Tina has always had an excessive devotion to her pets.

4. What details does Kya use to illustrate the topic of her first body paragraph?

   She discusses the money Aunt Tina spent on her cats' burial plots and medical insurance.

5. Kya's second body paragraph illustrates the money her neighbors have spent on their pet, yet Kya's paragraph has no topic sentence. Write a sentence that could serve as a topic sentence for this paragraph.

   Answers will vary. Possible answer: Some people go overboard in terms of the creature comforts they provide their pets.

6. Kya's third body paragraph has this topic sentence: *But it's people like Paris Hilton, Britney Spears, and Lindsay Lohan who have totally lost their perspective.* What are two details Kya uses to illustrate this topic sentence?

   Answers may vary. Possible answers: Paris Hilton's $350,000 Doggy Mansion and Oprah Winfrey's pet nanny

7. What sentence is not needed in the first body paragraph and can be taken out to improve paragraph unity?

   As a matter of fact, Kibbles was also the name of my best friend's cat.

8. A *simile* is a comparison between two unlike people or things using *like* or *as*. Find the simile Kya uses in paragraph 3.

   These bubble windows stick out from the fence like observation windows on a mini submarine.

9. Why is the phrase "in my opinion" not needed in the last sentence of the last body paragraph?

   Answers may vary. Possible answer: The phrase adds no meaning to the sentence. The sentence is an opinion with or without the phrase.

10. Kya's title is "I Love My Dog, but I'm Not 'In Love' with Her." In her essay, though, she does not discuss her own pet. What might be a better title for this essay?

    Answers will vary.

Professional Essay

Our View on Free Speech: Want to Complain Online? Look Out. You Might Be Sued.

An unsigned editorial published June 8, 2010, in USA Today

1 After a particularly painful visit to the dentist, San Francisco marketing manager Jennifer Batoon decided to **vent** on **Yelp.com**, a popular Internet rating site. "Don't go here," Batoon wrote, "unless u like mouth torture." She went on to describe her experience in excruciating detail.

2 The dentist, Gelareh Rahbar, responded on Yelp that Batoon's review was posted only after the dentist reported her to a credit bureau for a delinquent bill. She then went on to detail Batoon's dental problems.

vent to voice frustration

Yelp.com a Web site that shares evaluations of services provided by different businesses

3 Other patients weighed in, too, mostly praising Rahbar. But the free exchange, so typical of the Internet, didn't satisfy Rahbar. Last fall, she sued for **defamation**, charging that the review caused a drop in her revenue.

4 Batoon says she was shocked when she got served and "fearful about the prospect of paying tens of thousands of dollars in legal fees." Last November, though, thanks to a California law designed to protect public speech, a judge threw out the defamation counts. Last month, he ordered Rahbar to pay $43,000 for Batoon's legal fees.

5 Victory for Batoon, yes. Still, she didn't walk away **unscathed**—and neither did free speech. Once a **prolific** Yelp reviewer, Batoon now limits herself to occasional reviews and only positive ones. The Internet community lost one voice.

6 Such suits are becoming more common as **miffed** consumers who once warned friends not to patronize a business now publish their opinions on Web sites such as Yelp, Facebook pages, or personal blogs.

7 A towing firm in Kalamazoo, Michigan, for example, is suing a college student for **slamming** the firm on a Facebook page after it towed his car from a lot where he says he had permission to park. In February, a company that owns an Omaha knitting store sued a woman for negative comments on her personal blog. They sought $500,000 in damages, but later dropped the case.

8 Make no mistake: It's a certainty that some people lie for hidden motives, and people or businesses have every right to defend themselves against false accusations. But if **moneyed interests** can use the legal system to intimidate their critics, many honest reviewers will fear speaking their minds.

9 In theory, consumers would ultimately prevail. The Constitution guarantees a right to express opinions, even outlandish ones, as long as the facts are right. But the right has no meaning if people fear being bankrupted by the cost of defending themselves. In the process, the public stands to lose a useful source of information.

10 Which seems to be the whole idea.

11 It's a new twist on an old tactic, in which **well-heeled** landlords, developers and, most outrageously, government officials sued citizens who were causing them grief.

12 Often, the suits were meritless. The object wasn't to win but to intimidate, and they were dubbed SLAPP suits, for "strategic lawsuit against public participation."

13 In 1992, California came up with a solution: an "anti-SLAPP" law that made it easier for defendants to seek early dismissal of such suits at no cost. Twenty-six other states followed, though only a few laws are as strong.

14 Now two members of Congress are pushing a similar measure to govern federal courts. Good. They won't have an easy time protecting both honest critics and the falsely accused, but they should err on the side of protecting free speech.

defamation the act of harming someone's reputation

unscathed not hurt, unharmed

prolific frequent

miffed offended, upset, angered

slamming criticizing harshly

moneyed interests wealthy people or businesses

well-heeled having plenty of money

15 People who are unfairly attacked could still fight back. They'd just have to rely more on the **court of public opinion**, where they'd prevail only if the facts were on their side.

court of public opinion an idiom meaning "consensus of belief"

A CLOSER LOOK

Answer these questions about the essay:

1. The title of this editorial is a warning and is the thesis of the essay. Where in the text is the thesis reiterated?
   The thesis is reiterated in the first and last paragraphs.
2. How many examples of lawsuits similar to the one filed against Batoon are mentioned?
   Two examples are mentioned.
3. Why does the last sentence in paragraph 1 describe the "detail" in Batoon's review as "excruciating"?
   Answers may vary. Possible answer: because Batoon experienced excruciating pain during her dental visit
4. What transitions can you find in paragraphs 3 and 4?
   Transitions include *too, but, last fall, when, last November, last month.*
5. Paragraph 5 suggests that Batoon and free speech were harmed by this lawsuit. In what way was Batoon harmed? In what way was free speech harmed?
   Answers may vary. Possible answer: Batoon was physically harmed; free speech was harmed because Batoon now limits what she says online and the frequency with which she posts.
6. Paragraph 8 begins "Make no mistake" and speaks directly to you, the reader. What is the effect or impact of this point of view?
   Answers may vary. Possible answer: The writer's words are a powerful warning for the reader.
7. In paragraph 11, to what does the phrase "new twist on an old tactic" refer?
   Answers may vary. Possible answer: "New twist" refers to defamation lawsuits businesses file against consumers; the "old tactic" refers to the intimidation factor of those suits.
8. What point does the editorial present as the basis of people's fear?
   People fear being sued for defamation if they speak out about poor service.

Writing Your Illustration Essay

To write an illustration essay, you first need to discover your topic and expand on it with details, just as you do with an illustration paragraph. An illustration paragraph supports a topic sentence, and an illustration essay supports a thesis statement.

LEARNING GOAL
4 Write an illustration essay

Step One: Choose a Topic and Develop a Working Thesis Statement

See Chapter 2 to review thesis statements.

To begin your illustration essay, ask yourself these **starter questions**:

1. **What** point do I want to make?
2. **What** example(s) will best illustrate this?
3. **Why** do I want to illustrate this point? ("Because it's an assignment" isn't a good enough answer.)
4. **What** lasting impression do I want to leave with my readers?

For instance, in an illustration essay about the importance of community service, you might answer starter questions this way:

1. **What** point do I want to make?

   Being active in community service is rewarding.

2. **What** example(s) best illustrate this?

   My involvement and friends' involvement in volunteer organizations

3. **Why** do I want to illustrate this point?

   To get more people interested in volunteering

4. **What** lasting impression do I want to leave with my readers?

   That community service has helped people in my hometown and helped me personally—can see this through what I've gotten out of volunteering

Next, review the information above and then fill in the following **starter statement**:

I want to illustrate ______________________________ (the point I want to make), and I will do that by showing readers ______________________________ (example or impression that supports my point).

Review how the student writing about volunteering filled in the starter statement:

I want to illustrate that being active in community service is personally rewarding (the point I want to make), and I will do that by showing readers how volunteering has helped me and helped people in my hometown (example or impression that supports my point).

Answering Starter Questions
Kya Maalouf and *USA Today* editorialist

In answering the starter questions and forming her starter statement, student writer Kya Maalouf, the author of "I Love My Dog, but I'm Not 'In Love' with Her" (page 156) wrote:

1. The point I want to make is that too many pet owners go overboard in their appreciation for their pets.
2. The best examples to illustrate this are Aunt Tina and her cats, the Robinsons and the money they spent on Fitch, celebrities and the ways they treat their pets.
3. I want to illustrate this point because people are losing their perspective when it comes to animals.
4. The lasting impression I want to leave with my readers is that people too often go overboard with how they treat their animals.
5. I want to illustrate that pet owners lose their perspective, and I will do that by showing readers the lavish ways they treat their pets, the money they spend on their pets, and the unnecessary things they provide for their pets (example or impression that supports my point).

The *USA Today* editorialist might have addressed the starter questions and statement this way:

1. The point I want to make is that when opinions are voiced online, free speech is threatened.
2. The best examples to illustrate this are the cases of Jennifer Batoon's Yelp.com complaint about the dentist, the college student's Facebook complaint about the towing company, and the customer's personal blog complaint about the knitting store.
3. I want to illustrate this point because free speech should be protected.
4. The lasting impression I want to leave with my readers is that they should make sure they are fair and honest with how they use free speech platforms.
5. I will illustrate the point that free speech is being threatened, and I will do that by showing readers how some who publically complained were retaliated against with lawsuits (example or impression that supports my point).

Look at how two other writers chose to fill in the blanks for other topics.

I want to illustrate the point that today's youth don't deserve their slacker reputation, and I will do that by showing readers examples of young people involved in their schools, communities, and churches (example or impression that supports my point).

I want to illustrate the point that learning to cook can be humorous, and I will do that by showing readers experiences I've had in the kitchen (example or impression that supports my point).

8.8 Answer Starter Questions and Complete Starter Statement for an Illustration Essay

Directions: All essays have thesis statements. To help develop a thesis statement for your illustration essay, begin with the starter questions and a starter statement. Answer questions 1-4 and then fill in the blanks to compose your starter statement. Answers will vary.

(continued)

1. **What** point do I want to make through my illustration?
2. **What** example(s) best illustrate this?
3. **Why** do I want to illustrate this point?
4. **What** lasting impression do I want to leave with my readers?
5. I want to illustrate ___________________ (the point I want to make), and I will do this by showing readers ___________________ (example or impression that supports my point).

Once you've answered the starter questions and filled in the starter statement, you've begun to develop the focus of your essay. Return to your starter questions and starter statement and consider your purpose. From these, create your **working thesis statement**. Here is a working thesis statement created from the starter questions and statement above:

More people should take part in community service because doing so can be personally rewarding.

Derek wrote about interviewing techniques and filled in the starter statement this way:

I want to illustrate that someone being interviewed can make a good impression, and I will do that by showing readers tips for having a successful interview.

Derek reviewed this starter statement and also considered his purpose. Then he composed this working thesis statement:

Using several tips can help a person make a good impression at a job interview.

Javier wrote about random acts of kindness, and he filled in the starter statement this way:

I want to illustrate that random acts of kindness can change a person's outlook, and I will do that by showing readers several incidents that had a positive impact on the person receiving the kindness and the person offering it.

Javier reviewed this starter statement and also considered his purpose. Then he composed this working thesis statement:

Random acts of kindness are beneficial to both recipients of the acts and those who perform them.

8.9 Develop a Working Thesis Statement

Directions: Return to the starter statement you composed in Ticket to Write 8.8 and consider your purpose. From this, create your working thesis statement and share it with members of your writing group. Ask them if they understand what you will be illustrating in your essay and why it is important.

Step Two: Generate Ideas

Stay focused on the specific point you want to illustrate for your readers. To do this, you need to discover examples or illustrations that are detailed enough to

- clearly show the point you want to make
- catch and hold your audience's attention

Prewrite on your working thesis statement to discover such examples. Consider the following experiences you could use to illustrate your working thesis statement:

- **Personal** (examples of individual experiences)
- **Shared** (examples of common experiences)
- **Hypothetical** (examples of made-up experiences)

Look at these experiences and decide which details work well to help readers

- visualize the situation or condition
- understand the significance of the situation or condition

Complete Ticket to Write 8.10 to discover interesting experiences you can use as examples to illustrate the idea behind your working thesis statement.

See Chapter 1 to review activities and techniques for brainstorming and freewriting.

8.10 Generate Ideas

Directions: With your writing group, share your working thesis statement and any examples you generated in freewriting. Ask members to note which examples seem the most compelling or which most effectively illustrate your working thesis statement.

Step Three: Define Your Audience and Purpose

Your next step is to identify the audience for your essay. This helps you focus more on your purpose. Your audience choice also helps you decide the language to use and the type of illustration (personal, shared, or hypothetical) to use.

For instance, if your purpose is to draw attention to the negative effects of tuition increases at your school and your audience is a group of your peers, you will probably focus on personal and shared experiences. Your peers know you, so they might appreciate reading about your specific experience with this topic. Also, they already have their own experience with it.

However, if your audience is your college's administration, you will probably focus on personal and shared illustration. The administration may not know your experience or any other student's experience. Unlike your peers, they may have no concept of the harm the increase has had, so you will include your experience in addition to others' experiences.

8.11 Define Your Audience

Directions: Write your working thesis statement and identify your audience. Then answer the questions below. Share this with members of your writing group and ask if they agree that your working thesis statement will be satisfactory for your audience.

How familiar is the audience with my topic?

Would illustrating my personal experience help make my point?

Would illustrating shared experiences help make my point?

Would illustrating hypothetical experiences help make my point?

Step Four: Draft Your Essay

You have your topic, your working thesis statement, and ideas from prewriting to support your thesis statement. Plus, you have determined your audience. Now you should get your ideas in some kind of order and create a discovery draft. After completing this discovery draft, you might change your mind about the order of your ideas. That's the progress of your writing process. What is important now is to get your ideas down.

The details and examples you select should be characteristic of your topic and should not be unusual. Together, these examples and details should help your readers understand the point of your essay.

"Be able to draw an illustration at least well enough to get your point across to another person."
—Marilyn vos Savant

8.12 Write Your Discovery Draft

Directions: Use the material you have generated in Tickets to Write 8.8 through 8.11 to write your discovery draft. Refer to Chapter 2 to review discovery drafting.

When you have completed your discovery draft, share your work with members of your writing group. Ask them to note examples that work well to make your point and areas that may need more development.

Step Five: Organize Your Essay

In organizing your illustration essay, review your body paragraphs. To determine the type of order (*time, importance, space*) to use for them, think about how they best relate to your thesis statement.

Type of Order	Example
Time (Chronological) Order	If you're illustrating your transition from high school to college, your body paragraphs might deal with the time between leaving high school and enrolling in college.
Importance Order	If you're illustrating the advantages of your hometown, your body paragraphs might begin with details of one advantage and end with details highlighting the strongest advantage.
Space Order	If you're illustrating the layout of campus, you might begin at some point on campus and move in a certain direction to identify buildings, parking lots, and other landmarks.

Step Six: Apply Critical Thinking

Critical thinking means thinking and then rethinking about a topic, with the goal of discovering all the implications you can about it. In thinking critically about an illustration essay, you often do the following:

- examine why a situation or condition is important to share with others
- provide complete examples that illustrate the main point
- analyze or reflect on the examples that illustrate the main point
- reflect on the significance of a particular situation or condition
- discover a deeper appreciation for the topic

The following critical thinking questions are important when you write an illustration essay. After you write each of your drafts, ask yourself these questions and then apply your answers to your writing.

1. **Purpose:** Will my readers understand why my topic is significant to me? (This significance may have had a positive or negative impact.) Have I clearly stated or clearly implied my purpose?
2. **Information:** Have I given my readers enough details to appreciate the experiences I use as illustrating support? Would additional details enhance or further explain my topic?
3. **Reasoning:** Am I using an appropriate type of illustration (personal, shared, or hypothetical) in each body paragraph? Do I need to use more or fewer illustrations?
4. **Assumptions:** Have I made any assumptions that should be explained or justified? Have I used any vocabulary my audience may not understand?

8.13 Apply Critical Thinking

Directions: After revising for organization and responding to critical thinking questions, ask members of your writing group to read your latest draft and answer the critical thinking questions below.

Purpose

1. After reading my draft, why do you think this topic is important to me?
2. Did I clearly state or clearly imply my purpose? If not, where should I be clearer?

Information

3. After reading my draft, do you feel you read enough details to appreciate the experience I use as illustrating support?
4. Do you see any places where additional details would enhance or further explain my topic?

Reasoning

5. After reading my draft, do you feel all my facts are accurate? If not, where should I provide more accurate details?
6. Are all my facts clear? If not, where should I change my wording to be clearer?
7. Are all my facts relevant to my topic? If not, where did I stray from my topic?
8. Is my reasoning logical? If not, where do I seem to be illogical?

Assumptions

9. After reading my draft, do you feel I made any assumptions that need to be explained or justified? If so, what are these assumptions?
10. Did I use any vocabulary you don't understand? If so, what words or phrases need definitions?

Step Seven: Revise Your Essay

Creating your discovery draft is just your first step. When you revise and rewrite through subsequent drafts, you shape your discovery draft into a finished product of which you can be proud. First consult the general Review Checklist for an Essay in Chapter 3, page 61, and look for places in your essay that you need to polish. Then think about revisions that are specific to illustration essays: **examples**, **details**, and **topics**.

Heads Up!
Keep hard copies of all your drafts and number them so you'll know which is your latest draft.

Examples

When you revise, consider each example you use to support your thesis statement.

- Does each body paragraph have at least one example?
- Does each example illustrate your thesis statement?
- Do you have enough examples to support your thesis statement?
- Are your examples logically ordered?

Details

When you revise, look at your details.

- Are your details vivid or specific enough so readers identify with your impressions?
- Have you given too many details?
- Does each paragraph develop one example or multiple examples that illustrate your thesis statement?
- Have you written anything that flows away from a paragraph's focus?

"For example" is not proof.
—Yiddish proverb

Topics

When you revise, check for topic ideas.

- Does each topic idea directly support your thesis statement?
- Does each example support the topic idea of that paragraph?
- Do any topics flow away from your essay's focus (your thesis statement)?
- Do supporting sentences in body paragraphs connect to the paragraphs' topics?

8.14 Organize and Develop Your Essay

Directions: Read your latest draft and answer the following questions. When you have completed these questions, make any necessary revisions.

1. What is your thesis statement?
2. What is your first topic sentence?
3. What example does this topic sentence introduce?
4. How does this topic sentence illustrate your thesis statement?
5. What is your second topic sentence?
6. What example does this topic sentence introduce?
7. How does this topic sentence illustrate your thesis statement?
8. What is your third topic sentence?
9. What example does this topic sentence introduce?
10. How does this topic sentence illustrate your thesis statement?

MyWritingLab™

Writing Assignment

Consider the topics below or one that your instructor gives you. Choose one topic and expand it into an illustration essay.

1. Reasons people attend college
2. Dangers (or benefits) of living alone
3. Unusual celebrations
4. Respect (or lack of) in the workplace
5. Speaking your mind

TECHNO TIP

For more on illustration writing, conduct an Internet search for these sites:

How to Write 1 Well Developed Illustration Paragraph

5-1 Lecture: Exemplification

MyWritingLab™ Visit *MyWritingLab.com* and complete the exercises and activities in the **Paragraph Development-Illustrating** and **Essay Development-Illustrating** topic areas.

RUN THAT BY ME AGAIN

LEARNING GOAL
1. Get started with illustration writing

- **Illustration writing shows or proves . . .** a specific point through detailed and expressive examples.
- **Three possible purposes of an illustration paragraph or essay are . . .** creating interest in a particular situation, clarifying a position or idea about a particular situation, or informing your audience of something about a particular situation.
- **The three types of examples in illustration writing are . . .** personal, shared, and hypothetical.

LEARNING GOAL
3. Read and examine student and professional essays

- **Once you determine your topic for your illustration paragraph, you need to define . . .** an audience who would be interested in reading examples that illustrate your point.
- **The next step in writing an illustration paragraph is to think of . . .** your purpose in sharing each topic with the audience you define.

LEARNING GOAL
4. Write an illustration essay

- **Starter questions for an illustration essay include . . .** What point do I want to make? What examples best illustrate this point? Why do I want to illustrate this point? What lasting impression do I want to leave with my readers?
- **Completing the starter questions and statement can lead to . . .** your working thesis statement by focusing on the point you want to make.
- **In an illustration essay, identifying your audience can help you . . .** focus your purpose and determine the type of experiences you will use.
- **Details in illustration essays help readers . . .** visualize or understand the significance of the situation or condition.
- **The types of transition words and phrases you use in an illustration essay are determined by . . .** your purpose and how your body paragraphs relate to one another.
- **When thinking critically about an illustration topic, you should examine . . .** why a situation or condition is important to you.
- **When thinking critically about an illustration topic, you should analyze . . .** examples that illustrate your point.
- **When thinking critically about an illustration topic, you should reflect on . . .** the significance of a particular situation and on your examples.
- **When thinking critically about an illustration topic, you should provide . . .** concrete details.
- **When thinking critically about an illustration topic, you should discover . . .** a deeper appreciation for your topic.
- **When revising an illustration essay, you should look especially at . . .** examples, details, and topics you use to make your point.

ILLUSTRATION WRITING LEARNING LOG MyWritingLab™

MyWritingLab™

Answer the questions below to review your mastery of illustration writing. Answers will vary.

1. **What is the reason for writing an illustration paragraph or essay?**
   The reason for writing an illustration paragraph or essay is to offer examples that show the significance of a specific situation.
2. **What are three possible purposes for writing an illustration paragraph or essay?**
   Three purposes for writing an illustration paragraph or essay are to create interest in a particular situation, to clarify a position or idea about a particular situation, or to inform your audience of something about a particular situation.
3. **What are three types of experiences you could use to illustrate your thesis statement?**
   Three types of experiences you could use are personal, shared, or hypothetical.
4. **What are personal experiences?**
   Personal experiences are individual experiences.
5. **What are shared experiences?**
   Shared experiences are common experiences.
6. **What are hypothetical experiences?**
   Hypothetical experiences are made-up experiences.
7. **What are four starter questions that can help you discover a topic?**
   What point do I want to make? What example(s) will best illustrate this? Why do I want to illustrate this point? What lasting impression do I want to leave with my readers?
8. **For an illustration essay, to what will completing the starter questions and statement lead?**
   Completing the starter questions and statement will lead to your working thesis statement.
9. **What will defining your audience help you do in an illustration essay?**
   Defining your audience will help you focus your purpose, decide the language you will use, and decide the type of illustration (person, shared, or hypothetical) you will use.

10. **When details work well in an illustration essay, what should they help readers do?**
    Details should help readers visualize the situation (or condition) or help them understand the significance of the situation (or condition).

11. **What are three types of transitions that can help you organize your essay?**
    Three types of transitions are time order, importance order, and space order.

12. **When thinking critically about an illustration topic, what should you examine?**
    You should examine why a particular situation or condition is important to you.

13. **When thinking critically about an illustration topic, what should you analyze?**
    You should analyze examples that illustrate your point.

14. **When thinking critically about an illustration topic, on what should you reflect?**
    You should reflect on the significance of a particular situation or condition and on your examples.

15. **When thinking critically about an illustration topic, what should you provide?**
    You should provide concrete details.

16. **When thinking critically about an illustration topic, what should you discover?**
    You should discover a deeper appreciation for your topic.

17. **Identify four questions you should ask yourself when revising for examples.**
    Answers include: Does each body paragraph have at least one example? Does each example illustrate my thesis statement? Do I have enough examples to support my thesis statement? Are my examples logically ordered?

18. **Identify four questions you should ask yourself when revising for details.**
    Answers include: Are my details vivid or specific enough that readers will identify with my impressions? Have I given too many details? Does each of my paragraphs develop one example or multiple examples that illustrate my thesis statement? Have I written anything that flows away from a paragraph's focus?

19. **Identify four questions you should ask yourself when revising for topics.**
    Answers include: Does each topic idea directly support my thesis statement? Does each topic example support the topic idea of that paragraph? Do any topics flow away from the essay's focus? Do supporting sentences in body paragraphs connect to the paragraphs' topics?

Process Writing

LEARNING GOALS

In this chapter, you'll learn and practice how to

1. Get started with process writing
2. Write a process paragraph
3. Read and examine student and professional essays
4. Write a process essay

GETTING THERE

- In process writing, you explain the importance of a process (a procedure).
- In process writing, you relate steps in a specific order and you show their relevance.
- Your audience's familiarity with the procedure will guide your vocabulary choice and depth of details.
- Effective process writing allows your readers to understand your topic and adopt your procedure.

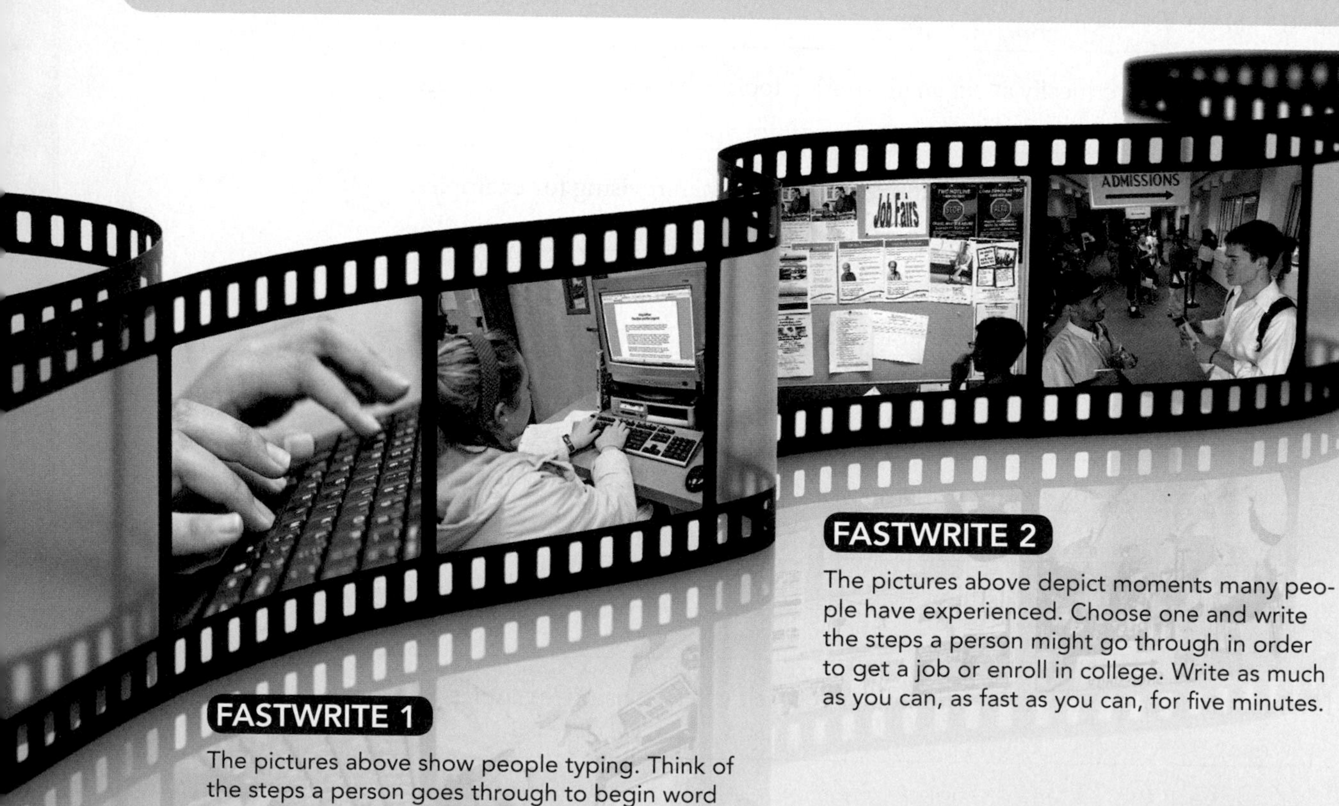

FASTWRITE 1

The pictures above show people typing. Think of the steps a person goes through to begin word processing, and then write the steps in order. If you think of a step you left out, write it and draw a line to where it should be inserted. Write as much as you can, as fast as you can, for three minutes.

FASTWRITE 2

The pictures above depict moments many people have experienced. Choose one and write the steps a person might go through in order to get a job or enroll in college. Write as much as you can, as fast as you can, for five minutes.

Complete these **Fastwrites** at **mywritinglab.com**

Getting Started in Process Writing

Simply put, **process writing** is *how-to* writing. You describe the steps necessary to do something (that is, the process carried out to reach an end).

Your **purpose** with process writing might be to

- instruct someone about how to achieve a certain goal
- analyze steps that lead to an outcome
- explain the importance of a process

Whatever your purpose, in process writing you carefully communicate the necessary steps and chronology.

Suppose you're brainstorming about the topic *how to manage your time*. Your list might include items like these:

- decide the time of day when you're the most productive
- prioritize the items on your to-do list
- buy or create some sort of personal organizer
- create a to-do list
- decide how to reward yourself after finishing a certain number of items on the list

If you looked at the steps in your brainstorming list and numbered and reordered them so they were chronological, they might look like this:

1. buy or create some sort of personal organizer
2. create a to-do list
3. prioritize the items on your to-do list
4. decide the time of day in which you're the most productive
5. decide how to reward yourself after finishing a certain number of items on the list

Next, you might review your list to see if any steps are more complex than others and can be broken down into even smaller steps. To show the relationship of any smaller steps to the larger steps, you could use a Roman numeral outline.

In that type of outline, each Roman numeral represents a large step in the overall process of how to manage time. The capital letters represent smaller steps that, combined with other smaller steps, make up a large step or stage in the overall process. If your process is showing how to create a personal organizer, your Roman numeral outline might look like this:

LEARNING GOAL

1. Get started with process writing

Heads Up!
You may be more familiar with this type of writing by its other names; in both writing and oral communications classes, process writing is sometimes called *how-to*, *instructional*, or *step-by-step* writing.

Create a personal organizer

I. Create a to-do list
   A. home
   B. school
   C. work
II. Prioritize the items on your to-do list
   A. family items
   B. school and work items
III. Decide the time of day when you're the most productive
IV. Decide how to reward yourself after finishing a certain number of items

To compose your process paragraph, develop a topic sentence and link each step. Your completed process paragraph might look like this:

If you want to manage your time more efficiently, then these easy-to-follow steps can help you. First, find an organizer that fits your needs. Organizers let you create lengthy lists that note chores at home, deadlines for school, responsibilities and duties for work, and other tasks you must do. You can use either a digital organizer or a paper organizer. Digital organizers are available from many software companies, and they come with multiple functions. Look at several digital organizers and decide which one would work best for you. If you have difficulty typing on a digital organizer, a traditional paper organizer might be a better choice. These are available for purchase, or you can create one yourself. After you have written your list, whether it's on a digital or paper organizer, you should next prioritize the items on your list. Lastly, put reminders for little rewards you can treat yourself to for finishing a certain number of tasks. Ice cream is always better when you're not pressed for time.

9.1 Fastwriting a Process

Directions: For three to five minutes, fastwrite the steps involved in getting your day started. Focus on those steps you take to move from waking up to getting to work or school each day. When you're finished, share your fastwrite with your writing group. As a group, discuss the fastwrites and note which steps are similar, different, or missing.

WRITING at WORK *Snapshot of a Writer*

Marvin Bartlett, Television Anchor

Marvin Bartlett has been a broadcast journalist for twenty-five years. He started as a reporter/videographer, then moved into producing, and became a full-time anchor in 1994. Marvin has won three regional Emmys for reporting and news writing and has taught broadcast writing at the University of Kentucky. In 2002, he was inspired to write a book, The Joy Cart, *about the subject of one of his feature reports.* The Joy Cart *is currently being made into a movie.*

Marvin's Writing Process

Choose Quotes First I write the story after I come back from conducting interviews and gathering video. A typical television news story runs less than ninety seconds. I want to find the most compelling sound bites I can from the interviewed subject. I'm looking for quotes that show emotion (joy, anger, concern, or humor). Then I write around the sound. I can fill in with the basic facts and tell them more succinctly than the subject. But it's the raw emotion that can really make a story memorable.

Look for the Human Element All broadcast stories are better if they are personalized. For example, a station may do a story on how school start times are being changed. The reporter could just spell out the facts—that buses will now run thirty minutes earlier and many people are upset—using video from the school board meeting where it was discussed. Or she could go to the bus stop with a mother and her children, showing how dark it is at the earlier hour and how drivers speeding past may not see the children waiting by the side of the road. I always look for ways to tell the story through the eyes of a "real person," not police officers or politicians.

Make It Conversational I have to keep in mind that I am writing for the ear. People are *listening* to what I write, rather than *reading* it. I am truly "telling" a story. So I use contractions and everyday phrases, and I try to shape a sentence the same way someone would if they were talking to a friend. We're told to write on a fifth-grade level because that's the average reading ability of our listeners. That may seem like "dumbing down" the news, but what's the use in writing things a large portion of the audience may not understand? Officials, such as police, use a lot of terms the average person does not, for example, saying "motorist" instead of "driver" or "premises" instead of "house." I rewrite police reports almost every day.

Make It Tighter Brevity is key, so I always go back through my copy and look for unnecessary words or phrases. We try to get as many stories into a newscast as possible, and saving words here and there can allow for additional stories. It's odd how often I can find three-word phrases to cut. For example

"The man was close to death when police arrived ~~at the scene~~."

"Police aren't releasing the victim's name ~~at this time~~."

"The lottery game is expected to get more popular ~~in the future~~."

(continued)

Ask "Who Cares?" Part of my job as an anchor is to write teases for the upcoming newscast—those ten- or fifteen-second spots you see in prime time that tell you what's coming up. In that amount of time, I'm trying to promise the viewers it will be worth their time to stick around. A bad tease would be

> "It's flu season. We'll show you people lined up to get a shot . . . tonight at ten."

Why would anyone tune in for that? We all know what it looks like to get a shot. A better tease would be

> "Flu season hits early this year. Tonight, at ten, we'll tell you where the first confirmed case has been reported and let you know where you can get your shot for free."

Now I've promised something that could be beneficial to know.

Read It Out Loud The best way for me to tell if I've written good broadcast copy is to read it out loud. This helps me catch tongue twisters, awkward phrases, and sentences that are too long. If I have to take a breath in the middle, it's not a sentence I want to read on air.

TRY IT!

Write a paragraph about the process you would follow to interview a student or faculty member about a controversial issue on campus or in your community. When you're finished, share your process with those in your writing group.

Getting Started with Your Process Paragraph

LEARNING GOAL
❷ Write a process paragraph

You may feel more comfortable writing a process paragraph before you try an essay. If that's the case, follow steps one through six below. If not, go on to Take 2 on page 184.

Step One: Choose a Topic

If you have the opportunity to choose the topic for your process paragraph, choose a process that

- you know well
- involves a number of easily explained steps
- your audience can re-create

Process topics usually concern subjects your audience may need to know about or could benefit from. Such topics could include

- installing a car battery
- changing a cell phone's ring tone
- protecting yourself from sun damage
- changing a bicycle tire
- finding a source in the library
- packing an overnight bag
- booking a flight online
- selling something on eBay
- figuring your GPA
- choosing the perfect pet

To begin searching for a topic, complete Ticket to Write 9.2 below.

TICKET to WRITE

9.2 Choose a Topic

Directions: Complete the following to start the process of choosing a topic. For each general area listed, list as many process-related tasks as you can. Answers will vary.

1. Arts and crafts
2. Cars
3. College and school
4. Computers
5. Cooking
6. Electronics
7. Hobbies and games
8. Home and garden
9. Parenting
10. Personal finance
11. Pets and animals
12. Relationships and family
13. Sports and fitness
14. Travel
15. Workplace

Step Two: Generate Ideas

Once you have decided on your topic (the process you will explain), you need to consider your purpose (the reason you're explaining it). Combined, these two ideas make up your topic sentence.

Topic: I want to explain how to prepare a pot roast.

Purpose: My purpose for explaining this is so you can feed your family something they'll love to eat.

Now compose your topic sentence, which should include

- letting your reader know **what** you're explaining
- giving the reason **why** you're explaining it (that is, stating your purpose)

What you're explaining — Purpose for explaining it

Topic sentence: Preparing a delicious pot roast is easy and will please your family.

Topic: I want to explain how to book a flight online.

Purpose: My purpose for explaining this is so a traveler can save money.

What you're explaining — Purpose for explaining it

Topic sentence: Purchasing airline tickets online can save you money.

Topic: I want to explain how to make a recycling compost bin.

Purpose: My purpose for explaining this is to reduce household garbage.

What you're explaining — Purpose for explaining it

Topic sentence: Making a recycle compost bin can reduce household garbage.

To define your possible process paragraph topics, complete Ticket to Write 9.3 below.

9.3 Generate Ideas: Creating Topic Sentences

Directions: Choose five possible topics from Ticket to Write 9.2 and complete the *topic* and *purpose* statements for each topic. Then use your responses to create topic sentences for each topic. Answers will vary.

Before selecting which process writing topic you will develop into a paragraph, make sure you are knowledgeable enough regarding the topic to instruct your readers. One way of doing this is to be sure you know which tools, utensils, materials, ingredients, or other necessary items will be needed to carry out this process.

To help determine what you will need to let your readers know, use Ticket to Write 9.4 and brainstorm a list of necessary materials for each of your possible subjects.

9.4 Generate Ideas: Listing Necessary Materials

Directions: Review the five topics you defined in Ticket to Write 9.3. List the tools, utensils, materials, ingredients, or other necessary items needed to complete each process.

Before you begin drafting your paragraph, freewrite on the steps that make up the entire process. Write down every step you can think of. Picture yourself as you perform each step, and imagine yourself completing the task.

Use Ticket to Write 9.5 to freewrite your list of steps.

9.5 Generate Ideas: Listing Steps

Directions: Choose one of the five topics you have been working with and write it, your purpose, and your topic sentence. Then freewrite all the steps you can think of that must be taken to complete this process. When finished, share your responses with your writing group. Ask them to note any steps they think are incomplete or missing.

Step Three: Define Your Audience

A written process offers readers necessary information in a step-by-step approach so they can

- achieve a certain goal
- analyze the steps that lead to an outcome
- understand the importance of a certain process

You should be aware of who your audience is for this process. As you draft your paragraph, keep the knowledge level of your audience in mind. You don't want to speak down to your audience or write over their heads.

For example, you probably wouldn't want to share with an audience of chefs how to make macaroni and cheese, just as you wouldn't want to explain how to make traditional risotto to an audience of fourth graders.

Consider terminology with which your audience is unfamiliar, and provide definitions for those terms. For example, in a process paragraph on how to cook macaroni and cheese, your directions might include "add pepper and salt to taste." For an audience of young people who are cooking for the first time, you would need to explain that this phrase means about a fourth of a teaspoon; conversely, you would not need to explain the phrase if you were writing for experienced cooks.

Step Four: Draft Your Paragraph

See Chapter 2 for more on chronological or time order.

As you draft your paragraph, focus on linking the steps you freewrote in Ticket to Write 9.4.

In process writing, transitional words and phrases that show **chronological order** are often used; these help your readers know when one step is completed and the next begins. Here are some you can draw on:

after (afterwards)	eventually	immediately before	meanwhile
at last	finally	initially	next
at the same time	first (second, third, etc.)	in the end	previously
before		in the future	simultaneously
concurrently	following this	in the meantime	soon after
currently	formerly	last	subsequently
during	immediately after	later	then

9.6 Draft Your Paragraph

Directions: Copy the topic sentence you chose in Ticket to Write 9.5 and then draft your paragraph by linking the steps you listed in 9.5. As you draft your paragraph, focus on chronological order and transitional words or phrases.

When finished, share your responses with your writing group. Discuss with your group who is a good audience for this paragraph and if the language you have used is appropriate for that audience.

Step Five: Revise Your Paragraph

After you've finished your paragraph, save it, print a hard copy, and read it out loud, looking and listening for errors in grammar and content. Refer to the general Review Checklist for a Paragraph in Chapter 3, page 60, to make sure you revise as completely as possible.

Use the following checklist for further revision of your process paragraph.

☑ REVIEW CHECKLIST for a Process Paragraph

- ☐ Is my topic sentence clear?
- ☐ Do I include relevant points of comparison and contrast?
- ☐ Do I hear any fragments or run-ons when reading out loud?
- ☐ Do all my complete thoughts have appropriate end punctuation?
- ☐ Do I use vocabulary appropriate for my audience?
- ☐ Have I reviewed my steps to make sure they are in the right order?
- ☐ Does my purpose (to instruct, analyze, or explain) remain constant?
- ☐ Do any sentences or ideas seem off-topic?

Heads Up!
See Chapter 20 to review fragments and run-ons.

Once you've made any necessary changes, save your work and leave it alone for a while. Then come back later, print out a new copy, and look and listen for errors again. Refer again to the general Review Checklist for a Paragraph in Chapter 3.

Step Six: Peer Review

You might think you've found all of your errors and made all of the improvements you can, but having someone else read your paragraph and offer suggestions will almost always improve your writing. This person is called your **peer reviewer**.

Heads Up!
See Chapter 4 for more on peer reviewing.

Your peer reviewer may use the review checklist above or the general Review Checklist for a Paragraph in Chapter 3, page 60, or your instructor may provide a different checklist. You might or might not agree with the suggestions you receive, but peer reviewers often find places for improvement or errors that slipped by you. Listen to or read closely your peer reviewers' suggestions, and don't be shy about asking questions to clarify their ideas and suggestions.

9.7 Peer Review

Directions: Once you have completed revising your paragraph, share it with your peer reviewers. Ask them to review your paragraph using the Review Checklist for a Process Paragraph above and the general Review Checklist for a Paragraph from Chapter 3, page 60. Then ask them to record their suggestions for revision. Review their suggestions and make any revisions you feel are necessary.

Student and Professional Essays

LEARNING GOAL

❸ Read and examine student and professional essays

Below are two process essays. The first was written by a student and the second by a professional. Read these two essays and answer the questions that follow them.

Student Essay

Choosing a College Major

by Zosima A. Pickens

1 "**Major**? Ahh! . . . I don't know!" These are words that a college freshman student normally answers to this question, "What is your major?" Most freshmen students enter college life without a plan or major in mind. Others cannot decide to pick their major until they are halfway through their college life. Considering certain steps might help freshmen students choose their college major through research, **volunteerism**, and gathering information from an expert.

2 First, do some research. The rapid growth and advanced technology in the world is one factor of choosing a college major or degree. List the courses you are interested in. Most students work on all the **prerequisite courses** before they pick their major. Here are examples of degrees you might consider: teaching, nursing, engineering, and computer science. These are very common and most in demand. Choose which degree is **suited** to your best interest. Picture yourself in a career with one of those degrees. Researching a particular field of work can give you a specific picture of what that career might be like. For example, someone researching the field of nursing might have many options within the field, such as an **OR nurse**, a surgical nurse, or a **nurse practitioner**.

3 Volunteering is another step in choosing a college major. Working as a volunteer can play a big role in choosing a college major. For example, working with children as a volunteer can be challenging. Suppose one time you were volunteering at a day care, and you had to deal with a child who was acting up. You didn't know how to handle the difficult behavior, so you decided to take an early childhood development class. This might be the way to work towards a college major. In this case, the background as a volunteer will help you choose your college major, teaching small children. I have a friend who volunteered at an animal shelter and with a veterinarian because she thought she was interested in majoring in veterinary science.

4 Lastly, consulting an expert is also a step to choosing a college major. Experts can be people in the field you're interested in or advisors and counselors at your college. Those in the work world can give you information based on their personal experience and career. Counselors at your school

major in college, the field a student concentrates on or specializes in

volunteerism the practice of working, without pay, on behalf of others for a particular cause

prerequisite courses required classes students must take before enrolling in other courses

suited appropriate, fit

OR nurse a nurse who works in an operating room

nurse practitioner a registered nurse with special training for providing primary health care

can lay out choices of possible college majors and help you discover interests you have. Make sure to ask questions like "If later I decide to change my major, what can I do?" Most experts are willing to help students every step of the way throughout college life.

5 "What do you want to do when you get out of college?" This question doesn't have to be scary. Sometimes choosing a college major can be very challenging. Careful choice of a major based on research, volunteering, and information from an expert is worth the effort to make sure you know what you're going into and that it's the right choice for you.

A CLOSER LOOK

Answer these questions about the essay:

1. Where is the thesis statement found?

   The thesis statement is in the last sentence of paragraph 1.

2. Which sentence in paragraph 2 needs to be eliminated for paragraph unity?

   The fourth sentence should be eliminated.

3. Which sentence in paragraph 3 could use more development to illustrate the paragraph's topic?

   The fifth sentence could use more development.

4. How many examples does the author develop in paragraph 3 to illustrate the paragraph's topic?

   The author uses two examples.

5. Which of the following is the topic sentence of paragraph 4?

   The first sentence is the topic sentence.

6. Which sentence in paragraph 2 needs further development to support the topic sentence?

   The second sentence needs more development.

7. What transition word is used to introduce the final step in the process of choosing a college major?

   Lastly

8. Which sentence in the conclusion reemphasizes the importance of the thesis statement?

   The third sentence reemphasizes the importance of the thesis statement.

9. How many steps does the writer include in her process?

three

10. If you were working with this student writer, what is an additional step you would suggest she include in her process?

Answers will vary.

Professional Essay

Finding Good Dining Away from Home

by Ed Shuttleworth

1 When away from home on vacation, dining out often becomes an integral part of the trip. To save time when you're driving, you may eat at fast-food restaurants, but once you get to your **destination**, you usually look forward to good food. When you're in a new area, you want to be sure you spend your money and your time wisely. The problem is you usually have many choices of eating establishments, and that can be frustrating and confusing. You could buy a guidebook that has a section about restaurants, but guidebooks are sometimes expensive. Finding a good restaurant is easy and free if you follow three suggestions.

2 Before you leave home, do some Internet research. In a search engine, type "restaurant reviews" and then the name of the city, and many sites will come up from which you can find a particular restaurant, type of food, or price range. Then you can see reviews others have posted about various places. Using **flikr.com** or **google.com/images**, you can find pictures of some restaurants or the foods they serve by typing in the name of the restaurant and the city. Another method to find opinions about restaurants is to type "restaurant **blog**" and then the name of the city. You'll get links to opinions written by fellow travelers and even professional reviewers.

3 If you're staying in a hotel or motel, use services there to help. Go to the front desk or the **concierge** counter and ask for copies of menus from local restaurants. Survey what the establishment offers and see if its menu fits your taste **palate** and if its prices fit your budget. Hotel employees can tell you if the restaurant is within walking distance, or they can give you driving directions. If you need to make a reservation, let someone from the hotel make it for you; hotel employees often have connections with certain restaurants and can ensure that you get a good table.

4 A third way to find a good restaurant is to ask the **locals** where they like to eat. **Striking up a conversation** is usually easy when you mention you're looking for a good place to eat, and most locals are quick and reliable with their opinions and recommendations. Locals also know the places that are **off the beaten path** but have really good food. They'll probably ask the type of restaurant you want or the type of food you're interested in. Are you looking for upscale dining or casual dining? Do you need for the restaurant to be family-friendly, or are you looking for something for adults? What kind of

destination the place to which a traveler is going

flikr.com an Internet site that hosts pictures

google.com/images an Internet site that hosts pictures

blog an Internet site with a personal journal featuring opinions and comments

concierge a staff member of a hotel who assists guests with various services

palate personal sense of taste

locals people from the area

striking up a conversation an idiom meaning "beginning to talk casually with someone"

off the beaten path an idiom meaning "in a location not widely known"

food do you like? What kind of food do you not like? If you want to experience something new, you might also ask locals about types of food or cooking styles that are found just in that area.

5 These three methods to finding a good restaurant when you're away from home are free and take little effort. Whatever method you choose, you can find a restaurant that suits your taste buds and your budget. **Bon appétit**!

bon appétit French for "enjoy your meal"

A CLOSER LOOK

Answer these questions about the essay:

1. Where is the thesis statement found?

   The thesis statement is in the last sentence of paragraph 1.

2. What is the topic sentence of paragraph 2?

   The first sentence is the topic sentence.

3. How many examples does the writer include to illustrate the step in paragraph 2?

   two

4. How many examples does the writer include to illustrate the step in paragraph 3?

   two

5. What is the topic sentence of paragraph 3?

   The second sentence is the topic sentence.

6. What transition word is used in the first sentence of paragraph 4?

   third

7. What is the topic of paragraph 4?

   "To find a good restaurant, ask locals where they like to eat."

8. List two of the examples the writer gives in paragraph 4 to illustrate the topic idea.

   Answers will vary.

9. How many steps does the writer include in his process?

   three

10. Given the style, language, and information in this piece, who is the intended audience?

    Everyday travelers are the intended audience.

Writing Your Process Essay

LEARNING GOAL
❹ Write a process essay

Essay writing is much like paragraph writing: you start with a topic and expand with details. Remember that just as a paragraph supports a topic sentence, an essay supports a thesis statement.

Step One: Choose a Topic and Develop a Working Thesis Statement

See Chapter 2 to review thesis statements.

To begin your process essay, ask yourself these starter questions:

1. **What** task do I want to explain?
2. **What** is my purpose in explaining this task? *or* **Why** is this task significant?

For instance, in a process piece about the task *how to parallel park*, Jacques answered the **starter questions** this way:

1. **What** task do I want to explain?

 the best way to parallel park

2. **What** is my purpose in explaining this task? *or* **Why** is this task significant?

 to make parallel parking simpler and safer

Answering Starter Questions
Zosima A. Pickens and Ed Shuttleworth

In answering the starter questions, student writer Zosima A. Pickens, the author of "Choosing a College Major," wrote:

1. The task I want to explain is how to choose a college major.
2. This task is significant because many college students struggle when trying to select the right major.

In answering the starter questions, professional writer Ed Shuttleworth, the author of "Finding Good Dining Away from Home" might have written:

1. The task I want to explain is how to find a good restaurant if you're out of town.
2. This task is significant because finding a restaurant when traveling can be frustrating and confusing.

As you learned in Chapter 2, all essays have **thesis statements**. To begin developing a working thesis statement for your process essay, create a **starter statement**. Fill in the blanks in this sentence:

I will write about ____________________(identify the task) so that ____________________(explanation of the purpose or significance).

In explaining how to parallel park, Jacques filled in the blanks like this:

I will write about how to parallel park so that drivers can use a safe and proper procedure.

Look at how three other writers chose to fill in the blanks for other topics:

I will write about how to roast a turkey so that cooks can prepare a delicious main dish.

I will write about steps involved in preparing, planting, and cultivating an organic garden so that novice gardeners can learn to go green.

I will write about how to read football plays so that my team can decode an opposing team's intentions.

9.8 Answer Starter Questions for a Process Essay

Directions: All essays have thesis statements. To help develop a thesis statement for your process essay, answer the starter questions and then fill in the blanks to complete a starter statement. Answers will vary.

1. What task do I want to explain?
2. What is my purpose in explaining this task? or Why is this task significant?

I will write about ______________________________ (identify the task) so that ______________________________ (explanation of the purpose or significance).

Once you've filled in the starter statement, you've begun to develop the focus of your essay. Review your starter statement, which centers on your topic and your purpose. From this starter statement, create your **working thesis statement**. These are working thesis statements created from the preceding starter statements:

Following eight steps will result in a delicious turkey.

If you want to grow an organic garden, then follow specific steps in preparing, planting, and cultivating your soil.

Learning how to read football plays can give a team's offense and defense an edge over their opponents.

9.9 Develop a Working Thesis Statement

Directions: Review the starter statement you composed in Ticket to Write 9.8. From this, create your working thesis statement and share it with members of your writing group. Ask them if they understand what process you will explain in your essay and the significance of the process.

Step Two: Generate Ideas

See Chapter 1 to review activities and techniques for prewriting.

Consider the process you are explaining and why you are explaining it. You are actually teaching your audience how to do something. Freewrite *why* your audience should learn this process, *how* they should learn it, or *what* could help them learn it.

Why Learn This Process

You express this idea in your working thesis statement, but gathering a few more ideas to support this statement can help when you create your introduction and conclusion. How can knowing this process help your audience? How can it make their lives better? Jacques wrote the following in his prewriting about parallel parking:

Parallel parking can help you when all the easy spots are gone.
sometimes the only option
can protect your car from others opening their doors and denting your car

How to Learn This Process

Because you are a teacher in this instance, you can draw from your learning experience or research and pass it on to others. What steps are important in this process? Here are the steps Jacques listed:

decide if car will fit in space
alert other traffic I intend to park
align rear tires with rear bumper of car in front
turn wheels all the way in direction of space I want
to go in
put car in reverse
back up slowly until I'm at forty-five-degree angle
stop
turn wheels away from curb
ease into spot until I'm parallel with curb
pull back and forth to adjust space between
other cars

What Could Help in Learning This Process

Brainstorming with learning aids like mnemonics or other resources may help your audience remember some of the pointers, devices, or other little extras. Jacques wrote the following in his prewriting about parallel parking:

found a couple of YouTube videos that were helpful
the online driver's manual from the DMV

9.10 Generate Ideas

Directions: Write your working thesis statement and any ideas you generated in prewriting. Then share these with your writing group and ask them to note which ideas need more development. Answers will vary.

Step Three: Define Your Audience

Your next step is to identify the audience for your essay. You have already determined the process you will explain, so your audience is people who would benefit from learning your process. As you draft your essay, be sure to use a level of language helpful to this audience.

- *Don't write over your readers' heads.* If you must use terminology the average reader would not recognize or understand, define those terms.
- *Don't speak down to your readers.* Consider the knowledge base of the topic your readers have, and don't needlessly explain any material.

Jacques described the audience for his process essay on parallel parking this way:

My audience is . . .

people who should know how to parallel park but don't

people who might not like to parallel park because they can't do it very well

college students, young drivers, older drivers, anyone who can drive

9.11 Define Your Audience

Directions: Write your working thesis statement and your intended audience. Share this with members of your writing group and ask if they agree that your working thesis statement is satisfactory for your audience.

Step Four: Draft Your Essay

To begin drafting your essay, review the information you created in Step Two and Ticket to Write 9.10. The sections of Step Two (*why learn this process*, *how to learn this process*, and *what could help in learning this process*) include information you may be able to use for various parts of your essay. When drafting your essay's introduction and conclusion, refer to what you created in response to *why learn this process*. When drafting your essay's body paragraphs, refer to what you wrote in response to *how to learn this process* and *what could help in learning this process*.

The body of your essay should be based on

- a list of tools, utensils, materials, ingredients, or other necessary items
- a set of step-by-step instructions
- an explanation of the special significance of the task

Lists of Materials

Written instructions often include a list of tools, utensils, materials, ingredients, or other necessary items. If you've ever put together a computer desk, entertainment center, swing set, or bike, you probably saw a parts and tools list at the beginning of the instructions; if you've ever read recipes in a cookbook, you probably saw a list of ingredients before the cooking instructions. The same goes for any written process: you need to let your readers know what they'll be working with before you start telling them how to use these items.

Step-by-Step Instructions

As you draft your essay, consider the number of steps in your process. Is each step listed separately or are any steps grouped together? If any steps are grouped, then you have what's called a **stage** in the process. A stage may be one body paragraph or more.

A **stage** in a process is also referred to as a *task*.

Explanation of Special Significance

As you present a step within the process, you may need to point out its significance. This could include drawing the audience's attention to any safety precautions or common problems related to that particular step or stage.

Jacques included a special warning about parking on a hill in this paragraph from his essay on parallel parking:

> Finally, pull your vehicle forward slightly, keeping enough distance between yourself and the car in front of you. Then back up your vehicle once more until the car is straight in the middle of the two cars. You can pull forward and back up

(continued)

until you are satisfied that your car is positioned evenly in its space. If you are parked on a hill, brace your wheels against the curb to guard against your vehicle's parking brake failing. To do this when facing downhill, turn your wheels toward the curb and pull forward until the wheel is against the curb. When facing uphill, turn your wheels away from the curb and then back up until the rear of the tire catches against the curb.

TICKET to WRITE

9.12 Write Your Discovery Draft

Directions: From the material you created in Ticket to Write 9.9 through 9.11, write your discovery draft. Refer to Chapter 2 to review discovery drafting.

When you have completed your discovery draft, share your work with members of your writing group. Ask them to note the following on your discussion of your process:

1. Steps that are detailed, specific, and easy to understand
2. Steps that are vague, unclear, or hard to follow
3. Any steps they believe are missing or are not necessary

Step Five: Organize Your Essay

In a process essay, the order of steps is important. To ensure the correct outcome, each step must logically lead to the next. Some steps may need more explanation or discussion than others, but you still must present them in the proper order. Also, when necessary, you should provide extra explanation for steps that may be confusing or need special attention.

Transition words and phrases help you show the relation and order between steps, and your essay flows much more smoothly when you use them. These transitions give hints to your readers of the direction your thoughts are taking, guiding them through your essay.

In a process essay, you might use words and phrases showing chronological order, such as the following:

after (afterwards)	eventually	initially	next
at last	finally	in the end	previously
at the same time	first (second, third, etc.)	in the future	simultaneously
before	following this	in the meantime	soon after
concurrently	formerly	last	subsequently
currently	immediately after	later	then
during	immediately before	meanwhile	

9.13 Organize Your Essay: Practice with Transitions

Directions: The sentences below form a paragraph detailing the steps in organizing a class reunion. Working in groups or individually, read the steps, put them in chronological order, add transitional words or phrases, and then compose a topic sentence to introduce the paragraph.

Alternately, with your writing group or individually, expand the details below into a process essay, adding transitional words or phrases, an introductory paragraph with a thesis statement, and a concluding paragraph.

Answers may vary.

1. Compile classmates' current information to either send out before the reunion or distribute at it
2. Form a committee of people interested in helping with organizing the reunion.
3. Enjoy getting reunited with friends from the past.
4. From the budget, figure out how much each graduate will pay in order to cover the total cost of the reunion.
5. Send out letters or e-mails to classmates, giving information about the time, place, and cost of the event, and to whom they should send their payment.
6. Ask those who have paid to attend to give current information about their lives.
7. Make name tags for the event so that people won't be wondering about what to call each other.
8. Determine a date and site for the reunion; the food, music, and decorations; and a budget.

Use Ticket to Write 9.14 to check your own draft for proper transitions.

9.14 Organize Your Essay: Check for Transitions

Directions: Print a copy of your draft. Then use a highlighter to mark each transition you use between steps and in discussions of specific steps that may take more development than others. When you are finished, read through your draft and look for places that

1. May be missing a needed transition
2. Have duplicate or redundant transitions
3. Have unnecessary transitions

Step Six: Apply Critical Thinking

Critical thinking means thinking and then rethinking about a topic, with the goal of discovering all the implications you can about it. In thinking critically about a process essay, you often

- examine why a particular process or task is significant
- analyze the importance and the order of the steps in a process
- reflect on the importance of being able to accomplish a certain task or process
- provide precise details

The following critical thinking questions are important when you write a process essay. After you write each of your drafts, ask yourself these questions and then apply your answers to your writing.

1. **Purpose:** Will my readers understand why my topic is important? (This importance may involve understanding how a specific process works or learning how to complete a specific process task.) Have I clearly stated or clearly implied my purpose?
2. **Information:** Have I given my readers enough details to understand the process or complete the process on their own? Do I see any place to add details that will enhance or further explain my topic?
3. **Reasoning:** Have I presented all the steps in the proper order? Are the steps clear? Is my reasoning logical?
4. **Assumptions:** Have I made any assumptions that need to be explained or justified? Have I used any vocabulary my audience may not understand?

In a process essay, critically thinking about specific details means analyzing details for their significance to a specific step.

9.15 Apply Critical Thinking

Directions: Ask members of your writing group to read your latest draft and answer the critical thinking questions below.

Purpose

1. Do you understand why my process topic is important?
2. Did I clearly state or clearly imply my purpose? If not, where should I be clearer?

Information

3. Do you feel you read enough details to understand the process and how it works?
4. Do you see any places where adding details would enhance or further support my topic?

Reasoning

5. Do you feel all the necessary steps are included, are appropriate, and are in the proper order? If not, what steps are out of order, missing, or redundant?
6. Are all my facts clear? If not, where should I change my wording to be clearer?
7. Are all my facts relevant to my topic? If not, where do I stray from my topic?
8. Is my reasoning logical? If not, where am I illogical?

Assumptions

9. Did I make any assumptions that need to be explained or justified? If so, what are these assumptions?
10. Did I use any vocabulary you don't understand? If so, what words or phrases need definitions?

Step Seven: Revise Your Essay

As you know, getting your essay on paper is just your first step. Revising and rewriting through subsequent drafts shape your first draft into a finished product of which you can be proud. First, consult the Review Checklist for an Essay in Chapter 3, page 61, and use it to look for places in your essay that need to be polished. Then think about revisions that are specific to process essays: **steps**, **discussions and details**, and **language**.

Heads Up!
Keep hard copies of all your drafts and number them so you'll know which is your latest draft.

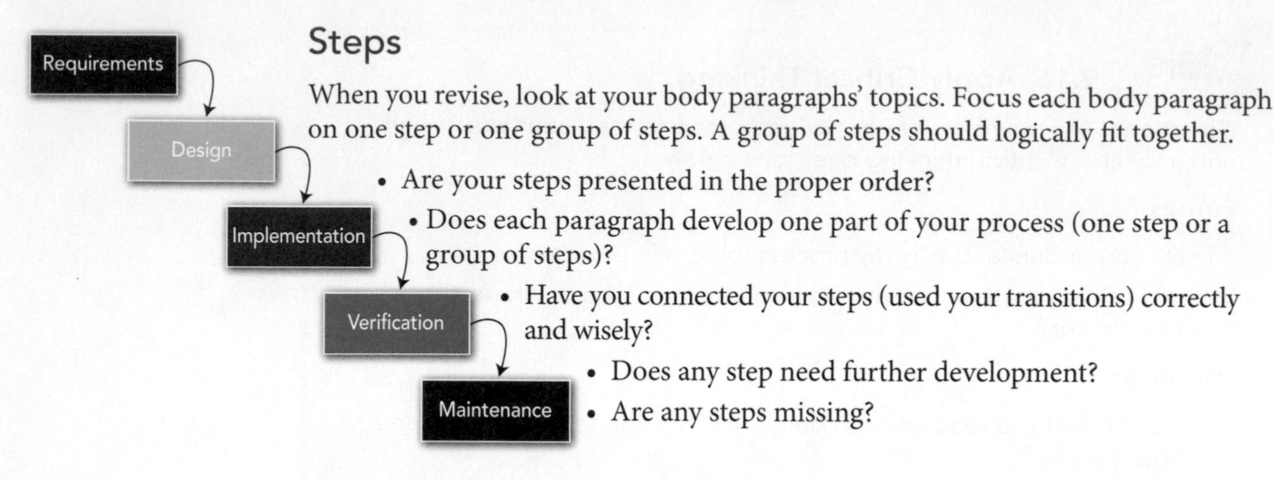

Steps

When you revise, look at your body paragraphs' topics. Focus each body paragraph on one step or one group of steps. A group of steps should logically fit together.

- Are your steps presented in the proper order?
- Does each paragraph develop one part of your process (one step or a group of steps)?
- Have you connected your steps (used your transitions) correctly and wisely?
- Does any step need further development?
- Are any steps missing?

Discussions and Details

When you revise, review the discussions of specific steps. Some steps may call for more explanation, some may require special warnings or cautions, and some may need detailed examples and illustrations.

- Does your title easily convey your topic?
- Are your details clear enough for readers to follow?
- Are your details complete enough to accomplish the task?
- Have you emphasized the importance of your topic?
- Have you given too many details?
- Have you written anything that flows away from your focus?

Depending on your process topic, you may need to include details that address the following:

- Do you identify whom, where, or what the process affects?
- Do you tell how long the process should take?
- Do you include any skills or equipment necessary for the process?
- Do you address any difficulties a person might encounter in the process?
- Do you identify specifically what should not be done in the process?
- Do you include any cautions or warnings?

Language

The audience for whom you're writing determines your language choice. You need to consider your audience's education level and their familiarity with the topic. Because your audience will probably not be specialists, make sure to adequately explain any terms that may be unfamiliar to them.

9.16 Revise Your Essay: Language Choice

Directions: Below are two lists of words and phrases that mean the same thing. Circle the word or phrase that would be appropriate for an audience who has no knowledge of the topic.

Topic	Choice A	Choice B
Cooking	sous chef	assistant
Motorcycles	panniers	luggage bags
First aid	lying face up	supine position
Computer skills	Universal Serial Bus device	flash drive
Lawn care	grubs and crickets	soil inhabitants

Writing Assignment

MyWritingLab™

Consider the topics below or one that your instructor gives you. Expand one into a process essay.

1. Shop for bargains
2. Give a haircut
3. Move from one apartment to another
4. Prepare a holiday meal
5. Dress a wound
6. Have a perfect wedding
7. Take memorable digital pictures
8. Be prepared to handle a natural disaster
9. Trace your genealogy

TECHNO TIP

For more on process writing, search the Internet for these videos:

Process Essay Video
Writing the Process Essay

MyWritingLab™ Visit *MyWritingLab.com* and complete the exercises and activities in the **Paragraph Development-Process** and **Essay Development-Process** topic areas.

RUN THAT BY ME AGAIN

LEARNING GOAL

❶ Get started with process writing

- **Process writing describes . . .** the steps necessary to do something.
- **Three possible purposes of a process paragraph or essay are . . .** instructing someone about how to achieve a certain goal, analyzing the steps that lead to an outcome, and explaining the importance of a process.

LEARNING GOAL
❸ Read and examine student and professional essays

LEARNING GOAL
❹ Write a process essay

- **When selecting a process topic, choose a topic that . . .** you know well, involves a number of easily explained steps, and your audience can re-create.
- **Before you begin drafting your process paragraph, freewrite on . . .** the steps that make up the entire process.
- **In a process paragraph, your topic and your purpose will combine to make up . . .** your topic sentence.
- **To compose your process paragraph, develop . . .** a topic sentence and link each step.
- **Starter questions for a process essay include . . .** What task do I want to explain? What is my purpose in explaining this task (or, why is this task significant)?
- **Completing the starter statement can . . .** lead to your working thesis statement by focusing on the reason you want to explain a specific process.
- **To generate ideas for your process, freewrite . . .** why your audience should learn this process, how they should learn it, and what could help them learn it.
- **The body of a process essay should be based on . . .** a list of tools, utensils, materials, ingredients, or other necessary items; a set of step-by-step instructions; an explanation of the special significance of the task.
- **In a process essay, identifying your audience can . . .** help you focus your purpose and determine the language you will use.
- **When writing a process essay, you should make sure your level of language . . .** is not written over your audience's head and does not speak down to your audience.
- **In a process essay, the types of transition words and phrases most often used are . . .** those that indicate chronological order.
- **When thinking critically about a process topic, you should examine . . .** why a particular process or task is significant.
- **When thinking critically about a process topic, you should analyze . . .** the importance and the order of steps in a process.
- **When thinking critically about a process topic, you should reflect on . . .** the significance of being able to accomplish a certain task or process.
- **When thinking critically about a process topic, you should provide . . .** precise details.
- **When revising a process essay, look especially at . . .** the steps in your process, the discussion and details about each step, and the language you use.

PROCESS WRITING LEARNING LOG MyWritingLab™

Answer the questions below to review your mastery of process writing. Answers will vary.

1. How is process writing like how-to writing?
   Process writing is like how-to writing because it describes the steps necessary to do something (that is, to carry out a process to reach its end).
2. What are three possible purposes for writing a process paragraph or essay?
   Three purposes for writing a process paragraph or essay are to instruct someone about how to achieve a certain goal, to analyze the steps that lead to an outcome, and to explain the importance of a process.
3. What should you make sure of before selecting a specific process topic?
   Before selecting which process topic you will develop into a paragraph or essay, you need to be sure you are knowledgeable enough regarding the topic to instruct your readers.
4. What two starter questions can help you discover a process topic?
   What task do I want to explain? What is my purpose in explaining this task?
5. In a process essay, to what will completing the starter questions and statement lead?
   Completing the starter questions and statement will lead to your working thesis statement.
6. Why should you know your audience's knowledge level of your process topic before you begin writing?
   Knowing your audience's knowledge level of your process topic can keep you from speaking down to your audience or over their heads.
7. On what three elements should the body of your process essay be based?
   The body of your process essay should be based on a list of tools, utensils, materials, ingredients, or other necessary items; a set of step-by-step instructions; and an explanation of the special significance.
8. When should you let your readers know of any tools, utensils, materials, ingredients, or other necessary items they will need to carry out the process you describe?
   You should tell your readers what materials they will be working with before you start listing the steps for which they'll need the materials.

9. **In a process essay, what is a set of grouped steps called?**
   A set of grouped steps is called a *stage.*

10. **Why might you need to explain any special significance of a particular step in a process essay or paragraph?**
    You might need to draw your audience's attention to any safety precautions or common problems associated with a specific step in the process.

11. **What types of transitional words and phrases are usually used with process writing?**
    Usually chronological order transitions are used with process writing.

12. **When thinking critically about a process topic, what should you examine?**
    You should examine why a particular process or task is significant.

13. **When thinking critically about a process topic, what should you analyze?**
    You should analyze the importance and the order of the steps in the process.

14. **When thinking critically about a process topic, on what should you reflect?**
    You should reflect on the significance of being able to accomplish a certain task or process.

15. **When thinking critically about a process topic, what should you provide?**
    You should provide precise details.

16. **What are five questions you should consider when revising the steps of your process essay?**
    Answers include: Are my steps presented in order? Does each paragraph develop one part of the process? Have I connected the steps using transitions? Does any step need further development? Are any steps missing?

17. **Identify two questions you should ask yourself when revising discussions and details.**
    Possible answers: Are my details clear enough for readers to follow? Are my details complete enough to accomplish the task? Does my title easily convey my topic? Have I emphasized the importance of my topic? Have I given too many details? Have I written anything that flows away from my focus?

18. **When revising for language in your process essay, for what should you look?**
    Look for any terminology that may be unfamiliar to your audience.

CHAPTER 10 Definition Writing

LEARNING GOALS

In this chapter, you'll learn and practice how to

1. Get started with definition writing
2. Write a definition paragraph
3. Read and examine student and professional essays
4. Write a definition essay

GETTING THERE

- Certain words or phrases have meanings that go beyond their dictionary definitions.
- Definition paragraphs and essays allow writers to explain a personal meaning of a word or phrase.
- Effective definition writing relies on details, analysis, and an understanding of the writer's audience and purpose.

FASTWRITE 2

The pictures above show meanings of the word *family*. Choose one of the pictures (or choose your situation) and fastwrite all that defines *family*. Write as much as you can, as fast as you can, for five minutes.

FASTWRITE 1

The definition of *college student* is "one who is enrolled in a college or university." This is true, but not all college students lead the same life. Consider your daily life while taking classes, and fastwrite all that defines the term *college student* as it applies to your experience. Write as much as you can, as fast as you can, for three minutes.

Complete these **Fastwrites** at **mywritinglab.com**

Getting Started in Definition Writing

LEARNING GOAL

1. Get started with definition writing

Your purpose in a **definition paragraph or essay** is to tell and show your audience what a specific word or phrase means to you. To be sure the audience understands your unique meaning, your definition paragraph or essay should

- identify the word or phrase you're defining
- tell what the word or phrase means to you
- provide supporting examples

Many words and phrases have concrete meanings that are easy to define. *Rain*, for example, is "water that is condensed in the atmosphere and falls to earth."

But words such as *ugly, cool*, and *relationship* can be highly abstract, and their definitions change depending on who is using them. The meaning of a word often depends not only on a person's specific point of view but also on a particular circumstance.

Heads Up!
Some instructors do not want you to include the dictionary definition of a word or phrase. Instead, they want to know what you can create and support on your own.

For instance, the word *awesome* can be used literally, as in expressing an emotion that is remarkable or even fear-inspiring. Outdoor editorialist Hal Borland uses *awesome* that way in this quote:

> "A woodland in full color is awesome as a forest fire, in magnitude at least, but a single tree is like a dancing tongue of flame to warm the heart."
>
> —Hal Borland

The word *awesome* can also be used figuratively, meaning *inspiring* or *exciting*, as in the song from a popular rock band:

> "Dude, that's freaking awesome!"
>
> —All Star United

Definition paragraphs and essays use different types of examples to explain meaning.

- **Details**—defining a word or phrase by focusing on its unique qualities

> The cuttlefish, a relative of the squid, is the world's stealthiest hunter.

- **Comparison**—defining a word or phrase by showing its similarities to other words or phrases

My patients are more at ease when I use layman's terms for their illnesses, like when I tell them they have a "rapid heartbeat" rather than "tachycardia."

- **Contrast**—defining a word or phrase by showing how it is unlike its commonly accepted definition

When my twelve-year-old son told me he was dating, I realized his definition of dating is much different now than it will be in five years.

- **Background**—defining a word or phrase by discussing its history

The graphic novel, a popular form of literature today, evolved from pulp fiction in the 1920s, comic book series in the middle of the century, and paperback novels that first appeared in the 1970s.

10.1 Fastwriting a Definition

Directions: For three to five minutes, fastwrite about the definition of *communication in the twenty-first century*. Think of the many ways people communicate today, and add the advantages and disadvantages of each. When you're finished, share this fastwriting with your writing group.

WRITING at WORK *Snapshot of a Writer*

Michael J. Minerva, Vice President, Corporate Real Estate, US Airways

Michael J. Minerva, Jr., is responsible for US Airway's portfolio of real estate at airports, offices, and support facilities. He also manages corporate security and, as an attorney, handles various legal matters for the company.

Mike's Writing Process

E-mail. It's ever-present, but as a genre it's unappreciated, in my opinion. On my busiest days at work, I might write as many as one hundred e-mails. Some

(continued)

are as a short as "OK" or "thx" or "FYI." Or even, "You CANNOT be serious!" Because e-mail is the most common form of writing white-collar workers use to communicate, the form deserves some attention.

Spelling Counts, as Do Punctuation and Grammar An e-mail that contains improper spelling—even a typographical error—or other mistakes makes the writer look sloppy, if not uneducated. Because e-mail is a casual form of communication, some writers mistakenly believe mistakes don't matter. But they do matter. The reader doesn't know if you made a typo or simply don't know the right spelling or usage. Yes, if you send an e-mail that says only "thx," the recipient will not assume you can't spell the word "thanks." However, if you send an e-mail that says, "Lets met wendsday," errors jump off the screen. Avoid them. Just before you hit "Send," reread your draft e-mail. It's well worth the extra few seconds.

In Addition to Being Written Correctly, E-mails Should Be Written Well Every e-mail gives you the opportunity to demonstrate not just your command of the subject matter but also your proficiency as a writer. Take pride in your e-mail writing. A well-drafted e-mail will catch the attention of the recipients. All of the tools you learn in a writing class—clarity, brevity, accuracy—will serve you well in a professional environment if you apply them to your e-mails. It's not enough to understand the subject matter of your chosen profession. You also have to be able to communicate that understanding to others. Words are the currency of ideas. Choose yours well.

Don't Be Afraid to Let Your Personality Show Through in Your Writing While excessive casualness is a problem in e-mail writing, so is its opposite: excessive stiffness. You may have occasions to write e-mails that require complete formality, but those are the exception. To be on the safe side in a professional setting, err on the side of formality. Remember that you write with your own style, so don't suppress style to the point of extinction. Some wit is welcome, especially if it helps make your point. Using wit is difficult because it requires balance and judgment, as well as an acute awareness of the situation and your audience. Start slowly. Use your style as seasoning for the writing and not as the main course.

Your E-mails, Other Than the Briefest Ones, Should Have a Beginning, a Middle, and an End Readers should understand why you are sending the e-mail and what action, if any, you expect them to take in response. Provide background, narrative, and clear direction.

Use a Greeting and a Closing Include a greeting and a closing, even if the greeting is just the recipient's name and the closing is just yours. Granted, if you are simply forwarding an attachment to your co-worker in the cubicle next door, this is probably not necessary. However, for any e-mail with a few sentences, the addition of "Diane" at the outset and "Justin" at the end personalizes a simple e-mail. Other times, you'll want to close with "Best Regards," "Warmly," or some other more formal closing. These little touches have an impact, just like using someone's name in conversation.

Here are two examples of an e-mail covering the same information. One is written in a professional (but not stuffy) fashion; the other in an overly casual fashion. I think you will see the difference.

All,

In a few days, you will receive from Barb an invitation to a meeting about a new project the Fixed Assets group will launch in 2013. Although the subject matter might strike you as uninteresting (and probably with good reason), this is a critical project for us. It was identified by our outside auditors as an area of concern. Therefore, I need you to be on the call or, if you are unavailable, to make sure your department is represented by someone with the background and authority to participate fully. Please direct any questions you have about the project to me or Barb.

Thanks,

Dave

Hey, everybody. BOLO for a conference call notice from Barb. It's about this Fixed Assets thing you might have heard about. Yeah, I know, not fun, but the auditors say we have to do it—what-EVER! Anyway, either be on the call or make sure somebody covers it for you. Call me or Barb with questions. —Dave

TRY IT!

Imagine you want to survey students and faculty concerning opinions about shortening classes by five minutes. Compose an e-mail to send to your peers, and compose a separate e-mail to send to faculty. When you're finished, share your e-mails with those in your writing group. Discuss the similarities and differences in your group members' e-mails.

TAKE I Getting Started with Your Definition Paragraph

LEARNING GOAL
❷ Write a definition paragraph

You may feel more comfortable writing a definition paragraph before you tackle a definition essay. If that is the case, follow steps one through six below. If not, go right on to Take 2 on page 212.

Generally your topic sentence will include the word or phrase you are defining. Sentences that follow prove, illustrate, or support your definition. For example, you might have this as your topic sentence:

> Workaholics are obsessed with their work, have no work-home boundaries, and have few activities they do just for pleasure.

In the rest of the paragraph, readers expect you, the writer, to develop or explain the details of the definition in your topic sentence.

Step One: Choose a Topic

Often instructors assign a particular word or phrase for a definition paragraph. If, however, you have the freedom to choose what you define, remember that your purpose is to make clear what a specific word or phrase means to you.

To begin discovering some possible topics, complete Ticket to Write 10.2.

10.2 Creating Possible Topic Sentences

Directions: Complete the following to start the process of choosing a topic. Answers will vary.

Step One

1. On the left-hand side of your page, create a list of people who have special meaning for you—close friends, relatives, mentors, or even those with whom you don't get along. On the right-hand side of the page beside each person's name, write a word or phrase that describes your relationship with that person.
2. On the left-hand side of your page, create a list of places that have a special meaning (positive or negative) for you. Beside each place, on the right-hand side write a word or phrase that describes the significance of the place.
3. On the left-hand side of your page, create a list of things that have a special meaning (positive or negative) for you. Beside each thing, on

the right-hand side write a word or short phrase that describes the significance of the thing.

Step Two

1. Review each list. Then from each list, circle one topic (word or phrase) in the right-hand column that attracts your attention.
2. For each of the three topics you circled, write your unique definition. Each of these definitions is a topic sentence.

Heads Up!
Remember, you may fine-tune your topic sentence as you develop your paragraph.

Step Two: Generate Ideas

In Ticket to Write 10.2, you created three topic sentences (your definitions). Now complete Ticket to Write 10.3 to begin generating ideas that will help you select the one topic sentence you want to develop into a paragraph.

10.3 Generate Ideas

Directions: Write each of the three topic sentences you created in Ticket to Write 10.2 . Then, freewrite answers to reporter's questions (*Who? What? Why? Where? When?* and *How?*) to produce support for each of your definitions. Form your own reporter's questions, or use some or all of the following. Answers will vary.

1. ***Person . . .*** Topic sentence (definition): ______________________________

   ____

   a. Who is an example of this definition?
   b. What does this person do that exemplifies this definition?
   c. Why does this person exemplify your definition?
   d. When is this person an example of your definition?
   e. Where is this person an example of your definition?
   f. How is this person an example of your definition?

2. ***Place . . .*** Topic sentence (definition): ______________________________

   ____

   a. To whom is this place an example of your definition?
   b. What about this place exemplifies your definition?
   c. Why does this place exemplify this definition?
   d. Where in (what part of) this place is an example of your definition?
   e. When is this place an example of your definition?
   f. How is this place an example of your definition?

(continued)

3. ***Thing . . .*** Topic sentence (definition): ______________________________

____

a. To whom is this thing an example of your definition?
b. What about this thing exemplifies your definition?
c. Why does this thing exemplify this definition?
d. Where is this thing an example of your definition?
e. When is this thing an example of your definition?
f. How is this thing an example of your definition?

Step Three: Define Your Audience and Purpose

You've written three tentative topic sentences (your definitions), and you have answers to reporter's questions that you can develop into support for your definitions. Now is the time to think about your **audience** and **purpose**.

Your audience—and therefore your language—will differ, depending upon your purpose. For example, when defining the word *recession*, you probably would not use the same language for your economics professor as you would when defining it for your roommate when discussing paying rent.

10.4 Discovering Your Audience and Purpose

Directions: Write each topic sentence you composed in Ticket to Write 10.2. Under each topic sentence, write a separate sentence that details your audience, your purpose, and what you want to illustrate.

When finished, share your topics and your audience and purpose information with your writing group. Discuss your group members' topic sentences. Use your group's feedback on your topic sentences to select one to develop into a paragraph. Answers will vary.

Step Four: Draft Your Paragraph

Now it's time to start drafting your details, examples, or illustrations that will support your definition.

Use the details you generated in Ticket to Write 10.3 to build your paragraph. You may find you don't need to include all your responses to reporter's questions; likewise, you may discover other examples of support you hadn't thought of before. Keeping your audience and purpose in mind will help you decide which details you need most. Complete Ticket to Write 10.5 when you are ready to draft your paragraph.

10.5 Compose Supporting Details

Directions: Using the topic sentence you selected from Ticket to Write 10.4 and the supporting ideas you generated in Ticket to Write 10.3 and 10.4, compose supporting details that prove or illustrate your definition.

Write your topic sentence and the details that support it. Then share this with your writing group. Ask members if you have provided enough details so you fulfill the purpose you express in your topic sentence.

Step Five: Revise Your Paragraph

After you've finished your paragraph, save it, print a hard copy, and read it out loud, looking and listening for errors in grammar and content. Refer to the general Review Checklist for a Paragraph in Chapter 3, page 60, to make sure you revise as completely as possible.

☑ REVIEW CHECKLIST **for a Definition Paragraph**

- ☐ Is my topic sentence (definition) clear?
- ☐ Do my supporting examples and statements illustrate or explain my meaning of the word or phrase?
- ☐ Do I hear any fragments or run-ons when reading out loud?
- ☐ Do all my complete thoughts have appropriate end punctuation?
- ☐ Is my paragraph appropriate for my audience?
- ☐ Does my purpose remain constant?
- ☐ Does my definition need any fine-tuning?
- ☐ Do any sentences or ideas seem off-topic?

Heads Up!
See Chapter 20 to review fragments and run-ons.

Once you've made any necessary changes, save your work and leave it alone for a while. Then come back later, print a new copy, and look and listen for errors again. Refer again to the general Review Checklist for a Paragraph in Chapter 3.

Step Six: Peer Review

You might think you've found all of your errors and made all of the improvements you can, but having someone else read your paragraph and offer suggestions will almost always improve your writing. This is person is called your **peer reviewer**.

Heads Up!
See Chapter 4 for more on peer reviewing.

Your peer reviewer may use the review checklist above or the general Review Checklist for a Paragraph in Chapter 3, page 60, or your instructor may provide a different checklist. You might or might not agree with the suggestions you receive, but peer reviewers often find places for improvement or errors that slipped by you. Listen to or read closely your peer reviewers' suggestions, and don't be shy about asking questions to clarify their ideas and suggestions.

TICKET to WRITE

10.6 Peer Review

Directions: Once you have completed revising your paragraph, share it with your peer reviewers. Ask them to review your paragraph using the Review Checklist for a Definition Paragraph above and the general Review Checklist for a Paragraph from Chapter 3, page 60. Then ask them to record their suggestions for revision. Revise your paragraph again, using the suggestions with which you agree.

Student and Professional Essays

LEARNING GOAL
3 Read and examine student and professional essays

Below are two definition essays. The first was written by a student and the second by a professional. Read these two essays and answer the questions that follow them.

Student Essay

Some Go Too Far

by Hassan Jordan

(1) All my friends are interested in sports. Some like team sports, including football, baseball, hockey, soccer, and basketball. Others prefer individual sports like skiing, bodybuilding, and golf. Most of them like sports because of the entertainment it gives them. Some like their sports for health reasons. But a few of them go too far. These people cross the line from sports fans to sports fanatics.

(2) A sports fanatic can be a person who spends too much money on a sport. This person might invest in equipment he or she says is necessary for the sport. For instance, a backpacking fanatic who could get by with a $50 sleeping bag might instead buy one that costs three times that much. A football fanatic might spend money he doesn't have on tickets to "the big game" even though that game might be on television. Eugenios, my neighbor, is a huge Iowa State fan, and has spent thousands on his **tailgating gear**. Eugenios recently bought a six-burner gas grill and new red-and-gold patio furniture in matching colors. Eugenios **spares no expense** when it comes to tailgating.

cross the line an idiom meaning "to behave in a socially unacceptable way"
tailgating gear cooking equipment for food served before a sports event; the food is often prepared and served from the tailgate of a vehicle
spares no expense an idiom meaning "does not care about the cost of something"

His wife told me that they are looking into buying an **RV** just so Eugenios has a place to stay dry and warm if it rains or snows at one of the games.

3 My friend Thom is another type of sports fanatic, one who spends way too much time watching or playing a sport. Thom devotes his entire weekend and many weeknights to watching as many baseball games as he can. He has the television turned to one game after another, and he often clicks over to watch a different game when a commercial comes on. He also **TiVos** games that are on at the same time. If he's not watching a baseball game, he's watching ***Sports Center*** or working on his **fantasy baseball team**. Last year Thom invited a few friends over to watch the final **AL** playoff. He made us all feel welcome, but he could have been just as content to watch the game alone. He also sacrifices time with his family just to watch a game.

4 Have you ever met one of those people who **puts somebody on a pedestal**? These sports fanatics are the worst because they think if someone can knock a few home runs or lead the league in rebounds, then that person is worthy of complete admiration. These fanatics believe their sports heroes can do no wrong. Even when the sports stars are **outed** for bad behavior, the fanatics still think their idols are above criticism and forgive them easily.

5 Sports are an important part of our American society. Following our favorite sport, team, or athlete can be a fun and harmless passion. But, too often, some take their sports too seriously and change from fans to fanatics.

RV the abbreviation for "recreational vehicle," such as a camper or motor home

TiVos uses TiVo, a company that provides a method to digitally record television programs

Sports Center an ESPN television sports program that offers highlights, analyses, and news of major sporting events

fantasy baseball team an imaginary baseball team made up of real-life players from different teams

AL the abbreviation for "American League," a division of teams in professional baseball

puts somebody on a pedestal an idiom meaning "thinks a certain person is more important than others"

outed disclosed or revealed, especially in a negative way

A CLOSER LOOK

Answer these questions about the essay:

1. What phrase does the student writer define?

   sports fanatic

2. Double underline the thesis statement of this essay.

3. List the three types of fanatics the writer discusses.

   A sports fanatic is someone who spends too much money, who spends too much time playing or watching a sport, or who puts sports figures on pedestals.

4. In the second sentence of paragraph 3, Jordan uses the word *many* twice. Reword this sentence to eliminate one use of the word. Answers will vary.

5. In paragraph 3, underline the sentence that needs to be either eliminated or developed further.

6. Which sentence in paragraph 4 unnecessarily restates an idea and can be eliminated without changing the meaning of the paragraph?

   These fanatics believe their sports heroes can do no wrong.

7. Which sentence in the conclusion reiterates the writer's thesis statement?

   But, too often, some take their sports too seriously and change from fans to fanatics.

8. In paragraph 2, the fifth, sixth, and seventh sentences all begin the same way. Rewrite these sentences, rewording them for sentence variety. Answers will vary.

9. Paragraph 4 mentions sports heroes who, in the eyes of fans, "can do no wrong." Think of a sports hero who fits in this category and write at least two sentences that give support to this example.

   Answers will vary.

10. Write a description of a sports fanatic you know who fits one of the categories in Hassan's essay.

    Answers will vary.

The professional essay below offers a unique definition of the word *wife*.

Professional Essay

I Want a Wife

by Judy Brady

(1) I belong to that classification of people known as wives. I am A Wife. And, not altogether **incidentally**, I am a mother.

(2) Not too long ago a male friend of mine appeared on the scene fresh from a recent divorce. He had one child, who is, of course, with his ex-wife. He is looking for another wife. As I thought about him while I was ironing one evening, it suddenly occurred to me that I too, would like to have a wife. Why do I want a wife?

(3) I would like to go back to school so that I can become **economically independent**, support myself, and if need be, support those dependent upon me. I want a wife who will work and send me to school. And while I

incidentally apart from the main subject
economically independent not reliant on another for money

am going to school I want a wife to take care of my children. I want a wife to keep track of the children's doctor and dentist appointments. And to keep track of mine, too. I want a wife to make sure my children eat properly and are kept clean. I want a wife who will wash the children's clothes and keep them mended. I want a wife who is a good nurturing **attendant** to my children, who arranges for their schooling, makes sure that they have an adequate social life with their **peers**, takes them to the park, the zoo, etc. I want a wife who takes care of the children when they are sick, a wife who arranges to be around when the children need special care, because, of course, I cannot miss classes at school. My wife must arrange to lose time at work and not lose the job. It may mean a small cut in my wife's income from time to time, but I guess I can tolerate that. Needless to say, my wife will arrange and pay for the care of the children while my wife is working.

4 I want a wife who will take care of my physical needs. I want a wife who will keep my house clean. A wife who will pick up after my children, a wife who will pick up after me. I want a wife who will keep my clothes clean, ironed, mended, replaced when need be, and who will see to it that my personal things are kept in their proper place so that I can find what I need the minute I need it. I want a wife who cooks the meals, a wife who is a good cook. I want a wife who will plan the menus, do the necessary grocery shopping, prepare the meals, serve them pleasantly, and then do the cleaning up while I do my studying. I want a wife who will care for me when I am sick and sympathize with my pain and loss of time from school. I want a wife to go along when our family takes a vacation so that someone can continue to care for me and my children when I need a rest and **change of scene**.

5 I want a wife who will not bother me with rambling complaints about a wife's duties. But I want a wife who will listen to me when I feel the need to explain a rather difficult point I have come across in my course of studies. And I want a wife who will type my papers for me when I have written them. I want a wife who will take care of the details of my social life. When my wife and I are invited out by my friends, I want a wife who takes care of the babysitting arrangements. When I meet people at school that I like and want to entertain, I want a wife who will have the house clean, will prepare a special meal, serve it to me and my friends, and not interrupt when I talk about things that interest me and my friends. I want a wife who will have arranged that the children are fed and ready for bed before my guests arrive so that the children do not bother us. I want a wife who takes care of the needs of my guests so that they feel comfortable, who makes sure that they have an ashtray, that they are passed the **hors d'oeuvres**, that they are offered a second helping of the food, that their wine glasses are **replenished** when necessary, that their coffee is served to them as they like it. And I want a wife who knows that sometimes I need a night out by myself.

6 I want a wife who is sensitive to my sexual needs, a wife who makes love passionately and eagerly when I feel like it, a wife who makes sure that I am satisfied. And, of course, I want a wife who will not demand sexual attention when I am not in the mood for it. I want a wife who assumes the complete responsibility for birth control, because I do not want more children. I want

attendant someone who waits on another

peers people with equal standing with others, as in rank, class, or age

change of scene an idiom for "a new place to go"

hors d'oeuvres appetizers served before a meal

replenished refilled, stocked up

a wife who will remain sexually faithful to me so that I do not have to clutter up my intellectual life with jealousies. And I want a wife who understands that my sexual needs may entail more than strict **adherence** to **monogamy**. I must, after all, be able to relate to people as fully as possible.

7 If, **by chance**, I find another person more suitable as a wife than the wife I already have, I want the liberty to replace my present wife with another one. Naturally, I will expect a fresh, new life; my wife will take the children and be solely responsible for them so that I am left free.

8 When I am through with school and have a job, I want my wife to quit working and remain at home so that my wife can more fully and completely take care of a wife's duties.

9 My God, who wouldn't want a wife?

adherence observance

monogamy the practice of having a single sexual partner for a period of time

by chance an idiom meaning "accidentally"

A CLOSER LOOK

Answer these questions about the essay:

1. What is the word Brady is defining?

   wife

2. Brady implies that the word *wife* is a synonym for *maid* and *cook*. List at least three more of Brady's implied synonyms for the word *wife*.

   Answers may vary. Possible answers: servant, housekeeper, secretary, personal assistant, nanny, driver, laundress, waitress, hostess, sex slave

3. Readers must infer Brady's thesis statement. How would you word her thesis statement?

   Answers may vary. Possible answer: The role a wife plays in a marriage unfairly has more responsibilities and is depended upon more than the role a husband plays.

4. Explain Brady's tone.

   Answers may vary. Possible answers: She is humorous and satirical (mocking). She pokes fun at the men (and women) who hold this sexist and outdated view of wives.

5. Brady repeats the phrase "I want a wife who" twenty-four times. How does this help her define her word?

   Answers may vary. Possible answers: Brady adds to her definition of the word *wife* each time she starts a sentence. In doing this, she emphasizes how many tasks a wife is expected to perform. She repeats this phrase to humorously make her point.

6. What is the effect Brady hopes to accomplish with the repetition of this phrase and other places where she repeats the pronouns *I* and *me*?

   Possible answer: She is demonstrating the selfishness of those who hold the view that wives are solely responsible for the duties she lists.

7. What types of responsibilities does Brady say she would like for her wife to take on?

   Answers may vary. Possible answers: responsibilities for the children's needs, household needs, personal needs, social needs, and sexual needs

8. Underline the topic sentence of paragraph 6.
9. This essay was written forty years ago. What beliefs or observations expressed in the essay are no longer relevant?

   Answers will vary.

10. Why does Brady ask the question, "My God, who wouldn't want a wife?" in the last line of the essay?

    Answers will vary.

Writing Your Definition Essay

Essay writing is much like paragraph writing: you start with a topic and expand on it with details. Remember that just as a paragraph supports a topic sentence, an essay supports a thesis statement.

LEARNING GOAL
4 Write a definition essay

Step One: Choose a Topic and Develop a Working Thesis Statement

To begin your definition essay, ask yourself these **starter questions**:

See Chapter 2 to review thesis statements.

1. **What** word or phrase do I want to define?
2. **What** is my particular definition?
3. **Why** do I want to define this, or **what** is my purpose?
4. **What** method of defining will work well for me to prove my definition?

"Write to be understood, speak to be heard, read to grow."
—Lawrence Clark Powell

Here is how Micah answered the starter questions in preparing to write her definition essay on the word *dedication*:

1. **What** word or phrase do I want to define? Dedication
2. **What** is my particular definition? Dedication is unwavering support of others.

3. **Why** do I want to define this, or **what** is my purpose? **My purpose is to show what some local volunteers do.**
4. **What** method of defining will work well for me? **I will use details, illustrating the unique qualities that characterize people who are dedicated to something or someone.**

Answering Starter Questions
Hassan Jordan and Judy Brady

In answering the starter questions, Hassan Jordan, the author of "Some Go Too Far," wrote:

1. I want to define the phrase sports fanatic.
2. My particular definition is sports fanatics are those who go too far and take their sports too seriously.
3. My purpose is to show what activities define sports fanatics.
4. To define my phrase, I will use the method of offering examples that illustrate my definition.

Judy Brady, the author of "I Want a Wife," might have addressed the starter questions this way:

1. I want to define what many wrongly believe the word wife means.
2. The particular (and false) definition I am showing is that wives are servants to their husbands.
3. My purpose is to show the unfair expectations often placed on married women.
4. To define my word, I will use the method of listing some common duties usually assigned to wives.

10.7 Answer Starter Questions and Complete Starter Statement for a Definition Essay

Directions: All essays have thesis statements. To help develop a working thesis statement for your definition essay, begin by answering the following starter questions. Answers will vary.

1. What word or phrase do I want to define? ____________________
2. What is my particular definition? ____________________
3. What is my purpose? ____________________
4. To define my word or phrase, I will use the method of ____________________.

Once you've answered the starter questions, you have the beginnings of a working thesis statement. (The second starter question can be especially helpful here.) Because this is a working thesis statement for a definition essay, the sentence you compose should

- identify the specific word or phrase you are defining
- offer your particular definition for the word or phrase

After reviewing her starter question responses, Micah wrote the following working thesis statement:

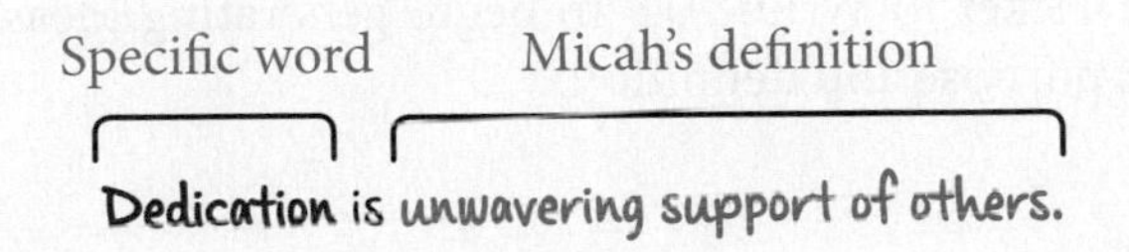

Here is how two other students phrased their working thesis statements:

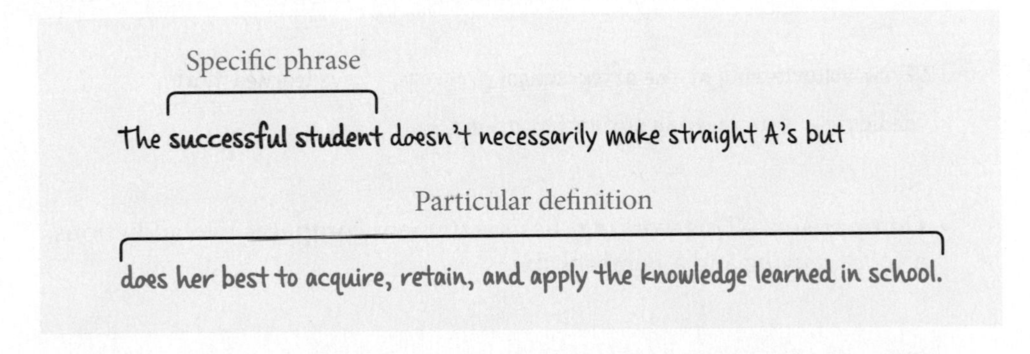

Specific word

In my family of seven, privacy means spending ten full minutes in the bathroom, eating a meal without someone getting angry, and watching a TV show without interruptions.

Particular definition

Once you've filled in the starter statement, you've begun to develop the focus of your essay. Return to your starter statement and consider your topic and your purpose. From this starter statement, create your **working thesis statement** (your word or phrase and definition).

10.8 Develop a Working Thesis Statement

Directions: Looking at your starter statement, consider the purpose for the term or phrase you are defining. Then create your working thesis statement. Share this with members of your writing group and ask them if they understand your particular definition and why it is important to you.

Step Two: Generate Ideas

See Chapter 1 to review activities and techniques for prewriting.

In Ticket to Write 10.8, you created a working thesis statement (your word or phrase and definition). Now complete Ticket to Write 10.9 to begin generating ideas to help support and illustrate your purpose and definition.

Type of Support

- **Details**—Micah will use **details** to illustrate and support her definition of *dedication* in her working thesis statement.

From volunteering at the after-school program, I have learned that dedication is unwavering support of others.

- **Comparison**—This working thesis statement **compares** two addictions, a *coffee craving* and a *Facebook fix*.

My spouse's morning coffee craving is an addiction, just like my daily need for a Facebook fix.

- **Contrast**—This working thesis statement defines its phrase, *the successful student*, with **contrast**. In this case, the writer begins by telling us what a successful student is not:

The successful student doesn't necessarily make straight A's but does her best to acquire, retain, and apply the knowledge learned in school.

- **Background**—In this working thesis statement, the writer gives background information to begin showing why the definition of *privacy* changed after the addition of triplets.

> Since adding triplets to our family of four, privacy means spending ten minutes alone in the bathroom, eating a meal without someone getting angry, and watching a TV show without interruptions.

10.9 Generate Ideas

Directions: Write the word or phrase you are defining, your working thesis statement, and any ideas you generated. Identify the type of support you will use, and freewrite answers to the reporter's questions (*Who? What? Why? Where? When?* and *How?*) that may produce support for your definition. Then share this with your writing group, and ask them to note which ideas seem the most compelling or which elaborate the most on the information in your working thesis statement. Answers will vary.

Step Three: Define Your Audience and Purpose

Your next step is to identify the audience for your essay. To decide this, consider who would benefit from learning your definition. Review your working thesis statement and your purpose.

After brainstorming on an appropriate audience for her topic of *dedication*, Micah decided the following:

> **Working Thesis Statement:** From volunteering at the after-school program, I have learned that dedication is unwavering support of others.
>
> **Purpose:** to show what local volunteers do
>
> **Audience:** people who aren't aware of the commitment some make to helping others

10.10 Define Your Audience

Directions: Write your working thesis statement, your purpose, and your intended audience. Share this with members of your writing group and ask if they agree that your working thesis statement will be satisfactory for your audience. If they do not agree, consider revising your working thesis statement.

Step Four: Draft Your Essay

To begin your definition discovery draft, look back at the ideas you generated. Consider which type(s) of support (*details, comparison, contrast,* or *background*) best supports your definition and accomplishes your purpose.

In generating ideas, you may have discovered information you now find you don't need. As you progress, you may discover other ideas that support your definition. Keeping your audience and purpose in mind as you compose your discovery draft will help you decide which details you need most.

Micah's method, shown in her working thesis statement, is to use details to define *dedication*. Micah reviewed all her prewriting and then used two topics and their supporting details when she composed her discovery draft.

Working Thesis Statement: From volunteering at the after-school program, I learned that dedication is unwavering support of others.

Topic: emotional support from volunteers

Detail: spend one-on-one time
Detail: listen and don't judge
Detail: are nurturing, always show patience

Topic: physical support from volunteers
Detail: drive kids different places
Detail: read and study with kids
Detail: play games with kids

When you are ready to compose your discovery draft, complete Ticket to Write 10.11.

10.11 Write Your Discovery Draft

Directions: From the material you compiled in Ticket to Write 10.8 through 10.10, write your discovery draft. Refer to Chapter 2 to review discovery drafting.

When you have completed your discovery draft, share your work with members of your writing group. Ask them to note the following about your essay:

1. Examples or details of support that are accurate and effective
2. Examples or details of support that are vague, unclear, or need more development

3. Other examples, details, or information you could use but haven't included

Review suggestions and consider where you wish to revise.

Step Five: Organize Your Essay

Sometimes looking at a subject from a different point of view can give you new insight or make you aware of different ideas. To discover all you can about your topic, take a different look at what you created in your drafting by splitting the draft into its parts. By doing this, you might find places where you can develop your supporting details and improve readers' understanding of your definition.

Below is a chart Micah used to help her take a different view of her definition essay discovery draft. Making this chart helped her organize ideas by

- showing particular points that need more details
- helping her see which details best illustrate her points
- letting her reexamine the order of details within body paragraphs

Thesis statement	From volunteering at the after-school program, I have learned that dedication is unwavering support of others.	
Topics	Emotional	Physical
Details of support	spend one-on-one time "soothing sessions" with kids who need someone to hear their problems tell stories to little ones always interested in what happened during day	drive kids different places take home when ride doesn't show field trips take across street to get ice cream
	listen and don't judge hear lots of problems never criticize	read and study with kids read books aloud math sessions homework help

(continued)

nurturing and always patient	play games with kids
kind voice	basketball
How can I help? attitude	board games
tolerant of juvenile outbursts	Wii

Complete Ticket to Write 10.12 to review support and development in your body paragraphs.

10.12 Organize Your Essay

Directions: Write your definition essay working thesis statement. Under that, make a chart of your essay's body paragraphs. Review the chart and make any changes as needed. Look for types of support you have used—details, contrast, comparison, and examples or illustrations—that need more development. Where needed, develop more details to complete the body paragraphs.

Step Six: Apply Critical Thinking

Critical thinking means thinking and then rethinking about a topic, with the goal of discovering all the implications you can about it. In thinking critically about a definition essay, you often do the following:

> Reflective thinking is examining decisions you've made or beliefs you've adopted. You consider both what sustains your decisions or beliefs and what conclusions you can draw from them.

- reflect on the importance of your particular definition
- reflect on the relationship between your definition and its supporting details, examples, and information
- analyze the subtle or obvious differences between your definition and traditional definitions of the word or phrase
- reflect on the relevance of details you provide

The following critical thinking questions are important when you write a definition essay. After each of your drafts, ask yourself these questions and then apply your answers to your writing.

> Critically thinking about specific details means analyzing how accurately they support the definition.

1. **Purpose:** Will readers understand why my definition is important? Does my thesis clearly state my definition? Will readers understand the connection between my particular definition and my supporting evidence?
2. **Information:** Are all my details accurate? Have I given readers enough details, examples, or illustrations to understand my specific definition? Do I see any place to add details that will enhance or further support my definition?
3. **Reasoning:** Are my details appropriate to my topic? Is my reasoning logical?

4. **Assumptions:** Have I made any assumptions that need to be explained or justified? Have I used any vocabulary my audience may not understand?

10.13 Apply Critical Thinking

Directions: After you have revised for organization and critical thinking, ask members of your writing group to read your latest draft and answer the questions below.

Purpose

1. Do you understand why my definition is important?

Information

2. Do you feel you read enough details to understand my particular definition?
3. Do you see any places where adding more support would further explain or clarify my definition of this word or phrase? If so, where should I add support and what kind of support do you suggest?

Reasoning

4. Do you feel all my facts are accurate? If not, where should I provide more accurate details?
5. Are all my facts clear? If not, where should I change my wording to be clearer?
6. Are all my details relevant to my definition? If not, where do I stray from my topic?
7. Is my reasoning logical? If not, where do I seem to be illogical?

Assumptions

8. Have I made any assumptions that need to be explained or justified? If so, what are these assumptions?

Step Seven: Revise Your Essay

Getting your essay on paper is just your first step. Revising through subsequent drafts shapes your discovery draft into a finished product of which you can be proud.

Consult the Review Checklist for an Essay in Chapter 3, page 61, and look for places in your essay you need to polish. Because your goal is to convince your readers of your definition's accuracy and legitimacy, you may want to focus revising efforts on your *topics*, *details*, and *reasoning*.

Heads Up!
Keep hard copies of all your drafts and number them so you'll know which is your latest draft.

Topics

Check that each body paragraph focuses on one specific topic.

- Does the topic in each body paragraph illustrate or explain some facet of your definition?
- Does each topic logically connect to the topics before and after it?

Details

When you revise, review your details.

- Have you given enough examples or illustrations in each body paragraph to support its topic sentence?
- Does each detail focus on the topic of its paragraph?
- Should any detail be eliminated or revised because it's unrelated to the topic?

Reasoning

Look at the reasoning you have presented for your definition and its support; then consider if your audience will think your essay is believable.

- Is your definition reasonable?
- Is each paragraph's topic logically linked to your definition?
- Are your examples or illustrations sensible?

Complete Ticket to Write 10.14 to review your essay's topics, supporting details, and reasoning.

10.14 Revise Your Essay—Peer Review

Directions: After your latest revision, ask members of your writing group to read your draft and answer the questions below.

Topics

1. Does each of my topics illustrate or explain some facet of my definition? If not, which ones need revising?
2. Does each of my topics logically connect to the topics before and after it? If not, which ones need revising?

Details

3. Have I given enough examples or illustrations in each body paragraph to support the topic sentence? If not, which topic needs more focus?
4. Do each of my details focus on the topic of its paragraph? If not, which needs more support?
5. Do I need to eliminate or revise any detail because it's unrelated to the topic? If so, which ones need eliminating and which need revising?

Reasoning

6. Is my definition reasonable? If not, what do you suggest to improve it?
7. Is each paragraph's topic logically linked to my definition? If not, which topic needs revision?
8. Are my examples and illustrations sensible? If not, which ones need revising?

Writing Assignment

MyWritingLab™

Complete this Exercise

Consider the words or phrases below or one that your instructor gives you. Choose one and expand it into a definition essay.

1. Slang
2. Victory
3. Defeat
4. Good music
5. Faith
6. Classwork that deserves an A
7. Thoughtlessness
8. Ideal working conditions
9. Healthy relationship
10. Unhealthy relationship

TECHNO TIP

For more on definition writing, search the Internet for these videos:

Definition Intro 1

CM-001-Definition Essay

MyWritingLab™ Visit *MyWritingLab.com* and complete the exercises and activities in the **Paragraph Development-Definition** and **Essay Development-Definition** topic areas.

RUN THAT BY ME AGAIN

- **Your purpose in a definition paragraph or essay is . . .** to tell and show your audience what a specific word or phrase means to you.
- **To explain meaning, definition paragraphs and essays use . . .** details, comparison, contrast, or background.

LEARNING GOAL
❶ Get started with definition writing

- **For a definition paragraph, instructors often assign . . .** a particular word or phrase.
- **In a definition paragraph, your audience and your language will differ, depending upon . . .** your purpose.
- **When you have completed your definition paragraph, check to see that your supporting examples and statements . . .** illustrate or explain your meaning of the word or phrase.
- **When you receive peer reviewers' suggestions . . .** listen to or read them closely, and don't be shy about asking questions to clarify them.

LEARNING GOAL
❷ Write a definition paragraph

- **For a definition essay, starter questions help lead you . . .** to composing a working thesis statement.
- **Your working thesis statement identifies . . .** the specific word or phrase you are defining and offers your particular definition for the word or phrase.
- **To decide your audience, consider . . .** who would benefit from learning your definition.
- **To discover all you can about your topic . . .** take a different look at what you created in your drafting by splitting the draft into its parts.
- **For critical thinking about your definition essay, focus on . . .** purpose, information, reasoning, and assumptions.
- **When revising your definition essay, focus on . . .** topics, details, and reasoning.

LEARNING GOAL
❸ Read and examine student and professional essays

DEFINITION WRITING LEARNING LOG MyWritingLab™

MyWritingLab™

Answer the questions below to review your mastery of definition writing. Answers will vary.

1. To be sure your audience understands your unique meaning, your definition paragraph or essay should accomplish what three tasks?
   It should identify the word or phrase you're defining, tell what the word or phrase means to you, and provide supporting examples.
2. How are details used in definition paragraphs or essays?
   Details define a word or phrase by focusing on its unique qualities.
3. How is comparison used in definition paragraphs or essays?
   Comparison defines a word or phrase by showing its similarities to other words or phrases.
4. How is contrast used in definition paragraphs or essays?
   Contrast defines a word or phrase by showing how it is unlike its commonly accepted definition.
5. How is background used in definition paragraphs or essays?
   Background defines a word or phrase by discussing its history.
6. In a definition paragraph, where will you generally place the word or phrase you are defining?
   You will generally place it in your topic sentence.
7. What are starter questions for a definition essay?
   (1) What word or phrase do I want to define? (2) What is my particular definition? (3) Why do I want to define this, or what is my purpose? (4) What method of defining will work well for me to prove my definition?
8. To decide who your audience is, what should you consider?
   Consider who would benefit from learning your definition.
9. In discovering all you can about your topic, why is splitting your draft into its parts often helpful?
   You might find places where you can develop your supporting details to improve readers' understanding of your definition.

10. In applying critical thinking to a definition essay, what do you often do?

Reflect on the importance of your particular definition; reflect on the relationship between your definition and its supporting details, examples, and information; analyze the subtle or obvious differences between your definition and traditional definitions of the word or phrase; provide precise details; reflect on the relevance of details you provide.

11. When revising your essay, you should keep your goal in mind. What is your goal in a definition essay?

Your goal is to convince your readers of your definition's accuracy and legitimacy.

CHAPTER 11 Compare-and-Contrast Writing

LEARNING GOALS

In this chapter, you'll learn and practice how to

1. Get started with compare-and-contrast writing
2. Write a compare-and-contrast paragraph
3. Read and examine student and professional essays
4. Write a compare-and-contrast essay

GETTING THERE

- Compare-and-contrast writing helps readers gain a deeper understanding and appreciation of two subjects through analyzing the relationships between them.
- Compare-and-contrast writing allows you to examine the advantages and disadvantages of two subjects.
- Compare-and-contrast writing helps you discover the unique similarities and differences between two subjects.

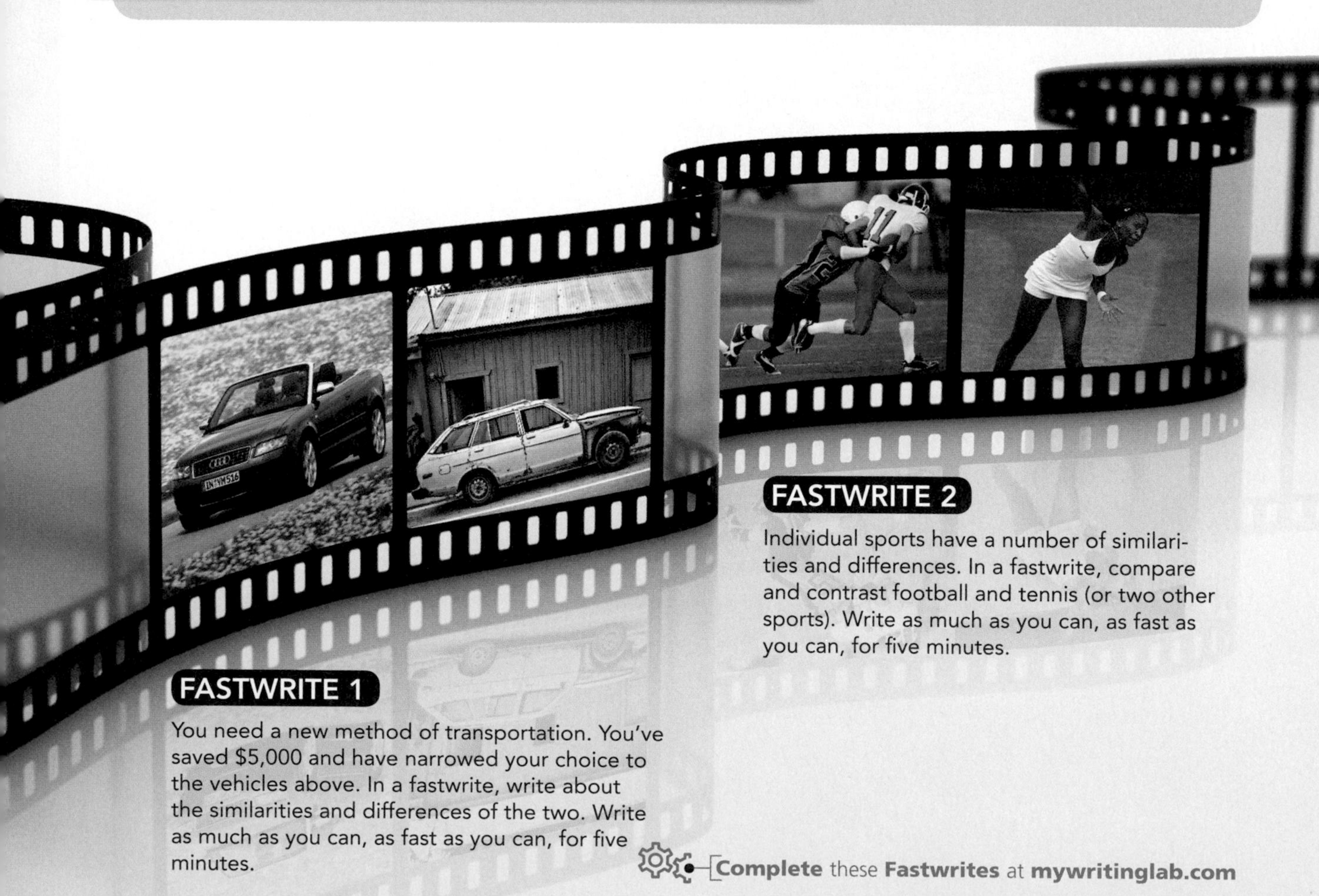

FASTWRITE 1

You need a new method of transportation. You've saved $5,000 and have narrowed your choice to the vehicles above. In a fastwrite, write about the similarities and differences of the two. Write as much as you can, as fast as you can, for five minutes.

FASTWRITE 2

Individual sports have a number of similarities and differences. In a fastwrite, compare and contrast football and tennis (or two other sports). Write as much as you can, as fast as you can, for five minutes.

Complete these **Fastwrites** at **mywritinglab.com**

Getting Started in Compare-and-Contrast Writing

In its simplest terms, compare-and-contrast writing details the *similarities* (the *comparisons*) and the *differences* (the *contrasts*) between two people, places, or things. Beyond reporting the similarities and differences of two subjects, compare-and-contrast writing may propose that one subject is superior to the other in specific ways, that both subjects are equal in importance, or that one subject is far less important.

LEARNING GOAL

❶ Get started with compare-and-contrast writing

Compare-and-contrast writing does the following:

- uses a particular pattern of organization
- highlights specific qualities or conditions
- offers a clearer understanding of two subjects

Compare-and-contrast writing can be objective and subjective—you're relating factual information about two subjects, and often you're relating your impressions of that information.

You make comparisons and contrasts every day. You might compare and contrast

- two cell phone service plans
- an associate degree in science and one in arts
- an in-home babysitter and a daycare facility

When you explore the similarities and differences between two subjects, you and your readers to gain a deeper understanding of the two.

Patterns of Development: Alternating and Divided

You may write compare-and-contrast essays or paragraphs using one of two patterns of organization: **alternating** pattern or **divided** pattern.

An **alternating pattern** is also known as a ***point-by-point*** pattern.

In the **alternating** pattern for essays or paragraphs, you compare and contrast the two subjects, point for point. Here is Brittney's paragraph that uses an alternating pattern:

Zippy Cell and Quick Call cell phone service plans are similar, but Quick Call is the right choice for me. Both require a two-year contract and have comparable monthly fees. Zippy's calling plan is $59.99 per month, compared to Quick Call's $49.99 per month. Zippy Cell's plan includes unlimited weeknight minutes after 8 p.m. and unlimited weekend minutes. Likewise, Quick Call has some unlimited night and weekend minutes, but its nights start at 9:00 p.m. Texting options with Zippy are fairly limited at either $13 for up to 1,000 messages per month

(continued)

or $24 for unlimited texting. The three texting plan options with Quick Call are $6 for 300 messages, $12 for 900 messages, or $26 for unlimited texting. Because I text more than I talk on the phone, I've opted for Quick Call with unlimited texting.

Outlining your paragraph sentence-by-sentence gives you a view of the supporting details in an alternating pattern and is a good method for checking the paragraph for balance.

B = both subjects Z = Zippy Q = Quick Call

B—Zippy Cell and Quick Call cell phone service plans are similar, but Quick Call is the right choice for me.

B—Each requires a two-year contract with comparable monthly fees.

Z—Zippy's calling plan is $59.99 per month compared to

Q—Quick Call's $49.99 per month.

Z—Zippy Cell's plan includes 1,000 anytime minutes and unlimited weeknight minutes starting at 8:00 p.m. and unlimited weekend minutes.

Q—Likewise, Quick Call has some unlimited night and weekend minutes, but its nights start at 9:00 p.m. and it offers only 900 anytime minutes.

Z—The texting options with Zippy are fairly limited at either $13 for up to 1,000 messages per month or $24 for unlimited texting.

Q—The three texting plan options with Quick Call are $6 for 300 messages, $12 for 900 messages, or $26 for unlimited texting.

B—Because I text more than I talk on the phone, I've opted for Quick Call with unlimited texting.

Sentence-by-sentence examination highlights the format. This lets you check for balance and find places that need more support.

A **divided pattern** is also known as a *side-by-side* pattern or *whole-to-whole* pattern.

Here is the same comparison and contrast in a divided pattern:

Zippy Cell and Quick Call cell phone service plans are similar, but Quick Call is the right choice for me. Adopting a Zippy Cell service plan requires a two-year contract at $59.99 per month. This plan includes 1,000 anytime minutes and unlimited weeknight minutes starting at 8:00 p.m. and unlimited weekend minutes. The texting options with Zippy are fairly limited at either $13 for up to 1,000 messages per month or $24 for unlimited texting. The Quick Call cell phone service also requires a two-year contract but is only $49.99 per month. Though Quick Call's plan has only 900 anytime minutes, it offers some unlimited weeknight minutes starting at 9:00 p.m. and unlimited weekend minutes. Subscribers also have three texting plan options to choose from: $6 for 300 messages, $12 for 900 messages, or $26 for unlimited texting. Because I text more than I talk on the phone, I've opted for Quick Call with unlimited texting.

Using a sentence-by-sentence outline gives you a view of the supporting details in a divided pattern.

B = both subjects   Z = Zippy   Q = Quick Call

B—Zippy Cell and Quick Call cell phone service plans are similar, but Quick Call is the right choice for me.

Z—Adopting a Zippy Cell service plan requires a two-year contract at $59.99 per month. This plan includes 1,000 anytime minutes and unlimited weeknight minutes starting at 8:00 p.m. and unlimited weekend minutes. Texting options with Zippy are fairly limited at either $13 for up to 1,000 messages per month or $24 for unlimited texting.

Q—The Quick Call cell phone service also requires a two-year contract but is only $49.99 per month. Though Quick Call's plan has only 900 anytime minutes, it offers unlimited weeknight minutes starting at 9:00 p.m. and unlimited weekend minutes. Subscribers also have three texting plan options to choose from: $6 for 300 messages, $12 for 900 messages, or $26 for unlimited texting.

B—Because I text more than I talk on the phone, I've opted for Quick Call with unlimited texting.

Sentence-by-sentence examination highlights the format. This lets you check for balance and find places that need more support.

"Don't try to figure out what other people want to hear from you; figure out what you have to say. It's the one and only thing you have to offer."
—Barbara Kingsolver

In the last sentence of the paragraph, Brittney explains why, for her, Quick Call is the better choice.

Using Venn Diagrams

Compare-and-contrast paragraphs and essays assist readers in understanding the differences and similarities between two subjects. Often, this allows readers a deeper understanding of each subject and a better understanding of each subject's significance.

In the topic sentence or thesis statement, you need to make clear

- *what* two subjects you are comparing and contrasting
- *why* their similarities or differences are important

Heads Up!
This is your purpose.

Keegan has this topic sentence:

While an associate's degree in science and one in arts both appeal to me, the degree in arts is better for me.

You can prewrite about areas of similarities and differences in several ways. In compare-and-contrast writing, some find that creating a Venn diagram is helpful in prewriting and in organizing the structure of their paper. A **Venn diagram** is a drawing of two intersecting circles; the circles note features that are either unique or common to two concepts.

In exploring which associate degree to pursue, Keegan created this Venn diagram:

Associate in Science
- High demand for most jobs
- Several classes not offered online
- High school science grades were B's & C's

both
- Will take 2 yrs @ full-time status
- Transfer to 4-year college
- Have interest in nursing
- Have interest in graphic arts

Associate in Arts
- Current job demand not strong
- Most classes are online
- Made A's in high school arts classes

11.1 Brainstorming with a Venn Diagram

Directions: Spend three to five minutes creating a Venn diagram that compares and contrasts your life as a fifteen-year-old student with your life as a student today. Use as many details as possible to note the similarities and differences. When you're finished, share your Venn diagram with your writing group. As a group, discuss each Venn diagram.

WRITING at WORK: *Snapshot of a Writer*

Deputy Chief Earl Brandon, Lt. Chip Stauffer

With over fifty years' experience between them, Deputy Chief Earl Brandon (left) and Lt. Chip Stauffer (right) are two police officers who know the importance of writing well. While many might think the job of a police officer is constantly filled with exciting action, that job is also filled with the action of writing. As Brandon and Stauffer note, precise and detailed writing is a crucial part of their job to protect and serve.

Police Writing Process

"Our goal is to keep people safe," says Deputy Chief Earl Brandon, "and to do this, we have to arrest criminals, keep them arrested, and get them convicted." Yet a lot goes into the process of arresting those who break the law, and part of that process is writing. The patrolman's and the investigator's writing is crucial in arresting and prosecuting criminals. Officers enjoy the excitement of getting criminals off the street, but as Lt. Chip Stauffer says, "Fifteen minutes of fun is two hours of paperwork to law enforcement."

Clarity The patrolman's report has to tell a story that anyone can pick up, read, and then visualize about what was taking place. Officers have to be careful with pronouns, and maybe follow each pronoun with a name or identifier. Also, verb tenses and order of events have to be accurate so readers know what happened, when it happened, and what actions follow what other actions. An office or a detective should always review and revise a report. Reviewing and revising are equally important in turning in the clearest report possible.

Description/Details An officer's notes should be as detailed as possible. Being vague or ambiguous won't help anyone—especially that officer. Always asking *who*, *what*, *why*, *where*, *when*, and *how* can help fill in the details. Not only should officers record names, addresses, jobs, and personal information, they should also note all that's going on around them.

When it is time for an officer to type up a report, the notes should be laid out chronologically. Then the report should be typed, proofread, revised, checked for accuracy, reviewed, and revised again. We stress repeatedly checking for accuracy throughout the revision process.

Following this process of drafting will help an officer remember the report. For example, an officer may have responded to a traffic accident, made an arrest, taken a suspect to jail, cleared a car wreck, and then responded to another crime or two before he or she can get back to the office to write that first report from his or her shift.

Accuracy Details are vital. They are what make a field report or a criminal investigation report clear. The more detailed your report, the more accurate your report. The more accurate your report, the better that report will serve you, your superiors, the investigation team, and possibly the prosecutor's office. There's nothing more embarrassing than going to trial and having a defense attorney question your report because it wasn't clear.

(continued)

A training video we've used shows Officer Buck Savage—a police officer whose reports are scrawled on the backs of matchbook covers and always lack detail—get hammered by a defense attorney because his report is one sentence on a scrap of paper.

Note-Taking When you can't remember what happened with a specific incident, you need your notes to help. If you don't have good notes, you open yourself up to questions about your report's reliability. Of course, you can't write down every word witnesses say or record in rich detail every observation, so using trigger words can help. Trigger words help you remember a specific piece of information, person, or scene. They also might identify a witness or remind you of some specific detail.

In the training video, Buck Savage's poor note-taking skills don't help him remember any details about an incident. After just a few questions from the defense attorney, the case is dismissed. Trainees always get a laugh when they see Buck Savage, but they get the point, too: poor note-taking could damage a police officer's credibility and cause others to question his integrity.

Revision Once a report is complete, the officer should compare information in the report to notes in the field. Each detail in the report should match some note or trigger word written in the field. If officers can put notes and the incident report side by side and tell which note or trigger words led to which bit of information in the report, they have an accurate and reliable report.

No matter what officers write, like a field report or an administrative memo, they should ask co-workers to read it over before they submit it. Co-workers should check for clarity and details, noting questions about any missing or confusing information. Anything officers submit should read well and flow correctly.

Heads Up!
Notice anything about this workplace peer review? It has the same purpose as your class peer review.

TECHNO TIP

For more on Buck Savage, search the Internet for this video: J.D. Buck Savage "Saw Drunk Arrested Same"

TRY IT!

For three to five minutes, write a description of what went on in the last meeting of this class. When finished, compare and contrast what you recorded with what those in your group recorded. Like police officers, compare both description and accuracy.

Getting Started with Your Compare-and-Contrast Paragraph

LEARNING GOAL
❷ Write a compare-and-contrast paragraph

You may feel more comfortable writing a compare-and-contrast paragraph before you tackle an essay. If so, follow steps one through six below. If not, go on to Take 2 on page 241.

Step One: Choose a Topic

If you have the freedom to choose the topic for your compare-and-contrast paragraph, remember that your purpose is to offer your readers a clearer understanding of two subjects, pointing out similarities, differences, or both. To begin searching for a topic, complete Ticket to Write 11.2 below.

TICKET to WRITE

11.2 Choose a Topic

Directions: Complete the following to start the process of choosing a topic. List as many subjects as possible for each of the following: Answers will vary.

methods of cooking
forms of exercise
types of television shows
types of natural disasters
kinds of pets

Step Two: Generate Ideas

Before you choose which topic and subjects you will develop into your paragraph, first generate ideas on a few topics to discover what you know about those subjects. Discovering similarities and differences for different subjects will help you get a feel for which topic you want to develop.

Comparing Subjects

When comparing, you usually have two subjects that seem as if they have nothing in common. For example, if you were to compare how you speak with your friends and how you speak with your boss, you might have a list of qualities like this:

Speaking with friends	Speaking with bosses
use slang	use workplace terms
don't say anything that hurts feelings	don't say anything that hurts job
really interested in what they say	pretend to be interested in what they say

11.3 Generate Ideas: Discovering Similarities

Directions: In Ticket to Write 11.2 you generated a list of subjects for five different topics. Choose two subjects from each topic and list their similarities. Answers will vary.

Contrasting Subjects

When contrasting two subjects, you focus on their differences. For example, if you were to contrast rural living with city living, your list might look like this:

Subject A: rural life	Subject B: suburban life
long, winding roads	busy intersections
few neighbors	people always outside
quiet living, birds, cattle	noisy life, cars, voices, lawn mowers
starry nights	light pollution

11.4 Generate Ideas: Discovering Differences

Directions: In Ticket to Write 11.2 you generated a list of subjects for five different topics. Choose two subjects from each topic and list their differences (contrasting qualities). Answers will vary.

Step Three: Define Your Audience and Purpose

When you compare and contrast two subjects, you give your readers a better understanding of the subjects. Your purpose could be to

- convince readers that one subject is better than the other
- show readers that the two subjects are equal
- inform readers of the unique qualities of each subject

Heads Up!
Sometimes these purposes overlap.

Once you determine your purpose, define an audience who would be interested in reading a compare-and-contrast paragraph on the pair of subjects you have chosen. Consider who, for example, might want to know the similarities and differences between an SUV and a four-door truck.

11.5 Define Your Audience and Purpose

Directions: Review the topics, subjects, and ideas you have generated, and choose one topic to become the focus of your compare-and-contrast paragraph. Write your topic and its two subjects, and answer the two questions below. Answers will vary.

Who is my audience?
What is my purpose for comparing or contrasting these two subjects?

Step Four: Draft Your Paragraph

Because paragraphs center on a topic sentence, review your purpose to focus your topic sentence. Include the purpose of the paragraph in your topic sentence.

11.6 Create a Topic Sentence

Directions: List your topic, purpose, and audience. Then, compose your topic sentence and share your work with members of your writing group. Ask if they agree that your topic sentence explains your purpose and is appropriate for your audience. Answers will vary.

After writing your topic sentence, use your list of similarities or differences to compose the supporting details of your paragraph. You may discover that you don't need to include all the qualities you listed, and you may discover other qualities you hadn't thought of before. Keeping your audience and purpose in mind will help you decide which details you need most.

11.7 Compose Supporting Details

Directions: Using the topic sentence you composed in Ticket to Write 11.6 and the ideas you generated in Tickets to Write 11.3 and 11.4, compose supporting details for your paragraph. Write the details and then share them with your writing group. Ask members if you have given enough details so that you fulfill the purpose you express in your topic sentence.

Step Five: Revise Your Paragraph

After you've finished your paragraph, save it, print a hard copy, and read it out loud, looking and listening for errors in grammar and content. Refer to the general Review Checklist for a Paragraph in Chapter 3, page 60, to make sure you revise as completely as possible.

Heads up!
See Chapter 20 to review fragments and run-ons.

☑ REVIEW CHECKLIST **for a Compare-and-Contrast Paragraph**

- ☐ Is my topic sentence clear?
- ☐ Do my points of comparison and contrast align with my purpose?
- ☐ Do I hear any fragments or run-ons when reading out loud?
- ☐ Do all my complete thoughts have appropriate end punctuation?
- ☐ Is my paragraph appropriate for my audience?
- ☐ Does my purpose remain constant?
- ☐ Do any sentences or ideas seem off-topic?

Once you've made any necessary changes, save your work and leave it alone for a while. Then come back later, print out a new copy, and look and listen for errors again, referring to the general Review Checklist for a Paragraph in Chapter 3.

Step Six: Peer Review

Heads Up!
See Chapter 4 for more on peer reviewing.

You might think you've found all of your errors and made all of the improvements you can, but having someone else read your paragraph and offer suggestions will almost always improve your writing. This person is called your **peer reviewer**.

Your peer reviewer may use the Review Checklist for a Compare-and-Contrast Paragraph above or the general Review Checklist for a Paragraph in Chapter 3, or your instructor may provide a different checklist. You might or might not agree with suggestions you receive, but peer reviewers often find places for improvement or errors that slipped by you. Listen to or read closely your peer reviewers' suggestions, and don't be shy about asking questions to clarify their ideas and suggestions.

TICKET to WRITE

11.8 Peer Review

Directions: Share your revised paragraph with your peer reviewers. Ask them to review your work using the Review Checklist for a Compare-and-Contrast Paragraph above and the general Review Checklist for a Paragraph from Chapter 3, page 60. Then ask them to record their suggestions for revision. Revise your paragraph, using any suggestions with which you agree.

Student and Professional Essays

Below are two compare-and-contrast essays. The first was written by a student and the second by a professional. Read these two essays and answer the questions that follow them.

LEARNING GOAL

❸ Read and examine student and professional essays

Student Essay

My Family Thanksgivings

by Frances Moret-Koerper

1 After being gone from home for fourteen years, I know that part of me is turning into a **Midwesterner**. One special day when this change shows the most is on Thanksgiving. Every other year, my husband and I have gone to my home in Louisiana for Thanksgiving. While I miss the culture that was a part of my growing up in the South, I have grown to appreciate the Thanksgiving tradition of my in-laws.

2 My family down south and my in-laws here to the north always have the same types of activities before we eat our Thanksgiving meal. Back home at my Uncle Gaston's farm, we enjoy horseshoes, **skeet shooting**, and, of course, football. The games and game-watching are always full of conversations about days gone by. Relatives pack into the den to watch the Saints. Meanwhile, outside, when the weather's good, we toss horseshoes, Frisbees, and footballs back and forth in the front yard while relatives line up for their turn at shooting skeet in the back field. My husband's family also gathers at his parents' home in Michigan, and TV and sports play a significant part in the Thanksgiving Day meal here. Instead of the New Orleans Saints, though, the Detroit Lions are playing on three televisions throughout the house. If the snow is right, some of the kids go outside and build forts and snowmen until supper time, but usually everyone stays inside where it's warm, visiting, **reminiscing**, and playing cards. We also listen to my husband's cousin Delinda playing tunes on the piano in the living room, inviting all to sing along.

3 At each place, my families even eat at certain times. Back home, we have "dinner" at 1:00. People start showing up as early as 10:00 in the morning at Uncle Gaston's so they can catch up with each other. Eating at 1:00 is convenient because the game usually doesn't start until 3:00 or 4:00. It's always a **heartwarming** day visiting my relatives, but especially my aunts and uncles who want to know about every detail of my life. Here in Michigan, on the other hand, the meal is referred to as "supper" and doesn't start until about 4:00 p.m. There are many more children at my in-laws' Thanksgiving and fewer older folks, so there's a lot more action with kids running in and out of the house, watching movies upstairs, and playing games all over the place.

Midwesterner someone from the north-central region of the United States

skeet shooting shotgun shooting in which clay targets are thrown into the air, simulating the flight of game birds

reminiscing remembering fondly and discussing past times

heartwarming inspiring sympathy or comfort

4 The Thanksgiving meals back home and at my in-laws' home have obvious similarities, but they also have pretty obvious differences. In both places, we enjoy the usual turkey, ham, cranberries, potatoes, stuffing, salads, and breads. Down home, the turkeys are always first smoke-roasted in Uncle Gaston's smokehouse and then in the kitchen oven. He also roasts a pig every year. My parents and aunts and uncles always bring many other dishes and entrées. Cranberry compote, sweet potato pies, collards, oyster dressing, cornbread, **gumbo**, red beans and rice, and gator are always on the buffet. At our Michigan celebration, the turkeys are all oven-roasted and are just as delicious as Uncle Gaston's turkey. A ham is baked, full of cloves, and smothered in mustard and brown sugar. Every year, my husband's family members, like mine back home, bring their special dishes: cranberry relish, sweet potato casserole, cornbread stuffing, yeast rolls, **Waldorf salad**, **brats** and red potatoes, and **venison**.

5 The travel home every other year grows more expensive, but I find something else I miss about home each Thanksgiving we have down there. Also, each year I find something new about my Michigan Thanksgiving that I look forward to every year we stay for Thanksgiving. If we have children one day, I hope they will recognize and be thankful for the differences and similarities that make up both sides of their family.

gumbo a thick soup, usually containing okra and seafood or other meats

Waldorf salad diced apples, chopped celery, and chopped walnuts, mixed with mayonnaise

brats short for *bratwurst*, a type of sausage

venison deer meat

A CLOSER LOOK

Answer these questions about the essay:

1. What two subjects are the focus of the essay?

   Thanksgiving celebrations with family in Louisiana and family in Michigan

2. Where is the thesis statement found?

   The thesis statement is in the last sentence of paragraph 1.

3. Which of the following is *not* a purpose of this essay? a

   a. to convince readers that one subject is better than the other
   b. to show readers that the two subjects are equal
   c. to inform readers of the unique qualities of each subject
   d. to provide readers a clear understanding of two subjects

4. List three points of contrast that are the basis of this essay.

   activities before the families eat, time the families eat, and food the families eat

5. List at least two differences (two contrasts) the writer details in paragraph 2.

   Differences include weather, games the families watch, and outside activities

6. What is the topic sentence of paragraph 3?

   At each place, my families even eat at certain times.

7. Does the essay follow an alternating or divided pattern of organization?

   alternating pattern

8. List at least two differences the writer details in paragraph 4.

   the way foods are prepared and some of the types of food

9. In paragraph 3, what transitional phrase does the author use to introduce the contrast?

   on the other hand

10. What sentence in the conclusion reiterates the thesis or main idea of the essay?

    If we have children one day, I hope they will recognize and be thankful for the differences and similarities that make up both sides of their family.

The professional essay below compares and contrasts couples' food preferences and eating habits.

Professional Essay

What's for Dinner, Sweetie? Heartburn

You Say Tomato, She Says "Ew!"

How Couples Cope When Their Cooking and Chowing Styles Clash

by Elizabeth Bernstein

(1) Ben Breeland slurps sauces, sucks on bones, smacks his lips and licks his fingers while eating. "You want to get the chipmunk effect," says the software consultant, of stuffing his cheeks full of peanuts, his favorite food. Eating this way is a pleasure to him: He grew up with five siblings on a farm in South Carolina, where mealtimes were chaotic affairs and the sounds of loud eating were a sign of appreciation. But how does his wife feel about it?

(2) "I struggle to keep my nerves **intact**," says Jocelyn Breeland, a communications and marketing director for a trade association that supports people with disabilities. "When he swallows, he makes a drain-flushing sound. And he can make grapes crunch."

(3) In the beginning of their 23-year marriage, Ms. Breeland tried to change her husband's eating habits by nagging or kicking his leg under the table. Now she drinks wine to calm down, dines in another room, or rushes through her own food so she can get away from his noises as quickly as possible. And she shoots him a look: "It's like a cartoon character, where her eyes bug

intact unbroken, in one piece

out and her mouth turns down," says Mr. Breeland. "You feel like the worst person ever."

4 Forget middle school. **Spats** over eating—where, when, how we do it—are just as likely to happen to grown-ups as children, especially grown-ups in a relationship, who eat together a lot. And in the adult world, the mess they leave tends to be emotional, rather than physical. Couples squabble over everything from how much mayo to put into the tuna salad to whether to order in or go out for dinner. Meat lovers vs. vegetarians? Organic vs. junk food? A spouse "gently" telling you to put down the Chunky Monkey Ben & Jerry's? The possibilities for food to go bad in a relationship are endless.

5 Heather Hills likes to eat dinner early around 5 p.m. Her husband, James, wants to eat later, around 9 or 10 p.m. Making matters worse, the two differ in their cooking styles: He loves to take his time creating beautiful entrées, with special sauces and carefully chosen side dishes. She throws ground meat, frozen vegetables, and cream of mushroom soup into a casserole.

6 The **nadir** of the Hills' battles? Chocolate-chip cookies. Mr. Hills prefers his flat and thin. His wife wants them cakey and thick. "There is always an argument," says Mr. Hills, a travel blogger. "It's usually resolved by the person who made them enjoying them and the other being ticked off." (Ms. Hills has been known to get so mad after a flat batch comes out of the oven that she's driven to the grocer to buy store-made cookies.)

7 When I asked people about the food fights they'd had with spouses or romantic partners, stories poured in. There were disputes over shopping lists, how closely to follow directions on a recipe, and exactly how brown a banana has to be before it becomes officially inedible.

8 One friend of mine told of her husband's "garbage pail" dinners, which she described as concoctions straight out of a trash can. "He opens the fridge and yanks whatever he can grab—beans, cheese, Indian or Mexican leftovers, pasta—puts it together in the microwave or a frying pan, and douses it with whatever kind of sauce is around, which is usually some kind of curry sauce or maybe ketchup," she says. This "nastiness" has made her wonder at times about the essence of her relationship, she says. "How can you not want to make someone you love happy with food?"

9 Sharing a meal—especially with candlelight and a bottle of rosé—can be loving and intimate. And, at least in the beginning of a relationship, we're typically on our best behavior when we eat. (Ms. Breeland has memories of her husband "cutting his food and taking dainty mouthfuls" when they were dating.)

10 So why all the bickering? We shouldn't need therapists to tell us that food cuts to a very basic issue of identity. It's no coincidence that one of the earliest ways we demonstrate our independence is by asserting our food preferences. By demanding that others respect what we eat, we are demanding that they see us as individuals. So maybe we should pay a bit more attention to people's eating habits when we first meet them. That's what Kathy Schwartz did. The Seattle resident once ended a relationship with a man

spats quarrels, arguments
nadir lowest point

because of the way he ate French onion soup. He had ordered a bowl one day at a restaurant, but found the typically stringy, melted **Gruyere** cheese to be a challenge. "After several attempts trying to twiddle the cheese into submission, he grabbed his knife and, samurai style, sliced through it," says Ms. Schwartz. "It dawned on me that this was his approach to dealing with life's challenges—to attack and **pummel** rather than negotiate, compromise, or find another less confrontational way." She declined further dates.

11 Sara Walker, an interior decorator from Birmingham, Ala., admits she grew up enjoying a very limited **palate**: chicken fingers, mac and cheese, pizza, and peanut-butter crackers. "I never even ate a sandwich," she says.

12 In college, she met her husband, Chris Walker, who hails from the **Mississippi Delta** and loves food: steak, tamales, catfish, game. He became the first person in her life to challenge her on her poor eating habits. A few months into their relationship, as the couple became more serious, Mr. Walker came up with a possible solution: He sent her to a therapist to get over her food **aversions**. The counselor had Ms. Walker make a list of the foods she refused to touch—her No. 1 offender lettuce, along with green beans, grapes, and spaghetti sauce—and helped her introduce them into her diet. How'd it go? Well, recently Ms. Walker ordered a salad to start. And then her entrée? Another salad. She's now a fan of green beans and asparagus. She has learned to love steak. There's just one problem: She's learned that her husband isn't really all that adventurous of an eater after all. For example, he likes his **quesadillas** plain—she throws beans, corn, salsa, and chicken into hers.

14 Now when it comes to eating habits, she says: "I am starting to pass him."

Gruyere a firm, nutty cheese
pummel strike, pound
palate taste, appetite
Mississippi Delta the northwest part of the state of Mississippi, lying between the Mississippi and Yazoo rivers
aversions dislikes, distastes
quesadillas tortillas filled with a mixture, folded, and usually fried

A CLOSER LOOK

Answer these questions about the essay:

1. *Onomatopoeia* is the use of words that echo the sound they indicate. In the first sentence, the author uses what onomatopoetic verbs?

   slurps, sucks, smacks

2. Why does Ben Breeland eat the way he does?

   eating that way "is a pleasure to him" (paragraph 1)

3. The thesis statement comes in paragraph 4. What is the thesis statement?

   Spats over eating—where, when, how we do it—are just as likely to happen to grown-ups as children, especially grown-ups in a relationship, who eat together a lot.

4. In what three ways are the food habits or preferences of Heather and James Hill contrasted?

   They like to eat dinner at different times, have different cooking styles, and like different types of chocolate-chip cookies.

5. According to the author, people are similar in "one of the earliest ways [that] we demonstrate our independence." What is that way?

   The author says that one of the earliest ways we demonstrate our independence is by asserting our food preferences.

6. Cite a way you have demonstrated your independence through such preferences.

   Answers will vary.

7. Paragraph 3 cites the way Ms. Breeland reacted to her husband's eating habits in the beginning of their marriage and the way she currently reacts. Describe both reactions.

   In the beginning, she nagged him or kicked his leg under the table; now she drinks wine to calm down, dines in another room, or rushes through her own food.

8. In paragraph 7, what examples of food fights does the author say she received?

   The author received stories of disputes over shopping lists, how closely to follow directions on a recipe, and how brown a banana has to be before it is inedible.

9. Bernstein quotes Sarah Walker in the last sentence of the essay. Explain what Walker means by "I am starting to pass him."

   Answers will vary.

10. The conclusion of the article is unusual because it is only one sentence, yet it wraps up the essay concisely. How does it remind the audience of the thesis?

    Answers will vary.

Writing Your Compare-and-Contrast Essay

LEARNING GOAL

4 Write a compare-and-contrast essay

Essay writing is much like paragraph writing: you start with a topic and expand on it with details. Remember that just as a paragraph supports a topic sentence, an essay supports a thesis statement.

Step One: Choose a Topic and Develop a Working Thesis Statement

To begin your compare-and-contrast essay, ask yourself these **starter questions**:

1. **What** two subjects do I want to compare or contrast?
2. **Why** do I want to compare or contrast these subjects, or **what** is my purpose?
3. To accomplish my purpose, **what** should I focus on—similarities or differences or both?

For instance, in a compare-and-contrast essay comparing places you might stay when traveling, you could answer starter questions this way:

1. **What** two subjects do I want to compare or contrast?

   staying in a hotel and staying in a relative's home

2. **Why** do I want to compare or contrast these subjects, or **what** is my purpose?

   I want to decide how to have the best vacation possible.

3. To accomplish my purpose, **what** should I focus on—similarities or differences or both?

   I will focus on both.

See Chapter 2 to review thesis statements.

"Writing comes more easily if you have something to say."
—Sholem Asch

Answering Starter Questions: Frances Moret-Koerper and Elizabeth Bernstein

In answering the starter questions, Frances Moret-Koerper, author of "My Family Thanksgivings," wrote:

1. I want to compare and contrast Thanksgiving with my family in Louisiana and Thanksgiving with my family in Michigan.
2. My purpose is to show appreciation for the way both my families observe Thanksgiving.
3. To accomplish my purpose, I will describe the unique qualities that make up both Thanksgiving celebrations.

Elizabeth Bernstein, author of "What's for Dinner, Sweetie? Heartburn," might have addressed the starter questions this way:

1. I want to compare and contrast adult spats about eating.
2. My purpose is to show that adults' eating habits and food preferences differ greatly.
3. To accomplish my purpose, I will focus on the differences adults show in their habits and food preferences.

After reviewing your answers to the starter questions, combine them into a single sentence to create a starter statement. For example, the writer who is comparing and contrasting staying in a hotel with staying with relatives might create a starter statement like this:

I will write about the similarities and differences of staying in a hotel and staying with relatives because I want to decide which will be better for me.

Look at how two other writers chose to create starter statements for their topics:

I will write about the similarities of taking an online class and a traditional class because I want to show readers that comparable tasks are required for success in either class.

I will write about the differences between tweeting and blogging because I want to use one of these to keep in touch with my friends back home.

TICKET to WRITE

11.9 Answer Starter Questions for a Compare-and-Contrast Essay

Directions: All essays have thesis statements. To help develop a thesis statement for your compare-and-contrast essay, begin by reviewing your answers to the starter questions. Then use those answers to complete the sentence below. Answers will vary.

I will write about *the similarities, the differences, the similarities and differences* (circle one) between ____________________ (list your two subjects) because I want to show readers ____________________ (list your purpose).

Once you've completed the starter statement, you've begun to develop the focus of your essay. Return to your starter statement and consider your topic and your purpose. From this starter statement, create your **working thesis statement**. These are working thesis statements created from the starter statements above:

> Staying with relatives when vacationing has wonderful advantages, but staying in a hotel has its own advantages.
>
> Online and traditional classes are not as different as many students might think.
>
> While tweeting is an excellent way to keep in touch with friends, blogging offers more choices in what I share and how I share it.

To develop your working thesis statement, complete Ticket to Write 11.10 below.

11.10 Develop a Working Thesis Statement

Directions: Return to the starter statement you composed in Ticket to Write 11.9 and consider your topic and your purpose. From this, create your working thesis statement and share it with members of your writing group. Ask them if they understand what you will be describing in your essay and why it is important to you.

Step Two: Generate Ideas

Consider your purpose for comparing and contrasting your two subjects.

Contrasting Subjects

If your purpose is to show that one subject is superior to the other or is a better choice than the other, focus on their differences. Think of the particular qualities that highlight those differences. Using reporter's questions (*Who? What? When? Where? Why? How?*) is often helpful in generating ideas.

For instance, if you're comparing and contrasting high school life with college life (your subjects) and you want to show that college life is better (your purpose), you might create ideas by using reporter's questions this way:

Who can attend high school?	Who can attend college?
students under 18	anyone with GED or diploma

(continued)

What can you study in high school?	What can you study in college?
mandatory subjects for a diploma	anything you want to
When can you go to high school?	**When can you go to college?**
must go from 7:30 a.m.–3:00 p.m.	any time; just find the right time for you
Where can you go to high school?	**Where can you go to college?**
in your school district	anyplace you apply and are accepted
Why do you go to high school?	**Why do you go to college?**
because you are made to go	because you want to go
How do you get into high school?	**How do you get into college?**
complete the eighth grade	apply, get accepted, register for classes

Comparing Subjects

If your purpose is to highlight two subjects' similarities, you probably have two subjects that are not usually thought of as being alike. For example, Zach is comparing and contrasting online and traditional classes (two subjects). He wants to show that both kinds of classes demand the same attention and study habits (purpose), so he might discover ideas by using the reporter's questions this way:

What types of assignments are in a traditional on-campus class?	What types of assignments are in an online class?
reading, writing, and group	reading, writing, and group
What class formats are used on campus?	**What class formats are used online?**
lecture, class discussion, small group, discussion, and project-based	lecture (recordings & notes), class and project-based
How often do I have to go to class?	**How often do I need to log in to class?**
go to all scheduled meetings	as often as possible/at least weekly
Whom can I contact for classwork help?	**Whom can I contact for online classwork help?**
My instructor, on-campus tutors	My instructor, on-campus tutors, or online tutors
Who can take on-campus classes?	**Who can take online classes?**
any student enrolled in the college	any student enrolled in the college

Once you have your two subjects, you need to discover what similarities or differences they have.

See Chapter 1 to review activities and techniques for brainstorming and freewriting.

Your purpose for comparing or contrasting subjects should be evident in your working thesis. With this in mind, **prewrite** about your two subjects. Use reporter's questions to discover similarities and differences between your two subjects.

11.11 Generate Ideas

Directions: Write your working thesis statement and any ideas you generated in prewriting. Then share this with your writing group and ask them to note which ideas seem the most compelling or which elaborate the most on the information in your working thesis statement. Answers will vary.

Step Three: Define Your Audience

Once you know *what* you want to write about two subjects, you need to figure out *who* would benefit from reading what you write. Defining your audience helps focus your purpose even more and helps you decide the language to use. Suppose you're comparing and contrasting two movies. For a review on your personal blog, you might use language that is less formal than for a review that is for a class assignment. You may relate the same details in both, but the way you relate them—your language—will be different because of your audience.

11.12 Define Your Audience

Directions: Write your working thesis statement and your intended audience. Share this with members of your writing group and ask if they agree that your working thesis statement will be satisfactory for your audience.

Step Four: Draft Your Essay

No matter what your purpose with your compare-and-contrast essay, you need to decide if you are going to use an alternating pattern or a divided pattern of organization.

An alternating pattern compares and contrasts the two subjects *within* the same body paragraph. The paragraph's topic is usually a specific quality or point of comparison. A divided pattern *divides* the two subjects into separate body paragraphs, addressing only one subject in one paragraph. Because the topics of the paragraphs are the subjects themselves, the paragraph usually discusses more than one quality or point of comparison at a time.

Alternating Pattern	Divided Pattern
Point 1	**Subject 1**
Subject A	Point 1
Subject B	Point 2
Point 2	Point 3
Subject A	**Subject 2**
Subject B	Point 1
Point 3	Point 2
Subject A	Point 3
Subject B	

Because you are exploring the similarities and differences between two subjects, allowing yourself and your readers to gain a deeper understanding of the two, you need to decide which pattern will work better for you.

Zach's essay compares and contrasts online classes with face-to-face classes and has the following thesis:

> While online classes demand more self-motivation than face-to-face classes, both types of classes have the same basic features.

Zach made the following outlines when deciding which pattern to select:

Alternating Pattern

Point 1 Class availability
- face-to-face
- online

Point 2 Participation requirements
- face-to-face
- online

Point 3 Types of assignments
- face-to-face
- online

Point 4 Class activities
- face-to-face
- online

Divided Pattern

Subject 1	face-to-face classes	Subject 2	online classes
	Class availability		Class availability
	Participation requirements		Participation requirements
	Types of assignments		Types of assignments
	Class activities		Class activities

Whether you're writing on paper or using a computer, the point is to get your ideas down. Know that these ideas will change—that's the progress of your writing process. Refer to Chapter 2 to review suggestions about discovery drafting.

In generating ideas, you may have discovered some information you now find you don't need. As you progress, you may discover other qualities you hadn't thought of before. Keeping your audience and purpose in mind helps you decide which details you need most.

11.13 Write Your Discovery Draft

Directions: Use the starter question, working thesis statement, ideas, and defined audience you compiled in Ticket to Write 11.9 through 11.12 to write your discovery draft. Refer to Chapter 2 to review discovery drafting. When finished, share your work with members of your writing group. Ask them to note the following on your two subjects: Answers will vary.

1. Points of comparison and contrast that are detailed and specific.
2. Points of comparison and contrast that are vague, unclear, or need more development.
3. Points of comparison you haven't already thought of that would be good to include in your draft.

Step Five: Organize Your Essay

Although you have already determined your pattern of organization, *alternating* or *divided*, you need to decide in what order to present your paragraphs, and in what order to present your support within those paragraphs. Consulting your working thesis statement and purpose will help you determine this.

Zach chose the alternating pattern for his essay comparing and contrasting online and face-to-face classes. Because anyone who meets the prerequisites can take online classes, Zach decided to make *class availability* the first point of the essay. The next three points of comparison progressed to what Zach believed is his strongest point of support for the essay's thesis.

Below is a chart Zach made to check that he was giving equal weight to the alternating points in his body paragraphs. Making this chart helped him organize his ideas by

- showing particular points that needed more details
- helping him see which details best illustrated his points
- helping him reexamine the order of the details within body paragraphs

THESIS: While online classes demand more self-motivation than face-to-face classes, both types of classes have the same basic features.			
		Details of support	
Subjects		Face-to-face Classes	Online Classes
Point 1	Class availability	Can take if pre-reqs are met Internet access at home	Can take if pre-reqs are met Internet access at home
Point 2	Types of assignments	Lecture: notes & ask prof questions Assigned readings discussed in class Group projects: meet with other students Essays: often work on in class & ask prof to look at drafts	Lecture: take notes from recorded lectures/can't ask questions Assigned readings not always discussed online Group projects: other members all over the state Essays: can't always have prof read drafts
Point 3	Participation requirements	Must attend class Work with groups in class Some profs can be talked into accepting late work	Must be self-motivated to log in daily Deadlines for assignments Assignments will lock you out after deadline passes

To stay reminded of his purpose, Zach wrote his thesis statement at the top of his chart.

Point 4	Class activities	Open discussions Small group work Presentations . . . PowerPoint, speeches, group discussions Service learning projects—hands-on . . . go into community	Discussion boards Group projects Presentations . . . PowerPoint or essays uploaded to site Service learning projects—virtual . . . all individual

Complete Ticket to Write 11.14 to review your body paragraphs' points of comparison and contrast and their supporting details.

11.14 Organize Your Essay

Directions: Using the example above, create a chart of your essay's body paragraphs. Review your chart for the following:

1. Look for points of comparison or contrast that need more developed details. Where needed, develop more details to complete the body paragraphs.
2. Determine which details are the strongest support in each body paragraph. Where needed, reorder the details of support so that body paragraphs end with their strongest details.

Step Six: Apply Critical Thinking

Critical thinking means thinking and then rethinking about a topic with the goal of discovering all the implications you can about it. In thinking critically about a compare-and-contrast essay, you often do the following:

- reflect on the relationship of two subjects
- analyze the similarities or differences between two subjects
- provide precise details
- reflect on the relevance of details you provide

Critically thinking about specific details means analyzing details for their significance in the relationship of the two subjects you're comparing and contrasting.

The following critical thinking questions are important when you write a compare-and-contrast essay. After you write each of your drafts, ask yourself these questions and then apply your answers to your writing.

1. **Purpose:** Will my readers understand why my comparison or contrast of these two subjects is important? (This importance may have helped you make a decision or come to a clearer understanding about the two subjects.) Have I clearly stated or clearly implied my purpose?
2. **Information:** Have I given my readers enough details to understand how these subjects relate to one another? Do I see any place to add details that will enhance or further explain my topic?
3. **Reasoning:** Are all my facts accurate? Are they clear? Are they relevant to my topic? Is my reasoning logical?
4. **Assumptions:** Have I made any assumptions that need to be explained or justified? Have I used any vocabulary my audience may not understand?

11.15 Apply Critical Thinking

Directions: After your revision, ask members of your writing group to read your latest draft and answer the critical thinking questions below. Answers will vary.

Purpose

1. Do you understand why my comparison or contrast of these subjects is important?
2. Did I clearly state or imply my purpose? If not, where should I be clearer?

Information

3. Did you read enough details to understand the relationship of these subjects? If not, where should I add details?
4. Do you see any places where adding details would enhance or further support my topic?

Reasoning

5. Do you feel all my facts are accurate? If not, where should I provide more accurate details?
6. Are all my details clear? If not, where should I change my wording to be clearer?
7. Are all my details relevant to my topic? If not, where did I stray from my topic?
8. Is my reasoning logical? If not, where do I seem to be illogical?

Assumptions

9. Have I made any assumptions that need to be explained or justified? If so, what are these?
10. Did I use any vocabulary you don't understand? If so, what words or phrases need definitions?

Step Seven: Revise Your Essay

As you know, getting your essay on paper is just your first step. Revising and rewriting through subsequent drafts shapes your first draft into a finished product of which you can be proud. First, consult the Review Checklist for an Essay in Chapter 3, page 61, and look for places in your essay you need to polish. Then think about revisions that are specific to compare-and-contrast essays: **topics**, **details**, and **transitions**.

Heads Up!
Keep hard copies of all your drafts and number them so you'll know which is your latest draft.

Topics

When you revise, look at your body paragraphs' topics. Focus each body paragraph on one topic. In an alternating pattern, each body paragraph's topic should be a specific point of comparison or contrast. In a divided pattern, each body paragraph's topic should be one subject.

Details

When you revise, review your details.

- Will readers understand the significance of the details of one subject when compared or contrasted to the other subject?
- Have you given too many details?
- Have you written anything that flows away from a paragraph's focus?
- Are your paragraphs arranged in the most logical or effective way?

Transitions

Also consider the language you use and word choices you make when developing the relationship between your two subjects. In compare-and-contrast essays, use words and phrases that show special relationships between your two subjects.

When comparing, use words and phrases that indicate equality or show similarity. This is called using *coordination*. These transition words and phrases show coordination:

along with	in addition	similarly
also	in the same way	thus
both . . . and	just as . . .	together with
by the same token	like	too
coupled with	likewise	
correspondingly	moreover	

When contrasting, use words and phrases that indicate inequality or show differences. This is called using *subordination*. These transition words and phrases show subordination:

although	in contrast	rather than
at the same time	in spite of	regardless
but	instead	still
by contrast	nevertheless	though
different from	on the contrary	unlike
even so (though)	on the other hand	yet
however	otherwise	

11.16 Recognizing Coordination and Subordination

Directions: In each sentence, underline the transition words or phrases. Then in the spaces provided, identify the type of transition (subordination or coordination) and explain the relationship between the two subjects. Answers may vary.

Example: Although football and boxing are dangerous sports, boxing is the more deadly sport.

Type of transition: subordination

Relationship: boxing is more dangerous than football

1. For some people, the need for caffeine can cause an addiction to chocolate in the same way it can to coffee or soft drinks.

   Type of transition: coordination

   Relationship: Chocolate can be as addicting as coffee or soft drinks.

2. Algebra and writing classes, along with all other basic core courses, can help students prepare for other more demanding classes.

   Type of transition: coordination

   Relationship: Algebra, writing, and all other core classes help prepare students.

3. The football coach was forced to resign because of five recruiting violations; on the other hand, the tennis coach committed the same violations but received only a warning.

   Type of transition: subordination

   Relationship: The two coaches were not treated equally.

4. Despite the disco music that I can't stand, I still listen to the local oldies station because it plays a lot of 70s progressive rock that I like.

   Type of transition: subordination

   Relationship: The writer prefers the 70s progressive rock to the 70s disco music.

5. Instead of spending extra money each week for gas, I decided to bike to class whenever the weather was good.

   Type of transition: subordination

   Relationship: The writer shows a preference for biking to classes over driving.

Writing Assignment

MyWritingLab™

Complete this Exercise

Consider the topics below or one that your instructor gives you. Choose one topic and expand it into a compare-and contrast essay.

1. two personal role models
2. a spring break vacation with your family and one with your friends
3. the main characters in two books you have read
4. two different movie action heroes
5. your first week as a college student and your current student life
6. two different places where you study
7. experiences of two people who have been in the military
8. single life and married life (or the way you think it will be)
9. local news coverage and national news coverage
10. the first job you held and your dream job

TECHNO TIP

For more on compare-and-contrast writing, search the Internet for these videos:

ECPI University English Compare and Contrast Essays

Compare/Contrast 1

Compare/Contrast 2

MyWritingLab™ Visit *MyWritingLab.com* and complete the exercises and activities in the **Paragraph Development-Describing** and **Essay Development-Describing** topic areas.

RUN THAT BY ME AGAIN

LEARNING GOAL
❶ Get started with compare-and-contrast writing

- **A compare-and-contrast paragraph or essay . . .** details the similarities (the comparisons) and the differences (the contrasts) between two people, places, ideas, events, or things.
- **A compare-and-contrast paragraph or essay . . .** may be written in an alternating or a divided pattern.
- **In the topic sentence of your compare-and-contrast paragraph, clearly state . . .** what two subjects you are comparing and contrasting and why their similarities or differences are important.

LEARNING GOAL
❷ Write a compare-and-contrast paragraph

- **When writing a compare and contrast paragraph, it is important . . .** to determine your purpose and identify your audience.
- **After you have completed your compare-and-contrast paragraph, check that your points of comparison and contrast align with . . .** your purpose.

LEARNING GOAL
❹ Write a compare-and-contrast essay

- **Starter questions for a compare-and-contrast essay include . . .** What two subjects do I want to compare or contrast? Why do I want to compare or contrast these subjects, or what is my purpose? To accomplish my purpose, what should I focus on—similarities or differences or both?
- **If your purpose is to show that one of your subjects is superior to the other or is a better choice than the other . . .** focus on their differences.
- **If your purpose is to highlight two subjects' similarities, you probably have . . .** two subjects that are not usually thought of as being alike.
- **Defining your audience helps . . .** focus your purpose even more and helps you decide the language to use.
- **An alternating pattern of organization . . .** compares and contrasts the two subjects within the same body paragraph.
- **A divided pattern of organization . . .** divides the two subjects into separate body paragraphs, addressing only one subject in one paragraph.
- **Consulting your thesis statement and purpose will help you determine the order to present . . .** your paragraphs and the support within your paragraphs.
- **In thinking critically about a compare-and-contrast essay, you often . . .** reflect on the relationship of two subjects, analyze the similarities or differences between two subjects, provide precise details, and reflect on the relevance of details you provide.
- **When revising a compare-and-contrast essay, look especially at . . .** topics, details, and transitions.

COMPARE-AND-CONTRAST WRITING LEARNING LOG MyWritingLab™

MyWritingLab™

Answer the questions below to review your mastery of compare-and-contrast writing. Answers will vary.

1. Beyond reporting the similarities and differences of two subjects, what might a compare-and-contrast paragraph or essay do?
   It may propose that one subject is superior to the other in specific ways, that both subjects are equal in importance, or that one subject is far less important.
2. What is a Venn diagram?
   A Venn diagram is a drawing of two intersecting circles that note features that are either unique or common to two concepts.
3. When contrasting two subjects, what do you focus on?
   You focus on their different qualities.
4. What could be your purpose in comparing and contrasting two subjects?
   Your purpose could be to convince readers that one subject is better than the other, show readers that two subjects are equal, or inform readers of unique qualities of each subject.
5. How should you create your starter statement?
   Combine the answers to your starter questions into a single sentence.
6. What should you create from your starter statement?
   You should create your working thesis statement.
7. What two elements should be evident in your working thesis statement?
   Your purpose for comparing or contrasting subjects should be evident.
8. After you have determined what you want to write about your two subjects, what is the next step?
   Determining who will benefit from reading what you write about the subjects is the next step.
9. In compare-and-contrast writing, what are two patterns of organization you must decide between?
   You must decide between an alternating pattern and a divided pattern.
10. In applying critical thinking to your essay, what questions help you in examining your purpose?
    Will my readers understand why my comparison or contrast of these two subjects is important? Have I clearly stated or clearly implied my purpose?

11. **In applying critical thinking to your essay, what questions help you in examining your information?**

    Have I given my readers enough details to understand how these subjects relate to one another? Do I see any place to add details that will enhance or further explain my topic?

12. **In applying critical thinking to your essay, what questions help you in examining your reasoning?**

    Are all my facts accurate? Are they clear? Are they relevant to my topic? Is my reasoning logical?

13. **In applying critical thinking to your essay, what questions help you in examining your assumptions?**

    Have I made any assumptions that need to be explained or justified? Have I used any vocabulary my audience may not understand?

14. **You have used an alternating pattern for your body paragraphs. When you are revising, you should check to see if you used what format?**

    Check to see that each body paragraph's topic gives a specific point of comparison or contrast.

15. **You have used a divided pattern for your body paragraphs. When you are revising, you should check to see if you used what format?**

    You should check to see that each body paragraph's topic gives only one subject.

16. **When you revise for details, what questions should you ask yourself?**

    Will readers understand the significance of the details of one subject when compared or contrasted to the other subject? Have I given too many details? Have I written anything that flows away from a paragraph's focus? Are my paragraphs arranged in the most logical or effective way?

17. **Name at least five transition words or phrases that show similarity (that use coordination).**

    Transition words or phrases that show similarity include *along with, also, both . . . and, by the same token, correspondingly, coupled with, in addition, in the same way, just as, like, likewise, moreover, similarly, thus, together with,* and *too.*

18. Name at least five transition words or phrases that show differences (that use subordination).

    Transition words or phrases that show differences include *although, at the same time, but, by contrast, conversely, despite, different from, even so (though), however, in contrast, in spite of, instead, nevertheless, notwithstanding, on one hand, on the contrary, on the other hand, otherwise, rather than, regardless, still, though, thus, unlike,* and *yet.*

CHAPTER 12 Classification Writing

LEARNING GOALS

In this chapter, you'll learn and practice how to

1. Get started with classification writing
2. Write a classification paragraph
3. Read and examine student and professional essays
4. Write a classification essay

GETTING THERE

- Effective classification writing sorts a large subject into more useful or understandable parts or categories, allowing you to better analyze it.
- Attention to detail makes your classification writing more critical and logical.

FASTWRITE 1

The pictures above depict typical restaurants, ones like those you might find in your area. Think of the types of food that could be served in either of these restaurants. Write as much as you can, as fast as you can, for five minutes.

FASTWRITE 2

The pictures above suggest general topics that can be broken into separate categories. Choose one (methods of exercise or locations or types of vacations). Write as much as you can, as fast as you can, for five minutes.

Complete these **Fastwrites** at **mywritinglab.com**

Getting Started in Classification Writing

In a **classification** paragraph or essay, you take a general topic and sort it into simpler or more helpful categories. Your purpose is to make a broad subject (your topic) easier to understand by creating divisions or categories that identify specific differences.

LEARNING GOAL

1. Get started with classification writing

This type of writing may not seem as overwhelming if you picture yourself doing a little sorting. Suppose you had all the notes from your various classes in a big pile. In order to study, you would separate the piles into notes from each class.

This is similar to writing a classification piece. The big pile of notes is like your topic. However, the topic or *big pile* is too general and includes too much information, so you need to separate it into smaller piles of notes. To make the smaller piles, you must first decide on an **organizing principle**. This is the method you use to determine your smaller sets of notes. Each smaller set is a category.

Within each category, you have a set of notes that belongs to a specific class. From here you might arrange each set of class notes by some specific detail, such as daily notes.

	Pile of Notes	(your *topic*)	
English	Study Skills	Math	(your *categories*)
notes from Day 1	notes from Day 1	notes from Day 1	(*explanations*
notes from Day 2	notes from Day 2	notes from Day 2	or *details*
notes from Day 3	notes from Day 3	notes from Day 3	of your
notes from Day 4	notes from Day 4	notes from Day 4	categories)

TICKET to WRITE

12.1 Fastwriting a Classification

Directions: For three to five minutes, write the names of as many television shows as you can (whether they're ones you enjoy or not). Next, use the organizing principle of genres (drama, comedy, reality, educational, etc.) to sort the shows, noting as many details as possible about each show and its genre. When you're finished, share this with your writing group.

When classifying material, you often need to take an assortment of examples or ideas and determine a category they share. In doing so, you not only look for ways the examples are similar, but you also look for ways they are different, excluding any examples or ideas from a category that do not ultimately fit with the others.

12.2 Classification Subject and Categories

Directions: In the groups below, every item except one could be included in the same classification essay. Working alone or in groups, underline the item that should be deleted. Then write the category into which the remaining items can be classified.

Example:

instant coffee stainless steel aerosol cans telegraph hula hoop

Category: inventions of the twentieth century

1. *Cloudy with a Chance of Meatballs* *Twilight* *Aladdin* *The Little Mermaid* *The Land Before Time*

   Category: animated films

2. England Ireland Canada Australia Germany

   Category: English-speaking countries

3. Rockies Missouri Green Ohio Mississippi

   Category: rivers in the United States

4. foxtrot waltz pop and lock logger mambo

   Category: dances

5. restaurant server elementary teacher TV anchor nurse paralegal

   Category: jobs that require college degrees

In classification writing, you frequently start with a topic and then look for an organizing principle to help you develop the categories (the kinds, groups, or types) through which you can separate examples. This approach allows you to develop the examples that share similar characteristics.

12.3 Classification Types

Directions: List at least three different classifications (kinds, groups, or types) from the topics below. Answers will vary.

Example:

Books

a. *How to Save for College Fund*
b. *30 Habits to Improve Your Sleep*
c. *Build Your Own Successful Marriage*

Organizing principle: Self-help books

1. College classes
2. Styles of dress
3. Traits in a friend
4. Communication
5. Medical conditions
6. Television comedy
7. Reading material
8. Childcare options
9. Music
10. Television drama

WRITING at WORK: *Snapshot of a Writer*

Renee LaPlume, Technical Writer and Editor

Renee LaPlume has been a technical writer and editor for over twenty years, primarily in the software development industry. She has worked both as an employee and contractor for a number of companies.

Renee's Writing Process

How I Use Writing I am a technical writer and editor. I solve business and organizational problems using communication skills and technical knowledge. In my work, I use writing to

- teach people about using technology, such as software—for example, using an online registration system
- inform people about procedures to follow or processes to understand—for example, payroll procedures
- inform members about upcoming events in an organization—for example, publishing an events calendar and articles in an organization's newsletter
- improve written products composed by others—for example, editing an author's book to correct errors, improve understandability, reduce wordiness, and expand on missing information

I like to help others solve business-related problems. Often those with the problem see just that—a problem—but I see problems as types. *Communication*, *procedural*, *self-evolving*, *visionary*, and *simple explicit* are five categories that most problems fall into. Though many problems can be in more than one category, thinking of a problem as one of these types helps me define the nature of the problem and the characteristics of the problem. Once I know the characteristics of a problem, I know how best to approach the issue and seek out a solution.

What My Writing Process Is My writing process is to follow these steps:

1. Define the communication problem or need, analyze the intended audience, and identify the timeline/schedule required for completion.
2. Design a communication solution to meet the project needs as defined.
3. Create the communication solution in a draft form.
4. Edit/test the solution.
5. Refine, revise, and implement the solution.

TRY IT!

Write a paragraph describing a communication problem at work, school, or home. Next, identify when and how the problem occurs, and then write a solution to address the problem. When finished, share your paragraph with those in your writing group.

Getting Started with Your Classification Paragraph

LEARNING GOAL

❷ Write a classification paragraph

You may feel more comfortable writing a classification paragraph before you tackle a classification essay. If that is the case, follow steps one through seven below. If not, go right on to Take 2 on page 273.

Step One: Choose a Topic

Remember *topic* and *topic sentence* are not the same.

When writing a classification paragraph, you'll often be assigned a particular topic to develop. If you have the freedom to choose a topic, however, keep in mind that your purpose in classification writing is to make a broad subject easier to understand. Your detailed examples should make clear your point or main idea.

To begin prewriting possible classification topics, complete Ticket to Write 12.4.

12.4 Choose a Topic

Directions: To begin choosing a topic, complete the following prompts. If necessary, freewrite for a few minutes on the subject and then fill in the prompt. Answers will vary.

1. List at least three instructors (K–12 or college) you like (or dislike).
2. List at least three places you're fond of visiting or would like to visit.
3. List at least three charitable organizations with causes you support.

Step Two: Generate Ideas

Once you've decided on your topic, the next step is to begin discovering details and examples that define the elements of each category. Return to Chapter 1 to review several prewriting methods that will help you generate ideas about your topic.

See Chapter 1 for prewriting techniques such as listing, clustering, fastwriting, reporter's questions, and journaling.

12.5 Generate Ideas and Choose a Topic

Directions: Choose one of the topics you worked on in Ticket to Write 12.4. Freewrite answers to reporter's questions (*Who? What? Why? Where? When?* and *How?*) for each category within that topic. Form your own reporter's questions or use the following:

1. Instructors

   Who is the instructor?
   Why is the instructor a favorite?
   When did you have this instructor's class?
   What subject does the instructor teach?
   Where does the instructor teach?
   How does the instructor capture your attention?

2. **Places to visit**

   Who would enjoy this place?
   Why is this one of your favorite places?
   When do you like to visit this place?
   What do you like to do at this place?
   Where is this place?
   How is this place unique?

3. **Charitable organizations**

   Who is involved in this organization?
   Why do you support this organization?
   When does this organization's work take place?
   What does this organization do?
   Where does this organization's work take place?
   How does this organization's work affect you or others?

Step Three: Define Your Audience and Purpose

Your purpose with a classification is to make a broad subject or topic easier to understand through categorizing types or kinds of it. After selecting a topic, you need to determine the organizing principle or the method you will use to create categories for the topic.

As you consider your topic and purpose, think of who would be interested in or benefit from reading about your topic and the categories you create for it.

For example, Tavon was classifying places to study on campus, so he considered new students as his audience. For this audience, he used an organizing principle of looking at the qualities various study areas had, and he created these categories:

- quiet study area
- study area with background noise
- distracting study area
- nature lover's area

12.6 Discovering Your Audience and Purpose

Directions: Write your topic. Under that, identify your audience, organizing principle, purpose, and categories.

Here is how Tavon filled in the Ticket to Write 12.6:

Topic: campus study spots

My audience is new students.

My organizing principle is qualities of various study areas.

My purpose is to help students find the right study area.

My categories are quiet study area, study area with background noise, and distracting study area.

Step Four: Create a Topic Sentence

Because paragraphs center on a topic sentence, review

- who your readers are
- what topic you want to classify
- what your purpose is in classifying this topic

Focusing on these three ideas will help you compose a topic sentence your audience will understand. Here is Tavon's topic sentence:

Our college has four types of study areas students can choose from.

Sometimes a classification topic sentence names the categories the paragraph will develop. This is called a **plan of development**. Tavon considered an alternative top ic sentence that included a plan of development:

Your instructor may require you to use a topic sentence that includes a plan of development.

The choices of study areas on our campus are the quiet areas, areas with background noise, distracting areas, and the nature lover's area.

12.7 Create a Topic Sentence

Directions: Write your topic, purpose, audience, and categories. Using these, create your topic sentence. Then share your work with members of your writing group to see if they agree that your topic sentence explains your purpose and is appropriate for your audience. Answers will vary.

Step Five: Draft Supporting Details

After writing your topic sentence, build your examples from the fastwriting you did to compose the supporting details of your paragraph. As you write, keep your audience and purpose in mind.

Work with the prewriting and freewriting you created as part of Tickets to Write 12.3, 12.4, and 12.5. Select the details that work best to show your categories and their differences. Remember, you do not have to use all the ideas from your prewriting. Instead, choose the examples that help you make your point the best way you can.

12.8 Draft Supporting Details

Directions: Using the topic sentence you composed in Ticket to Write 12.7 and the ideas you generated in Ticket to Write 12.5, draft categories for your classification paragraph. Write your categories, describing each of them in detail. When finished, share your work with your writing group, checking that each category has enough description so your readers understand fully your topic sentence.

Step Six: Revise Your Paragraph

See Chapter 4 to review proofreading techniques.

After you've finished your paragraph, save it, print a hard copy, and read it out loud, looking and listening for errors in grammar and content. To make sure you revise as completely as possible, refer to the general Review Checklist for a Paragraph in Chapter 3 (p. 60). Then look at the review checklist below for specific elements that may need revision.

☑ REVIEW CHECKLIST **for a Classification Paragraph**

- ☐ Is my topic sentence clear?
- ☐ Do my categories represent my topic well?
- ☐ Are my categories sufficiently detailed?
- ☐ Do I hear any fragments or run-ons when reading out loud?
- ☐ Do all my complete thoughts have appropriate end punctuation?
- ☐ Is my paragraph appropriate for my audience?
- ☐ Do any sentences or ideas seem off-topic?

Heads Up!
See Chapter 20 to review fragments and run-ons.

Once you've made any necessary changes, save your work and leave it alone for a while. Then come back later, print out a new copy, and look and listen for errors again. Refer again to the general Review Checklist for a Paragraph in Chapter 3.

Step Seven: Peer Review

Peer reviewers bring a fresh perspective. Consider them valuable resources as audience members and fellow writers.

You might think you've found all of your errors and made all of the improvements you can, but having someone else read your paragraph and offer suggestions will almost always improve your writing. This person is called a **peer reviewer**.

Your peer reviewer may use the review checklist above or the general Review Checklist for a Paragraph in Chapter 3, or your instructor may provide a different checklist. You might or might not agree with the suggestions you receive, but peer reviewers often find places for improvement or errors that slipped by you. Listen to or read closely your peer reviewers' suggestions, and don't be shy about asking questions that clarify their ideas and suggestions

Heads Up!
See Chapter 4 for more on peer reviewing.

12.9 Peer Review

Directions: Once you have completed revising your paragraph, share it with your peer reviewers. Ask them to review your paragraph using the Review Checklist for a Classification Paragraph above and the general Review Checklist for a Paragraph from Chapter 3. Then ask them to record their suggestions for revision.

Student and Professional Essays

Below are two classification essays. The first was written by a student and the second by a professional. Read these two essays and answer the questions that follow them.

LEARNING GOAL
❸ Read and examine student and professional essays

Student Essay

Vacation Types for Everyone

by Skyy Laughlin

(1) I love to travel, and growing up in central Illinois, I daydreamed of vacations on sandy ocean shores and on mountain overlooks. My parents were very **outdoorsy**, so our family vacations usually involved tents, hiking boots, boat paddles, or bikes and were most often only an hour or two from home. Based on these and other experiences, I have learned that vacations come in three types: the **jam-packed**, the adventure-filled, and the laid-back.

(2) The jam-packed vacation is usually the most expensive and tiring. This is the vacation that includes a car full of kids, overstuffed suitcases, and an **itinerary** that includes everything but a break. The jam-packed vacation is usually expensive and has a schedule crammed full of things to do and places to see. Once my aunt, my cousins, my mom and dad, and I drove for

outdoorsy fond of nature and the open air
jam-packed crammed full
itinerary schedule of events or activities, route plan

three days to spend five days and nights at Disney. By the time we got to Orlando, we had seen Lincoln's Boyhood Home in Indiana, Mammoth Cave in Kentucky, the **Grand Ole Opry** in Tennessee, and the Coca-Cola Museum in Georgia. My mom said that by the time we got to Disney, we had spent as much on roadside attractions and other stops as we had on Disney tickets for all seven of us. We were so tired when we got to our hotel at Disney World that we all stayed in our hotel rooms and slept almost the whole first day there. Jam-packed vacations are fun, of course, because they usually have plenty of activities that are once-in-a-lifetime events and memory-making material.

3 The adventure-filled vacation is the type that my parents taught me to appreciate. Every summer, fall, and spring break, my parents would take me hiking, camping, fishing, or on another outdoor **escapade**. We seemed always to have some experience that made me grateful for nature. Once, when we were hiking in southern Illinois, my parents and I were resting on a small **outcropping** that looked over a valley and two hummingbirds started buzzing around us. Dad told us to be real still as the birds flew around Mom's head and seemed very interested in her. We later learned from the park ranger that the birds were probably attracted to Mom's bright red and yellow tee shirt. There are 250 different types of birds in the Shawnee National Forest's 280,000 acres. Another awesome adventure was when we went canoeing on the Green River. Because of a drought that year, we canoed eight miles on the river and carried the canoes for two miles. I learned much that year about early settlers and the **trailblazing** they had to do to establish settlements and towns. I liked these outdoor adventures more as I got older and still appreciate the outdoors today. In fact, I can't wait to take my kids hiking one day so they too can grow their own appreciation for the outdoors.

4 As much as I enjoy the adventure-filled vacation, the type of vacation I enjoy the most is the laid-back vacation. When I was twenty-five, a friend of mine and I drove up to Chicago and stayed in a **chic** hotel downtown. We mostly stayed in the hotel and took advantage of room service, indoor pools, and saunas. We even ordered an in-room massage. We had no itinerary or plans other than to relax for a few days. We found time to take in a Cubs game and a jazz concert. It was one of the most relaxing and peaceful vacations I've ever had—indoors or out. Of course, the laid-back vacation can be expensive like this one was, but it doesn't have to be. I once went to Moline and spent a week on my Aunt Doris's farm. I took three books and a hammock. By the time I came home, I was rested, relaxed, and looking for another novel to start reading. The same goes for camping. I still enjoy going out in the woods, pitching my tent, and relaxing in nature for a few nights.

5 If you're in the mood for a vacation, then consider the type of vacation you want and what you can afford to do. For me, although my wallet might determine the **frills** that go along with my vacation plans, my mood will determine the type of vacation I take. No matter if I'm **flush** or broke, I can always grab my camping gear and hit the woods for a couple of days; the cost is minimal and the rewards are great.

Grand Ole Opry a famous country music stage concert attraction in Nashville that began as a radio show in 1925

escapade adventure or experience

outcropping a rock formation that thrusts out and away from the earth

trailblazing making a new way; forging a path through the wilderness

chic stylish, fashionable

frills added extras; attractive but unnecessary embellishments

flush having money

A CLOSER LOOK

Answer these questions about the essay:

1. Where is the thesis statement found?

   The thesis statement is the third sentence of the first paragraph.

2. What topic does the student writer classify?

   types of vacations

3. What three categories does the writer develop for the topic?

   the jam-packed vacation, the adventure-filled vacation, and the laid-back vacation

4. What is the topic sentence of paragraph 3?

   The first sentence is the topic sentence.

5. In paragraph 3, which sentence does not support the topic and could be removed?

   There are 250 different types of birds in the Shawnee National Forest's 280,000 acres.

6. What details does the writer use in paragraph 4 to develop that paragraph's type or category?

   room service, indoor pools, and saunas; in-room massage; Cubs game and a jazz concert; three books and a hammock; pitching a tent, relax in nature

7. List the transitions and transition phrases the writer uses in paragraph 4.

   As much as, after, even, Of course, but, By the time

8. In the second sentence of paragraph 2, the writer repeats the word *includes*. How might you rewrite this sentence to avoid the repetition?

   Answers may vary. Sample answer: This is the vacation that *usually has* a car full of kids, overstuffed suitcases, and an itinerary that includes everything but a break.

9. The third, fourth, and fifth sentences in paragraph 4 all begin the same way. How might you rewrite some of these sentences to vary the sentence beginnings?

   Answers may vary. Sample answer: Mostly we stayed in the hotel and took advantage of room service, indoor pools, and saunas. We even ordered an in-room massage. With no itinerary or plans other than to relax for a few days, we found time to take in a Cubs game and a jazz concert.

10. Look at body paragraphs, 2, 3, and 4, which describe various types of vacations. Compose a body paragraph of your own as a substitute for one of the writer's body paragraphs. You may use the same or a similar topic sentence, or you may go in a different direction. This may be a real vacation you have taken or it may be all from your imagination. Answers will vary.

Professional Essay

The 12 Most Annoying Types of Facebookers

by Brandon Griggs

(1) Facebook, for better or worse, is like being at a big party with all your friends, family, acquaintances, and coworkers. There are lots of fun, interesting people you're happy to talk to when they stroll up. Then there are the other people, the ones who make you cringe when you see them coming.

(2) Sure, Facebook can be a great tool for keeping up with folks who are important to you. Take the status update, the 160-character message that users post in response to the question, "What's on your mind?" An artful, witty, or newsy status update is a pleasure—a real-time, tiny window into a friend's life.

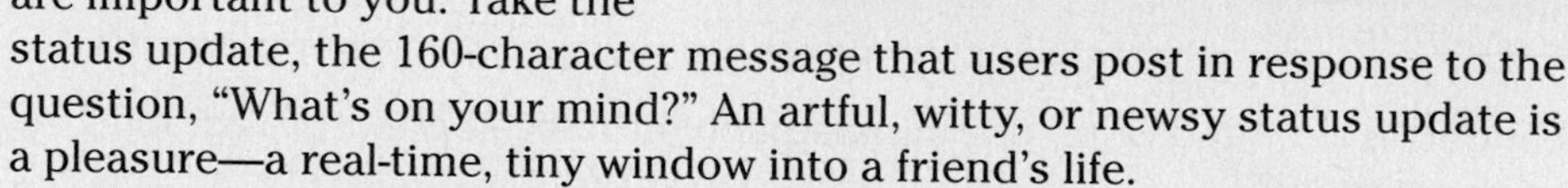

(3) But far more posts read like navel-gazing diary entries, or worse, spam. A recent study categorized 40% of Twitter tweets as "pointless babble," and it wouldn't be surprising if updates on Facebook, still a fast-growing social network, break down in a similar way. Combine dull status updates with shameless self-promoters, "**friend-padders**," and that friend of a friend who sends you quizzes every day, and Facebook becomes a daily reminder of why some people can get on your nerves. Here are 12 of the most annoying types of Facebook users:

(4) The Let-Me-Tell-You-Every-Detail-of-My-Day Bore. "I'm waking up." "I had Wheaties for breakfast." "I'm bored at work." "I'm stuck in traffic." You're kidding! How fascinating! No moment is too mundane for some people to broadcast unsolicited to the world. Just because you have 432 Facebook friends doesn't mean we all want to know when you're waiting for the bus.

(5) The Self-Promoter. OK, so we've probably all posted at least once about some achievement. And sure, maybe your friends really do want to read the fascinating article you wrote about beet farming. But when almost EVERY update is a link to your blog, your poetry reading, your 10k results or your art show, you sound like a bragger or a self-centered **careerist**.

(6) The Friend-Padder. The average Facebook user has 120 friends on the site. Schmoozers and social butterflies—you know, the ones who make lifelong pals on the subway—might reasonably have 300 or 400. But 1,000 "friends"? Unless you're George Clooney or just won the lottery, no one has that many. That's just showing off.

"friend-padders" a coined term meaning "people who exaggerate the number of friends they really have"

careerist a person whose chief aim in life is the quest of his or her professional progress

7 The Town Crier. "Michael Jackson is dead!!!" You heard it from me first! Me, and the 213,000 other people who all saw it on **TMZ**. These **Matt Drudge wannabes** are the reason many of us learn of breaking news not from TV or news sites but from online social networks. In their rush to trumpet the news, these people also spread rumors, half-truths, and **innuendo**. No, Jeff Goldblum did not plunge to his death from a New Zealand cliff.

8 The **TMIer**. "Brad is heading to Walgreens to buy something for these pesky hemorrhoids." Boundaries of privacy and decorum don't seem to exist for these too-much-information updaters, who **unabashedly** offer up details about their sex lives, marital troubles, and bodily functions. Thanks for sharing.

9 The Bad Grammarian. "So sad about Fara Fauset but Im so gladd its friday yippe". Yes, I know the punctuation rules are different in the digital world. And, no, no one likes a spelling-Nazi schoolmarm. But you sound like a moron.

10 The Sympathy-Baiter. "Barbara is feeling sad today." "Man, am I glad that's over." "Jim could really use some good news about now." Like anglers hunting for fish, these sad sacks cast out their hooks—baited with vague tales of woe—in the hopes of landing concerned responses. Genuine bad news is one thing, but these manipulative posts are just pleas for attention.

11 The Lurker. The Peeping Toms of Facebook, these **voyeurs** are too cautious, or maybe too lazy, to update their status or write on your wall. But once in a while, you'll be talking to them and they'll mention something you posted, so you know they're on your page, hiding in the shadows. It's just a little creepy.

12 The Crank. These **curmudgeons**, like the **trolls** who spew hate in blog comments, never met something they couldn't complain about. "Carl isn't really that impressed with idiots who don't realize how idiotic they are." [Actual status update.] Keep spreading the love.

13 The **Paparazzo**. Ever visit your Facebook page and discover that someone's posted a photo of you from last weekend's party—a photo you didn't authorize and haven't even seen? You'd really rather not have to explain to your mom why you were leering like a drunken hyena and French-kissing a bottle of **Jagermeister**.

14 The **Obscurist**. "If not now then when?" "You'll see . . ." "Grist for the mill." "John is, small world." "Dave thought he was immune, but no. No, he is not." [Actual status updates, all.] Sorry, but you're not being mysterious—just nonsensical.

15 The Chronic Inviter. "Support my cause. Sign my petition. Play Mafia Wars with me. Which 'Star Trek' character are you? Here are the 'Top 5 cars I have personally owned.' Here are '25 Things About Me.' Here's a drink. What drink are you? We're related! I took the 'What President Are You?' quiz and found out I'm Millard Fillmore! What president are you?"

16 You probably mean well, but stop. Just stop. I don't care what president I am—can't we simply be friends? Now excuse me while I go post the link to this story on my Facebook page.

TMZ a television show that centers on celebrity gossip

Matt Drudge wannabes people who would like to model themselves after Matt Drudge, a conservative online and print columnist

innuendo indirect remark or hint

TMIer a coined term meaning "someone who gives too much information"

unabashedly in a shameless or bold way

sad sacks incompetent or unskilled people

voyeurs people who derive pleasure from secretly observing other people

curmudgeons grumpy or disagreeable people

trolls those who post fiery or off-topic messages online

paparazzo a freelance photographer who takes unexpected photos of celebrities, politicians, and other public figures

Jagermeister a 70-proof imported German liquor

obscurist a coined term meaning "someone who is deliberately vague or ambiguous"

A CLOSER LOOK

Answer these questions about the essay:

1. Which sentence is the thesis of the essay?

   The third or fourth sentence of paragraph 4.

2. How many categories does the author define?

   12

3. What organizing principle does Griggs use to establish his categories?

   His organizing principle is annoying Facebook users.

4. How many examples of posts does Griggs use to illustrate the topic idea of paragraph 10?

   three

5. Griggs opens the essay with a simile (a comparison of two unlike things using *like* or *as*) comparing Facebook users to people at a party. Underline three other similes in the essay.

6. Paragraphs 4 through 15 have the same pattern of development. The author begins by stating the type of Facebook user. What do the next few sentences in each of these paragraphs accomplish?

   Answers will vary. Possible answers: The sentences provide examples of posts; then they convey the author's evaluation or thoughts on those posts and posters.

7. The author comments on each type of "Facebooker" except the last one, The Chronic Inviter. In a sentence or two, compose your own comment evaluating this type of user.

   Answers will vary.

8. *Irony* is the difference between what is expected to happen and what does happen. What does Griggs state in the conclusion that is ironic?

   Answers will vary. Sample answer: By posting a link to his story, the author takes on the qualities of one of the annoying Facebook posters he is criticizing in his article.

9. Which of these types of social network users do you find most annoying and why?

   Answers will vary.

10. The author gives a number of examples for each type of Facebook user. Choose two types and compose additional examples of your own.

    Answers will vary.

Writing Your Classification Essay

Writing a classification essay takes the same tack as writing a classification paragraph. You first need to discover your topic, purpose, and audience. A classification essay effects the same purpose as a classification paragraph: it takes a broad topic and divides it into categories so that it is easier to comprehend.

LEARNING GOAL
4 Write a classification essay

Step One: Choose a Topic and Develop a Working Thesis Statement

To begin your classification essay, ask yourself these **starter questions**:

See Chapter 2 to review thesis statements.

1. **Whom** or **what** do I want to classify?
2. **What** is my purpose?
3. **What** divisions or categories identify specific differences of the subject?

For instance, in a classification essay about college classrooms, Melissa answered the starter questions this way:

1. **Who** or **what** do I want to classify?

   types of classrooms at my college

2. **What** is my purpose?

   to help students understand the benefits of different classrooms types

3. **What** divisions or categories identify specific differences of the subject?

   the features certain classrooms offer that other classrooms don't

Answering Starter Questions:
Skyy Laughlin and Brandon Griggs

In answering the starter questions, Skyy Laughlin, author of "Vacation Types for Everyone," wrote:

1. I want to classify different types of vacations I've experienced.

2. My purpose is to show the value people can get out of each type of vacation.
3. The divisions or categories I identify are jam-packed, adventure-filled, and laid-back.

Brandon Griggs, author of "The 12 Most Annoying Types of Facebookers," might have addressed the starter questions this way:

1. I want to classify various types of Facebook users.
2. My purpose is to show the different types of pointless, unwanted information that some people share.
3. The divisions or categories I identify are the Let-Me-Tell-You-Every-Detail-of-My-Day Bore, the Self-Promoter, the Friend-Padder, the Town Crier, the TMIer, the Bad Grammarian, the Sympathy-Baiter, the Lurker, the Crank, the Paparazzo, the Obscurist, and the Chronic Inviter.

Using information from these starter questions helps you complete your **starter statement**:

I want to classify (point I want to make), and I will do that by showing readers (divisions or categories).

Melissa, whose topic is *types of classrooms at her college*, filled in the starter statement like this:

Point

I want to classify the benefits of different kinds of classrooms, (and I will do that by showing readers) features certain classrooms offer that others don't.

Categories

12.10 Answer Starter Questions and Complete Starter Statement for a Classification Essay

Directions: To help develop a working thesis statement for your classification essay, answer the starter questions and then fill in the blanks for your starter statement: Answers will vary.

1. **Whom** or **what** do I want to classify?
2. **What** is my purpose?
3. **What** divisions or categories identify specific differences of the subject?
4. I want to classify ________________ (point I want to make), and I will do that by showing readers ________________ (divisions or categories).

From these starter questions and statement, create your **working thesis statement**. Here is the working thesis Melissa created from her starter questions and starter statement:

To deal with the needs of different courses, our college has several types of classrooms.

To further develop her categories, Melissa used clustering to prewrite on her subject. Here is Melissa's clustering that helped her discover details of her categories:

See Chapter 1 to review activities and prewriting techniques.

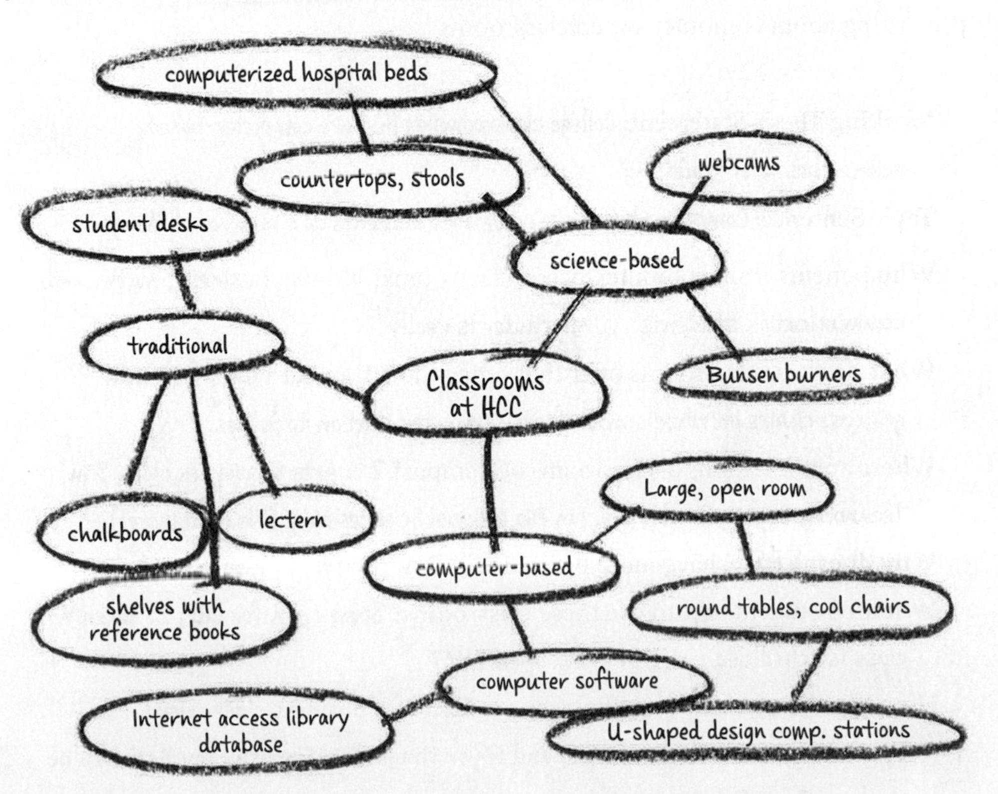

Looking over her clustering, Melissa discovered she thought of the classrooms in three distinct types: *traditional*, *computer-based*, and *science-based*.

Depending on how Melissa composes her thesis statement, she may have a thesis statement that

- lists the categories she will develop in her essay

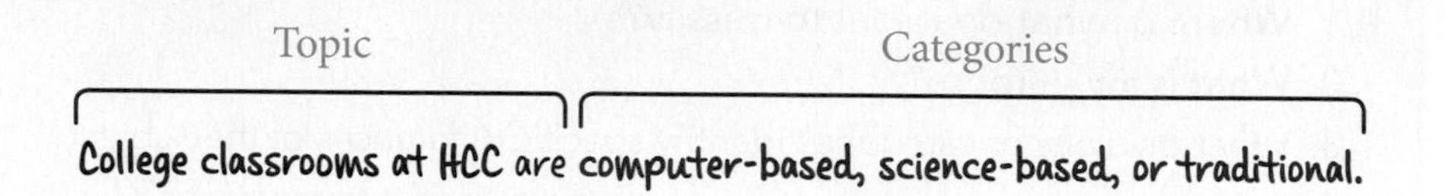

- or implies that categories will be developed in her essay

Topic Implied categories

College classrooms at HCC are available three distinct types.

Step Two: Generate Ideas

To classify a topic, discover details unique to each category. After clustering, Melissa used reporter's questions to discover more details about her categories. Below is her prewriting about computer-based classrooms.

Working Thesis Statement: College classrooms at HCC are computer-based, science-based, or traditional.

Topic Sentence: Computer-based classrooms let students do a number of things.

Who benefits from computer-based classrooms? Writing, business, math, and communication students . . . all students really

What do these classrooms offer that others don't? Round tables for small groups, chairs on wheels move from computer station to tables

Where are these kinds of classrooms on campus? 2 in Arts & Sciences bldg., 3 in Technology Center, 2 in library, 1 in Bio & Social Sciences bldg. (We need more!)

Why doesn't HCC have more on campus? Don't know . . . money maybe

When do students get to use these classrooms? Some open for lab use when no class is scheduled . . . 2 of them open 24/7

How have these classrooms helped me? Excel for business class, word processing, research databases, and MyWritingLab in English, MyMathLab in algebra, PowerPoint in speech

In a classification essay, description is essential. Details you offer about different categories should provide information to help readers understand each category and the larger topic. Melissa developed her answers to reporter's questions into this body paragraph:

Computer-based classrooms let students do a number of activities. Because they have tables instead of desks, students interact more freely. The rooms have wheeled chairs so students can get in small groups pretty quietly. The computers have all kinds of software students can use for different classes. For example, Internet access lets students do research in class and helps them access programs like MyWritingLab and MyMathLab. Word-processing programs help writing students; spreadsheet programs help with business and math classes; PowerPoint software enables students to create colorful presentations for any class.

After prewriting on your categories, complete Ticket to Write 12.11 to discover interesting details and examples that illustrate your working thesis statement.

12.11 Generate Ideas

Directions: Write your working thesis statement and categories. Describe any examples you generated in prewriting for each category. Then share this with your writing group and ask them to note which examples illustrate most effectively your categories.

Using the feedback from your writing group, review the supporting details (examples, explanations, and illustrations) you generated for Ticket to Write 12.11. Then develop further any supporting details you can for each category. Next go through what you have written and check that your details are separated by categories.

12.12 Generate Ideas

Directions: You are writing a classification essay concerning Third World countries. Here is your thesis statement: Answer: Para 1:11, 8, 2, 9, 3, 10; Para 2: 12, 15, 4, 5, 13; Para 3:1, 14, 6, 7

Today, countries are classified as being Third World if they rank low in political rights, rank high in poverty, and lack freedom of the press.

(continued)

Look at the sentences below and determine how you should separate them into three body paragraphs. Each paragraph should flow smoothly according to its topic sentence.

1. Besides few political rights and a high poverty level, a lack of freedom of the press is a characteristic of a third world country.
2. Myanmar (formerly known as Burma) is governed by a military junta.
3. In 1987, Zimbabwe's prime minister Robert Mugabe declared himself president of the country.
4. African countries that are among the poorest include Angola, Chad, and Ethiopia; six Asian countries also rank among the world's poorest.
5. In the Pacific, people living in Kiribati, Samoa, Tuvalu, Vanuatu, and the Solomon Islands have a GNI of under $750.
6. The northern African country of Eritrea always ranks at the bottom or near the bottom for freedom of the press.
7. Other countries with dismal records for freedom of the press include North Korea, Cuba, and Vietnam.
8. Countries around the world, including Myanmar, North Korea, and Zimbabwe, are plagued by the absence of political rights.
9. Although Myanmar boasts two major political parties, the parties are allowed little freedom and are heavily regulated and suppressed.
10. Mugabe and his ZANU party have won every election held since then, in spite of national and international condemnation that all of the elections have been rigged.
11. The Economist Intelligence Unit, which profiles countries around the world, ranks North Korea as having the most authoritarian regime in all the 167 countries it assesses.
12. A second issue that factors into a country's third world status is its poverty level.
13. Currently, citizens of these countries earn under $750 as their GNI.
14. Reporters Without Borders publishes an annual index of countries of the world, ranking them according to their respect for press freedom.
15. The world's most impoverished countries suffer from a low gross national income (GNI).

Step Three: Define Your Audience and Purpose

Your next step is to define the audience for your essay, which will help you focus more on your purpose. Defining your audience will help you focus more on your purpose. Your audience choice will also help you decide the language you use to name your categories, the type of details you select, and the language you use to illustrate or explain those details.

12.13 Define Your Audience

Directions: Write your working thesis statement, categories, audience, and purpose. Share this with members of your writing group and ask if they agree that your working thesis statement, categories, and purpose are appropriate for your audience.

Step Four: Draft Your Essay

Now you have a topic, a working thesis statement, categories, examples, illustrations, purpose, and an audience. Next you should get your ideas in some kind of order—that is, you should create a discovery draft.

The examples you select for your categories should

- explain, clarify, or illustrate their specific category
- be realistic or practical for your topic
- be appropriate for your audience
- help your readers understand the category and broad topic

Whether you're writing on paper or computer, the point is to get your ideas down. Remember, this is a discovery draft, so your ideas and thoughts may change as you write or later—that's the progress of your writing process.

12.14 Write Your Discovery Draft

Directions: Use the material you compiled in Tickets to Write 12.10 and 12.11 to write your discovery draft. Refer to Chapter 2 to review discovery drafting.

When you have completed your discovery draft, share your work with members of your writing group. Ask them to note details and examples that work well to make your point and areas that may need more development.

Step Five: Organize Your Essay

Your essay flows much more smoothly when you use transitional words and phrases. These give hints to your readers, guiding them through your essay.

In a classification essay, you might use transitional words and phrases such as

also	further(more)	next
as a result	in addition	similarly
because of	in brief	specifically
besides	in conclusion	then
consequently	in other words	therefore
due to	in particular	thus
finally	in summary	to conclude
first (second, etc.)	last(ly)	whereas
for example	likewise	
for instance	moreover	

In addition to using simple transitions, consider if your categories need to be presented in a specific order. Sometimes the importance of your body paragraphs is equal; in that case, the order in which they appear does not matter. Other times, following a particular method of organization helps you make your point. Look at your body paragraphs and decide if your categories need to be presented in a specific order. You might consider these types of organization: *emphasis*, *frequency*, *nationality*, *age*, *personal preference*, *price*, *value*, or *quality*.

Step Six: Apply Critical Thinking

Critical thinking means thinking and then rethinking about a topic, with the goal of discovering all the implications you can about it. In writing a good classification essay, several concepts of critical thinking are important. In a classification essay, you often

- examine why the parts of a topic are important
- analyze the categories that comprise your topic
- provide precision in details
- discover a deeper appreciation for your topic

After you write each of your drafts, ask yourself the critical thinking questions below, and then apply your answers to your writing:

1. **Purpose:** Will readers understand why your topic is important? For example, if your thesis statement asserts that crackers come in three categories, readers will probably be bored. However, if your thesis statement includes your purpose—that, say, the categories are determined by the crackers' nutritional benefits—then you have a focused purpose your readers will understand and appreciate.

2. **Information:** Have you given readers enough evidence or details to understand the various aspects of the categories? Is all of your information relevant to your topic? Do you see any place to add details that will enhance or further explain your categories?
3. **Reasoning:** Are all your facts accurate? Are they clear? Are they relevant to your topic? Is your reasoning logical?
4. **Assumptions:** Have you made any assumptions that need to be explained or justified? Have you used any vocabulary that your audience may not understand?

12.15 Apply Critical Thinking

Directions: Ask members of your writing group to read your latest draft and answer the critical thinking questions below.

Purpose

1. Why is this topic important to me?
2. Did I clearly state or clearly imply my purpose? If not, where should I be clearer?

Information

3. In what places would adding details further enhance or explain my topic?
4. In what places would adding details further enhance or explain each category?

Reasoning

5. Where should I provide more accurate details?
6. Are my examples and categories relevant to my topic? If not, where did I stray from my topic?
7. Is my reasoning accurate? If not, where am I illogical?

Assumptions

8. Did I use any vocabulary you don't understand? If so, what words or phrases need definitions?

Step Seven: Revise Your Essay

Getting your essay on paper—creating your discovery draft—is your first step. Revising through subsequent drafts shapes your discovery draft into a finished product of which you can be proud. First, consult the general Review Checklist for an Essay in Chapter 3, page 61, and look for places in your essay you need to

Heads Up!
Keep hard copies of all your drafts and number them so you'll know which is your latest draft.

polish. Then think about revisions that are specific to classification essays: **broad topic**, **categories**, and **illustrations** or **examples.**

1. **Broad topic:** When you revise, check that you make clear in the introduction what your broad topic is and why you are classifying it.
2. **Categories:** When you revise, consider each category.
   - Is each category a logical part of the broad topic?
   - Does each category have at least one example you have elaborated on?
   - Does each category explain or illustrate some aspect of your broad topic?
   - Have you included enough categories to fully cover your broad topic?
   - Are your categories logically ordered?
3. **Illustrations or examples:** When you revise, look at your illustrations and examples.
   - Are your illustrations vivid or specific enough so readers will understand your categories?
   - Have you given too many illustrations or examples?
   - Have you written anything that flows away from a paragraph's focus?

12.16 Revise Your Essay

Directions: From your latest draft, write your thesis statement, the topic sentence for each body paragraph, the category each topic sentence introduces, and a short description of how each topic sentence illustrates your thesis statement. Ask members of your writing group to review this material and note any places where . . .

- categories do not prove topic sentences
- topic sentences stray from your thesis statement
- sentences in body paragraphs stray from their topic sentences

Review suggestions from your writing group members and revise your draft accordingly.

MyWritingLab™

Writing Assignment

Consider the topics below or one that your instructor gives you. Expand one into a classification essay:

1. current fads
2. housing options
3. important (local, state, national, or international) events in the last year
4. presents you have received

5. ways you spend (or would like to spend) free time
6. features that are important when purchasing a vehicle
7. people who are in the news
8. physical attraction
9. animals in your area
10. clothes in your closet

TECHNO TIP

For more on classification writing, visit this site:
Guide to Grammar and Writing Classification and Analysis

MyWritingLab™ Visit *MyWritingLab.com* and complete the exercises and activities in the **Paragraph Development-Classification** and **Essay Development-Classification** topic areas.

RUN THAT BY ME AGAIN

- **In classification writing, you . . .** take a general topic and then sort it into simpler or more helpful categories.
- **In classification writing, you frequently start with a topic and then . . .** look for an organizing principle to help develop categories through which you can separate examples.
- **Your purpose in classification writing is . . .** to make a broad subject easier to understand.

LEARNING GOAL
❶ Get started with classification writing

- **After you select a topic for your classification paragraph or essay, you need to determine . . .** your organizing principle or the method you will use to create categories for the topic.
- **In a plan of development topic sentence for a classification paragraph, the sentence names . . .** the categories the paragraph will develop.

LEARNING GOAL
❷ Write a classification paragraph

- **When writing a classification essay, answer the starter questions . . .** Whom or what do I want to classify? What is my purpose? What divisions or categories identify specific differences of the subject?
- **To classify a topic, you need to discover . . .** details unique to each category.
- **Your audience choice will help you decide the language you use . . .** to name your categories, the type of details you select, and the language you use to illustrate or explain those details.
- **If body paragraphs are of equal importance . . .** the order in which they appear in your essay does not matter.
- **In classification writing, critical thinking means you often . . .** analyze categories that comprise your topic, provide precision in details, discover a deeper appreciation for your topic, and reflect on the importance of your topic.
- **Types of revisions specific to classification essays include . . .** broad topic, categories, and illustrations or examples.

LEARNING GOAL
❹ Write a classification essay

CLASSIFICATION WRITING LEARNING LOG MyWritingLab™

MyWritingLab™

Answer the questions below to review your mastery of classification writing.
Answers will vary.

1. In classification writing, what is your purpose?
   Your purpose is to make a broad subject (your topic) easier to understand by creating divisions or categories that identify specific differences.
2. In classification writing, what is the advantage of starting with a topic and then looking for an organizing principle?
   This approach allows you to develop the examples that share similar characteristics.
3. After considering your topic and purpose, what should you think of?
   You should think of who would be interested in or benefit from reading about your topic and categories.
4. After you answer your starter questions and compose your starter statement, what should you create?
   You should create your working thesis statement.
5. In order to classify a topic, what do you need to discover?
   You need to discover the details unique to each category.
6. Details in different categories should provide information that helps readers understand what areas?
   Details should help readers understand each category and the larger topic.
7. Defining your audience will help you focus more on what area?
   Defining your audience will help you focus more on your purpose.
8. Examples you select for your categories should have what four qualities?
   Examples should explain, clarify, or illustrate their specific category; be realistic or practical for the topic; be appropriate for the audience; and help readers understand the category and broad topic.
9. What types of organization might be helpful or necessary for body paragraphs?
   These types of organization might be helpful or necessary: emphasis, frequency, nationality, age, personal preference, price, value, or quality.
10. In critically thinking about the purpose of your classification writing, addressing what question is helpful?
    Will readers understand why my topic is important?

11. In critically thinking about information in your classification writing, what questions are helpful?

    Have I given readers enough evidence or details to understand the various aspects of the categories? Is all my information relevant to my topic? Do I see any place to add details that will enhance or further explain my categories?

12. In critically thinking about reasoning in your classification writing, what questions are helpful?

    Are all my facts accurate? Are they clear? Are they relevant to my topic? Is my reasoning logical?

13. In critically thinking about assumptions made in your classification writing, what questions are helpful?

    Have I made any assumptions that need to be explained or justified? Have I used any vocabulary that my audience may not understand?

14. In revising the broad topic of your essay, what areas should you check?

    Check that you make clear in the introduction what your broad topic is and why you are classifying it.

15. In revising your essay, what areas should you check when looking at your categories?

    Check that each category is a logical part of the broad topic, has at least one example that you have elaborated on, explains or illustrates some aspect of your broad topic, has enough categories to fully cover your broad topic, and is logically ordered.

16. In revising your essay, what areas should you check when looking at your categories' illustrations or examples?

    Check that illustrations are vivid or specific enough that readers will understand your categories, that you have not given too many illustrations or examples, and that you have not written anything that flows away from a paragraph's focus.

CHAPTER 13

Cause-and-Effect Writing

LEARNING GOALS

In this chapter, you'll learn and practice how to

1. Get started with cause-and-effect writing
2. Write a cause-and-effect paragraph
3. Read and examine student and professional essays
4. Write a cause-and-effect essay

GETTING THERE

- Cause-and-effect writing lets you discover and analyze the origins and results of events, actions, or influences.
- Successful cause-and-effect writing helps readers understand connections through your analysis.
- Relevant details give authenticity and support to cause-and-effect writing.

FASTWRITE 1

The pictures above show two businesses that were not successful. Choose one of the businesses and write about the reason(s) why it closed. Be creative and consider what cause(s) might have led to this effect. Write as much as you can, as fast as you can, for three minutes.

FASTWRITE 2

These pictures capture life-changing moments. Choose one, and write about what might have been the cause that led to the moment depicted in the picture. Then elaborate on what effects this moment might have for the person or people. Write as much as you can, as fast as you can, for five minutes.

Complete these **Fastwrites** at **mywritinglab.com**

Getting Started in Cause-and-Effect Writing

Cause-and-effect writing identifies and examines the reasons for and consequences of an action, event, or idea. Your purpose might be to

- show why an event occurred
- identify the reasons something changed
- explain the consequences of an event or action

LEARNING GOAL

❶ Get started with cause-and-effect writing

Whatever your purpose, cause-and-effect writing carefully explains the relationship between the cause(s) and the effect(s). For instance, when Ella listed the causes and effects of texting during a meeting, she wrote this:

Your topic may have one cause and many effects, one effect and many causes, or any combination.

Cause: was really bored by the discussion
Cause: wanted to know what friends were doing
Cause: needed to get update about sister's health
Effect: made boss angry
Effect: missed information about what my next project would entail
Effect: was relieved to learn that sister got an "all clear" from doctor

In a cause-and-effect essay, body paragraphs may focus on causes or effects or both.

Next, Ella examined the causes and effects of her texting. From her list, she chose to examine the following relationship:

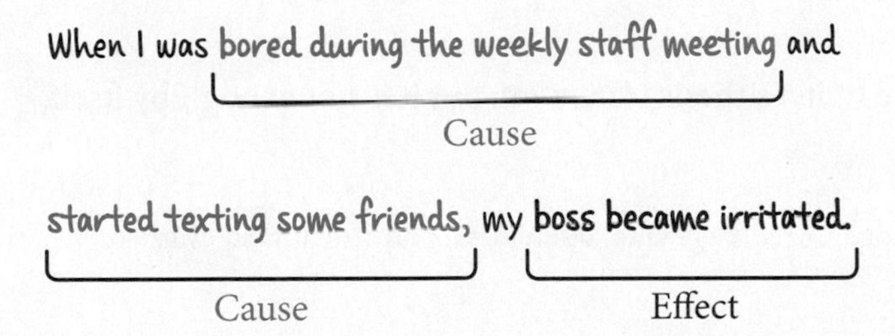

Ella elaborated on the cause-and-effect of texting like this:

The staff meeting had already lasted forty-five minutes, and I was really bored. To keep from falling asleep, I texted my bff. "OMG!!! BOOMS here. How RU?" I wrote. Just as I hit "send," Mr. Park, my boss, said my name. I took from his tone he was repeating a question he'd already asked me. Someone kicked me under the table. Looking up, I was shocked to see Mr. Park standing over my

Heads Up! Academic writing does not allow abbreviations and slang common in text speak, unless the writer is directly quoting.

(continued)

shoulder, hands on hips, glaring down at my cell phone. Then I heard a few nervous laughs from other staffers.

Here Ella built on her listing with more complete details of both the causes (lengthy meeting, boredom, texting) and the effects (irritated boss, concerned coworker, amused coworkers) of her actions.

When drafting cause-and-effect writing, make sure your causes are true to their effect. That is, focus on developing relevant and reasonable causes. Two problems often arise:

1. **Writers sometimes don't cite enough details to prove their effect.**

> The principal at my daughter's school unfairly suspends students. He needs to be reassigned.

In this example, the writer hasn't provided enough details to prove that the principal is unfair when suspending students. While the suspensions may be a relevant cause to reassign the principal, the writer needs to provide more details to prove the suspensions were unfair.

> The principal suspended my daughter and two of her friends for being five minutes late to school.

This supports the writer's thesis statement, but it is not enough by itself.

> He also suspended three boys who laughed during the spring choral concert.

Again, this supports the thesis statement, especially when combined with the previous reason.

> In fact, he has suspended more than 33% of the student population in the last two months.

With three examples, the writer has sufficient evidence (causes) to support the effect in the thesis statement.

2. Writers sometimes cite details that are irrelevant to their effect.

> Fourteen people in class wore blue jeans on the day of the test. The people who wore blue jeans all scored an 80% or above on the test. Therefore, wearing blue jeans leads to good grades.

Obviously, the cause (wearing blue jeans) does not lead to the effect (earning a good test grade). This was merely a coincidence, so this detail should not be included.

13.1 Matching Causes and Effects

Directions: In 1–5, match the causes with their effects.

	Causes	Effects
d	1. trips to gym, vitamin regimen, less junk food	a. didn't study
c	2. low performance ratings, frequent absences, bad attitude	b. no summer travel
a	3. napped, partied, played video games	c. lost job
e	4. increased spam, server interruptions, more pop-ups	d. better health
b	5. increased gas prices, no pay raises, high insurance rates	e. changed ISP

In 6–10, match the effects with their causes.

	Effects	Cause
c	6. faster communication, less paperwork, more identity theft	a. global warming
d	7. insecurity about academic abilities, shyness	b. volunteering
e	8. relaxed, headache is soothed, forgot about work	c. use of Internet
b	9. self-esteem, feeling of reward, community pride	d. lack of self-confidence
a	10. spread of disease, more hurricanes, melting ice caps	e. listening to jazz

13.2 Cause-and-Effect Fastwriting

Directions: For three to five minutes, fastwrite about the causes and effects of stress in your life. Use as many details as possible. Try to identify the same number of effects and causes. When you're finished, share this with your writing group.

WRITING at WORK: *Snapshot of a Writer*

Judy Carrico, Advanced Practice Registered Nurse

As an advanced practice registered nurse (APRN), Judy Carrico has additional education and training, allowing her to provide expert health care similar to that of a physician. APRNs have completed at least a master's degree in a specialty field in nursing. Carrico works for an endocrinologist, and Carrico is concerned with controlling the problems her patients experience as a result of diabetes.

Writing at Work

As a medical professional, I use writing in several ways. First, I use it as a collection of data describing a patient's visit and reminding the health-care provider (HCP) of what has previously been done. Writing also serves to meet the criteria for insurance billing, and to communicate with other HCPs.

The day-to-day documentation is an office note that includes subjective data, objective data, reviews of various tests, effects of treatment, and plan of care. Additional documentation includes letters to other HCP and insurance companies. In general, the letters I write are just that—detailed notes explaining a patient's office visit or request for a needed service.

I write office notes in very simple and to-the-point wording involving short statements that may not even be complete sentences. I write these notes in various ways, but they generally follow a format called a SOAP note. SOAP is short for Subjective data, Objective data, Assessment, Plan.

Subjective Data First, I list a reason for a patient's visit. Because I work for an endocrinologist, the patients I see are diabetics, and their visits are related to problems associated with diabetes.

Objective Data Second, I offer a detailed description reviewing the patient's problem. This area must include statements regarding severity, quality, location, duration, context, timing, modifying factors, and associated symptoms of the problem. Information here comes from what the patient tells me in his or her visit. For my diabetic patients, this involves their latest blood sugar readings, along with their diet and the amount of exercise they fit in each day.

Assessment Generally at this point I provide some reflection on the patient's medical, social, and family history. Next, I give a review of symptoms not addressed thus far in the documentation. Then I record the actual physical exam, followed by a list of one- or two-word items in my diagnosis.

Plan Last, I describe the plan of care. This includes any testing, teaching, medications, and treatments, as well as patient responses and a follow-up plan. For many of my patients, this includes attending classes to learn more about diabetes and blood glucose management. We may also talk about adjustments in the patient's diet and exercise routine.

In years past, medical notes served only as a record and reminder of the patient's history and what had been done. Because of billing issues, today insurance and Medicare drive the process of documentation. Others in my profession also use their writing to provide continuity of care for patients; in addition, their written documentation helps obtain approval of payment for medicines, testing, and procedures.

TRY IT!

Recall a time you went to or took someone to a doctor's office or an emergency room. Following the SOAP format, freewrite the subjective data, objective data, assessment, and plan that might have been used. When you're finished, share your work with those in your writing group.

Getting Started with Your Cause-and-Effect Paragraph

LEARNING GOAL

❷ Write a cause-and-effect paragraph

You may feel more comfortable writing a cause-and-effect paragraph before you tackle a cause-and-effect essay. If that is the case, follow steps one through six below. If not, go on to Take 2 on page 303.

Usually, a cause-and-effect paragraph begins with a topic sentence that explains the effect that came about as a result of a cause or causes. Sentences that follow the topic sentence explain the cause(s) that led to the effect. For instance, you might have this effect:

Finally, after many attempts, I quit smoking for good.

In the rest of the paragraph, readers expect you to discuss reasons why your final attempt was successful—that is, the causes that led to this effect. You might cite several causes (e.g., you were experiencing shortness of breath, you realized that you were setting a poor example for someone who looked up to you, the price of tobacco had risen too much for your budget). Or, you might cite one cause that was dramatic enough to lead to this effect (e.g., a loved one died from lung cancer).

Remember that "e.g." (Latin for *exempli grati*) means "for example."

Step One: Choose a Topic

Heads Up!
Substituting the word "result" for "effect" and "reasons" for "causes" might help you better understand your topic.

Often instructors will assign a particular topic. If you have the freedom to choose a topic for your cause-and-effect paragraph, remember that your goal is to explain the relationship between a consequence and its reason.

See Chapter 1 for additional prewriting techniques such as listing, clustering, freewriting, reporter's questions, and journaling.

TICKET to WRITE

13.3 Choose a Topic

Directions: Complete the following to start the process of choosing a topic. If necessary, freewrite for a few minutes on the subject and then on your lists. Answers will vary.

1. List at least three occasions when you had an extremely bad day.
2. List at least three occasions when you had an especially good day.
3. List at least three times when you were forced to tell a "little white lie."
4. List at least three times when you lost your temper.

Step Two: Generate Ideas

"I keep six honest serving-men
(They taught me all I knew);
Their names are What and Why and When
And How and Where and Who."
—Rudyard Kipling

Now determine which topics you can develop. Do this by generating supporting ideas for them. One effective method of generating this support is creating and answering reporter's questions (*Who? What? When? Where? Why?* and *How?*).

First, write a statement that identifies the effect you want to illustrate, and then ask questions about that effect. Your answers to reporter's questions will help you generate ideas you may use later to write your paragraph.

Here is how Gregg generated ideas to support his topic regarding *a bad day at work.*

Topic: my bad day at work

Statement identifying effect: I had to apologize to coworkers because of my bad day.

Generating ideas: To whom did I apologize? — To three coworkers.

What did I do that led to the apology? — I was rude with them.

<u>Why</u> was I rude to them? — I had a run of bad luck before work.

<u>Where</u> was I when things were not going my way? — I was not even at work yet.

<u>When</u> did I apologize to them — At the end of my shift.

<u>How</u> did they react? — They appreciated my apology.

13.4 Generate Ideas

Directions: From each list of topics you generated in Ticket to Write 13.3, circle one topic that attracts your attention. Then write a sentence identifying the effect you will illustrate. Next, freewrite answers to reporter's questions to name causes that bring about the effect. Form your own reporter's questions or use the following: Answers will vary.

1. **Had bad day**
   a. What were the causes of your bad day? (Why did you have a bad day?)
   b. Who else was involved?
   c. How did you react?
   d. Where did this happen?
   e. When did this happen?
2. **Had good day**
   a. What were the causes of your good day? (Why did you have a good day?)
   b. Who else was involved?
   c. How did you react?
   d. Where did this happen?
   e. When did this happen?
3. **Told white lie**
   a. Why did you lie?
   b. Who else was involved?
   c. What were the repercussions? (How did others react?)
   d. Where did this happen?
   e. When did this happen?
4. **Lost temper**
   a. Why did you lose your temper?
   b. Who else was involved?
   c. What were the repercussions? (How did others react?)
   d. Where did this happen?
   e. When did this happen?

Step Three: Define Your Audience and Purpose

"Life is a perpetual instruction in cause-and-effect."
—Ralph Waldo Emerson

Once you have written a statement identifying your effect and you have answers to reporter's questions that elaborate on your causes, you should think about your **audience** and **purpose**.

Your purpose for writing determines your audience, and your audience helps you decide your language. You probably would not use the same level of language in a company newsletter about workplace etiquette as you would in a presentation to elementary school students about being kind to others.

After reviewing his responses to reporter's questions on the topic of *a bad day at work*, Gregg completed Ticket to Write 13.5 to discover his audience and purpose.

Topic: Bad Day at work

My audience is anyone who's taken out frustrations on others.

I want to illustrate what caused me to treat my coworkers poorly.

My purpose is to show others how a little bad luck can ruin the day for others.

TICKET to WRITE

13.5 Discover Your Audience and Purpose

Directions: To discover your audience and purpose, review the material you created in Ticket to Write 13.4. For each possible topic, list the audience, purpose, and effect you want to illustrate.

When finished, share this with your writing group. Ask group members to identify which topic they feel you could best develop into a cause-and-effect paragraph. Consider their feedback as you decide which topic to develop. Answers will vary.

Step Four: Draft Your Paragraph

Because paragraphs center on a topic sentence, you need to review your topic, audience, and purpose to develop your topic sentence. Your topic sentence should focus on the topic of the paragraph and the point you want to make about that topic. Use this information to help focus your purpose and create a topic sentence your audience will understand.

Here is Gregg's review of his audience and purpose, and the topic sentence he created:

Topic: *What* do I want to illustrate?
my bad day at work

Audience: *Who* are my readers?
people who've taken out their frustrations on others

Purpose: *Why* do I want to illustrate this?
to show others how a little bad luck can ruin the day for others

Topic Sentence: At the end of my shift yesterday, I apologized to three of my coworkers for having been grumpy with them.

13.6 Create a Topic Sentence

Directions: Write your topic, its purpose, and your audience. Then create your topic sentence. Ask members of your writing group if they agree that your topic sentence explains your purpose and is appropriate for your audience. Answers will vary.

After writing your topic sentence, use examples from your prewriting to compose the supporting details of your paragraph. Remember, you do not have to use all the ideas from your prewriting, but you should include all the important points necessary. Choose examples that help make your point, and use details you created to build your paragraph. You may discover you don't need to or can't include all your prewriting, and you may discover other supporting ideas you hadn't thought of before. Keeping your audience and purpose in mind helps you decide which details you need most.

"Shallow men believe in luck, believe in circumstances . . . Strong men believe in cause-and-effect."
—Ralph Waldo Emerson

In the example below, Gregg illustrates the causes (the reasons he shares in the paragraph) and the effect he expresses in his topic sentence (why a bad day led him to apologize).

At the end of my shift yesterday, I apologized to three of my coworkers for having been grumpy with them. I intended to run several errands before my shift started at 3:00 p.m. However, a number of events soured both my day and then me. First, I overslept. Then, Freddie, my coworker who was supposed to help me load some boxes in my truck for delivery, called to say he had to go into work early and couldn't help me. So, I had to load the truck by myself, which took twice as long. By the time I was finished, I had just ten minutes to get to work. Then, I ran into unexpected construction on the road. I could see the plant and parking

(continued)

lot, but I was stuck in traffic, not moving for fifteen minutes. My nerves were shot by the time I got to work, and I snapped at anyone who talked to me. I was rude to Freddie when he apologized for not helping me, to my boss when he joked about my being two minutes late, and to Cheryl, who offered me some of her birthday cake. By the end of the day, I had calmed down and realized I'd been a jerk, so I made a point of apologizing to everyone.

13.7 Compose Supporting Details

Directions: Using the topic sentence you created in Ticket to Write 13.6 and the supporting ideas you generated in Tickets to Write 13.4 and 13.5, compose supporting details that illustrate your cause-and-effect topic. Then share them with your writing group. Ask members if you have given enough details to fulfill the purpose you express in your topic sentence.

Step Five: Revise Your Paragraph

See Chapter 4 to review proofreading techniques.

After you've finished your paragraph, save it, print a hard copy, and read it out loud, looking and listening for errors in grammar and content. To be sure you revise as completely as possible, refer to the general Review Checklist for a Paragraph in Chapter 3 and the Review Checklist for a Cause-and-Effect Paragraph below.

Heads up!
See Chapter 20 to review fragments.

☑ REVIEW CHECKLIST **for a Cause-and-Effect Paragraph**

- ☐ Is my topic sentence clear?
- ☐ Have I included sufficient details to illustrate my causes and effects?
- ☐ Do I hear any fragments when reading my paragraph out loud?
- ☐ Do all my complete sentences have appropriate end punctuation?
- ☐ Is my paragraph appropriate for my audience?
- ☐ Does my purpose remain constant?
- ☐ Are any sentences or details unessential?

Once you've made any necessary changes, save your work and leave it alone for a while. Then come back later, print out a new copy, and look and listen for errors again.

Step Six: Peer Review

You might think you've found all of your errors and made all of the improvements you can, but having someone else read your paragraph and offer suggestions will almost always improve your writing. This person is called your **peer reviewer**. Your peer reviewer may use the review checklist above, or your instructor may provide a different checklist. You might or might not agree with the suggestions you receive, but peer reviewers often find places for improvement or errors that slipped by you. Listen to or read closely your peer reviewers' suggestions, and don't be shy about asking questions to clarify their ideas.

Peer reviewers bring a fresh perspective. Consider them valuable resources as audience members and fellow writers.

13.8 Peer Review

Directions: Once you have revised your paragraph, share it with your peer reviewers. Ask them to review your paragraph using the Review Checklist for a Cause-and-Effect Paragraph above and the general Review Checklist for a Paragraph from Chapter 3. Then ask them to record their suggestions for revision. Use these suggestions as you revise your paragraph again.

Student and Professional Essays

Below are two cause-and-effect essays. The first was written by a student and the second by a professional. Read these two essays and answer the questions that follow them.

LEARNING GOAL

❸ Read and examine student and professional essays

Student Essay

A Cause for Pinching Pennies

by Richard Ervin

(1) Last year, I traded in my 8-cylinder, 4-wheel-drive truck for a compact car. The new vehicle took a while for me to **get used to**, but the money I saved on gas didn't. Now because of the **recession**, I have taken more steps to save money.

pinching pennies an idiom meaning "spending as little as possible"

get used to an idiom meaning "become accustomed to"

recession an extended decline in general business activity

2 A household with two children and two adults spends a great deal on food. When my wife and I reviewed our checkbook and our weekly spending habits, we discovered we were spending almost $200 a month going out to eat on certain nights. Between Rick Jr.'s Monday night **FFA** meetings and Friday night soccer games, and Reagan's Tuesday night softball games, we easily slipped into some unhealthy and expensive habits. Monday nights we had pizza delivered; Tuesday nights we had burgers and hotdogs at the ball fields; and Friday nights we went out for a late dinner with other soccer families at Mr. B's, a family restaurant known for its pizza, chicken wings, and onion rings. We found that with preplanning, we could prepare full meals and freeze them for later in the week. Breaking the old habits took a few months, but once we started seeing the money we saved, we loved our new eating habits. We were also spending between $120 and $150 per week for our children to buy lunch at school and for the two of us to purchase lunch at work. So, we came up with a list of **brown-bag** lunch options for all four of us. Just like limiting our pizza and burger nights, we discovered our nutrition had improved and we saved up to $70 a week by taking our lunches.

3 Another decision we made to save money was to make our home more **energy efficient**. After reading an article on how to conduct our own **home energy audit**, we made a few alterations to the house. First, we replaced the washers in the kitchen faucet. The cold water had been dripping for almost six months. I'm sure this didn't help the water bill any. Next, we lowered our water heater thermostat from 140° to 120° and we put a timer on it. According to the Department of Energy's Web site, the timer will pay for itself in one year. Then, we **insulated** the house for drafts. This included weather stripping all the doors and installing foam insulators in all our light switches and electrical outlets. We also checked out all our **ductwork** for leaks and taped three holes. The cost of the weather stripping, outlet and switch insulators, and duct tape was $19.75. Since our utility bill averages $250 per month, we should save $300–$600.

4 One of my family's best money-saving efforts was changing our summer and spring break vacations. Last spring break, we went to Disney World for five days. The year before, we enjoyed a four-day cruise to the Bahamas. This year we made a couple of weekend trips in our home state. I always wanted to see the world's largest pumpkin, which was just a two-hour drive from home. Plus, we love baseball.

5 Don't believe you can't save money in a recession. In just one year, and with only a little time and money, my family saved over $6,000. I am sure we'll return to see Mickey and Minnie one day, but now we are going to enjoy baseball and camping in our own backyard.

FFA Future Farmers of America, a national youth organization promoting agriculture

brown-bag lunch brought to work, often in a small brown paper bag

energy efficient using little gas, electricity, or other resources

home energy audit a check of conditions of heating and air systems, lighting, water, and appliances

insulated prevented the loss of heat by surrounding with non-conducting material

ductwork a system of pipes used in heating and air conditioning

A CLOSER LOOK

Answer these questions about the essay:

1. What is the general *cause* behind the problems the author details?

   The general cause is the recession.

2. What is the general *effect* the author identifies in response to the cause?

   He has taken a number of cost-saving steps.

3. Where is the thesis statement in this essay?

   The thesis statement of the essay is in the last sentence of paragraph 1.

4. The topic sentence of paragraph 2 is not clearly stated. Write a topic sentence for that paragraph.

   Answers will vary. Possible answer: The first step we took to save money was to change our eating habits.

5. List the details in paragraph 2 that illustrate the expense of the family's eating habits.

   spending $200 per month eating out; pizza delivery; family restaurant; $120–$150 per week for lunches

6. What details illustrate the family's responses to the food budget problem?

   brown-bag lunch alternatives; limiting pizza and burger nights

7. List the transitions in paragraph 3.

   Transitions include *another, after, first, next, lastly,* and *also.*

8. Which sentence in paragraph 4 is not fully developed?

   Plus, we love baseball.

9. Suppose you were the author of this essay. List details that would fully develop the idea from paragraph 4 that the author does not develop.

   Answers will vary. Possible answer: Plus, we love baseball. The summer months are full of inexpensive minor league baseball games we can see locally. Minor league games are usually family-oriented and always have something for everyone to enjoy in addition to the game itself.

10. In paragraph 5, the author states the overall *effect* of his family's money-saving efforts. Write it below.

    In one year, the author's family will save over $6,000.

The professional essay below describes a winter storm that the author experienced.

Professional Essay

Life Interrupted

by Laura Winspear

1 We should have known. Newscasts had warned us, and, after all, it was January. In spite of advance notice that a winter storm was approaching, most people had no idea of what was to come. We went to bed that Monday night expecting to see several inches of snow the next morning, but we didn't expect to wake up during the night wondering if we were hearing gunshots.

2 As it turned out, what we had heard was the result of a two-part winter storm that packed the expected snow and also dumped an unprecedented amount of frozen precipitation. This ice and sleet led to the strange nighttime sounds of branches and trees crackling and falling, and transformers blowing, as the great ice storm of 2012 marched through. When the storm was finally over, everything was **encrusted** in over an inch of ice, and about five inches of snow lay on the ground. All of this had a crippling effect on our entire area; in fact, our everyday lives turned upside-down for quite some time.

3 When the storm had been raging for just an hour, officials cautioned that road conditions were extremely dangerous; a number of vehicles had already slid into ditches. The slick roads, however, weren't the only problem, as destruction caused by the storm led to other hazards. The ice was so heavy that it brought down thousands of branches and even entire trees, which lay across roads, **rendering** the roads **impassable**. When the branches and trees came down, they often toppled nearby wires and utility poles; in fact, more than twenty thousand downed wires and one thousand broken poles were reported by the storm's end. As more power lines fell, county fire departments quickly became overburdened. On a normal night, these departments make no more than ten runs; on the night of the storm, the crews were called out almost three hundred times. Because of these **impediments**, city and county officials declared a Level 1 state of emergency, meaning life-threatening conditions were present and only essential travel was advised. This state of emergency lasted thirty-six hours.

encrusted covered, coated
rendering leaving, causing to be
impassable blocked, closed
impediments obstacles, obstructions

(4) Another problem that accompanied downing of wires and poles was the loss of electricity, cable service, Internet service, and telephone service (landline and cellular) in about 90% of the county. Power went off in the middle of the night, so not until the next morning did most realize the temperature in their homes had **plummeted**. Thousands were without power, leaving those without alternate heat—that is, most of them—in houses that were chilled and getting chillier. With outside temperatures reaching the low twenties during the day, the only way to get warm was to put on many layers of clothing. Electrical teams from the city and county worked frantically to restore power, but they were hampered by the length and **severity** of the storm and by problems with removing limbs and trees. Cable and telephone companies faced the same problems in restoring their services. Because of losses in these areas, most residents had little contact with the outside world, and the only way to receive any information was via a battery-powered or hand-cranked radio.

(5) As a result of the problems, routine life was disrupted in many ways. Businesses tried to reopen as soon as the state of emergency was lifted; however, since most were still without power, their efforts were futile. Some of the businesses with electricity had no telephone service, so paying for a purchase with a credit or debit card was out of the question. Customers who wanted to pay cash were unable to get money out of their ATMs. Schools, too, were shut for a number of days. The community college closed its doors the rest of the week; area public and private schools stayed **shuttered** eight school days, and some daycare centers did not reopen for more than two weeks.

(6) In spite of all of the hardship the ice storm brought, stories of courage and kindness quickly emerged. For weeks after the storm, local media were **engulfed** with tales of neighbor helping neighbor. Those without power were invited into the homes of others—even complete strangers—fortunate enough to have electricity. Children shoveled and scraped the sidewalks of their elderly neighbors. Youth groups from local churches delivered hot meals and space heaters. The Red Cross, the Salvation Army, and private businesses opened shelters. Boy Scout **Venture Crew #21** organized a shuttle service, taking those in need to these area shelters and assisting the National Guard, who checked on every county resident.

(7) County rescue teams performed a number of services. They delivered prescriptions to those unable to drive to their pharmacies; they took healthcare workers to their jobs; they transported donated cots and blankets to shelters; they drove **dialysis** patients to the local hospital. In some cases, they even carried animals to area veterinarians.

(8) For thirty-six hours we lived in a state of emergency, and for a longer time we lived without everyday conveniences like electricity and telephone service. But we also lived with the comfort of seeing neighbors and strangers caring for each other. Because of all these consequences—good and bad—the ice storm of 2012 will remain a memorable event for all who experienced it.

plummeted fallen dramatically

severity harshness, brutality

shuttered closed

engulfed overwhelmed

Venture Crew #21 a scouting group of youth aged fourteen to twenty

dialysis a machine that provides artificial replacement for lost kidney function

A CLOSER LOOK

Answer these questions about the essay:

1. Underline the thesis statement of this essay.
2. Paragraph 3 details *causes* that led to the state of emergency being declared. List these causes.

   roads were impassable, power lines fell, utility poles were broken
3. Slick roads led to what *effect* (paragraph 3)?

   A number of vehicles slid into ditches.
4. Cite problems *caused* by heavy ice on trees (paragraph 3).

   Ice toppled branches and trees, which lay across roads and made roads impassable; branches and trees often toppled wires and utility poles.
5. Cite transition words and phrases in paragraph 4.

   another, so, though, because
6. Cite the *effects* of the downed wires and poles (paragraph 4).

   Possible answers: the loss of electricity, cable service, Internet service, and telephone service (landline and cellular); houses that were chilled and getting chillier; electrical teams from the city and county worked frantically to restore power
7. Which sentence in the conclusion sums up the writer's view of the ice storm's effects?

   Because of all these consequences—good and bad—the ice storm of 2012 will remain a memorable event for all who experienced it.
8. Cite ways in which "routine life was disrupted" (paragraph 5).

   Businesses could not reopen because of lack of power or telephone service (which meant no credit or debit card service); customers could not access their ATMs; public and private schools and the community college were closed; day care centers were closed.
9. In paragraph 6, what phrase signals a contrast in listing hardships to listing "stories of courage and kindness"?

   In spite of
10. Which statement best describes this essay? d

    a. It cites two effects of a single cause.
    b. It cites a single effect of a single cause.
    c. It cites only the negative effects of a cause.
    d. It cites a major cause that led to several effects.

TAKE 3 Writing Your Cause-and-Effect Essay

LEARNING GOAL

❹ Write a cause-and-effect essay

Essay writing is much like paragraph writing: you start with a topic and expand with details. Remember that just as a paragraph supports a topic sentence, an essay supports a thesis statement. Your cause-and-effect essay will do one of the following:

- illustrate how something came to be
- show the consequences of an event or action
- identify the reasons something changed

See Chapter 1 to review thesis statements.

Keep in mind that the causes behind your effect must be events that you can elaborate on or explain. A statement like "I drank just a cup of coffee for breakfast, so I was hungry at lunch" would be too narrow to develop as a thesis statement. Unless your story has interesting details for the reasons you skipped breakfast, you don't have a basis for a good cause-and-effect essay.

Step One: Choose a Topic and Develop a Working Thesis Statement

To begin your cause-and-effect essay, choose a topic that focuses on the relationship between a consequence and its cause. Then ask yourself these **starter questions**:

1. **What** result or effect am I writing about?
2. **How** and **why** did this result or effect come about? **What** caused it?
3. **What** details should I concentrate on or make readers aware of?
4. **How** will my readers benefit?

For instance, Randa chose to write a piece about a recently passed ordinance that banned smoking in public places in her hometown.

1. **What** result or effect am I writing about?

   Mt. Pilot adopting a no-smoking ordinance

2. **How** and **why** did this result or effect come about? **What** caused it?

   - new city council members
   - aggressive antismoking ad campaign
   - organization of health-care professionals

3. **What** details should I concentrate on or make my readers aware of?

- new city council members had campaigned on "progressive policies"
- aggressive antismoking ad campaign financed by local man who had been diagnosed with lung cancer
- health-care coalition advanced cause in several ways (spoke to commission, mailed antismoking brochures to residents, handed out pamphlets in offices)

4. **How** will my readers benefit?

purpose—to analyze the causes of the ordinance

Answering Starter Questions: Richard Ervin and Laura Winspear

In answering the starter questions, Richard Ervin, author of "A Cause for Pinching Pennies," wrote:

1. I am writing about **how my family responded to the recession.**
2. This came about because my family needed to change its spending habits.
3. I will concentrate on **what led up to the money-saving steps we took.**
4. Readers will benefit by **learning ways they can save money.**

Laura Winspear, author of "Life Interrupted," addressed the starter questions this way:

1. I am writing about **the hardships and acts of kindness local people experienced.**
2. This came about because of a recent ice storm.

3. I will concentrate on what led to the state of emergency being declared, the loss of power and other services, and acts of kindness.
4. Readers will benefit by learning negative and positive effects of a natural phenomenon.

Cathi is a nontraditional student who had not been around young people in quite some time. She had formed an opinion that local youth were unmotivated and deserved a slacker reputation. However, certain observations caused her to change her mind, and this is what she chose as the topic of her cause-and-effect essay.

Here is how Cathi answered the starter questions.

1. I am writing about the undeserved slacker reputation many young people have.
2. This came about because I saw positive activities many local young people are involved in.
3. I will concentrate on examples of young adults involved in their schools, communities, churches, and civic organizations.
4. Readers will benefit by learning reasons why many young adults do not deserve the slacker reputation.

13.9 Answer Starter Questions for a Cause-and-Effect Essay

Directions: To develop the focus of your cause-and-effect essay, answer these starter questions. Answers will vary.

1. *What* result or effect am I writing about?
2. *How* and *why* did this result or effect come about? (*What* was its cause?)
3. *What* details should I concentrate on or make my readers aware of?
4. *How* will my readers benefit?

As you learned in Chapter 1, all essays have **thesis statements**. Once you've answered the starter questions, you've begun to develop the focus of your essay. Return to your starter questions and consider your purpose. Then condense the information into a single sentence, using the format below. This becomes your **starter statement**.

Cause

I will write about ______________(cite conditions, events, or behaviors)
because I want to show readers ______________

Effect

(cite results of conditions, events, or behaviors).

Here is how Cathi filled in the starter statement:

I will write about the school, community, and church activities many young adults are involved in because I want to show readers **why I no longer believe that all young adults are slackers.**

Look at how three other writers chose to fill in the blanks for other topics to create their starter statements:

I will write about events that led to a drastic change in my hometown because I want to show readers **why Mt. Pilot adopted a nonsmoking ordinance.**

I will write about abuses of government bailout funds because I want to show readers **how these brought about American taxpayers' frustrations.**

I will write about methods I used to reduce my carbon footprint because I want to show readers **that everyone can have a green impact.**

13.10 Create a Starter Statement for a Cause-and-Effect Essay

Directions: Review your responses to the starter questions in Ticket to Write 13.9. Then consider your purpose and condense the information into a single sentence, using the starter statement format.

Next, rework your starter statement to form your working thesis statement. Eliminate any language not essential to your purpose, like *I will write about* and *because I want to show readers*.

Cathi reworked her starter statement into this working thesis statement:

Because I noticed young adults' involvement in school, community, and church activities, **I realized that many of them do not deserve a slacker reputation.**

Other writers reworked their starter statements to form these working thesis statements.

After pressure from various organizations and citizens, **the Mt. Pilot City Commission adopted a nonsmoking ordinance.**

Loopholes available only to the rich, bonuses given to those responsible for their firms' problems, and tax money underwriting perks to businesspeople **are three reasons American taxpayers are frustrated.**

By changing how I consume energy at home, at work, and in travel, **I have discovered some methods everyone can use to reduce their carbon footprints.**

13.11 Develop a Working Thesis Statement

Directions: Review your starter statement in Ticket to Write 13.10. Then reword it to form your working thesis statement.

Step Two: Generate Ideas

In a cause-and-effect essay, you have two primary goals:

- to tell your readers of a result (the effect)
- to show what brought about the result (the cause)

The key to a cause-and-effect essay is explaining and illustrating the link between certain events (causes) and their result (effect). For example, if the effect you want to discuss is that you're out of money, you would cite reasons that led to that problem. Maybe you spent too much at the mall, had an unexpected car repair bill, or didn't get the overtime pay you'd counted on. These are legitimate causes that resulted in your effect.

However, you probably couldn't identify other events, like window shopping downtown, staying up late watching a TV movie, or reading about treatment for the common cold as causing your financial problems. They wouldn't have been the reason you spent money, so you can't link them to your being in a financial bind.

To begin generating ideas, **prewrite** about your topic. When you think about the conditions, events, or behaviors that led to your effect, ask yourself these questions:

- What sensory details would offer readers a clear picture of what led to your effect?
- What personal experiences or examples support your effect?
- What statistical, factual, or research-based details support your effect?

Cathi's freewriting on her topic looked like this:

Before going back to school, the only place I saw young adults was on the streets around my neighborhood. These kids did nothing all day long. They weren't in high school anymore and they weren't working. My next door neighbor's 21-year-old son says he's in a band, but the only place they play is in the garage and usually during the day. He says he can't find a job. His dad says he doesn't try hard enough! Because of these kids, I thought all young people were lazy and deserved the reputation of slackers. But since I've gone back to college, I've seen so many active young people—most of whom are younger than 25. These college kids run the Soup and Social Program. They work in the downtown soup kitchen all week and they feed and interact with the clients. Many of them go to school and work part-time jobs. They're in clubs and groups that sponsor charities & other events. The student Nursing club offers free blood pressure checks. The Young Independents Club holds the annual Turkey Trot 10k race to raise money for the humane society. Fraternities and sororities raise money for the campus Feed the Hungry campaign. On most warm weekends, church youth groups from all over town hold car washes, and the proceeds support all kinds of programs in town. Until I became a student, I guess I never really looked closely at what so many young people were doing.

Cathi wondered how many young adults actually helped out in their communities. Here is what she found:

I read online at VolunteeringInAmerica.gov that 55 % of our nation's youth volunteer. The site also said more college students are volunteering every year.

Cathi reviewed her freewriting and researching to see what causes supported her effect:

Effect: **I realized that young adults don't deserve the rep of slackers.**

Cause: They run the Soup and Social Program

Cause: Most volunteers in the Soup and Social Program are under 25

Cause: Student Nursing club offers free blood pressure checks every semester

Cause: Young Independents Club holds the annual Turkey Trot 10k race to raise money for the local animal shelter

Cause: Fraternities & sororities raise money for Feed the Hungry campaign

Cause: On most warm weekends, church youth groups hold car washes

Cause: VolunteeringInAmerica.gov states 55% of US youth volunteer

Cause: More college students volunteering every year

13.12 Generate Ideas

Directions: Write the working thesis statement you created in Ticket to Write 13.11. Then freewrite as many details as you can about your cause-and-effect topic. Next share this with your writing group and ask them to note which ideas seem the most compelling or which best elaborate on the information in your working thesis statement.

Step Three: Define Your Audience and Purpose

Now that you have your topic, think of the reason you selected it. This is the time to put that purpose to paper. Decide *why* you want to write about your topic. Consider the following:

- **What** do I want to write about (**what** is my topic)?
- **What** do I want to illustrate or show about my topic?
- **Why** do I want to write about that topic?

Answering these questions helps you narrow your *purpose*; narrowing your purpose helps you pinpoint your audience. To determine your audience, ask yourself

- **Who** will *benefit* from reading my cause-and-effect essay?
- For **whom** is my cause-and-effect essay *not intended?*

Answering these questions helps you define your *audience*. Identifying your audience guides you in choosing the language and details that will communicate your purpose to this specific audience.

In defining her purpose and audience, Cathi reviewed her starter statement and working thesis statement.

Starter statement: I will write about the school, community, and church activities many young adults are involved in because I want to show readers why I no longer believe that all young adults are slackers.

Working thesis statement: Because of their involvement in school, community, and church activities, many young adults do not deserve a slacker reputation.

Then Cathi answered the purpose and audience questions this way:

- *What* do I want to write about? activities young adults are involved in
- *What* do I want to illustrate or show about that topic? show that many young adults are in lots of activities
- *Why* do I want to write about that topic? prove the slacker rep isn't right
- *Who* will benefit from reading my cause-and-effect essay? over thirty crowd
- For *whom* is my cause-and-effect essay not intended? young adults

13.13 Define Your Purpose and Audience

Directions: To discover your purpose and audience, review the material you created in Ticket to Writes 13.10 through 13.12. Write your starter statement and your working thesis statement. Then answer the questions about purpose and audience.

What do I want to write about?
What do I want to illustrate or show about that topic?
Why do I want to write about that topic?
Who will benefit from reading my cause-and-effect essay?
For **whom** is my cause-and-effect essay not intended?

When finished, share your information with your writing group. Use the group's feedback to make any necessary revisions. Answers will vary.

Step Four: Draft Your Essay

As you draft, keep your purpose in mind. If your working thesis statement tells about a specific effect, the purpose of your essay is showing and explaining the causes of that effect. If your working thesis statement focuses on a particular cause, the purpose of your essay is showing and explaining the effects of that cause.

Develop each supporting cause or effect by showing and explaining its significance to your working thesis statement. For example, because Cathi wanted to show that not all young adults were slackers, she focused her essay on showing and explaining the reasons she believed this (the causes that led her to believe this).

Working thesis statement: Because of their involvement in school, community, and church activities, many young adults do not deserve a slacker reputation.

Cause: College students everywhere are always busy on their campuses with charitable organizations, clubs, and other extracurricular activities.

Cause: Young adults take on volunteer roles.

Cause: Youth church groups always active in our community.

Cathi can develop each of her topic ideas (causes) to support her effect by describing young adults who are not slackers but are active and involved citizens. Although Cathi discovered other causes in her fastwriting, she eliminated those that were too vague or insignificant.

13.14 Write Your Discovery Draft

Directions: Use the material you compiled in Ticket to Write 13.10 through 13.13 to write your discovery draft. Refer to Chapter 2 to review discovery drafting.

When you have completed your discovery draft, share your work with members of your writing group. Ask them to note the following: Answers will vary.

1. What causes or effects are especially detailed and specific?
2. What causes or effects are vague, unclear, or need more development?
3. What additional causes should be included?
4. What additional effects should be included?

Consider group members' responses and revise your essay.

Step Five: Organize Your Essay

Your body paragraphs may focus on causes or effects.

In a cause-and-effect essay, the order of your body paragraphs and the details in them can be as important as the order of steps in a set of instructions. The organization of your cause-and-effect essay is based on the relationship of your topic and its support.

Gino is writing an essay that explains why his math grade dropped from an 80 to a 70. Here is Gino's list of details for one of his body paragraphs:

left my books in the car

teacher announced test would be in one week, on Friday

noticed Thursday night my books were gone

left top down all week because of good weather

put off studying until the last minute

When explaining the causes why his grade dropped, Gino needs to place his details in the proper order. When he does this, the reasoning behind the cause becomes clear. Here is Gino's paragraph with the details in order:

Putting off studying is one reason my math grade fell. A couple of weeks ago, on a Friday, our teacher announced we would have a test the next Friday. This gave us seven days to study. The day she announced the test, I put my books in the backseat of my car, put the top down, and went cruising around the lake after school. The weather was perfect all week, so I went to the lake almost every day, and I never put the top up. Thursday night around 9:00, I decided it was time to study for my math test. When I went to the garage to get my books, I couldn't find them. Someone had stolen them from the backseat. It was too late to do much of anything about replacing my math book or notes. I realized immediately I should never have put off studying until the last minute.

13.15 Organizing Your Supporting Details

Directions: Read through your draft and check that the details in your body paragraphs are logical and properly ordered. Make any revisions necessary so each body paragraph flows smoothly from its topic sentence to its concluding idea. When finished, ask members of your writing group to review your draft, checking for information that is missing or out of order. Ask them to write any suggestions they have for improving the order of details in each body paragraph and any suggestions they have about missing details.

Step Six: Apply Critical Thinking

Critical thinking means thinking and then rethinking about a topic, with the goal of discovering all the implications, results, or consequences. When you think critically about a cause-and-effect essay, you often

Critical thinking is also examining decisions you've made or beliefs you've adopted. You consider both what sustains your decisions or beliefs and what conclusions you can draw from them.

- examine the origins of a particular decision, action, or event
- analyze the relationship of particular events or actions that brought about a specific result
- discover a deeper appreciation for your topic

In writing a cause-and-effect essay, several concepts of critical thinking are important. After you write each of your drafts, ask yourself the critical thinking questions below, and then apply your answers to your writing:

1. **Purpose:** Have I clearly stated the cause(s) and effect(s) I plan to discuss? Will readers understand the connection between the cause(s) and effect(s)?
2. **Information:** Have I given readers details that are relevant in explaining the cause, the effect, and the relationship between the two? Should I add any details to further explore this relationship?
3. **Reasoning:** Are all of my facts accurate? Are they clear? Is my reasoning logical?
4. **Assumptions:** Have I made any assumptions that need to be explained or justified? Have I used any vocabulary my audience may not understand?

13.16 Apply Critical Thinking

Directions: Apply answers to the critical thinking questions and revise your essay again. After your revision, ask members of your writing group to read your latest draft and answer the critical thinking questions below. Answers will vary.

(continued)

Purpose

1. Do you understand why my cause-and-effect topic is important?
2. Did I clearly state or clearly imply my purpose? If not, where should I be clearer?

Information

3. Did I include enough details for you to understand the cause-effect relationship of my topic?
4. Do you see any places where adding details would enhance or further support my topic?

Reasoning

5. Are all my facts accurate? If not, where should I provide more accurate details?
6. Are all my facts are clear? If not, where should I change my wording to be clearer?
7. Are all my facts relevant to my topic? If not, where did I stray from my topic?
8. Is my reasoning logical? If not, where do I seem to be illogical?

Assumptions

9. Did I make any assumptions that need to be explained or justified? If so, what are these assumptions?
10. Did I use any vocabulary you don't understand? If so, what words or phrases need definitions?

Step Seven: Revise Your Essay

Heads Up!
Keep hard copies of all your drafts and number them so you'll know which is your latest draft.

Getting your essay on paper is just your first step. Revising through subsequent drafts shapes your first draft into a finished product of which you can be proud.

To begin, consult the Review Checklist for an Essay in Chapter 3 and look for places in your essay you need to polish. Then think about revisions that are specific to compare-and-contrast essays: **topics**, **details**, and **transitions**.

Topics

When you revise, look at your body paragraphs' topics. Focus each body paragraph on one topic. If you are illustrating the causes (reasons) of an effect, each body paragraph's topic will probably be a specific *cause*. Similarly, if you are illustrating the effects (results) of a specific cause, each body paragraph's topic will probably be a specific *effect*.

Details

When you revise, review your details.

- Will readers understand the importance of each cause and effect?
- Have you given enough details?
- Have you given too many details?
- Have you written anything that flows away from a paragraph's focus?
- Are your paragraphs arranged in the most logical or effective way?

Transitions

Your essay flows much more smoothly when you use transitional words and phrases. These show the relationship between causes and effects, linking your ideas logically. In this type of essay, you might use words and phrases that **emphasize** or **clarify**, such as those in the following chart:

Transitions Used to Show Cause and Effect		
accordingly	for this reason	probably
as a result	hence	result (from, in)
because (of this)	if . . . then	since
beside	in spite of	then
certainly	leading to	therefore
consequently	may	thus
created from	perhaps	unquestionably
due to	possibly	what followed

Writing Assignment

MyWritingLab™

Complete this Exercise

Consider the ten effects below or one that your instructor gives you. Choose one and expand it into a cause-and-effect essay:

1. Dropping a class
2. Applying for a personal loan
3. Breaking up with someone
4. Switching Internet or cell phone service providers
5. Finally speaking your mind
6. Realizing the need to economize
7. Signing up for identity theft protection
8. Unrest in (name and area of the country or the world)
9. The rise (or decline) in people suffering from (name a disease)
10. The rise in viewership of cable TV channels

Some of these topics may need outside support. Consult newspapers, magazines, and various online sites such as CNN.com, MSNBC.com, USA-TODAY.com, and THEWEEK.com.

TECHNO TIP

For more on cause-and-effect writing, search the Internet for these videos:
Academic Writing Tips: How to Write a Cause & Effect Essay
Chapter 6: Cause & Effect, Process Analysis

MyWritingLab™ Visit *MyWritingLab.com* and complete the exercises and activities in the **Paragraph Development-Cause and Effect** and **Essay Development-Cause and Effect** topic areas.

RUN THAT BY ME AGAIN

LEARNING GOAL
❶ Get started with cause-and-effect writing

- **Cause-and-effect writing identifies and examines . . .** reasons for and consequences of an action, event, or idea.
- **When drafting cause-and-effect writing, make sure your causes are true . . .** to the effect.

LEARNING GOAL
❷ Write a cause-and-effect paragraph

- **A cause-and-effect paragraph usually begins with . . .** a topic sentence that explains the effect that results from a cause or causes.

LEARNING GOAL
❹ Write a cause-and-effect essay

- **The purpose of a cause-and-effect essay could be to . . .** show why an event occurred, identify the reasons something changed, or explain the consequences of an event or action.
- **The causes behind your effect must be . . .** events you can elaborate on or explain.
- **The key to a cause-and-effect essay is . . .** explaining and illustrating the link between certain events (causes) and their result (effect).
- **When determining your purpose, consider . . .** what you want to illustrate or show about your topic and why you want to write about that topic.
- **Narrowing your purpose helps you . . .** pinpoint your audience.
- **Develop each supporting cause or effect by . . .** showing and explaining its significance to your thesis statement.
- **In a cause-and-effect essay, the order of body paragraphs and the details in them can be as important as . . .** the order of steps in a set of instructions.
- **When you think critically about a cause-and-effect essay, you often . . .** examine the origins of a particular decision, action, or event; analyze the relationship of particular events or actions that brought about a specific result; or discover a deeper appreciation for your topic.
- **Revisions that are specific to compare-and-contrast essays include . . .** topics, details, and transitions.

CAUSE-AND-EFFECT WRITING LEARNING LOG MyWritingLab™

Answer the questions below to review your mastery of cause-and-effect writing. Answers will vary.

Complete this Exercise
MyWritingLab™

1. **What might be your purpose in cause-and-effect writing?**
   Your purpose might be to show why an event occurred, to identify the reasons something changed, or to explain the consequences of an event or action.

2. **What two problems do writers often face when drafting their cause-and-effect writing?**
   Writers sometimes don't cite enough details to prove their cause, and they sometimes cite details irrelevant to their cause.

3. **In a cause-and-effect paragraph, what usually follows the topic sentence?**
   Sentences that follow the topic sentence explain the cause(s) that led to the effect.

4. **What starter questions should you use to begin your cause-and-effect essay?**
   What result or effect am I writing about? How and why did this result or effect come about? What caused it? What details should I concentrate on or make my readers aware of? How will my readers benefit?

5. **How can you compose your starter statement?**
   Consider your purpose and answers to your starter statements. Condense that information into a single sentence, using the format "I will write about (cite conditions, events, or behaviors) because I want to show readers (cite results of conditions, events, or behaviors)."

6. **In developing your working thesis statement from your starter statement, what kind of language should you eliminate?**
   Eliminate language not essential to your purpose, like *I will write about* and *because I want to show readers.*

7. **What questions should you consider to determine your purpose?**
   What do I want to illustrate or show about that topic? Why do I want to write about that topic?

8. **What questions should you consider to determine your audience?**
   Who will benefit from reading my cause-and-effect essay? For whom is my cause-and-effect essay not intended?

9. **On what is the organization of your cause-and-effect essay based?**

   The organization is based on the relationship of your topic and its support.

10. **In critically thinking about the purpose of your cause-and-effect essay, what questions are important?**

    Have I clearly stated the cause(s) and effect(s) I plan to discuss? Will readers understand the connection between the cause(s) and effect(s)?

11. **In critically thinking about information in your cause-and-effect essay, what questions are important?**

    Have I given readers details that are relevant in explaining the cause, the effect, and the relationship between the two? Should I add any details to further explore this relationship?

12. **In critically thinking about reasoning in a cause-and-effect essay, what questions are important?**

    Are all of my facts accurate? Are they clear? Is my reasoning logical?

13. **In critically thinking about assumptions in a cause-and-effect essay, what questions are important?**

    Have I made any assumptions that need to be explained or justified? Have I used any vocabulary my audience may not understand?

14. **In revising, what should you be certain of regarding the focus of each body paragraph?**

    The focus of each body paragraph should be on one topic.

15. **In revising, what questions should you ask about the details in body paragraphs?**

    Will my readers understand the importance of each cause and effect? Have I given enough details? Have I given too many details? Have I written anything that flows away from a paragraph's focus? Are my paragraphs arranged in the most logical or effective way?

16. **In revising, what types of transitional words or phrases are helpful in a cause-and-effect essay?**

    Words and phrases that emphasize or clarify are helpful.

CHAPTER 14 Persuasive Writing

LEARNING GOALS

In this chapter, you'll learn and practice how to

1. Get started with persuasive writing
2. Write a persuasive paragraph
3. Read and examine student and professional essays
4. Write a persuasive essay

GETTING THERE

- Persuasive writing presents your strong, distinct position on a specific issue.
- In persuasive writing, you offer valid support for your view and you refute opposing views.
- Effective persuasive writing reflects a full understanding of the topic.
- In successful persuasive writing, you change readers' opinions or move them to action.

FASTWRITE 1

Many cities ban skateboarding and rollerblading, citing that these activities damage property and cause hazards for pedestrians and drivers. Suppose your town is considering such a ban. Take a stand either for or against the proposed ban. Then support your position, writing as much as you can, as fast as you can, for three minutes.

FASTWRITE 2

Some colleges and universities have mandatory attendance policies—your school may be one that does. Choose a position either for or against a school-wide attendance policy, and fastwrite your support for your position. Write as much as you can, as fast as you can, for five minutes.

Complete these **Fastwrites** at **mywritinglab.com**

Getting Started in Persuasive Writing

LEARNING GOAL
1 Get started with persuasive writing

Persuasive writing seeks a reaction from its audience. In **persuasive writing**, you may try to change the way your audience feels about an issue or you may try to move them to action. Such action could be supporting a cause by voting a certain way, writing letters to government officials, stopping or starting certain behaviors, or signing a petition.

Your purpose as a persuasive writer is to prove to your audience that your position on an issue is correct, so the support for your argument must be convincing and trustworthy:

- To be **convincing**, you must be knowledgeable about the issue and understand your position and opposing positions.
- To be **trustworthy**, you must be truthful and focus on the real concern and not minor concerns.

Anytime you write persuasively, you should accomplish three tasks:

- Clearly state your position
- Offer rational and sound support for your position
- Counter opposing views

Tatiana's topic was *the campus parking fee*, and her position was that this fee should be lowered. In order to discover her support, Tatiana used freewriting in this way:

The parking fee is already too high. Now they want to raise the fee to $120. If my math is right, this is a 33% increase. What else goes up that much all at once? They expect us in class and on time but I can't afford the parking fee. I have to park at least a mile from my bldg & it might take me 20 mins to walk that far. I got a new car last semester and try hard to take good care of it. Rain or snow . . . it takes longer than I've got to walk back. Profs and staff don't pay parking fees and they get paid to come here. If I can't afford $90 I can't afford $120. More I spend on parking now, less I have for tuition and books next semester. The fee should be lowered to $50. A welcome-back e-mail said the fee will cover additional parking lot upkeep costs. It doesn't need any upkeep—was paved & painted last year.

Next, Tatiana reviewed her fastwrite to see if it achieved the three persuasive tasks, and this is what she found:

1. Clearly state your position

   The parking fee should be lowered.

2. Offer rational and sound support for your position

   Parking fee went up 33% at once.
   I can't afford the parking fee . . . so I have to park at least a mile away.
   Profs and staff don't pay parking fees.
   The more I spend on parking, the less I have for tuition and books.

3. Identify and counter an opposing view

   The administration says it needs the fee hike for upkeep of the lot, but it doesn't need any upkeep since it was paved and painted just last year.

Always review your early prewriting for the three persuasive tasks. While it might be too soon to achieve all three completely, you could find ideas you can further develop to accomplish the three tasks.

14.1 Persuasive Fastwriting

Directions: State your position on a controversial issue affecting your campus. Then, prewrite for three to five minutes, discovering at least two specific reasons supporting your position. Next, prewrite for three minutes on the opposing views and how you would respond to them. When you're finished, share this prewriting with your writing group.

WRITING at WORK: *Snapshot of a Writer*

Breck Norment, Law Student

Breck Norment is in his third year of law school, and one of the classes he took in his first year was legal writing. The purpose of the class is to familiarize students with the types of writing they will use once they become practicing attorneys.

Breck's Writing Process

Learning Legal Writing Two types of legal writing we've learned so far are memos and appellate briefs. Law firms use memos (legal memorandums) to help them decide whether or not to take on cases. Memos analyze problems in potential cases and offer legal opinions on the cases. In the real world, memos are written by law firm associates and presented to the senior members of the firm, who study the memos and decide if the firm should take the case—that is, if the case is viable for the firm. The associate looks at material presented by a client and then writes a memo that (1) states the case law that would back up the potential client, (2) presents strategies and formulates arguments for the case, and (3) predicts what the court would rule in the case.

In writing memos for class, we must follow specific rules about formatting and structure (these are modeled after formatting and structure actually required by law firms). The language we use must be objective, and we need to present the strengths and the weaknesses of both our client and his or her opponent in the potential case. In our conclusion, we must present which side is more likely to win, based on the reasoning and analysis we provide.

Heads Up!
This is the **audience** for whom the attorney is writing.

We start by writing closed memos. For these, we're given research about a particular case, and then we write the memo and submit it. Next, we progress to writing open memos. These are different from closed memos in that we do all the research. This is the kind of memo we'll write when we've graduated and are employed in law firms.

We also write an appellate brief, another type of writing important to practicing attorneys. An appellate brief is a document filed after an unsuccessful lawsuit. It is read by judges in a higher court, and it is used to prove that the decision of the first court was based on faulty application of the law.

Heads Up!
This is the **purpose** of the piece.

How I Write Generally, I write at home on a computer because this is quicker for me than going to the library. I need to use case studies, and I can find almost all of them at one of two online sites, LexisNexis or Westlaw. I usually do my research first and then create an outline for my material (outlines are required by our instructor). When I research a case, I often find other cases that provide additional support or that turn out to be more pertinent to the argument I'm presenting. The cases presented by LexisNexis have summaries at the beginning of the entry, so I read the summary and see if the case seems to be helpful to my client and our case.

In addition to using my computer for research, I also use it for writing. Unless I'm taking notes in class, I rarely write in longhand. For me, writing on a computer is easier and faster.

How I Revise When the time has come to revise my written work, I read it over on the computer screen many times. Then I print it out and reread it—and I always find additional spots that need correction or clarification. After

I make those revisions, I send my work to my father, a practicing attorney. He offers suggestions for changes in content or language that has correct legal phrasing. For instance, I recently had written that a prosecutor "did not provide evidence," but my dad said that the correct wording should be "did not present evidence." That type of precision is important.

In legal writing, you learn to structure legal analysis by writing about IRAC (Issue, Rule, Apply, Conclusion). First, you write about the Issue (the facts of a case or the law behind a case), then you look for the Rule (the governing law) that applies to that issue, then you Apply the rule to the facts of your issue, and finally you write your Conclusion (how those rules logically affect your facts). Throughout your legal papers, you use the IRAC format over and over.

Heads Up!
Language is **adapted** to the purpose and audience.

TRY IT!

Think of a controversial issue on campus or in your community. Freewrite about that issue and the rule (it may not be a law) behind it. What is the problem? Why is it a problem? When you're finished, share your work with those in your writing group.

Getting Started with Your Persuasive Paragraph

LEARNING GOAL
❷ Write a persuasive paragraph

You may feel more comfortable writing a persuasive paragraph before you tackle a persuasive essay. If that is the case, follow steps one through six below. If not, go right on to Take 2 on page 335.

Your persuasive paragraph usually begins by stating your position, the specific idea or opinion the rest of the paragraph will support through

- descriptive details
- examples
- logical, rational reasoning

You combine these three writing methods, attempting to convince your audience that your position is one they should believe or that your opinion or idea is one they should adopt. Darnell, whose topic was *student smokers*, stated his position this way:

Students who smoke should have to pay for smoking areas on campus.

Because not all of your audience may be aware of the issue you are addressing, you may need to explain it through *descriptive details*. Darnell knew this was the case with his issue, so he explained with these descriptive details:

When our campus became smoke-free inside the buildings, the school administration built seven outdoor shelters just for smokers. Each of these designated smoking areas is 12′ x 16′ and has a roof, concrete picnic table, and two smokeless ashtrays.

The descriptive details helped Darnell further define the issue by providing a visual impression with an *example*. Once he was sure the audience understood the issue, Darnell moved from details about his position to support for his position—his *logical reasoning*.

The student newspaper said each of the seven shelters costs the school $8,800. Nonsmokers should not have to pay for these shelters or picnic tables because most will never use them.

See Chapter 1 for additional prewriting techniques such as listing, clustering, fastwriting, reporter's questions, and journaling.

Remember that support for your position also needs to counter opposing views. So, as you prewrite, address the position opposite to yours. Opposing views will probably have more than one supporting point; refute the strongest or most popular one. Here is how Darnell addressed an opposing view:

Smokers say they should not be responsible for any costs. They feel their rights are being abused since they have been forced to move to certain areas. However, this is not the case, since these people make a conscious decision to smoke. If they didn't smoke, they wouldn't be responsible for the fees.

Step One: Choose a Topic

Heads Up!
Stay focused when you counter opposing views. Your aim is to reject the view, not the people who hold it.

Instructors assign a particular persuasive topic, offer a list of possible topics to choose from, or let students choose their own. If you have the opportunity to choose your persuasive topic, remember your three tasks:

1. Clearly state your position
2. Counter opposing views
3. Offer rational and sound support for your position

14.2 Choose a Topic

Directions: Prewrite to find at least three areas of concern related to the following: (1) your campus, (2) your hometown or college town, and (3) national or international issues. Answers will vary.

Step Two: Generate Ideas

Once you have some possible topics, determine which ones you have strong opinions about and can develop by generating supporting ideas. Prewriting can help you discover not only your position but also the details supporting it.

14.3 Generate Ideas

Directions: From each list of topics you generated in Ticket to Write 14.2, first circle one topic that attracts your attention. Prewrite on this topic to discover what you know about it. Then write a sentence that expresses your position on it. Repeat this process for the other two topics you selected. When finished, share these with your writing group and get their feedback on each possible topic.

Step Three: Define Your Audience and Purpose

For the topic you chose, clearly state your position so your **audience** knows exactly what your **purpose** is. Often with persuasive writing, your purpose determines your audience.

A strong position usually asks the audience to do something, like adopt a certain stance or support a specific cause. Tatiana's freewriting stated that the campus parking fee was too high, so her audience included people responsible for setting the campus parking fee. Her audience also included other students, so she could enlist their support of her position.

Topic: campus parking fees

My position is the campus parking fee is too high.

My purpose is to make people aware of how the fee is unfair to students.

My audience is students and school administrators.

"To be persuasive, we must be believable; to be believable, we must be credible; to be credible, we must be truthful."
—Edward R. Murrow

14.4 Discover Your Audience and Purpose

Directions: To discover your audience and purpose, write your topics from Ticket to Write 14.3. Under each topic, write a sentence that identifies your position, a sentence that identifies your purpose, and a sentence that identifies your audience. When finished, share your topics and your audience and purpose information with your writing group. Use your group's feedback to help select one topic to develop into a paragraph.

Step Four: Draft Your Paragraph

Next, you need to compose your topic sentence. Your prewriting should give you a firm handle on your purpose and position, so composing your topic sentence should be fairly simple. Your paragraph's topic sentence should clearly state your position and what you want your audience to do—support a certain cause, adopt a specific position, or commit to a particular action.

Often in persuasive writing, composing your topic sentence in the form of a *should statement* makes clear what you want from your audience or what you want them to know. A *should statement* declares what you believe *should* happen. *Should* is an auxiliary verb; other auxiliary verbs that also work well in persuasive writing include

- ought to
- need(s) to
- must
- has (have) to

After reviewing her prewriting, Tatiana composed her topic sentence as a *should* statement:

The campus parking fee should be lowered.

14.5 Create a Topic Sentence

Directions: Choose one of the three topics you have been working with. Write your topic, audience, and purpose. Using these, compose your topic sentence and write it under your purpose. Then share your work with members of your writing group to see if they agree that your topic sentence explains your purpose and is appropriate for your audience.

After writing your topic sentence, build your examples from your prewriting to compose supporting details of your paragraph. These details provide evidence that proves your position. As you write, keep your audience and purpose in mind. Work with your prewriting and select examples that work best to illustrate your topic and support your position.

You may discover that you don't need to or can't include all your prewriting, and you may discover other supporting ideas you hadn't thought of before. Keeping your audience and purpose in mind will help you decide which details you need most.

In reviewing her prewriting, Tatiana chose details that best *explained* and *illustrated* her purpose. Here is her paragraph:

The campus parking fee should be lowered. The administration wants to raise the fee to $120. This is a 33% increase for a fee many students already cannot afford. No service should go up that much. Because I cannot afford the current $90 parking fee, I have to park off campus. Sometimes I park over a mile away from my building, and it can take me almost twenty minutes to walk that far. When the weather is bad, the walk takes even longer. The parking fee is also unfair because faculty don't have to pay a fee and they get paid to be on campus. Their parking lots are closer to the buildings. We were told in an e-mail that the recent fee increase would cover parking lot upkeep. I don't see that it needs any upkeep. It was paved and painted just last year.

Tatiana clearly states her position in the first sentence, uses descriptive details to explain the issue, and provides examples to show the impact of the problem.

14.6 Compose Supporting Details

Directions: Use the topic sentence you created in Ticket to Write 14.5 and the supporting ideas you generated in Tickets to Write 14.3 and 14.4 to compose a paragraph that supports your position. Write your paragraph and then share it with your writing group. Ask members if you have given enough details to fulfill the purpose expressed in your topic sentence.

Step Five: Revise Your Paragraph

After you've finished your paragraph, save it, print a hard copy and read it out loud, looking and listening for errors in grammar and content. Refer to the general

See Chapter 4 to review proofreading techniques.

Review Checklist for a Paragraph in Chapter 3, page 60, to make sure you revise as completely as possible.

Use the checklist below for further revision of your persuasive paragraph.

✓ REVIEW CHECKLIST **for a Persuasive Paragraph**

- ☐ Does my topic sentence clearly state my position on my topic?
- ☐ Do I include sufficient details to support my position?
- ☐ Do I identify and counter an opposing view?
- ☐ Do I hear any fragments when I read my paragraph out loud?
- ☐ Do all my complete thoughts have appropriate end punctuation?
- ☐ Is my paragraph appropriate for my audience?
- ☐ Does my purpose remain constant?
- ☐ Are any sentences or details unessential?

Once you've made any necessary changes, save your work and leave it alone for a while. Then come back later, print a new copy, and look and listen for errors again.

Heads Up!
See Chapter 20 to review fragments.

Step Six: Peer Review

You might think you've found all your errors and made all the improvements you can, but having someone else read your paragraph and offer suggestions will almost always improve your writing. This is person is called your **peer reviewer**. Your peer reviewer may use the review checklist above, or your instructor may provide a different checklist.

Peer reviewers bring a fresh perspective. Consider them valuable resources as audience members and fellow writers.

Listen to or read closely your peer reviewers' suggestions, and don't be shy about asking questions to clarify their ideas. You might or might not agree with the suggestions you receive, but peer reviewers often find places for improvement or errors that slipped by you.

14.7 Peer Review

Directions: Once you have completed revising your paragraph, share it with your peer reviewers. Ask them to read your paragraph, using the Review Checklist for a Persuasive Paragraph above and the general Review Checklist for a Paragraph from Chapter 3. Then ask them to record their suggestions for revision. Revise your paragraph again, using the suggestions with which you agree.

Student and Professional Essays

Below are two persuasive essays. The first was written by a student and the second by a professional. Read these two essays and answer the questions that follow them.

LEARNING GOAL
❸ Read and examine student and professional essays

Student Essay

Keeping a Watchful Eye
by T. W. Burnette

1 Shortly after 5:00 a.m. one morning in May, I was on my way home when I got a call asking me to return to work, **stat**. One of our patients, Mrs. Betty, was missing, so our manager needed all hands on deck to search for her. The nursing home where I work has a **residential** neighborhood to its east, an access road and highway to its south, and farmland to its north and west. When a patient wanders off in this area, the results can be deadly. We've only had three **elopements** in four years, and my supervisor says that's pretty good. Still, I believe electronic **monitoring** devices should be required for all **wander-risk** patients.

2 Mrs. Betty was not required to have a personal monitoring device because she is a voluntary patient and did not have doctor's orders to be monitored. Luckily, we found Mrs. Betty about a mile away, walking by a busy highway. Picking fresh wildflowers for her room. The next day her doctor wrote an order for her to be given a monitoring bracelet.

3 Wanderers can cost an arm and a leg. My supervisor says the **liability** of elopement is why our facility manager is willing to pay time-and-a-half to all off-shift workers who come in to search for a wanderer. Nine of us came in to search for Mrs. Betty. Since we all searched for two hours, at $16 an hour, the facility paid us over $280 to look for her. That may not seem like much, but when you think of paying the four city police officers and the two state police officers who were dispatched to search for Mrs. Betty, the money adds up. I read online in the ***Occupational Outlook Handbook*** that the average police officer's salary is over $52,000. I'm not going out on a limb when I say that two hours of all six of their paychecks would more than pay for a monitoring device for every resident at the facility.

stat in medical terms, at once or with no delay

residential having to do with houses, apartments, and other household dwellings

elopement the act of a patient leaving a care facility without supervision

monitoring checking, supervising

wander-risk in danger of walking off without supervision

liability legal responsibility

Occupational Outlook Handbook a manual compiled every two years by the US Bureau of Labor Statistics; it provides career information and job search tips

4 Wanderers often meet a tragic end. People with **dementia** are prone to leaving their home and getting lost. According to the **Alzheimer's** Association, almost half of all nursing home residents have Alzheimer's or some other form of dementia. The Alzheimer's Association estimates that more than 34,000 Alzheimer patients wander from their homes or nursing homes each year and some with deadly results. Last month I read about an elderly Michigan couple, both of whom had just been diagnosed with Alzheimer's. They were walking home from the grocery store. Evidently, they turned the wrong way, got lost, and took **refuge** in an abandoned building. Their son couldn't reach them by phone, so he drove to their house. After waiting for a while, he called the police. Hours later his parents were discovered in the abandoned building a few blocks from their home. They had both frozen to death.

5 Providing Mrs. Betty an electronic monitoring device would have kept her family from worry, and providing the Michigan couple with those devices would have saved their lives. The simple solution of using monitoring devices can solve many problems.

dementia decline in memory, concentration, or judgment
Alzheimer's a progressive form of dementia
refuge shelter, protection

A CLOSER LOOK

Answer these questions about the essay:

1. Which sentence in the first paragraph is the thesis statement of the essay?

   Still, I believe all nursing homes should use electronic monitoring devices for all wander-risk patients.

2. Why does the writer begin the essay by relating the story of Mrs. Betty?

   Possible answer: The story is an attention-getter that illustrates the problem.

3. What is the writer's purpose with this persuasive essay?

   The writer wants to support the use of monitoring devices for certain patients.

4. Many groups of people might be part of the audience for this essay. Identify two.

   Possible answers: family members of wander-risk patients, nursing home personnel, physicians, legislators

5. The first body paragraph does not have a topic sentence. Instead the paragraph elaborates on the thesis statement by giving an example. Do you think this is an effective method of support? Explain your answer.

   Answers will vary.

6. Persuasive writing should address and refute an opposing view. The author does not do that. Compose a body paragraph that (1) addresses a reason

to oppose requiring electronic monitoring devices and (2) refutes that reason.

Answers will vary.

7. The topic of the third body paragraph is the tragic end met by some wanderers. How does the use of statistics in this paragraph strengthen its topic?

   Answers will vary.

8. The writer uses the same sentence structure and even the same first word to begin paragraphs 3 and 4. How might he reword the beginning sentence in paragraph 4?

   Possible answer: Tragedy occurs in many cases of elopement.

9. A *cliché* is an overused expression that usually does not have a place in academic writing. In this essay, the writer uses three clichés. Identify each and suggest how the writer might reword the sentence to eliminate the cliché.

   Paragraph 1 cliché: "all hands on deck"; *possible change*: everyone

   Paragraph 3 cliché: "cost an arm and a leg"; *possible change*: become very expensive

   Paragraph 3 cliché: "going out on a limb"; *possible change:* exaggerating

10. Paragraph 2 includes a sentence fragment. Find and repair this fragment.

    Possible answer: Luckily, we found Mrs. Betty about a mile away, walking by a busy highway and picking fresh wildflowers for her room.

One controversial topic facing many colleges today is whether or not to allow concealed weapons to be legally carried on campuses. When the Texas legislature was considering this issue, the essay below appeared in the *Dallas Morning News*.

Professional Essay

Texas Lawmakers' Guns-On-Campus Bills Still a Bad Idea

from The Dallas Morning News

1 The gun agenda is back in the Legislature with the **velocity** of a speeding bullet. Proponents are determined, finally, to make it possible for a university student to legally pack a Glock for class along with textbook and iPad. What an awful idea.

2 State law contains a list with the heading "Places Weapons Prohibited." Schools and educational institutions are listed first, ahead of courts, polling places, airports, and prisons. And there's a common-sense reason for that. Colleges and universities are places of learning, where young people converge, soak up knowledge, evolve, and mature. Inviting weaponry into the atmosphere recklessly introduces a risk factor into an already higher-risk age

velocity speed

group. University presidents struggle with ways to **tamp** down the excesses of a youthful community, from **hazing** to substance abuse. Fostering responsible gun ownership is far too much to ask of a university administration.

(3) Authors of the legislation say they want to give students and staff members the ability to use a concealed handgun to defend themselves or possibly cut short a **Virginia Tech–type massacre**. Since 1995, trained, licensed Texans have had the right to carry concealed handguns, except where prohibited, and this newspaper respects that right and the permit-holder's interest in self-defense. Still, it's a merely **theoretical notion** that the permit holder, after a few hours in the classroom, plus time on the firing range, has the training or nerve needed to stop a rampage-in-progress without causing **collateral damage**.

(4) Further, since the **carnage** that occurred at Virginia Tech, while horrific, is exceedingly rare, it makes no sense to allow a campus to be populated with guns based on that level of danger. The threat would seem to be higher from an accidental shooting or from an argument that escalated into gun violence.

(5) The state has more than 1.4 million students at 38 public universities, 39 private universities, and 50 community college campuses. They are now essentially gun-free zones, but the author of the guns-on-campus bill in the Senate, Jeff Wentworth, prefers to call them "victim zones." That's a sad, pessimistic rendering of today's campus life. Wentworth, R-San Antonio, easily got a similar measure through the Senate two years ago, and there's little question he can do it again with his current version (**SB 354**).

(6) One salvation for private schools is an "out" clause in the legislation. The bill says that after "consulting with students, staff and faculty of the institutions," university leaders may adopt rules barring concealed firearms. That's one thing that makes a horrendous proposal a little less horrendous. Lawmakers can improve it a **hair more** by extending the "out" clause to every campus in Texas, public school or private, professional or community college. It's called local control, something lawmakers say they honor. This legislation would be a good place to **bend** to that principle.

Source: Dallas Morning News, February 28, 2011. Reprinted with permission of *The Dallas Morning News*.

tamp pack, push
hazing forcing a new fraternity member to perform humiliating or dangerous acts
Virginia Tech–type massacre a reference to the April 2007 shooting at Virginia Tech University, when thirty-two people were killed and twenty-five were wounded
theoretical notion idea that has not been proven
collateral damage unintended injuries or deaths
carnage killing, bloodshed
SB 354 SB stands for "Senate Bill"; 354 is the number of a specific bill proposed in the Texas Senate
hair more an idiom meaning "small amount"
bend to direct or turn to a particular course

A CLOSER LOOK

Answer these questions about the essay:

1. What is the thesis statement for this essay?

   What an awful idea.

2. The author begins this essay with a metaphor (a comparison of two unlike things, without using *like* or *as*) comparing the quickness of the reappearance of the gun agenda to the velocity of a speeding bullet. Why is that metaphor appropriate?

   Possible answer: The metaphor use of "speeding bullet" plays off the topic of guns.

3. The author refutes arguments presented by those who support the bill. In paragraph 3, what argument do those supporters give?

   Possible answer: Supporters want to give students and staff members the ability to use a concealed handgun to defend themselves.

4. In paragraph 3, what does the author say to refute the supporters' argument?

   Possible answer: Still, it's a merely theoretical notion that the permit holder, after a few hours in the classroom, plus time on the firing range, has the training or nerve needed to stop a rampage-in-progress without causing collateral damage.

5. What transitional word shows the author will present another piece of evidence to refute the opposition?

   The author uses the word *further.*

6. In paragraph 4, what is the evidence the author presents to refute the opposition?

   Possible answer: The threat would seem to be higher from an accidental shooting or from an argument that escalated into gun violence.

7. In paragraph 5, the author describes the state's campuses in two different ways, reflecting the author's side and the opposition's side. What are these two descriptions?

   The author calls the campuses "gun-free zones"; Senator Wentworth calls them "victim zones."

8. What does the author say is "one thing that makes a horrendous proposal a little less horrendous"?

   Possible answer: The author says the "out" clause that allows private schools to adopt rules barring concealed firearms is one salvation.

9. What is the effect of the statistics the author presents in paragraph 5?

   Possible answer: Statistics show how many people would be affected by the passage of this bill.

10. In paragraph 3, the author introduces one point of the opposition's argument. Suppose you are an advocate of this legislation and compose a paragraph to support your point of view.

    Answers will vary.

Writing Your Persuasive Essay

LEARNING GOAL
❹ Write a persuasive essay

Essay writing is much like paragraph writing: you start with a topic and expand with details. Remember that just as a paragraph supports a topic sentence, an essay supports a thesis statement.

See Chapter 1 to review thesis statements.

Your persuasive writing is most convincing when you

- state your position clearly
- offer rational and sound support for your position
- counter opposing views

To prove your position to your audience, your support must be convincing and trustworthy.

- To be **convincing**, you must be knowledgeable about the issue and understand both your position and opposing positions.
- To be **trustworthy**, you must be truthful and focus on the real concern and not minor concerns.

Step One: Choose a Topic and Develop a Working Thesis Statement

Heads Up!
You might recognize these starter questions as reporter's questions.

To begin your persuasive essay, ask yourself these **starter questions**:

1. **What** issue or problem am I writing about?
2. **What** position do I hold on this issue?
3. **Why** do I hold this position?
4. **What** details support my position?
5. **Who** or **what** benefits from my position?
6. **How** will my readers benefit from adopting my position?
7. **What** is an opposing view of my position?
8. **How** can I refute that opposing view?

Answering Starter Questions:
T. W. Burnette and *The Dallas Morning News*

In answering the starter questions, T. W. Burnette, author of "Keeping a Watchful Eye," wrote:

1. I am writing about the dangers of failing to monitor certain patients.

2. My position is that electronic monitoring devices should be required for wander-risk patients.
3. I hold this position because wandering patients can cost lots of money, can cause worry, and can even be hurt or killed.
4. Details that support my position are Mrs. Betty's incident, the Michigan couple, and statistics from the Alzheimer's Association.
5. Those who benefit from my position include nursing homes, caregivers, family members, and patients.
6. Readers will benefit from adopting my position because they may have family who are wander-risk patients, or they may work at a nursing home or similar facility.
7. The opposition says electronic monitoring devices can be unreliable.
8. I refute the opposition by noting if batteries are checked and changed regularly, the devices are dependable.

The author of "Texas Lawmakers' Guns-On-Campus Bills Still a Bad Idea" might have addressed the starter questions this way:

1. I am writing about a bill that would allow handguns on campuses.
2. My position is that I strongly oppose this bill.
3. I hold this position because having a permit to carry a gun doesn't mean someone has enough training to use a gun correctly.
4. Details that support my position are almost anyone can get a permit to carry a gun after spending just a few hours in a classroom and a little time on a firing range.
5. Those who benefit from my position include students at state campuses.
6. Readers will benefit from adopting my position because they will know that state campuses are far safer if the legislation is not passed.
7. The opposition says carrying a concealed weapon will let people defend themselves and might also prevent a massacre.
8. I can refute the opposition by noting with only minimal training, permit holders may not have the preparation or the courage to stop someone else.

Yasmin chose to write an essay about a high school in her area that was considering adopting a school uniform policy. Here is how she answered the starter questions:

1. **What** issue or problem am I writing about?

   I'm writing about the possible adoption of a school uniform policy.

2. **What** side or position do I hold on this issue?

   I think the high school should adopt a uniform policy.

3. **Why** do I hold this position?

   The current policy allows clothing that is too revealing.

   The current policy allows clothing that has inappropriate messages.

   Schools with uniform policies have better learning environments.

4. **What** details support my position?

   Many boys wear their pants below their waist.

   Many girls wear tops that are too low-cut.

   Many students wear t-shirts that promote drug use or underage drinking.

   Many students wear t-shirts with curse words or insulting messages.

   Students who wear uniforms aren't distracted with looks, so they can focus on studying.

   Students who wear uniforms don't face problems of peer pressure about clothing styles.

   Uniforms all cost the same. Buying them could save money compared to the cost of clothing fads.

5. **Who** or what benefits from my position?

   Students, teachers, administrators, and students' parents

6. **How** will my audience benefit from adopting my position?

   The board will understand that a school uniform policy benefits students and their school.

7. **What** is an opposing view of my position?

   Opponents say uniforms deny students their freedom of expression.

8. **How** can I refute that opposing view?

   Freedom of expression means speaking your mind, not baring your skin.

Persuasive thesis statements are strongest when they declare the writer's desire for a change to take place. This change could be as simple as people becoming aware of an issue or as drastic as a law needing to be abolished.

To develop your working thesis statement, review your answers to the starter questions and use the following format to organize them:

Topic

I will write about ________________ (state concern, problem, or issue) to convince readers to support my position that ________________ (state your viewpoint).

Position

Composing your thesis statement in the form of a *should statement* makes clear what you what you want from your readers or what you want them to know. *Should* is an auxiliary verb; auxiliary verbs that also work well in persuasive writing include

- ought to
- need(s) to
- must
- has (have) to

After she answered the starter questions, Yasmin considered her purpose and filled in the starter statement like this:

Yasmin uses the auxiliary verb *should* in her starter statement.

Look at how two other writers chose to fill in the blanks for other topics:

I will write about **cash advance businesses** to convince readers to support my position that the interest rates those businesses charge should be capped at 10%.

I will write about **professional sports eligibility** to convince readers to support my position that college athletes should graduate before they enter professional sports.

14.8 Answer Starter Questions and Complete a Starter Statement for a Persuasive Essay

Directions: To help develop a working thesis statement for your persuasive essay, answer each of the following starter questions. Then complete the starter statement.

1. **What** issue or problem am I writing about?
2. **What** side or position do I hold on this issue?
3. **Why** do I hold this position?
4. **What** details support my position?
5. **Who** or what benefits from my position?
6. **How** will readers benefit from adopting my position?
7. **What** is an opposing view to my position?
8. **How** can I refute that opposing view?

Review your responses to the starter questions. Then consider your purpose and condense the information into a single sentence, filling in the blanks of the starter statement below:

> I will write about ________________ (state a concern, problem, or issue) to convince readers to support my position that ________________ (state your viewpoint).

Next you need to rework your starter statement to form your working thesis statement. Eliminate the phrases that are not essential to your purpose, like *I will write about* and *to convince my readers to support my position that.*

Yasmin reworked her starter statement into this working thesis statement:

The proposed school uniform policy should be adopted in local public schools.

The other writers created these working thesis statements from their starter statements:

The interest rates charged by cash advance businesses should be capped at 10%.

College athletes should graduate before they enter professional sports.

14.9 Develop a Working Thesis Statement

Directions: Review your starter statement in Ticket to Write 14.8. Then, reword it using an auxiliary verb (*should, ought to, must, has to, have to, need to, needs to*) to form your working thesis statement. Share this with members of your writing group, and ask them if they believe you clearly state your strong position.

Step Two: Generate Ideas

The working thesis statement you create establishes your commitment to an issue by clearly stating your position about it. This meets the first of the three persuasive writing tasks.

Next, build a strong case supporting your position by fulfilling the other two tasks: offering rational and sound support and countering opposing views (see p. 349). All supporting ideas should serve one of these two tasks.

Offering Rational and Sound Support

To begin generating ideas of support, review your answers to the starter questions. Your responses to some of the starter questions probably contain key ideas you could use to support your position.

Yasmin reviewed her responses to the starter questions and compiled a list of supporting ideas for her position (see page 342). After reviewing this list, she found she had obvious reasons that supported her position. Yasmin then used reporter's questions to develop these reasons.

My working thesis statement: The proposed school uniform policy should be adopted in local public schools.

What is one reason **why**? One reason why the policy should be adopted is the inappropriate outfits so many kids wear to school these days. So much of it is too revealing. I don't want to see any guy's underwear, and some girls wear skimpy clothes that are just tasteless. Sleeveless and backless shirts, short shorts, and miniskirts all are unnecessary in high schools.

What is another reason **why**? Another reason the policy should be adopted is offensive clothing. T-shirts with disgusting messages and images, gang symbols, and other types of clothes are meant to be hateful. School should not be a place that promotes hate and rudeness. These days, some t-shirts look more like bumper stickers than clothing.

What is another reason **why**? Probably the most important reason local students need school uniforms is so students can focus more on learning and less on who's wearing what. The competition in high school to be popular is enormous. Also, sometimes those who can afford the name brand popular styles make fun of those who can't afford those types of clothes. Uniforms would make everyone a bit more equal at school. This would put more focus back on learning and less on who's wearing the latest designer clothes.

14.10 Generate Ideas

Directions: Review the list of details supporting your position. Write your working thesis statement; under it, fastwrite detailed responses to each question below. Share this with your writing group and ask them to note which ideas seem the most compelling or which elaborate the most on the information in your working thesis statement. Answers will vary.

What is one reason why I hold this position?

What is another reason why I hold this position?

What is another reason why I hold this position?

Step Three: Define Your Audience

"At the end of reasons comes persuasion."
—Ludwig Wittgenstein

Your next step is to identify the audience for your essay. To decide who your audience is, review your purpose and then answer the following:

1. **Who** would benefit from adopting your position?
2. **Who** can help you accomplish your goal?
3. **Whose** support would you like to have?

After considering her purpose, here is how Yasmin responded to these questions about audience:

1. Who would benefit from adopting my position? students, teachers, administrators, parents
2. Who can help me accomplish my goal? administrators, parents, and school board members
3. Whose support would I like to have? administrators, parents, students, and school board members

14.11 Define Your Audience

Directions: Write your working thesis statement. Under that, write a sentence about who would benefit from adopting your position, a sentence about who can help you accomplish your goal, and a sentence about whose support you would like to have. Share this information with members of your writing group and ask if they agree that your working thesis statement will be satisfactory for the audience you describe.

Step Four: Draft Your Essay

To begin discovery drafting your persuasive essay, review the ideas you generated from your prewriting and starter questions.

State Your Position Clearly

At the beginning of your essay, clearly stating a strong position in your working thesis statement establishes your sincerity on the issue and your commitment to your position. By presenting your position in the first paragraph of your essay, readers will know exactly what you want from them.

Provide Rational and Sound Support

The support you provide for your position is what you use to persuade your readers to adopt your position. To continue drafting your body paragraphs, review the support you generated in prewriting and select the points you believe best support your position. Each body paragraph should have a topic sentence that focuses on one point of support for your position.

Yasmin reviewed all her prewriting and then used these topics and supporting details or evidence when she composed her discovery draft.

Working Thesis Statement: The proposed school uniform policy should be adopted in our public schools.

Topic: resolves problem of inappropriate outfits

evidence: guys' pants too low

evidence: girls' clothes show too much

Topic: resolves problem of offensive clothing

evidence: t-shirts with hateful messages

evidence: gang clothing causes fights

Topic: resolves problem of students distracted by clothing

evidence: some students compete through what they wear

evidence: some are bullied because of what they wear

Counter Opposing Views

To make a convincing argument, you must be able to defend it against opposing views; because of this, you need to refute or disprove the other side. Challenging the opposing view shows your audience that you

- know your topic well
- have taken into account an alternate position

Counter the opposing view early in your essay. That way you can focus your efforts on supporting your position. Yasmin examined a position opposite hers by pretending to support it. She developed support for that position the same way she did for her own: by considering reasons someone would have for supporting the position.

Opposition's View

Position The proposed school uniform policy should not be adopted in our public schools.

Why? School uniforms stifle individuality.

Why else? School uniforms can cost more money than everyday clothes.

Why else? School uniforms are not very stylish.

Yasmin looked over the ideas she generated that supported an opposing view and decided to address what she considered to be the strongest of them. She addressed the point that uniforms stifle individuality in the first body paragraph so she could acknowledge the position, disprove it, and then move on to support her position.

Here is Yasmin's first body paragraph. In it she addressed possible opposition to her argument:

> Opponents of the proposed policy claim school uniforms stifle individuality. To agree with this idea suggests adults teach kids to express who they are through what they wear instead of through what they say and the way they act. I would rather be recognized for who I am than for what I wear. The individuality I see consists of blue jeans and different color tops. Where's the individuality in that?

Yasmin's paragraph is consistent and effective. It doesn't spend too much time discussing the opposition and quickly moves into arguing against the opposition.

As you draft your essay, you may want to

- address the opposition in the first body paragraph; then support your position in subsequent body paragraphs
- state your position first and then take on the opposition; that way you have provided credibility for your argument before bringing in opposing views

14.12 Write Your Discovery Draft

Directions: Use the material you compiled in Tickets to Write 14.8 through 14.11 to write your discovery draft. Refer to Chapter 2 to review discovery drafting. When you have completed your discovery draft, share your work with members of your writing group. Ask them to note the following on your subject: Answers will vary.

1. Does the first body paragraph successfully disprove the opposition? What additional statements could be made to refute the opposition?
2. What supporting point has especially detailed and effective supporting details or evidence?
3. What supporting point needs more specific or clearer supporting details or needs more development?
4. What additional evidence could be included?

Step Five: Organize Your Essay

The organization of your persuasive essay should focus on winning the reader over to support your position. You already know the main three parts that make up the persuasive essay:

1. State clearly your position
2. Use rational and sound support
3. Counter opposing views

Don't repeat yourself. Word your conclusion differently from your thesis statement.

Now that you've completed your introduction and body paragraphs, reaffirm your strong position in your conclusion. Here is Yasmin's conclusion, which echoes her thesis statement in the first sentence.

Adopting the proposed school uniform policy will improve life for all our public school students. Imagine a school where, from 7:30 a.m. to 3:00 p.m., no one would have to see offensive shirts with hateful messages, or see exposed armpits; imagine a school where no students were picked on because they couldn't afford the latest fashions; imagine a school where you could focus on what you're learning and not on what someone else is wearing.

To check the organization of her ideas, Yasmin made a chart of the ideas in her draft.

Paragraph	Topic of Paragraph	Purpose—Description
Introduction	The proposed school uniform policy should be adopted in our public schools.	Thesis statement
Body paragraph #1	Opponents of the proposed policy claim school uniforms stifle individuality.	Counter opposing view
Body paragraph #2	One reason the policy should be adopted is the inappropriate outfits kids wear to school.	Supports position

Body paragraph #3	Another reason the policy should be adopted is because of the offensive clothing.	Supports position
Body paragraph #4	Probably the most important reason is so students can focus more on learning and less on who's wearing what.	Supports position
Conclusion	Adopting the proposed school uniform policy will improve life for all local public school students.	Reminds audience of position

14.13 Organizing Your Essay

Directions: Make a chart in which you outline your draft. Use the chart to review your draft, paragraph by paragraph, and look for areas that need revision. Review your chart and your draft to see that you have

1. Stated your position clearly
2. Countered your opposition
3. Provided sufficient points of support
4. Offered plenty of supporting details or evidence for each supporting point
5. Reminded your readers of your position

Your chart will highlight areas to strengthen in your next revision.

Step Six: Apply Critical Thinking

Critical thinking means thinking and then rethinking about a topic, with the goal of discovering all the implications you can about it. In writing a persuasive essay, several concepts of critical thinking are important.

In a persuasive essay, you

- state your position
- refute opposing positions

- examine the reasons behind your position
- analyze the evidence supporting your position

After you write each of your drafts, ask yourself the critical thinking questions below, and then apply your answers to your writing:

1. **Purpose:** In your thesis statement, have you clearly stated your position? Will your readers recognize that your position is realistic? Is your thesis narrow enough, or have you claimed too much?
2. **Information:** In explaining your position, have you given your readers relevant details? Will readers find your supporting evidence reasonable? Do you see anyplace to add details that will further support your position? Have you presented an opposing view?
3. **Reasoning:** Are all of your facts accurate? Are they clear? Is your reasoning logical? Have you logically refuted the opposing view?
4. **Assumptions:** Have you made any assumptions that need to be explained or justified? Have you used any vocabulary that your audience may not understand?

14.14 Apply Critical Thinking

Directions: After examining your answers to the critical thinking questions, revise your draft again. Ask members of your writing group to read your latest draft and answer the critical thinking questions below. Answers will vary.

Purpose

1. Did I state my position clearly? Is my position realistic? If not, why not?
2. Did I state or imply my purpose clearly? If not, where should I be clearer?

Information

3. Are my details relevant to my position? Is my supporting evidence reasonable? If not, why not?
4. Where would adding details enhance or further support my position?
5. What opposing view have I presented?

Reasoning

6. Are all my facts are accurate? If not, where should I provide more accurate details?
7. Are all my facts clear? If not, where should I change my wording to be clearer?
8. Are all my facts relevant to my topic? If not, where do I stray from my topic?
9. Is my reasoning logical? If not, where do I seem to be illogical?
10. Did I logically refute the opposing view? If not, how can I improve my reasoning?

Assumptions

11. Did I make any assumptions that need to be explained or justified? If so, what are these assumptions?
12. Did I use any vocabulary you don't understand? If so, what words or phrases need definitions?

Step Seven: Revise Your Essay

Getting your essay on paper is just your first step. Revising through subsequent drafts shapes your first draft into a finished product of which you can be proud.

Heads Up!
Keep hard copies of all your drafts and number them so you'll know which is your latest draft.

Consult the Review Checklist for an Essay in Chapter 3, page 61, and look for places in your essay you need to polish. Because your goal is to convince your readers of your position, you may want to focus revising efforts on your **body paragraphs**, **details**, and **reasoning**.

Body Paragraphs

Check that each body paragraph focuses on one specific topic.

- Does the topic in each body paragraph support your position by introducing some point of your position?
- Does each topic rationally and reasonably support your position?
- Does a body paragraph include information that counters your opposition?

Details

When you revise, review your details.

- Have you given enough examples or illustrations in each body paragraph to support the topic sentence?
- Does each detail focus on the topic of its paragraph?
- Should any detail be eliminated or revised because it's unrelated to the topic?

Reasoning

Look at the reasoning you have presented for your position and its support, and consider if your audience will think your essay is convincing.

- Is your position reasonable?
- Is each paragraph's topic reasonable and realistic support for your position?
- Are your examples or illustrations sensible?

14.15 Revise Your Essay—Peer Review

Directions: Revise again, looking especially at body paragraphs, details, and reasoning. Then ask members of your writing group to read your revision and answer the questions below. Answers will vary.

Reasons

1. Early in the essay, do I include a paragraph that counters my opposition?
2. Does each body paragraph support my position by introducing some specific point of my position? If not, which ones need revising?
3. Does each body paragraph rationally and reasonably support my position? If not, which ones need revising?

Details

4. Have I given enough examples or illustrations in each body paragraph to support the topic sentence? If not, which body paragraphs need more examples or illustrations?
5. Does each of my details focus on the topic of its paragraph? If not, which ones need more support?
6. Do I need to eliminate or revise any details because they are unrelated to the topic? If so, which need eliminating and which need revising?

Reasoning

7. Is my position reasonable? If not, what suggestions do you have to improve my position?
8. Is each paragraph's topic logically linked to my position? If not, which topic needs revision?
9. Are my examples and illustrations sensible? If not, which ones need revising?

MyWritingLab™

Writing Assignment

Consider the topics below or one that your instructor gives you. Choose one topic and expand it into a persuasive essay:

1. Nutritional information should (should not) be printed on bags in fast food restaurants.
2. Curse words should (should not) be banned from primetime television.
3. Paparazzi should (should not) be required to stay at least ten feet from celebrities.
4. Marriage counseling should (should not) be required for a couple filing for divorce.
5. A class in economics should (should not) be mandatory for college students.
6. Smoking should (should not) be banned on college campuses.

To acquaint yourself with various positions on these issues, consult newspapers, magazines, and online sites such as CNN.com, MSNBC.com, USATODAY.com, and THEWEEK.com.

7. Sex education should (should not) be taught in middle school.
8. Steroids should (should not) be allowed in professional sports.
9. Drivers over age seventy should (should not) be required to take annual driving tests.
10. College writing classes should (should not) be pass-fail.

TECHNO TIP

For more on persuasive writing, search the Internet for these videos:

Sweet Spot: A Lesson in Persuasive Writing with Barry Lane

The Art of Persuasive Writing: A Few Pointers

How to Write a Good Argumentative Essay: Logical Structure

MyWritingLab™ Visit *MyWritingLab.com* and complete the exercises and activities in the **Paragraph Development-Argument** and **Essay Development-Argument** topic areas.

RUN THAT BY ME AGAIN

- **Persuasive writing seeks . . .** a reaction from its audience.
- **Any persuasive writing should . . .** clearly state your position, offer rational and sound support for your position, and counter opposing views.
- **Your persuasive paragraph usually begins by . . .** stating your position.
- **In a persuasive essay, the thesis statement is strongest when it declares . . .** the writer's desire for a change to take place.
- **In composing your working thesis statement . . .** rework your starter statement and eliminate phrases not essential to your purpose.
- **The working thesis statement you create . . .** establishes your commitment to an issue by clearly stating your position about it.
- **To find support for your working thesis statement, look at . . .** responses to your starter questions.
- **To begin discovery drafting your persuasive essay . . .** review the ideas you generated from your prewriting and starter questions.
- **The organization of your persuasive essay should focus on . . .** winning readers over to support your position.
- **After you write each of your drafts, ask yourself the critical thinking questions about . . .** your purpose, the information you provided, your reasoning, and your assumptions.
- **In revising your persuasive essay, look especially at . . .** your body paragraphs, details, and reasoning.

LEARNING GOAL
❶ Get started with persuasive writing

LEARNING GOAL
❷ Write a persuasive paragraph

LEARNING GOAL
❹ Write a persuasive essay

PERSUASIVE WRITING LEARNING LOG MyWritingLab™

MyWritingLab™

Answer the questions below to review your mastery of persuasive writing. Answers will vary.

1. **What is your purpose as a persuasive writer?**
   Your purpose is to prove to your audience that your position on an issue is correct.
2. **In a persuasive paragraph, what do you use to support your position?**
   You may use descriptive details, examples, and logical, rational reasoning.
3. **In a persuasive essay, answering eight starter questions will help you develop what?**
   Starter questions help you develop a working thesis statement.
4. **When composing your working thesis statement, what should you eliminate?**
   You should eliminate the phrases that are not essential to your purpose, like *I will write about* and *to convince my readers to support my position that.*
5. **Ideas that support your working thesis statement serve which two tasks?**
   Supporting ideas (1) offer rational and sound support and (2) counter opposing views.
6. **To determine your audience, what questions should you answer?**
   Who would benefit from adopting my position? Who can help me accomplish my goal? Whose support would I like to have?
7. **Why should you clearly state your position in your thesis statement at the beginning of your essay?**
   This establishes your sincerity on the issue and your commitment to your position.
8. **What do you use to persuade readers to adopt your position?**
   Support you provide for your position is what you use to persuade readers to adopt your position.

9. **What does challenging the opposition show your audience?**
   Challenging the opposing view shows your audience that you know your topic well and you have taken into account an alternate position.

10. **In organizing your essay, where should you reaffirm your strong position?**
    You should reaffirm your strong position in your conclusion.

11. **What critical thinking questions help you look at the purpose of your essay?**
    In your thesis statement, have you clearly stated your position? Will your readers recognize that your position is realistic? Is your thesis statement narrow enough, or have you claimed too much?

12. **What critical thinking questions help you look at the information in your essay?**
    Have I given readers details that are relevant in explaining my position? Will they find my supporting evidence reasonable? Do I see any place to add details that will further support my position? Have I presented an opposing view?

13. **What critical thinking questions help you look at the reasoning in your essay?**
    Are all my facts accurate? Are they clear? Is my reasoning logical? Have I logically refuted the opposing view?

14. **What critical thinking questions help you look at the assumptions in your essay?**
    Have I made any assumptions that need to be explained or justified? Have I used any vocabulary that my audience may not understand?

15. **In revising your essay, what should you check about each body paragraph?**
    You should check that (1) each body paragraph focuses on one specific topic, (2) you include a body paragraph with a topic that counters your opposition, (3) the topic in each body paragraph supports your position by introducing some point of your position, and (4) each topic rationally and reasonably supports your position.

16. **In revising your essay, what should you check about your details?**
    You should (1) check that you have given enough examples or illustrations in each body paragraph to support the topic sentence, (2) check that each detail focuses on the topic of its paragraph, and (3) look for any detail that should be eliminated or revised because it's unrelated to the topic.

17. **In revising your essay, what should you check about your reasoning?**
    You should check to see if (1) your position is reasonable, (2) each paragraph's topic is reasonable and gives realistic support for your position, and (3) your examples or illustrations are sensible.

Writing Situations

PART 3

On or off campus, you'll face a variety of writing situations. From in-class writing to blogging, all writing situations have two elements in common: *audience* and *purpose*. Your audience—those who will read your writing—helps you determine what kind of language you will use. Your purpose—the reason or cause for your writing—shapes how you convey your message. Part 3 will introduce you to various writing situations many college students experience.

"A writer . . . has the duty to be good, not lousy; true, not false; lively, not dull; accurate, not full of error. He should tend to lift people up, not lower them down. Writers do not merely reflect and interpret life, they inform and shape life."
– E. B. White, *Writers at Work*, Eighth Series

Resource-Based Writing

LEARNING GOALS

In this chapter, you'll learn and practice how to

1. Use information properly from another source by quoting
2. Use information properly from another source by paraphrasing
3. Use information properly from another source by summarizing
4. Cite sources accurately using MLA style
5. Cite sources accurately using APA style

GETTING THERE

- A quotation is another person's exact words.
- A paraphrase is another person's ideas presented in your words and style of writing.
- A summary is an overview of a source, focusing only on its main points.
- In academic writing, all quotations, paraphrases, and summaries must be attributed to their original source.

FASTWRITE 1

Think of something you have recently seen (movie, TV drama or sitcom episode, viral video, or news piece). Recap what you saw, as if you were explaining it to a friend. Write as much as you can, as fast as you can, for seven minutes.

FASTWRITE 2

Think of something you have recently read (newspaper or magazine article, blog entry, short story, or book). Recap what you read, as if you were explaining it to a friend. Write as much as you can, as fast as you can, for seven minutes.

Complete these **Fastwrites** at **mywritinglab.com**

Integrating Sources

You include other people's work in your college writing in three ways: *quoting*, *paraphrasing*, and *summarizing*. Used properly, each of these skills helps you

- establish credibility with your readers
- emphasize specific points and topics
- support your ideas and opinions

The more you read other people's ideas and perspectives, the more your ideas and perspectives develop. Your instructors want you to distinguish for your readers which ideas are yours and which come from others. To do this, you *cite* sources by identifying the author and title. **Citing** sources shows readers that

Citing is also called *referencing* or *documenting*.

- you understand what you have read
- you can apply what you have read to your ideas

Always cite sources for other people's ideas, theories, opinions, facts, statistics, or any other information that is not considered *common knowledge*. Any idea that is generally accepted or widely known is considered common knowledge. For example, if you write that Oregon borders the Pacific Ocean, you do not need to cite a source for this widely-known piece of information.

"I quote others only to better express myself."
—Michel de Montaigne

Quoting

LEARNING GOAL
1 Use information properly from another source by quoting

When **quoting**, you present another person's exact words with quotation marks around those words. You should

- quote material that supports, explains, or proves your point
- credit the source (the person or text) that provided the original material
- use correct formatting and punctuation to incorporate quoted material into your writing

Identifying the source adds credibility to your writing and adds trustworthiness to you as a writer.

Look at how one student uses a source in the following example:

The squeeze at the pump not only causes us to second-guess where we vacation but even where we live and work. "Distance is now an enemy," writes environmentalist Bill McKibben, author of The End of Nature, a book that details the damage humans have caused on Earth.

Notice the information from the source is used to support a point the writer makes. The source is credited, and quotation marks show what information comes from the source.

Here are some reasons for using quotations in your writing.

You might quote to . . .	Because . . .
show the source's tone	the source is ironic, hopeful, or passionate
add credibility to your opinion or point	the source is an authority on the subject or a respected public figure
highlight a particularly well-written sentence or passage	the sentence or passage is written in a unique style or makes a point so well that paraphrasing would not do it justice

Short Quotations

Use quotation marks only if you quote the exact words a person said, in exactly the order the person said them.

To incorporate a **complete sentence** into your writing

- Place quotation marks around the words you quote.
- Capitalize the first word of the quoted sentence.
- Place a comma between the signal phrase (like *Clinton said,* or *Marlotte asserted,*) and the quotation mark.
- Place closing quotation marks after the punctuation mark and last word of the quoted sentence *if* your sentence ends with the last word of the quotation.

In Dracula, Bram Stoker wrote, "It seems to me that the further east you go, the more unpunctual are the trains."

Full sentence ends, so the period precedes the quotation marks

- Place a comma and closing quotation marks after the last word of the quoted sentence *if* your sentence continues after the last word of the quotation.

"It seems to me that the further east you go, the more unpunctual are the trains," Bram Stoker wrote in Dracula.

Full sentence continues, so the comma precedes the quotation marks

- Check to see if you interrupted a quoted sentence with a phrase like *Stoker wrote*. If you did, place a comma and then closing quotation marks at the end of the first segment you quote; place a comma and opening quotation marks where the quoted material begins again. When the quoted material restarts, do not capitalize the first word.

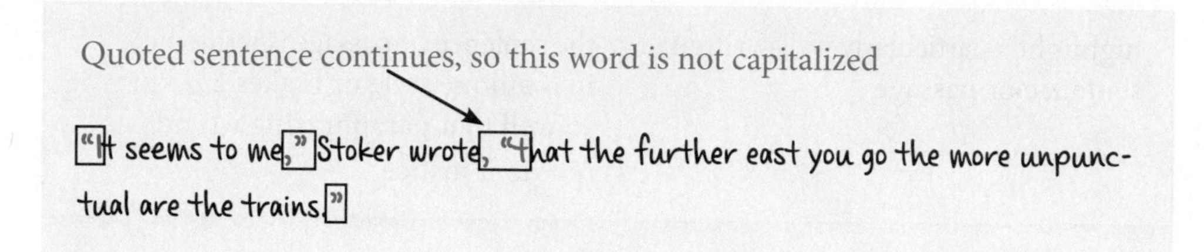

To incorporate a quote of a **partial sentence**, place quotation marks around the words you quote, but do not use capital letters.

For more on using quotation marks, see Chapter 30.

Bram Stoker wrote that trains are more unpunctual "the further east you go."

15.1 Quoting Others' Ideas

Directions: Working alone or in groups, return to the essays in Chapters 6 and 7. Choose a sentence (or more) to quote from each essay. Identify the source and author, and use correct punctuation. Answers will vary.

Paraphrasing

LEARNING GOAL

❷ Use information properly from another source by paraphrasing

When you **paraphrase**, you take another person's material and present it in your words. The skill of paraphrasing is part of the skill of summarizing but doesn't focus on squeezing a work down to its main idea and supporting points. A paraphrase does the following:

- credits the source (the person or text) that provided the original material

- restates information from a source and uses your words, not just synonyms (words with the same meaning)
- simplifies complicated language
- is roughly the same length as the original source

Identifying the source adds credibility to your writing and adds trustworthiness to you as a writer.

Compare the original excerpt of Dr. King's speech to the paraphrased version below it.

Original

"We must forever conduct our struggle on the high plane of dignity and discipline. We must not allow our creative protest to degenerate into physical violence. Again and again we must rise to the majestic heights of meeting physical force with soul force."

—Dr. Martin Luther King, Jr., "I Have a Dream"

Paraphrased

For more on paraphrasing, search the Internet for the video **Stop Thief! Avoiding Plagiarism by Paraphrasing.**

In his "I Have a Dream" speech, Dr. King said that those seeking equality for black Americans had to do so with respect for others. To gain the rights they wanted, black Americans had to avoid hatred and brutality and be sure all their actions toward gaining equality reflected how they wanted to be treated.

Now compare the original excerpt of Don Knabe's commencement address to the paraphrased version on the following page.

Original

Today you join an elite group of the 35% of Americans who hold an associate's or higher degree. Only one of every three Americans has attained what you will attain today. And many of you, I suspect, are among the first in your families to attend college. That is exclusive company, and membership does have its privileges.

—from the 2010 Cerritos (CA) College commencement address by Don Knabe

Paraphrased

In his commencement speech at Cerritos College, Don Knabe encouraged graduates to recognize their unique position in earning a college degree, in achieving an educational goal that 65% of Americans will never know, and, for many, in being the first in their families to join the ranks of college graduates.

15.2 Paraphrasing Others' Ideas

Directions: Working alone or in groups, paraphrase each of the following passages. Use your words and style of writing. Remember that titles of essays or articles go inside quotation marks; titles of books and newspapers are italicized or underlined. Answers will vary.

1. ***Original*** MLB should schedule at least one World Series game in the daytime, which hasn't happened since 1987. Kids—and many adults—who don't live in the Pacific time zone have trouble staying up late to watch night games, so the Series has lost a considerable percentage of its viewership. Plus, if the Series goes to seven games, the action would stretch into November, with the potential of playing in freezing temperatures or even snow. —Jay Goulooze, "Bring Back the Day," in the *Freeport News*
2. ***Original*** The goal of ecotourism is to offer vacations that are enjoyable but that also preserve and protect the ecosystem. In the United States, resorts involved in ecotourism often emphasize recycling, conservation, and volunteerism. Travelers frequently study the carbon footprints humans leave in the surrounding environment and how those footprints can be reduced or even eliminated. —Paula Fowler, *Ecotours for Everyone*
3. ***Original*** Although conscription has not been used in the United States since 1973, it has a long history. During the Revolutionary War, individual states sometimes drafted men into military service. The first national draft came into effect during the Civil War, with both sides drafting recruits. During WW I, conscription began in 1917, two months after the United States entered the war. History changed in 1940 when a peacetime draft was put into effect. This draft continued through World War II and, with a short hiatus, was in effect until near the end of the Vietnam War. Although the United States has had an all-volunteer army for almost forty years, men who are between ages 18 and 25 are still required to register with the Selective Service. —Reagan Abbott, "Conscription in the U.S."

Summarizing

When you **summarize**, you condense someone else's work. To write an effective summary you should

- include the name of the author and the title of the source (article, book, movie, etc.) you are summarizing
- focus on only the work's main idea and main points
- present the main points in the order they appear in the source
- present the material in your words
- not include your opinion of the work or its ideas
- keep your writing shorter than the original work

LEARNING GOAL

3 Use information properly from another source by summarizing

Summarizing is an essential skill in every corner of education. You might summarize a week's worth of lecture notes before you study for a biology test. You might summarize a movie for a film class, a short story for a literature class, a news article for a government class, a specific battle for a history class, or a news conference for the school's newspaper. In fact, you could write a summary of any number of topics for almost any class.

A good rule of thumb is to limit your summary so it's no longer than half the length of the original source.

For instance, if you're writing a summary of *Eclipse*, one of the *Twilight Saga* novels, you might write something like this:

For more on summarizing, search the Internet for this video title: **Video: How to Write a Summary**

In Stephenie Meyer's vampire novel Eclipse, the Cullens and their former enemies the werewolves join to battle Victoria and her army of newborns. The newborns are feeding on Seattle citizens to hunt down and kill Bella, Edward Cullen's human girlfriend. Victoria's drive to kill Bella is for revenge. Because Edward killed Victoria's lover, she has vowed to kill his. Shortly after Bella agrees to marry Edward, Bella's close friend Jacob, a werewolf with the La Push pack, overhears their plans to marry and for Bella to become a vampire. Jacob, who is also in love with Bella, threatens to join the fight against the newborns to escape his heartache. Bella professes her love for Jacob and yet tells him she loves Edward more.

"A word is not the same with one writer as with another. One tears it from his guts. The other pulls it out of his overcoat pocket."
—Charles Peguy

In her essay about single-parent families, Jill summarized the following excerpt from President Obama's June 21, 2010, Father's Day speech. She found the transcript on the **Whitehouse.gov** Web site, on the **Briefing Room** page, in the **Speeches and Remarks** database.

Original

But we also know that what too many fathers being absent means—too many fathers missing from too many homes, missing from too many lives. We know that when fathers abandon their responsibilities, there's harm done to those kids. We know that children who grow up without a father are more likely to live in poverty. They're more likely to drop out of school. They're more likely to wind up in prison. They're more likely to abuse drugs and alcohol. They're more likely to run away from home. They're more likely to become teenage parents themselves. And I say all this as someone who grew up without a father in my own life. He left my family when I was two years old. And while I was lucky to have a wonderful mother and loving grandparents who poured everything they had into me and my sister, I still felt the weight of that absence. It's something that leaves a hole in a child's life that no government can fill.

Summarized

Signal phrase identifies text and author →

Word count is less than one-half length of the original text

Source identified in parenthetical notation →

In his 2010 Father's Day speech at the Town Hall Education Arts Recreation Campus, in Washington, DC, President Obama reminded listeners that too many homes are fatherless in the United States. He listed the negative experiences—from dropping out of school to becoming teen parents—suffered by many children who grow up without a father's presence. Growing up without his own father's influence, President Obama said he will always know the emptiness such children feel. ("Remarks by the President at a Father's Day Event")

15.3 Writing Summaries

Directions: Return to an essay that you have read. Working alone or in groups, summarize the essay. When you have completed the summary, compare yours to that of another student or group in the class. Remember that titles of essays or articles go inside quotation marks. Answers will vary.

1. Underline the main idea in your peer's summary.
2. Double underline the main points in your peer's summary.

3. At the bottom, write any main points you feel should have been included in the summary.
4. At the bottom, write any words or phrases that were in the original author's words and should have been converted into your peer's words.

TECHNO TIP

For more on quoting, paraphrasing, and summarizing, search the Internet for these sites:

Quoting	Paraphrasing	Summarizing
Integrating Quotations into Sentences	Paraphrase: Write It in Your Own Words	Writing a Summary
Writing: Using Source Materials	How to Paraphrase a Source	Process for Writing a Summary

MLA Citation

LEARNING GOAL

4 Cite sources accurately using MLA style

As you advance in college classes, your instructors will require you to use particular methods, or styles, of identifying and crediting the sources you consulted when writing a paper (this is called **documentation**). MLA style, used chiefly in English classes, requires citations in two places: a short, in-text citation in parentheses immediately following quoted, paraphrased, or summarized information, and a full listing of a source's information on a Works Cited page at the end of your paper.

Chapter 5 has more information about MLA, APA, AMA, Turabian, and CMS styles.

In-Text Citations

In-text citations provide the page number(s) of the material you quote, paraphrase, or summarize, and the name of the author, if it is not included in your text. Suppose your topic is the women's movement, and you quote from Tom Brokaw's book *Boom!: Voices of the Sixties: Personal Reflections on the '60s and Today.*

- **If you use the author's name in your text**, put the page number in parentheses after the quotation and before your final punctuation mark:

> Brokaw says that, while the women's movement has gained steam in the years since its inception, it "still is making its way across uneven terrain" (237).
>
> Citation comes between quotation marks and final punctuation

- **If you do not include the author's name in your text,** place the author's last name and the page number of the text (with no punctuation between the two) in parentheses after the quotation and before the end punctuation:

> While the women's movement has gained steam in the years since its inception, it "still is making its way across uneven terrain" (Brokaw 237).
>
> Citation comes between quotation marks and final punctuation

In MLA, a block quotation consists of more than four typewritten lines.

- **If you use a block quotation**, place a colon (not a comma) at the end of the signal phrase before the quotation. Indent the quotation ten spaces from the left margin, double space it, do not use quotation marks at the beginning or end of the quotation (the indentation tells readers where the quotation begins and ends), and place the punctuation of the quotation at the end of the quotation's final sentence (not after the parenthetical citation). Place the page number—and author's name, if it is not mentioned in your text—in parentheses *after* the end punctuation:

> In his chapter about attorney Carla Hills and author Joan Didion, Brokaw wrote: ← Colon here
>
> No quotation marks
>
> Two women I came to know and greatly admire during my pilgrimage through the Sixties were native-born Californians, one from the South, the other from the North, one a Stanford graduate, the other a Berkeley alum, both Republicans, one a champion athlete and hard-driving lawyer, the other a frail woman of letters, and neither of them baby boomers. (224) ← Citation after final punctuation

The examples above show you how to cite from a *book* in a *printed* format, written by *one author*. Your reading and research, however, may lead you to information presented

- in other formats (e.g., article, short story, poem, drama)
- in other media (e.g., online, television, radio, lecture, newspaper, magazine)
- by more than one author or by an anonymous author

MLA and other styles, such as APA and CMS, have particular requirements for citing any source you use. (See Chapter 5 for more details.)

Works Cited Page

The **Works Cited** page contains an alphabetical listing of all the sources in a paper. In-text citations give readers enough information so they can consult the Works Cited page if they want to read more from a cited text or check the accuracy of information.

The Works Cited page comes at the end of the paper and starts on a new page. Entries are arranged in alphabetical order. Each entry starts on a new line, and if an entry runs longer than one line, subsequent lines are indented; this is called a **hanging indent**. A citation of a book by one author includes the following information in the exact order listed below.

- **Author name** (last name, comma, first name, followed by a period)
- **Complete title** (in italics, followed by a period)
- **Edition of the book**, if indicated (followed by a period)
- **Place of publication** (followed by a colon)
- **Name of the publisher** (followed by a comma)
- **Date of publication** (followed by a period)
- **Medium of publication** (followed by a period)

Using this format, the entry for the Brokaw book should look like this:

> Brokaw, Tom. *Boom!: Voices of the Sixties: Personal Reflections on the '60s and Today*. New York: Random House, 2007. Print.

MLA Handbook

Additional information on MLA and other documentation styles is available in Chapter 5. For more detailed information on MLA format for in-text citations and works cited entries, you can use the *MLA Handbook for Writers of Research Papers*, seventh edition; a current writer's handbook; or the MLA Web site (**www.mla.org**). These online sites generate in-text and Works Cited MLA citations:

citationmachine.net/ KnightCite

APA Citation

LEARNING GOAL

5 Cite sources accurately using APA style

Classes in social and behavioral sciences usually require students to use citations in APA style. APA requires citations in two places: a short, in-text citation immediately following quoted, paraphrased, or summarized information, and a full listing of a source's information on a References page at the end of the paper. Every in-text citation should be noted in full on the References page. Readers use a References page if they want to locate the original text to read beyond what you cited or to verify that you cited material correctly.

In-Text Citations

In-text citations come within sentences and paragraphs. You cite the author's last name, the year of publication, and in most cases the page number.

If you are writing about the women's movement and you quote from Tom Brokaw's 2007 book *Boom!: Voices of the Sixties: Personal Reflections on the '60s and Today*, APA style requires citations in these ways:

- **If you use the author's name in your text**, cite the year of publication of the work, in parentheses, after his or her name. Use a signal phrase to introduce the quotation. After the final the quotation mark, provide in parentheses the page number of the reference (preceded by "p.") followed by the final punctuation mark.

Year the book was published

While the women's movement has gained steam in the years since its inception, Brokaw (2007) says the movement currently "still is making its way across uneven terrain" **(p. 237).**

Citation comes between quotation marks and final punctuation

- **If you do not use the author's name in your text**, place the author's last name, the year of publication, and the page number in parentheses after the quotation. Separate each with a comma.

The women's movement "still is making its way across uneven terrain" **(Brokaw, 2007, p. 237)** although it has gained steam in the years since its inception.

Citation comes after quotation ends

- **If your quotation is more than forty words long**, use a block quotation. In a block quotation, use no quotation marks unless they are used in the original text. Begin the quotation on a separate line, indented ten spaces from both the left and right margins, and double space throughout.

In his chapter about attorney Carla Hills and author Joan Didion, Brokaw (2007) wrote:

No quotation marks

Two women I came to know and greatly admire during my pilgrimage through the Sixties were native-born Californians, one from the South, the other from the North, one a Stanford graduate, the other a Berkeley alum, both Republicans, one a champion athlete and hard-driving lawyer, the other a frail woman of letters, and neither of them baby boomers.

(p. 224) ← Citation comes after final punctuation

The examples above show how to cite from a *printed work* written by *one author*. Your reading and research, however, may lead to information presented

- in other formats (e.g., article, short story, poem, drama)
- in other media (e.g., online, television, radio, lecture, newspaper, magazine)
- by more than one author or by an anonymous author

References Page

The **References** page contains an alphabetical listing of all sources used in a paper. In-text citations give readers enough information so they can consult the References page if they want to read more from a cited text or if they want to check the accuracy of information.

The References page is at the end of your paper and starts on a new page. Entries are arranged in alphabetical order. Each entry starts on a new line, and if an entry runs longer than one line, subsequent lines are indented; this is called a hanging indent.

A citation of an in-print book by one author includes the following information in the exact order listed below.

- **Author name** (last name, comma, first initial, followed by a period)
- **Year of book's publication** (in parentheses, followed by a period after the parentheses)
- **Title of book** (in italics, followed by a period; capitalize only the first word of a title, first word of a subtitle, and all proper nouns)
- **Place of publication** (followed by a colon)
- **Name of the publisher** (followed by a period)

Using this format, an entry for the Brokaw book should look like this:

Brokaw, T. (2007). *Boom!: Voices of the sixties: Personal reflections on the '60s and today*. New York: Random House.

Note that Brokaw's book has two subtitles.

APA Manual

For more detailed information on the APA format for in-text citations and reference entries, use the *Publication Manual of the American Psychological Association*, sixth edition. You can also use a current writer's handbook or the APA Web site (**www.apastyle.org/**). These online sites generate in-text and Reference page APA citations:

citationmachine.net/
KnightCite

MyWritingLab™ Visit *MyWritingLab.com* and complete the exercises and activities in the **Research Process** topic area.

RUN THAT BY ME AGAIN

LEARNING GOAL
❶ Use information properly from another source by quoting

- **Quotations are . . .** other people's exact words.
- **Your reasons for quoting are . . .** (1) crediting the source (the person or text) that provided the original material; (2) adding credibility to your opinions, topics, and ideas; (3) highlighting well-written material.

LEARNING GOAL
❷ Use information properly from another source by paraphrasing

- **Paraphrasing is . . .** presenting someone else's ideas in your words.
- **Your reasons for paraphrasing are . . .** (1) crediting the source (the person or text) that provided the original material; (2) adding credibility to your opinions, topics, and ideas; (3) simplifying complicated language.

LEARNING GOAL
❸ Use information properly from another source by summarizing

- **Summarizing is . . .** condensing someone else's work to its main points.
- **Your reasons for summarizing are . . .** (1) crediting the source (the person or text) that provided the original material; (2) adding credibility to your opinions, topics, and ideas; (3) simplifying complicated language; (4) highlighting the main ideas of a source.

LEARNING GOAL
❹ Cite sources accurately using MLA style

- **Documentation is a method of . . .** identifying and crediting sources consulted in a paper.
- **MLA style requires citations . . .** in text and on a Works Cited page.
- **MLA in-text citations provide . . .** the page number(s) of the material quoted, paraphrased, or summarized and the name of author, if it is not included in text.
- **In MLA style, use a block quotation if you quote . . .** more than four typewritten lines.
- **At the end of a signal phrase before a block quotation . . .** place a colon.
- **With a block quotation, you show readers where the quotation begins and ends by . . .** indenting ten spaces from the left margin.

- **When you include a block quotation . . .** do not use quotation marks and double space the quotation.
- **A Works Cited page contains . . .** an alphabetical listing of all sources used in a paper.
- **A Works Cited page is located . . .** at the end of a paper, on a new page.
- **APA style requires citations . . .** in text and on a Reference page.
- **APA in-text citations provide . . .** an author's last name and the year of publication of material cited.
- **In APA style, use a block quotation if you quote . . .** more than forty words.
- **At the end of a signal phrase before a block quotation . . .** place a colon.
- **With a block quotation, show readers where the quotation begins and ends by . . .** indenting five spaces from the left margin and double spacing.
- **When you include a block quotation . . .** do not use quotation marks and double space the quotation.
- **A References page contains . . .** an alphabetical listing of all sources used in a paper.
- **A References page is located . . .** at the end of a paper, on a new page.

LEARNING GOAL

5 Cite sources accurately using APA style

RESOURCE-BASED WRITING LEARNING LOG MyWritingLab™

MyWritingLab™

Answer the questions below to review your mastery of resource-based writing.
Answers will vary.

1. **What is a quotation?**
   A quotation is another person's exact words.
2. **How can you use quotations in your papers?**
   You can use quotations to show a source's tone, add credibility to your point, or highlight a particularly well-written sentence or passage.
3. **What is a paraphrase?**
   A paraphrase is another person's material presented in your words.
4. **How can you use paraphrases in your papers?**
   You can use paraphrases in your papers to simplify complicated language and to support your points.
5. **What is a summary?**
   A summary is an overview of a source, focusing only on its main points.
6. **How can you use summaries in your papers?**
   You can use summaries to support a topic or idea in your papers.
7. **What is MLA documentation?**
   MLA documentation is a specific method or style of identifying and crediting sources cited in a paper.
8. **In MLA, when should you use a block quotation?**
   Use a block quotation if you are quoting more than four typewritten lines.
9. **In MLA, how should you introduce a block quotation?**
   A block quotation should be preceded by a signal phrase that ends with a colon.
10. **Where are MLA-style citations required?**
    MLA-style citations are required in text and on a Works Cited page.
11. **In MLA, what information do in-text citations provide?**
    In-text citations provide the page number(s) of the material quoted, paraphrased, or summarized and the name of author, if it is not included in text.
12. **What information does the Works Cited page include?**
    The Works Cited page includes an alphabetical listing of all sources used in a paper.

13. **Where is a Works Cited page located?**

    A Works Cited page is located at the end of a paper, on a new page.

14. **What types of classes usually require students to use citations in APA style?**

    Classes in social and behavioral sciences usually require students to use citations in APA style.

15. **In APA, when should you use a block quotation?**

    Use a block quotation if your quotation is more than forty words long.

16. **In APA, where is the References page located?**

    The References page is at the end of your paper and starts on a new page.

CHAPTER

16

In-Class Writing

LEARNING GOALS

In this chapter, you'll learn and practice how to

1. Interpret an essay question and respond to a writing prompt
2. Adapt the writing process for in-class writing situations

GETTING THERE

- Directives are verbs found in essay questions and writing prompts; they instruct writers about how to organize their responses.
- Key words in essay questions and writing prompts are the ideas on which writers must focus.

FASTWRITE 1

Think of a class you attended recently (earlier this week or last week) in which you learned a new concept. Then pretend you have come to the next class meeting and have a pop quiz in which you are asked to "*explain* the concept as clearly as possible without consulting your text or notes." Write as much as you can, as fast as you can, for five minutes.

FASTWRITE 2

Pretend you are in a sociology class and have been asked to "*compare* and *contrast* your life now and your life five years ago." Write as much as you can, as fast as you can, for five minutes.

Complete these **Fastwrites** at **mywritinglab.com**

Reading and Understanding Essay Questions to Determine Form and Purpose

In-class writing, like out-of-class writing, has two key elements: audience and purpose. In one respect, all written exams have the same audience and purpose: to demonstrate to your instructor what you know about a subject. However, in-class writing carries the purpose of your writing to another level, as it requires you to express your knowledge in a particular way. Because of this, reading the questions, directions, instructions, or writing prompts thoroughly is crucial. Essay questions are worded carefully and always include specific instructions that are your keys to a successful in-class writing experience.

LEARNING GOAL

1. Interpret an essay question and respond to a writing prompt

Form: How to Organize Your Answer

Read closely the instructions for any exam question or prompt and you will find **directives** (key verbs) that will help you organize your response. These directives give you a more specific instruction than to just share your knowledge with the teacher; instead, they tell you *how* to share your knowledge.

For more ideas about identifying directives, search the Internet for this video: **How to Write Good Timed Essays (2 of 6)**

Directive (Key Verb)	Definition
Analyze (Discuss, Review)	Examine all the parts of a topic, commenting briefly on the topic's major points or problems.
Compare (Relate)	Detail the similarities between two or more people, places, or things.
Contrast (Relate)	Detail the differences between two or more people, places, or things.
Define (Classify)	Offer a unique meaning for a specific term by explaining what the term is and is not.
Evaluate (Justify)	Offer a judgment (opinion) supported by other points of view and any related course-specific information.
Explain (Describe, Illustrate, Interpret)	Provide a detailed answer to the topic. Depending on the topic, offering examples, opinion, or analysis may be appropriate.
List (Enumerate, State)	Create a concise inventory of points that elaborate on the topic, but omit minor points.

(continued)

Prove (Argue)	Confirm or verify a statement or assertion by providing evidence based on logical reasoning or experiments.
Summarize (Outline, Trace)	Systematically (e.g., chronologically or by order of importance) highlight the main points or important facts.

16.1 Identifying Directives

Directions: Underline the directives in the sentences below. Then write the directive for each essay prompt and use the chart above to find its definition. Using the definitions, answer each essay question in one paragraph or more. Share your responses with those in your writing group to check that you have correctly followed each directive. Answers will vary.

1. Compare what you are wearing with what another person in the room is wearing.
2. Explain why you are taking this class.
3. Define "conscientious student."
4. Summarize four concepts you have learned in a class you are taking or have taken.
5. State the main duties of a job you have now or have had.

Purpose: How to Determine Your Focus

In addition to deciding how to organize your answer, you must also determine what to focus on. To decide this, read the question or writing prompt thoroughly and underline the key words, the ideas on which you must concentrate. Identifying key words helps you focus your answer by guiding you to the specific ideas that are the basis of the examination and the purpose of your response. Some students find using reporter's questions (*Who? What? Why? Where? When?* and *How?*) helpful in identifying the key words in essay questions and writing prompts.

The following essay questions have the directives marked. The key words are labeled with reporter's questions Amelia used to fully understand the essay questions. Amelia wanted to make sure she understood the purpose of the prompts, so she mentally restated their purpose based on the directives and key words.

Directive — Who — Who — What

Outline John Quincy Adams' and Thomas Paine's contributions to the

What — Where

abolitionist movement in the United States.

My purpose is to list the important contributions these two men made to the abolitionist movement.

Directive — What — Who — What

Explain three ways in which humans have had a negative impact on the

Where

Louisiana coast.

My purpose is to provide three detailed ways people have harmed the Louisiana coast.

Directives — Who

Compare and contrast the characters of Sammy from Updike's "A & P" to

Who

the narrator in Joyce's "Araby."

My purpose is to show the similarities and differences between the main characters in these two short stories.

TICKET to WRITE

16.2 Identifying Key Words

Directions: Read the essay questions below and underline the key terms. Then compare what you identified as key terms to what your classmates identified. Discuss why you selected the terms you did. Answers will vary.

1. List three major results from the stem cell research funding ban being lifted in 2009.

(continued)

2. Evaluate the governor's inaugural address for style, relevance, length, and impact.
3. Prove that the New Deal transformed American federalism.
4. Outline the major events in the Trail of Tears relocation of Native Americans.
5. Explain why in 2010 President Obama accepted the resignation of General McChrystal.

As you read questions or prompts, think about what you are marking and why. Jotting down brief notes can help you before you begin writing. Use the following checklist when you read questions or prompts.

Checklist for Reading Essay Questions

- Read the question or prompt all the way through.
- Circle the directives.
- Underline the key words.
- Label the key words using reporter's questions.
- Restate mentally the question or writing prompt in simpler terms.

16.3 Identifying Directives and Key Words

Directions: Read the essay questions below and circle the directive and underline the key words. Then compare what you identified as directives and key words to what your classmates identified and discuss why you selected the terms you did. Answers may vary.

1. Analyze Arthur Lewis's theory of economic growth.
2. Compare the main characters in *The Adventures of Huckleberry Finn* and *To Kill a Mockingbird*.
3. Describe the uniform of a Union soldier during the American Civil War.
4. Summarize the main points in Chapter 2.
5. Evaluate the contribution of Run DMC to rap music.

Using the Writing Process to Answer Essay Questions

LEARNING GOAL

❷ Adapt the writing process for in-class writing situations

In-class writing is much like out-of-class writing because it relies on the writing process. The only difference is that your time is limited within the stages of the process. With a pop quiz, for example, you may have only fifteen minutes, and with an in-class essay, you may have two or three hours. Still, you must decide how best to use your time.

You need to adapt the stages of the writing process to fit your in-class writing situation, as shown in the following chart.

Stage of Writing Process		In-Class Writing Situation
Prewriting	becomes	*Exploring the Topic* (List or fastwrite any ideas related to the quiz, exam, or writing prompt topic.)
Discovery Drafting	becomes	*Drafting* (Use the quiz, exam, or writing prompt to form your topic or thesis statement.)
Revising, Editing, and Publishing	become	*Revising and Editing* (Read through your writing and check for topic focus and accuracy in spelling, punctuation, and mechanics.)

TECHNO TIP

For additional information on writing in-class essays, essay exams, and other timed writing situations, search the Internet for these sites:

Inside the Classroom: College Board

The Learning Center: Taking Essay or Short-Answer Exams

RUN THAT BY ME AGAIN

LEARNING GOAL

❶ Interpret an essay question and respond to a writing prompt

- **Always reread an essay question and writing prompt for . . .** directives (verbs) or key words.
- **Directives tell you . . .** how to organize an essay.
- **Key words identify . . .** your focus.
- **After you read an essay question or writing prompt . . .** (1) circle directives; (2) underline key words; (3) label the words using reporter's questions; (4) mentally restate a question or prompt in simpler language.

LEARNING GOAL

❷ Adapt the writing process for in-class writing situations

- **In adapting the writing process for in-class writing, prewriting becomes . . .** exploring the topic.
- **In adapting the writing process for in-class writing, discovery drafting becomes . . .** drafting.
- **In adapting the writing process for in-class writing, revising, editing, and publishing become . . .** revising and editing.

IN-CLASS WRITING LEARNING LOG MyWritingLab™

Complete this Exercise

MyWritingLab™

Answer the questions below to review your mastery of in-class writing.

1. **What is a directive?**
   A directive is a verb found in essay questions and writing prompts; it instructs writers on how to organize their responses.
2. **What is a key word?**
   A key word in essay questions and writing prompts is an idea on which writers must focus.
3. **How can you use reporter's questions with in-class writing?**
   Reporter's questions can help identify key words in essay questions and writing prompts.
4. **How can you adapt the writing process for an in-class writing situation?**
   Prewriting becomes *exploring the topic* (listing or fastwriting any ideas related to quiz, exam, or writing prompt topic). *Discovery drafting* becomes *drafting* (using the quiz, exam, or writing prompt to form your topic or thesis). *Revising, editing, and publishing* become *revising and editing* (reading through your writing and checking for topic focus and accuracy in spelling, punctuation, and mechanics).

CHAPTER 17 Personal and Business Writing

LEARNING GOALS

In this chapter, you'll learn and practice how to

1. Keep a personal journal
2. Format business letters

GETTING THERE

- Personal journaling is an activity that can improve both your writing and your critical thinking skills.
- Personal journaling is a useful way to track ideas as they develop.
- Letter writing is still an important part of the business world today, and it is useful to know how to format a business letter and compose a letter of request and a business thank-you letter.

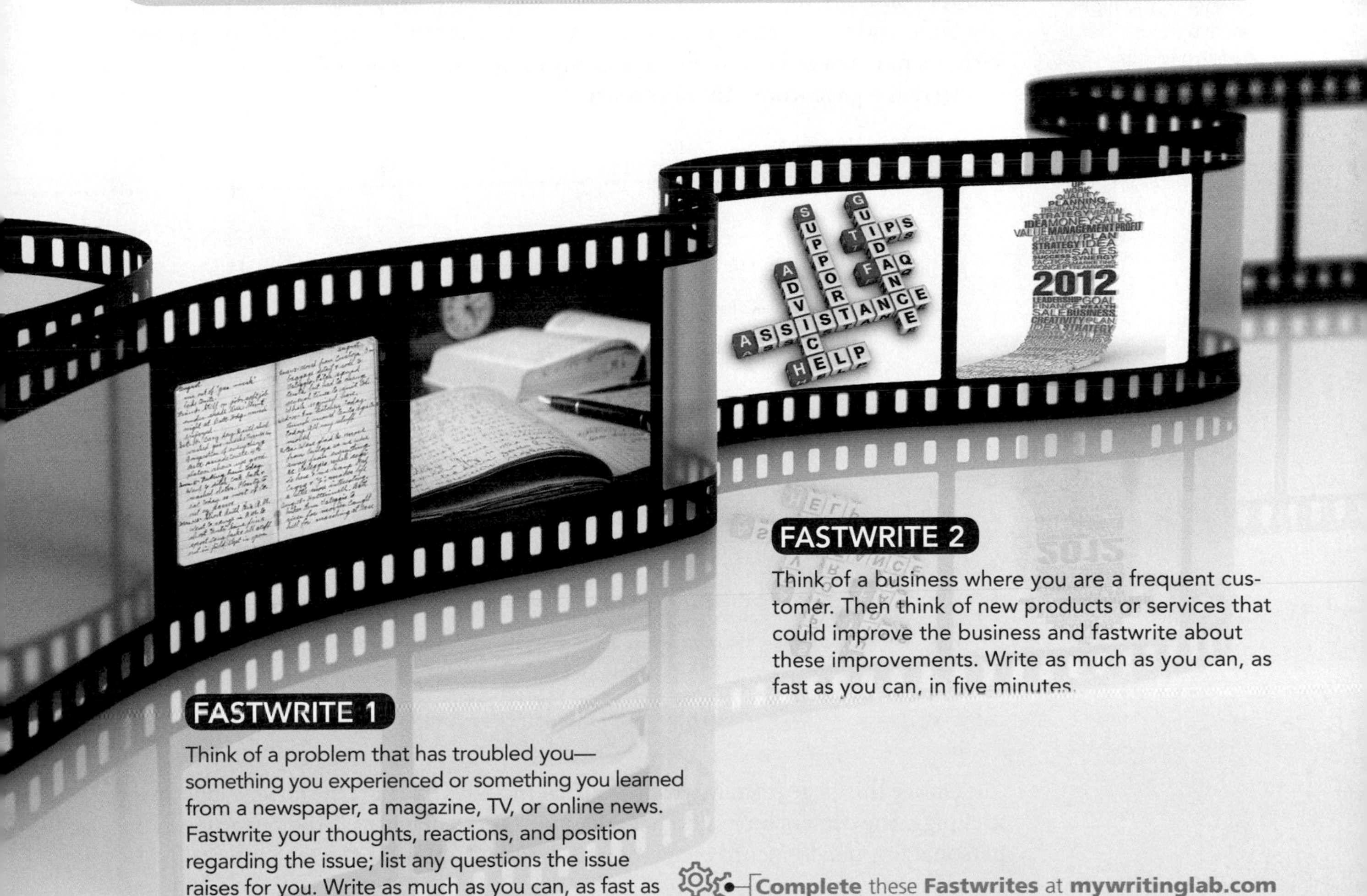

FASTWRITE 1

Think of a problem that has troubled you—something you experienced or something you learned from a newspaper, a magazine, TV, or online news. Fastwrite your thoughts, reactions, and position regarding the issue; list any questions the issue raises for you. Write as much as you can, as fast as you can, in five minutes.

FASTWRITE 2

Think of a business where you are a frequent customer. Then think of new products or services that could improve the business and fastwrite about these improvements. Write as much as you can, as fast as you can, in five minutes.

Complete these **Fastwrites** at **mywritinglab.com**

Personal Journaling

LEARNING GOAL

1. Keep a personal journal

For more on journaling, search the Internet for these videos:
The Benefits of Keeping a Journal
Writing Lessons: How to Keep a Personal Journal

"Keep a grateful journal. Every night list five things that happened this day . . . that you are grateful for. What it will begin to do is to change your perspective of your day and your life."
—Oprah Winfrey

In Chapter 1 you learned about journaling as a form of prewriting. Keeping a journal, however, can serve a more personal purpose. **Personal journaling** is both record keeping and collecting. A journal is your collection of experiences, opinions, questions, feelings, memories, and musings. It is a record you keep just for yourself, a record of thoughts that mean something to you one moment and may mean something more (or less) to you another time. Because the journal has no intended audience but yourself, spelling, grammar, mechanics, and even neatness do not matter. And neither does subject matter. Simply keep a record of ideas that interest you, new ideas you've learned, controversial issues you're wrestling with, and anything else you want to include.

Why Journal?

For both novice and professional writers, journaling is exercise. It develops your writing muscle (your brain) and helps you grow as a writer and a thinker. To build and tone muscle, you have to be dedicated to an exercising plan. In the same way, you build and tone your writing skills if you are dedicated to a journaling regimen. Write as much or as little as you like at any moment; in other words, just do it.

Here is a page from Miyoshi's journal

TGIF!! Yesterday was long. Thought it'd never end. Conquered math final. Felt good abt Soc101, Speech and Comm finals too. So glad theyre over.
Im upset at Arturo. Can't believe he won't go w/ us to work on the clean up at Sunny Lake. His so-called "boycott" is ridiculous. Doing nothing? Really? More fish will die More birds and turttles and who knows what else will die. Why not do something?! What if he's right What if Gov. does come and see the eyesore will he do anything anyway
Garbage pollution do-nothings politicians
I want to be proud of my town. I am proud of moss Bluff. Shouldn't I pitch in?
Arturo says let governer see what happened to the lake and then see what happens.
But who's to say the city commission will do anything about it after the governer leaves?
Maybe Arturo's right...

Notice this page from Miyoshi's journal includes fragments, run-ons, and misspellings. She centers on issues that are important to her. This is the purpose of any personal journal: to record concerns and thoughts important to the writer.

Three Journaling Categories

Journal entries are commonly sorted into one of three categories, though one journal entry can cross over into another category:

This Category . . .	Means
Me	The focus is on you—your values, your sense of identity, how you see yourself, how you believe others see you.
My World	The focus is on events, places, and people in your life.
Larger World	The focus is on events, places, and people who are otherwise not a part of your life.

17.1 Keeping a Personal Journal

Directions: Keep a journal for one week, writing in it at least once a day. Don't worry about spelling, grammar, or mechanics; you're writing for yourself. Below are three journal writing prompts you may choose. After a week, share your journal with your writing group. Discuss which entries are most meaningful to you and why you chose those topics. Answers will vary.

1. *Me* —What value do I have that others around me seem not to have? Why do others not have this value?
2. *My World* —Our school should offer ____________________ for its students.
3. *Larger World* —The crisis in _____________ doesn't seem to be ending. What could be done to change this? How might this affect me?

Business Letters

Today, letter writing is frequently replaced by cell phone, e-mail, and text message communication. These electronic modes are not only becoming more popular in private lives, but they are also becoming more common in business lives. Still, the traditional letter has an established foothold in business and is the preferred method of communication in a number of business writing situations.

LEARNING GOAL
❷ Format business letters

The standard business letter has eight parts. All these parts should be left-justified and aligned on the same vertical point.

In the business world, the accepted standard size and font is 12 point Times New Roman.

Letter Part	Description
Sender's Address	Optional. If included, place your address (street address, city, state, and zip code) in the upper left-hand corner on the first two lines of the page. Your name is not needed because it is in the closing.
Date	This is the date you write your letter. The date should be written in the convention of month, day, and year (e.g., August 10, 2012) and should appear on the line below your address or on the first line if you do not include your address.
Inside (Recipient's) Address	Write the recipient's address two lines below the date. Include the recipient's first and last name and appropriate title (e.g., Mr., Ms., Dr., or military title) on one line; office or room number (if applicable) on the next line; street address on the next line; and city, state, and zip code on the last line.
Salutation (Greeting)	Place your salutation two lines below the recipient's address. Use the greeting *Dear,* followed by the recipient's title and last name, and then a colon.
Body (Message)	Single space your paragraphs. Do not indent the first line of each paragraph; instead, leave a blank line between paragraphs.
Closing and Signature	Skip a line after the last body paragraph. Capitalize only the first word of the closing and place a comma at the end of the closing (e.g., *Thank you,* or *Sincerely,*); leave three lines between the closing and your name (typed) for your signature. Sign your name between the closing and your typed name.
Enclosure	If you include another document, skip one line after your typed name and type the word **Enclosure**. If you include more than one enclosure, type the word **Enclosures**, followed by a colon and the number of documents you include. For example, type **Enclosures: 3** if you include three other documents.
Courtesy or Carbon Copy (cc:)	If you distribute copies of your letter to people other than the addressed recipient, type **cc:** and list alphabetically those on your distribution list.

Some software programs have ready-to-use letter templates; however, these templates do not always use a standard format style. The following business letter includes all eight parts. It is properly formatted with one-inch margins, single spacing, block-style paragraphs, and an easy-to-read font style and size.

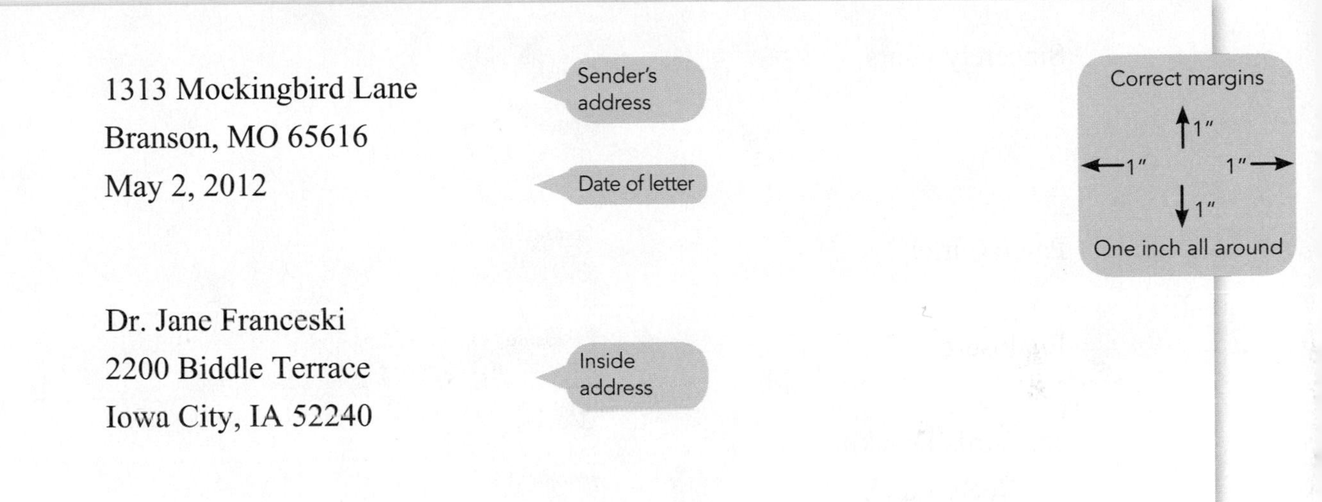

1313 Mockingbird Lane
Branson, MO 65616
May 2, 2012

Dr. Jane Franceski
2200 Biddle Terrace
Iowa City, IA 52240

Dear Dr. Franceski:

Salutation

Thank you for your offer to assist my school's Civic Leadership Team with the layout and design of our proposed community center. We hope our proposal is approved and funded by August so the necessary remodeling can begin.

As you requested, I have enclosed a drawing my team put together using a computer-aided design program. This plan includes all the features we believe those in the neighborhood want and would benefit from in a community center. The snack bar, game room, exercise/weight room, and television rooms are situated to best use the space we have available in the old hardware store we are renovating. We look forward to any suggestions you have that would help us use the space more economically.

Body

Our deadline for submitting the proposal to the city commission is July 2, so we would appreciate hearing from you by June 19. This would give us two weeks to make any necessary changes.

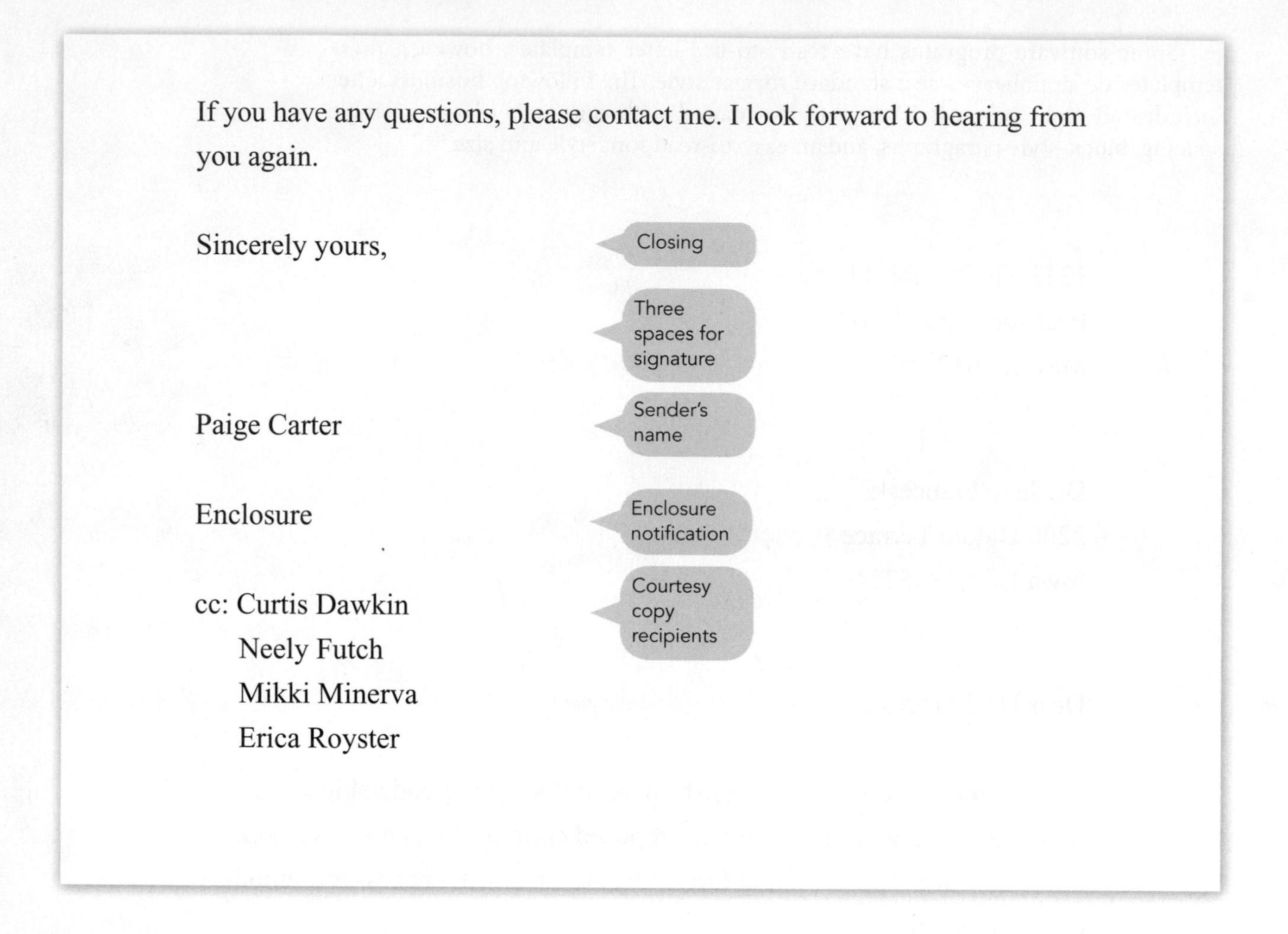
If you have any questions, please contact me. I look forward to hearing from you again.

Sincerely yours,

Paige Carter

Enclosure

cc: Curtis Dawkin
Neely Futch
Mikki Minerva
Erica Royster

Inquiry or Request Letters

One of the most common reasons people write business letters is to ask about something or to ask for something. When you ask in writing for information, products, or any type of service, you are writing **inquiries** or **requests**.

"Never write a letter while you are angry."
—Chinese Proverb

Letters of inquiry or request have five requirements:

- **Be polite.** If you've ever heard the saying, "You catch more flies with honey than with vinegar," you know the importance of being polite. Often the recipient did not ask you to write to him or her, so reading and fulfilling the request could be an imposition.
- **State your purpose quickly.** A reader will appreciate immediately knowing what you want, so state your purpose in the first paragraph.
- **Be specific.** Using as few words as possible but being as precise as possible helps keep you and your reader on track.

- **Explain why you are making the request.** Stating why you need the information, material, or service helps clarify your situation for your reader.
- **Provide more than one way to respond to you.** Offering your e-mail address, postal address, and even phone number might speed the response, so include at least two ways the recipient can reach you.

For more on writing request letters, search the Internet for this video:
Teaching Tips: How to Write an Inquiry Letter

17.2 Writing a Request Letter

Directions: Using the information below, write a request letter that meets the five requirements. The focus of your letter should be to receive a full refund for your purchase and to be let out of your two-year contract. Be sure to format your letter correctly. Answers will vary.

You ordered and received the CyberSurfeit-100 (CS-100), the latest in cell phone technology, from a service provider you've been with for the last six years. This phone is advertised as having the best connectivity, reception, sound, visual attributes, and memory. The company also hypes the CS-100 as being the only waterproof and shock-resistant phone on the market. However, once you received your new phone, you discovered the CS-100 was not what its advertising led you to believe.

1. Address your letter to the following:

   Ms. Hazel Ratchett, Manager
   Customer Service
   Cyberdyne Cellular
   1111 Cyberdyne Way
   Moss Bluff, FL 32179

2. Check your letter for all five business letter requirements.
3. Proofread and edit your letter.
4. Print your letter and share it with those in your writing group. Compare and discuss how your group members stated their purpose, were polite and precise, explained the reason for their request, and concluded their letter.

Business Thank-You Letters

Thank-you letters are often the first letters many people write, and in business and personal life thank-you letters can often be the most important letters they write.

Like request letters, thank-you letters should be honest, concise, and detailed. If you are writing a business or personal thank-you letter, you should also

- **Send it promptly.** Write a thank-you letter quickly; your recipient will appreciate your swift attention.

For more on writing thank-you letters, search the Internet for this video:
How to Write a Thank-You Letter

- **Be personal.** Your thank-you letter should be a sincere response for some thoughtful consideration. It should not gush with emotion but should simply express your appreciation.
- **Remind the reader why you are thanking him or her.** Make clear in the first sentence what you are thanking the recipient for.
- **Be specific and brief.** Include details that stand out from the event or situation, or specific thoughts about why you are thanking the recipient.

TICKET to WRITE

17.3 Writing a Business Thank-You Letter

Directions: Write a business thank-you letter to either someone who interviewed you for a job or someone who wrote you a recommendation for a job. Be sure your letter addresses the requirements and is formatted correctly. Answers will vary.

1. Look up the mailing address of your recipient or make up an address.
2. Check for all thank-you letter requirements.
3. Proofread and edit your letter.
4. Print and share your letter with your writing group. Compare and discuss how your group members stated their purpose, were polite and precise, explained the reason for their request, and concluded their letter.

TECHNO TIP

To review additional information on journal and letter writing, search the Internet for these sites:

Top 10 Thank-You Letter Tips

How to Keep a Journal

How to Write a Formal Business Letter

RUN THAT BY ME AGAIN

LEARNING GOAL
❶ Keep a personal journal

- **A Personal journal is . . .** a collection of your experiences, opinions, questions, feelings, memories, and musings.
- **Three categories of personal journaling are . . .** Me, My World, and Larger World.
- **In personal journaling, the category *Me* focuses on . . .** your values, your sense of identity, how you see yourself, and how you believe others see you.
- **In personal journaling, the category *My World* focuses on . . .** events, places, and people in your life.
- **In personal journaling, the category *Larger World* focuses on . . .** events, places, and people who are otherwise not a part of your life.

LEARNING GOAL
❷ Format business letters

- **All parts of a business letter should be . . .** left-justified and aligned on the vertical point.

- **The eight parts of a business letter are . . .** sender's address, date, inside (recipient's) address, salutation (greeting), body (message), closing and signature, enclosure, courtesy or carbon copy.
- **Five requirments of a request or inquiry letter are to . . .** be polite, state your purpose quickly, be specific, explain why you are making the request, and give more than one way to respond to you.
- **When writing a thank-you letter . . .** send it promptly, be personal, remind the recipient why you are thanking him or her, and be specific and brief.

PERSONAL AND BUSINESS WRITING LEARNING LOG MyWritingLab™

MyWritingLab™

Answer the questions below to review your mastery of personal and business writing.

1. **A personal journal is a collection of what?**
   A personal journal is a collection of experiences, opinions, questions, feelings, memories, and musings.
2. **What are the three categories of personal journaling?**
   The three categories are Me, My World, and Larger World.
3. **What does each personal journaling category help you focus on?**
   Me focuses on your values, sense of identity, how you see yourself, and how you believe others see you. *My World* focuses on events, places, and people in your life. *Larger World* focuses on events, places, and people that are otherwise not a part of your life.
4. **How should the eight parts of a business letter be justified and aligned?**
   All parts should be left-justified and aligned on the same vertical point.
5. **What are the eight parts of a business letter?**
   The eight parts are sender's address, date, inside (recipient's) address, salutation (greeting), body (message), closing and signature, enclosure, and courtesy copy or carbon copy.
6. **What are five requirements of a request or inquiry letter?**
   Be polite, state your purpose quickly, be specific, explain why you are making the request, and give more than one way to respond to you.
7. **What are the steps you should take when writing a thank-you letter?**
   Send it promptly, be personal, remind the recipient why you are thanking him or her, and be specific and brief.

CHAPTER 18 Electronic Writing and New Technologies

LEARNING GOALS

In this chapter, you'll learn and practice how to

1. Write e-mail
2. Post and respond to blogs
3. Participate in discussion groups
4. Instant message or text

GETTING THERE

- E-mail is a useful tool for both personal and professional communication.
- Blogs are electronic journals through which people share thoughts and interests.
- Discussion groups are Web sites that allow individuals interested in a specified topic to share ideas.
- Instant messaging (texting) is another immediate form of electronic communication.

FASTWRITE 2

Conduct an Internet search for Popular Blogs. Choose one of the blogs, read the most recent five blog posts, and then fastwrite a response to the post. Write as much as you can in five minutes.

FASTWRITE 1

You need to send an e-mail to a member of your class who has been absent for the last two class sessions. You are concerned about the student's health and you also want to let him or her know what has been happening in class. Write as much as you can in five minutes.

Complete these **Fastwrites** at **mywritinglab.com**

E-mail

LEARNING GOAL
1 Write e-mail

E-mail is a shortened version of the phrase *electronic mail.* As a noun, *e-mail* refers to a specific type or system of Internet communication or a piece of communication similar to a letter sent through the post office. As a verb, *e-mail* refers to the action of sending that communication.

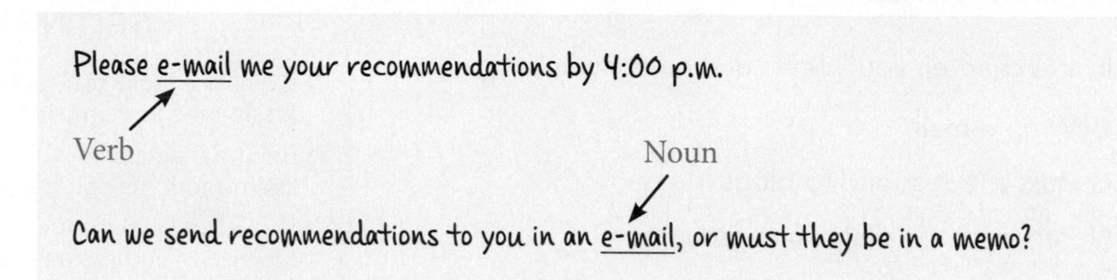

For an introduction to e-mail and for free e-mail sites, visit any of these Web sites:
yahoo.com
mail.google.com
hotmail.com
mozilla.com
mail2web.com

E-mail is like letter writing, but, unlike letters, which can be written with pen and paper, e-mail relies on a keyboard and software. Letters and e-mails are both sent to the recipient's address, but an e-mail goes through a type of service made up of e-mail servers. An e-mail is similar to a business letter or office memo because it usually contains the sender's address, the recipient's address, the date the e-mail was written, and a body message.

Your e-mail's intended audience determines the level of language formality you use. Look at the e-mail messages below. Both have the same writer and the same topic but different recipients and a different level of language.

The first e-mail Moira sent is to the director of a company in her hometown. It expresses her concerns with an environmental issue.

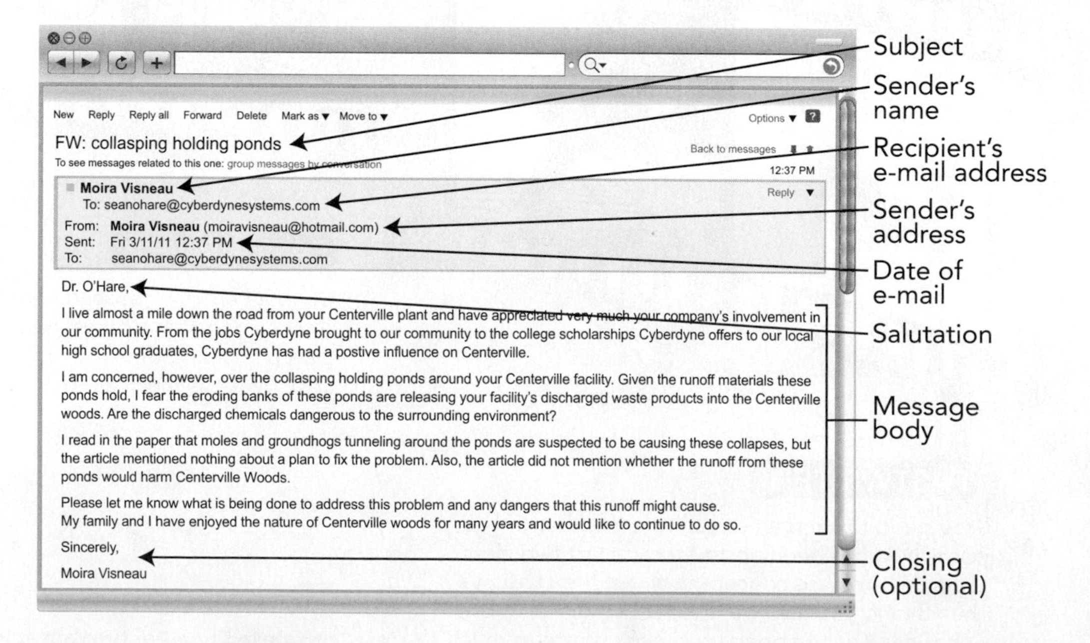

Moira sent this email to a friend. It addresses the same subject.

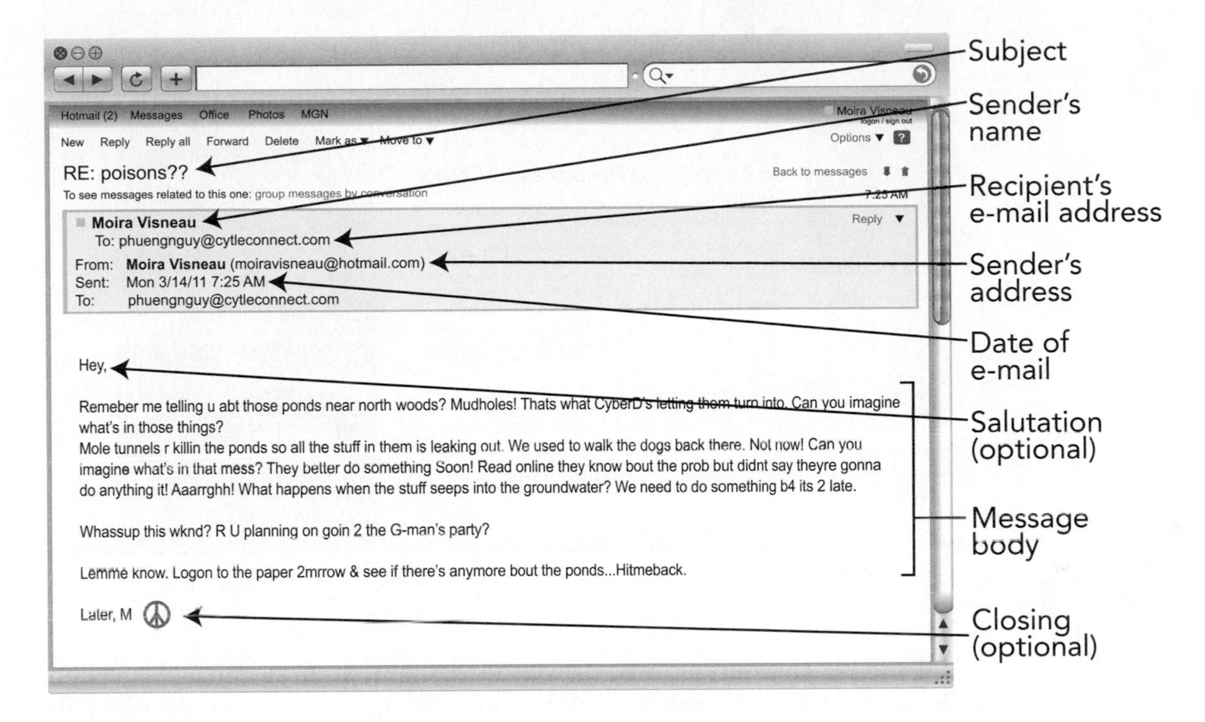

In both e mails, Moira makes the same point about her dissatisfaction with the holding ponds. The way she expresses her dissatisfaction, however, changes. She uses a formal approach when writing to someone she doesn't know; she uses a casual approach when writing to her friend. If she had used casual language in her e-mail to the director of the company, she would have risked the director not taking her seriously. Similarly, if she had used formal language in writing to her friend, the friend may have thought Moira didn't sound as friendly as usual.

Blogs

LEARNING GOAL

❷ Post and respond to blogs

If e-mail is the parent, its child is blogging. E-mailing is similar to writing letters, and blogging is similar to writing articles.

In Chapters 1 and 17 you learned how journaling can help you discover writing topics and improve writing and critical thinking skills. A **blog** is an electronic journal, a collection of a **blogger**'s (blog writer's) thoughts and ideas. Bloggers share not only writing, but also photography, videos, music, voice recordings, or other multimedia.

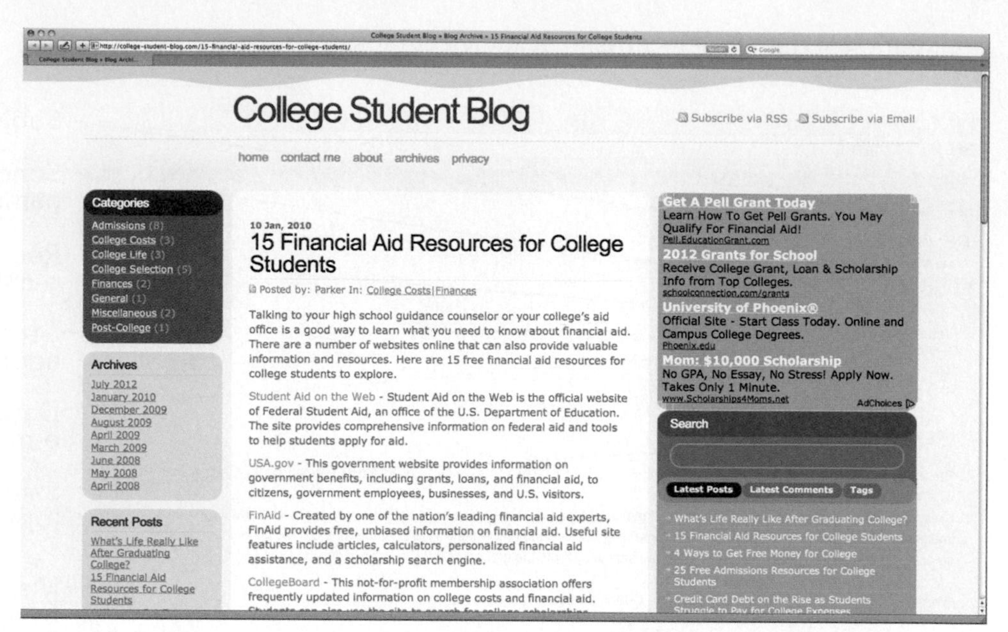

College Student Blog, copyright © 2009 College Student Blog. Reprinted with permission.

> **TECHNO TIP**
> For an introduction to blogging, search the Internet for this site: Create a blog. It's free.

Unlike a personal journal, a blog is intended to be shared with an audience and usually invites feedback from its readers. Any writing, video, sound recording, or image placed on the blog is called a **post**, and blog readers (members) are usually welcome to comment on any blog post.

Discussion Groups

> **LEARNING GOAL**
> ❸ Participate in discussion groups

A **discussion group** is a community of Internet users who take part in online talks regarding various topics of interest common to the group. Discussion groups are also referred to as *newsgroups, bulletin boards*, and *forums*. Group members post blocks of text, comments, and questions for other members to read and respond to. Becoming a part of one of these groups can aid you in research for your classes. Discussion members may be experts in various fields or may be able to quickly direct you to helpful resources. Usually, each member's response to a post becomes part of a conversation that all group members can see.

Instant Messages

> **LEARNING GOAL**
> ❹ Instant message or text

Instant messaging (texting) is conversation that takes place immediately, just as spoken conversation does. Unlike an e-mail, in which the writer types a letter-like message and then sends it to another person, an instant message usually consists of only a sentence or two or a single question. The recipient can respond to the message instantly and then wait for a reply. The conversation is as formal or informal as the writers want it to be. Instant messaging began via the Internet. Today, however, instant messages are more commonly sent with cell phones and are often referred to as *text messages*.

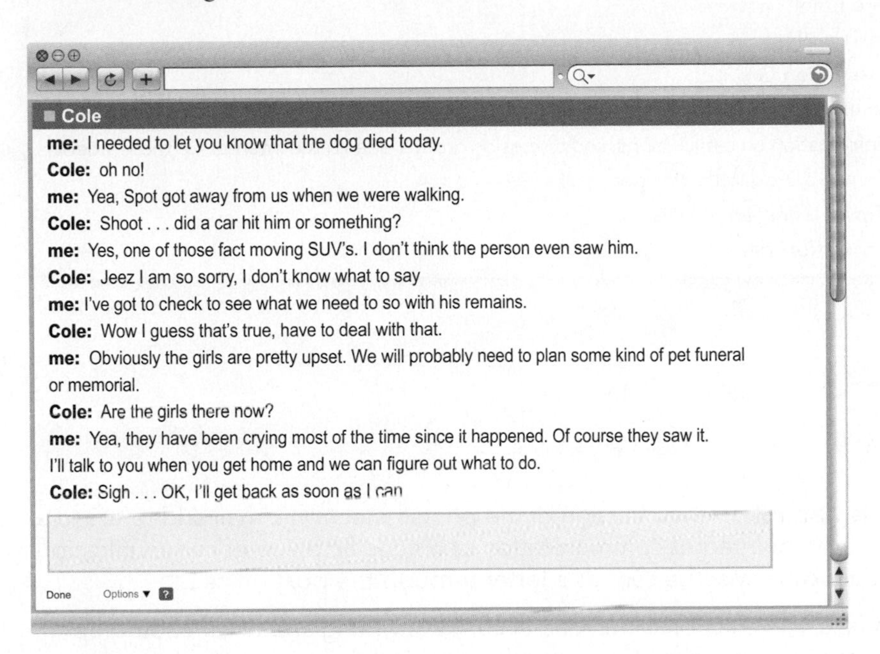

TICKET to WRITE

18.1 Electronic Writing

Directions: Conduct an Internet search for Popular Blogs. Choose one of the sites, and then select a blog from that site. Then read a few recent posts until you find one that interests you. The post may cause you to have questions, to want to share your opinion, to want to show your appreciation to the author, or to express your disagreement with something the author wrote.

Draft a response to the post. Then describe the original post to your writing group or show them the post. Next share your response with your group.

The group should discuss each member's blog response and answer the following questions.

1. Does the writer's response stay focused on the post's main idea?
2. Do you believe the writer would get a response from another reader?
3. Do you suggest the writer add or delete anything from the response? If so, what and why?

TECHNO TIP

For more information on e-mail, blogs, and discussion groups of interest, search the Internet for these titles:

Top 19 Free Email Services
How to Start a Blog
Google Discussion Groups
Yahoo Discussion Groups

For more information on e-mail, blogs, and discussion groups, search the Internet for these videos:

How to Create a Blog with Blogger
Internet Tips: Definition of Listserv
Google Group Tutorial

RUN THAT BY ME AGAIN

LEARNING GOAL
❶ Write e-mail

- **E-mail is . . .** a shortened version of the phrase *electronic mail* and refers to a specific type of Internet communication or a specific piece of communication that would otherwise be sent as a letter through the post office.

LEARNING GOAL
❷ Post and respond to blogs

- A **blog is . . .** an electronic journal that contains writing and often photography, videos, music, voice recordings, or other multimedia.

LEARNING GOAL
❸ Participate in discussion groups

- A **discussion group is . . .** a community of Internet users who take part in online discussions regarding topics of specific interest to the group by posting blocks of text, comments, and questions for group members.

LEARNING GOAL
❹ Instant message or text

- **Instant messaging is . . .** electronic conversation that takes place in real time, just as spoken conversation does.

ELECTRONIC WRITING AND NEW TECHNOLOGIES LEARNING LOG

MyWritingLab™

Answer the questions below to review your mastery of electronic writing.

MyWritingLab™

1. How is e-mail similar to a business letter or office memo?
   E-mail is similar to a business letter or office memo in that it usually contains the sender's address, the recipient's address, the date the e-mail was written, and a body message.
2. How is a blog similar to a personal journal?
   Like a personal journal, a blog is a collection of the writer's thoughts and ideas.
3. How is a blog different from a personal journal?
   Unlike a personal journal, a blog is meant to be shared and often invites feedback from readers.
4. What is the focus of an online discussion group?
   An online discussion group focuses on topics of interest common to the group's members.
5. How is an instant message different from an e-mail?
   An instant message usually contains only a couple of sentences or a question; an e-mail is a longer message, similar to a letter.

Writing Newspaper Articles and Examining Journal Articles

LEARNING GOALS

In this chapter, you'll learn and practice how to

1. Recognize and write various types of newspaper articles
2. Find and examine articles in professional journals
3. Discover and evaluate works in literary journals

GETTING THERE

- News articles, editorials, and feature articles are basic categories of newspaper writing.
- Articles in professional journals provide current research-based information in a particular field of study.
- Literary journals provide writers outlets for fiction, poetry, and creative nonfiction.

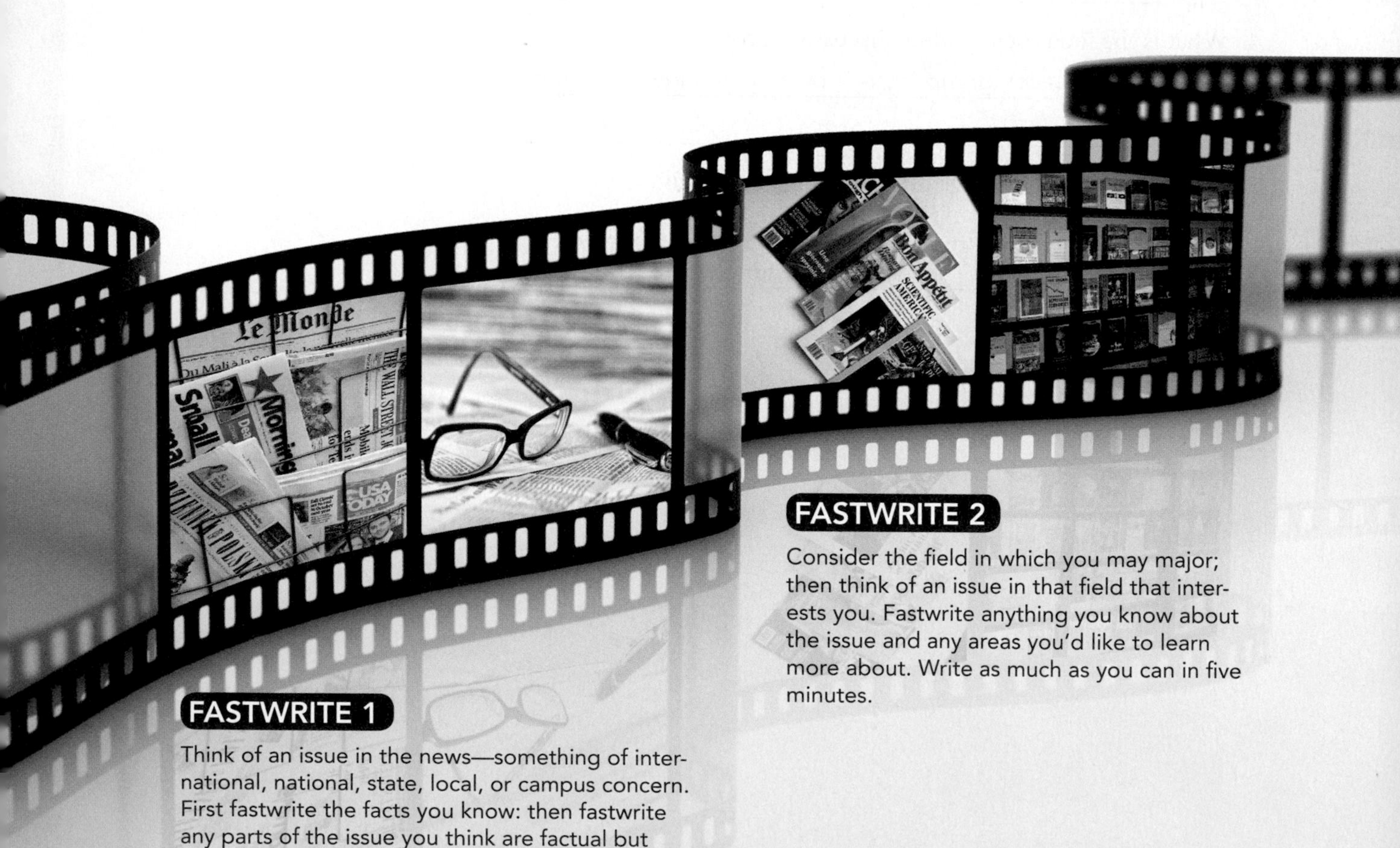

FASTWRITE 2

Consider the field in which you may major; then think of an issue in that field that interests you. Fastwrite anything you know about the issue and any areas you'd like to learn more about. Write as much as you can in five minutes.

FASTWRITE 1

Think of an issue in the news—something of international, national, state, local, or campus concern. First fastwrite the facts you know: then fastwrite any parts of the issue you think are factual but that you need to check. Write as much as you can in five minutes.

Complete these **Fastwrites** at **mywritinglab.com**

Newspapers

Newspapers include three main types of articles: *news articles*, *editorials*, and *feature articles*. In every major newspaper, you will find examples of each.

LEARNING GOAL
1. Recognize and write various types of newspaper articles

News Articles

News articles are information pieces found on the front pages of papers, as well as in many inside pages. Writers of news articles focus solely on factual information, so opinions are allowed only in quotations in the article.

News articles are written in a particular format, called an *inverted pyramid*. In this format, the writer presents the most important points of the story first. Subsequent facts come in descending importance.

"Were it left to me to decide whether we should have a government without newspapers or newspapers without a government, I should not hesitate a moment to prefer the latter."
—Thomas Jefferson

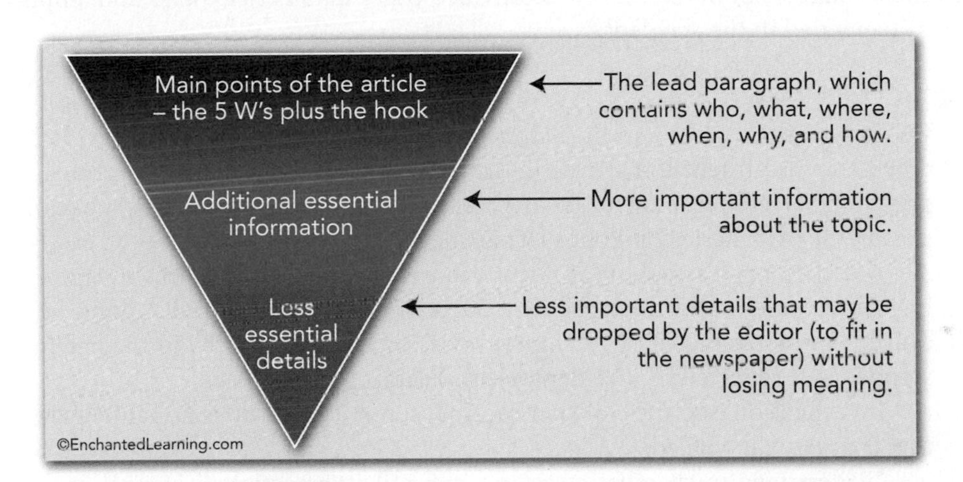

The first paragraph of the article (called the **lead**) presents the article in a nutshell. Information in the lead, which may be only one sentence long, gives readers short answers to reporter's questions (*Who? What? When? Where? Why?* and *How?*).

For a story he was writing for the campus newspaper, Pedro collected these notes for his lead:

Brianna Conweroun is a student at Rowland College. She is 20 years old. She was in an accident. The accident happened on Saturday night. Brianna was crossing Third Street. She was near the campus ball field. She was hit by a car. She was taken to Salyer's Hospital the night of the accident. She has a concussion.

In news articles, ages of key individuals are usually noted.

Pedro turned these notes into a lead that summarized the most important facts and answered reporter's questions:

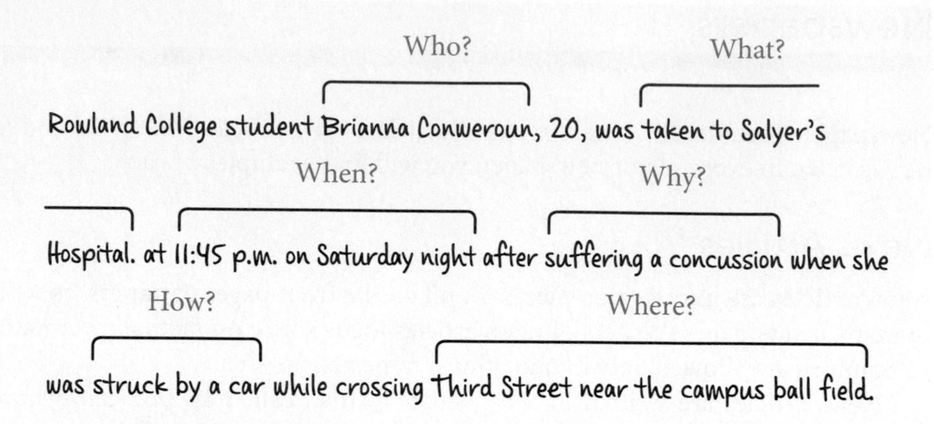

In the remainder of his story, Pedro provided other details to explain and amplify what happened in the accident.

In news articles, sources are clearly identified.

Charleston firefighters and the Salyer's Hospital ambulance service provided emergency treatment at the scene of the accident. Conweroun was then transported to Salyer's Hospital for further treatment, said Sgt. Jeff Henkel, spokesperson for the Charleston Police Department.

Investigators learned the accident occurred when Conweroun was crossing Third Street and was struck by a 2010 Chevrolet heading eastbound on Third. The driver of the car, Kammy Davis, 36, of Martinsville, immediately stopped and reported the accident, Henkel said.

Investigation into the collision continues, but Henkel does not anticipate that charges will be filed.

"Pedestrians must cross streets at marked intersections," Henkel said. "Drivers may not see someone walking in the middle of the road, especially late at night."

TECHNO TIP

For more on how to write a lead, search the Internet for these videos:

How to Write News Articles: How to Write a News Lead: Part 1

How to Write News Articles: How to Write a News Lead: Part 3

For more on how to write in an inverted pyramid format, search the Internet for this title:

How to Write News Articles: The Inverted Pyramid News Article Style

19.1 Composing a Lead for a News Article

Directions: Use the information below to compose a lead for a news article. Include answers to reporter's questions (*Who? What? When? Where? Why?* and *How?*) in the lead. Answers will vary.

A man was arrested. He was arrested at his home. The man was Herman G. Ursalle. He is 55 years old. He lives in New Albans. He was arrested October 2. Police arrested him at his home. This happened at 3:45 p.m. He was arrested for second-degree arson. The arson was in connection with burning down a building at 389 Oak Street.

Possible answer: Herman G. Ursalle, 55, of New Albans, was arrested at his home on October 2 at 3:45 p.m. He was charged with second-degree arson in connection with a building fire at 389 Oak Street.

19.2 Examining a News Article from a US Newspaper

Directions: The site **newslink.org** contains links to newspapers across the United States. On that site, under US newspapers, click on "Dailies." Next click on a state and then a newspaper in that state. Read a news article in that paper. Use information from the article to identify the following:

Name of newspaper
Title of article
Whom the article is about
What the article is about
When the events took place
Where the events took place
Why the events took place
How the events took place

19.3 Examining a News Article from a Foreign Newspaper

Directions: Use an online search engine to find a newspaper from another part of the world. Read a news article in that paper. Use information from the article to identify the following:

Name of newspaper
Title of article
Whom the article is about
What the article is about
When the events took place
Where the events took place
Why the events took place
How the events took place

Editorials

Every major newspaper has an assortment of **editorials**, articles that state writers' opinions on current topics. Editorials

Sometimes an editorial page is called an opinion page.

- are usually unsigned
- often state the perspective of those who run the newspaper
- run on the same page and same section in each issue of a newspaper

An editorial shares characteristics with a persuasive essay, in which you gear writing to your purpose and audience. Your purpose in both a persuasive essay and an editorial is the same: to present arguments that convince readers you are correct on a certain issue. However, your audience in an editorial for your student newspaper is different than your audience for, say, the *Wall Street Journal*, and so the topic you address may be different. Other important points to remember in writing an editorial include these:

Think of your position sentence as your thesis statement.

- **Take one side of an argument or the other, but not both.**
- **Be clear about your position.** Relate it in your first paragraph, perhaps even in your first sentence.
- **Be certain of your facts.** While readers understand that an editorial is your opinion, your support must be solid and factual. If necessary, use research, statistics, or other data as support.
- **Watch your language choice.** An editorial is not the place for long sentences or complex language. Keep your writing as simple as possible, and use strong and well-defined language.
- **Because readers know an editorial expresses your opinion, expressions like "I think that . . ."** or "My opinion is that . . ." are unnecessary.

~~I think that~~ students should have access to more ATMs across campus.

~~My opinion is that~~ the process for getting extra tickets to ball games needs to be changed.

Mikayla composed this editorial for her campus newspaper:

Students should have access to more ATMs across campus. Frequent use of ATMs is a way of life for many, if not most, students at RCC, but they are inconvenienced by the small number of ATMs available.

Mikayla begins by stating her position

In a recent informal poll of fifty students, 72% said that they found using campus ATMs was far easier than making a trip to a local bank or other location for an ATM.

Part of Mikayla's research supporting her position

In the same poll, 66% of those who responded were in favor of having more ATMs on campus. Comments posted by those who responded included statements like "I don't have time to search for an ATM" and "Sure would be nice not to have to trek almost a mile to get some money."

Another part of Mikayla's research supporting her position

(4) RCC has twenty-five buildings on campus, including three dorms, but only six buildings have an ATM. None of the dorms has one. Students do not need to spend their time tracking down which buildings can accommodate their banking needs. Since the only cost of more ATMs is in their installation, the administration and local banks should correct the problem immediately and put an ATM in each building on campus.

The strongest part of Mikayla's research supporting her position

19.4 Examining an Editorial

Directions: To find an editorial, select one of the following newspapers and type its title into an online search engine. At the homepage for that newspaper, click on "Opinion." Alternately, use an online search engine to find an editorial from another current newspaper.

Chicago Tribune | *The New York Times*
Los Angeles Times | *The Washington Post*

After you read the editorial, write the name of the paper and title of the editorial; then respond to the following:

1. Topic of the editorial
2. Writer's position on the topic
3. Support the writer gives for his or her position
4. Your opinion about the writer's strongest piece of support for his or her position
5. Your opinion as to whether or not the writer gives enough support to persuade readers
6. Your opinion as to where support is given or needed

Now that you have studied and read editorials, use Ticket to Write 19.5 to compose your own.

19.5 Composing an Editorial

Directions: Choose an issue (on campus, in your campus town or your hometown, in your state, in the country, or in the world) important to you. Compose an editorial about your position on the issue. When you have finished, share your editorial with your writing group and ask them to use the questions below to evaluate how effectively you supported your position.

1. What is my topic?
2. What is my position on the topic?
3. What support do I give for my position?
4. What is my strongest piece of support for my position?
5. Do I give enough support to persuade readers? If not, where do I need additional support?

Feature Articles

Feature articles are also used in magazines, TV news shows, newsletters, blogs, and zines.

A **feature article**, often found in inside sections of newspapers or in newspaper supplements, educates or entertains readers by providing well-researched, in-depth information on events, issues, trends, or people. Like news articles, feature articles are factual, timely, and cover an interesting angle about a person or story. Feature articles, however, differ from news articles in several aspects. A feature article

- is not centered only on "breaking news"
- is not written in an inverted pyramid form
- is usually longer than a news article
- does not answer reporter's questions in its lead

Newspapers and other outlets run many different types of feature articles, including the following:

Type	Purpose	Sample Title
Human Interest	relates how people cope with problems or realize accomplishments	"Losing Weight While Working as a Chef"
Commemorative	highlights past events or celebrations of annual event	"Bicentennial Celebrations Beginning in 2014"

Developing Trend	informs about increased interest in or use of something	"Laptops and Tablets Replacing Desktops"
Review	analyzes a product, place, performance, work of art, etc.	"One Student's View of On-Campus Eateries"
How-To/Service	gives step-by-step directions	"How to Use the New Campus E-mail System"
Informational	provides facts about what, when, or why to do something	"What to Do to Prepare for Next Semester"
First-Person	presents writer's experiences and comments	"My Week-Long Vacation from Computers"
Interview/Profile/ Personality	offers in-depth look at a certain person	"A Front Porch Chat with Carrie Underwood"

As in a news article, the first paragraph of a feature article is called the *lead*. In a feature article, the purpose of the lead is to draw readers in so they will be eager to read more. Here are some common types of leads in feature articles:

"I look at leads as my one frail opportunity to grab the reader. If I don't grab them at the start, I can't count on grabbing them in the middle, because they'll never get to the middle."
—N. Don Wycliff, *Chicago Tribune*

Type	Purpose	Sample
Question	a tease to get readers to ponder the answer	Who knew a spade and a shovel could have such quirky uses?
Revelation	a shocking or revealing statement or statistic	Today, almost eleven thousand people died of hunger.
Quote	a thoughtful quote that promises insight of what is to come	"There is always an easy solution to every human problem—neat, plausible and wrong." —H. L. Mencken

Narrative	a short account of experiences providing interesting action central to the article	James entered the room and heard a soft noise behind him. Looking back over his shoulder, he saw . . . nothing.
Description	a picture painted with words	Straining my sunburned neck upward, all I saw were the steep, red rocks of the canyon rising on both sides of me.

In writing a feature article, keep in mind

A main idea in a feature article is similar to a thesis statement in an essay.

- the audience you are writing for
- the main idea of your article
- the type of lead you will use to draw your reader in
- the focus and purpose of your article, which should be apparent in the lead and first paragraphs
- details in your body paragraphs, which should elaborate on both the slant and focus of your article
- supporting details, which may include quotes, anecdotes, vignettes, charts, graphs, photos, bulleted lists, headings, statistics, timelines, sidebars, or other text features

Writers use text features to add authenticity or depth to their piece.

19.6 Analyzing Pulitzer-Prize Feature Writing

Directions: The Pulitzer Prize is a prestigious US award given every year for achievements in various categories of journalism, literature, and music. To access recent Pulitzer winners in the feature writing category, enter **Pulitzer .org** in an online search engine. Next click on "Past winners & finalists by category" and then "Feature writing." Read one of the articles and identify

1. The title
2. The author
3. The source
4. The type of feature writing
5. The writer's purpose in the article
6. The details the writer provides to help readers understand the purpose
7. Why the title is or is not appropriate for the piece
8. The type of lead the author uses
9. How the lead draws you in and makes you want to read the rest of the article
10. The topic of the article
11. The topics presented in the body paragraphs
12. Why the conclusion is or is not appropriate for the piece

In each issue, Anthony's school newspaper runs a feature article titled "Who Knew?" The article highlights what someone on campus—a faculty or staff member, or another student—did before entering college. Anthony interviewed his history professor and then wrote this feature article:

Who Knew?

by Anthony King

Who knew that, for the last two months, I've been hobnobbing with the founder and past president of the Job-a-Month Club?

Anthony used a question as his lead.

We don't exactly pal around outside of class, and we haven't bonded over a cup of joe. But since classes began, I've spent three days a week with her, so I've been in the presence of greatness with Dr. Lindsay Byrne. She is the founder and former president of the Job-a-Month Club and now, I'm proud to say, my history professor.

When eighteen-year-old Lindsay Byrne graduated from high school, she had no idea what she wanted to do. "Some of my teachers had talked with me about going to college, but I didn't pay any attention. Honestly, when I was a teenager I was more caught up in the social scene than the academic world," Dr. Byrne said.

Anthony used a number of quotes throughout the article.

As a result, when she graduated in June she had no plans for her future. She quickly picked up a job selling purses and jewelry in a kiosk at the local mall. To her, that job became boring very quickly, and she immediately began looking for another one.

Four weeks later, in July, she was hired at one of the mall's clothing stores. "The manager and I didn't see eye-to-eye about my approach with customers. That's an indirect way of saying that I didn't have the best customer service record," Dr. Byrne admitted. "A month later, I was out of a job."

See a trend here?

Next came jobs at the local McDonald's (August), a consignment store (September), a grocery (October), an auto-repair shop (November), and a mall kiosk selling holiday items (December).

"I wasn't fired from those jobs," Dr. Byrne laughed, "but for one reason or another none of them worked out for me. In December, when I was working at the kiosk, I knew I would again be out of work—again!—at the end of the month since the job was seasonal. When I started thinking about my life after high school, I realized that I'd had a different job every month. That's when I gave myself the title of founder and president of the Job-a-Month Club."

It's also when Lindsay Byrne decided to do something more substantial with her life. In January of the next year, she enrolled in her local community college. "I had no idea what I wanted to study or to do with my life. But I sure knew what I didn't want to do—to continue being in the Job-a-Month Club."

While completing her basic classes, student Lindsay Byrne found that she enjoyed her studies in history the best. "I really hadn't connected with history when I was younger. Maybe I had to have a little history of my own before I could appreciate the subject," she says. Byrne went on to finish her work at the community college and then transferred to Benson State, where she completed her bachelor's, master's, and doctoral work.

"I've been teaching here for almost sixteen years," Dr. Byrne says, "so I think I can safely say that the Job-a-Month Club is—at least for me—a thing of the past."

19.7 Writing a Feature Article: Interview

Directions: Interview someone on your campus and write your version of an interview/profile type of feature article, like the "Who Knew?" article Anthony wrote. You might find these questions helpful in your interview:

- What is an interesting or unusual job you have held?
- What is the background of your first name or your family name?
- What is something about you that would surprise readers?
- What is something about you that readers would find interesting?
- What is something you've done that you're proud of?
- How did you come to live here?
- What is something about your childhood that was unusual?
- As a child, what kind of work interested you?
- In past schooling, what classes were your best? Why did you excel at them?
- In the past, were you interested in sports? Which ones? Were you a participant or a spectator?
- Do you have an interest in music? Do you play an instrument or sing?
- Do you like pets? Describe one of the pets you've felt especially close to.
- Describe any friends with whom you have a special bond. What makes those people so close to you?
- What world events have had an impact on you? What was that impact?
- What hobbies do you have? How do you participate in those hobbies now?
- What family stories have come down to you from any of your ancestors?
- If you had time to save just five items in your current home, what would you save and why would you pick those items?
- Have you ever lived anywhere else? If so, what was the most appealing part of living in that location?
- If you are married, how did you meet your spouse?

- If you are a parent, what can you tell about your relationship with your children?
- If you could thank one of your parents or another relative for something, what would you say and why?

19.8 Writing a Feature Article of Your Choice

Directions: Choose any type of feature article to write. Use one of the sample titles below or a variation of it (substitute other nouns, verbs, or adjectives to fit your needs), or use a title approved by your instructor.

- The Unsung Hero
- Why Math Classes Are So Difficult
- What Every College Freshman Should Know before Classes Begin
- The Places to Be: Student Hangouts
- Shh! What I'll Never Tell My Kids
- The Worst Break-Up Lines Ever Used
- What Makes Our City Tick?
- Smoking Used to Be Hip; Now It's Hopped
- How to Deal with Peer Pressure in College
- Oh! The Places You'll See: Recalling Dr. Seuss
- The Best Movie of the Year (So Far!)
- My Most Heartbreaking Time
- Fifty Years: Half-Century of Campus Life

Professional Journals

LEARNING GOAL
❷ Find and examine articles in professional journals

As you advance in your studies, you will research information from a number of different **periodicals**, publications issued at regular intervals. Some of your research may come from popular magazines you're familiar with, like *Time*, *Good Housekeeping*, or *Sports Illustrated*.

You may find other reasearch in professional journals, which differ from popular magazines in several ways. Articles in **professional journals** include authoritative information devoted to the latest trends or research in a specific occupation or industry, and they are authored by leaders in their field. Because these authors assume their readers understand the field, the language they use often includes technical terms and other specialized vocabulary specific to their discipline. Authors must prove the information in their texts, so supporting evidence (e.g., charts, graphs, and tables) is usually included. Other features often found in professional journals but not in popular magazines include the following:

Professional journals are also called scholarly journals, academic journals, refereed journals, or peer-reviewed journals.

An abstract, sometimes called an executive summary, helps readers determine if an article contains information or other research they seek.

- an **abstract** (a short summary of important points of the text)
- descriptions of methods authors used in their research
- conclusions authors reach as a result of their research
- a bibliography

Comparing Popular Magazines and Professional Journals

	Popular Magazine	Professional Journal
Purpose	inform, entertain, or persuade readers on general interest topics	provide original research articles, or book reviews; all devoted to a specific field
Audience	general public	other scholars and professionals in a field
Authors	staff writers, freelance authors, guest authors	experts in a field
Documentation	not used	bibliographies or references mandatory
Language and Graphic Aids	informal, easy-to-read language; photographs or images common; tables, charts, and other graphic aids infrequent	specialized formal language, including vocabulary common to the field; tables, charts, equations, and other graphic aids to support text
Layout	various formats; no abstract	standard formats (e.g., APA or MLA) common to a discipline; abstract often included
Examples	*Newsweek*, *Car and Driver*	*Minority Psychology*, *Comparative Economic Studies*

Finding Professional Journals

Like publication schedules of popular magazines, publication schedules of professional journals vary (e.g., weekly, monthly, quarterly).

In the reference section of your college library, you'll find hard copies of professional journals that relate to fields of study offered in your school. If you are studying dental hygiene, for instance, your library may subscribe to professional journals like these:

Dental Assistant
Dental Association
Dentistry Today
Journal of the American Dental Association
Journal of Dental Education
Journal of Public Health Dentistry

Or, if your institution offers classes in education, you might find these professional journals:

American Educator
Dimensions of Early Childhood
Educause Review
Education Digest
Exceptional Children
Journal of Education
Journal of Educational Research
Phi Delta Kappan

Your library may also subscribe to online databases that provide articles from a wide assortment of professional journals. Some databases cover many fields of study, and some concentrate on particular disciplines. Databases commonly found in college libraries include these:

Discipline	Database
Interdisciplinary	EBSCO Academic Search Elite, Expanded ASAP, ProQuest
Social Sciences and Education	ECONbase, Education Abstracts, JSTOR, Project MUSE, PscyINFO, ERIC
Science	American Chemical Society Publications, JSTOR, Science Direct, PubMed, Healthnet, MedlinePlus
Business	ABI/Inform, ECONbase, JSTOR
Humanities and Arts	JSTOR, Project MUSE

If you searched in your library and in online databases but were unable to find a copy of an article you need, you can request that your library obtain a copy of it through interlibrary loan. This service allows patrons of one library to borrow books or obtain copies of documents housed in another library.

Academic Writing and Writing for a Professional Journal

Articles in professional journals share many qualities that are also important in your academic writing.

Audience and Purpose

In some professions, being published is mandatory in order to retain a position.

In essay writing, you always focus on both audience and purpose. Likewise, authors in professional journals focus on their audience and purpose. Authors gear their topics to their audience. For instance, authors know the *audience* of the *Political Review* expects topics that reflect the world of politics, so that is what these authors provide.

Equally important is the *purpose* of an article. In professional journals, the purpose is to keep readers up-to-date on original research or experiments regarding a specific occupation or industry. Authors always focus on this purpose as they write their articles.

Peer Review

Heads Up!
Peer-reviewed articles are also known as refereed articles.

In your paragraph and essay writing, you often use peer review when you share your brainstorming and drafts with members of your writing group, asking them for input about ways you might improve your writing. Authors of articles in professional journals also ask peers to check their material before they submit articles for possible publication. Like you, they seek peer suggestions about ways to improve what they write.

After this, though, comes a formal process, also known as *peer review*. In a peer review for a professional journal, the finished article is sent to a group of scholars in the field, and that group decides if the material in the article is worthy of publication. The peers look at the article and judge if

Heads Up!
Professors may require you to use only peer-reviewed articles in your research.

- its content is relevant and significant to the field
- its data is well researched and well documented
- the article is written clearly and logically

Readers know that peer-reviewed articles have been meticulously researched and that the information in them is highly reliable and respectble.

TECHNO TIP

For more on peer-reviewed articles, search the Internet for this video: What Are Scholarly and Peer-Reviewed Articles?

TICKET to WRITE

19.9 Examining Professional Journal Articles

Directions: Many professional journals offer some or all of their articles online. For instance, *The New England Journal of Medicine* has articles available at **nejm.org/** and the *Journal of Family Psychology* has articles available at **apa.org/pubs/journals/fam/index.aspx.**

Choose one of those journals or one listed below; use links from the journal's home page to access an article. Alternately, use any online database of professional journals available through your college library. Read the article and note the article's author(s) and information about whether or not the article includes an abstract, footnotes, and a bibliography. Then write a summary of the article.

Education	*Current Issues in Education*
	Practical Assessment, Research & Evaluation
Firefighter	*Fire Safety Journal*
Management and Entrepreneurship	*Journal of Applied Management and Entrepreneurship*
Medicine	*Journal of the American Medical Association*
Peace and Conflict	*Peace and Conflict Studies*
Real Estate	*Journal of Real Estate Research*

Guidelines

As a writing student, you probably must adhere to guidelines your instructor gives, including topics you may address and formats you must use when presenting your paper in its final draft (that is, when you *publish* it in your class).

Articles submitted to professional journals must follow specific guidelines as well, and often these guidelines are far more rigorous than those in classrooms. Examining guidelines for writing in a professional journal will give you a greater understanding of the meticulous process a writer must undertake before submitting an article for possible publication, as well as the requirements the author and the article must meet.

TICKET to WRITE

19.10 Examining Professional Journal Guidelines

Directions: Most professional journals offer writer's guidelines (sometimes called "submission guidelines" or "call for papers") describing how to format and submit an article to that journal. For instance, writer's guidelines for *The College Mathematics Journal* are available at **maa.org/pubs/cmj_info.html**; guidelines for the *Journal of College Writing* are available at **la-cc.org/jcw-cfp.html**.

Choose one of those journals or one listed below and enter its title in an online search engine. Look on the site for information regarding writer's guidelines, submission guidelines, or call for papers. After finding that journal's writer's guidelines, answer the questions. (Write n/a if the information is not available.)

Action in Teacher Education
Administration
American Journal of Economics and Business
American Journal of Environmental Sciences
American Journal of Nursing
American Secondary Education
Current Research in Psychology
Facts & Findings—The Bimonthly Journal for Paralegals
Fire Safety Journal
International Journal of Automotive Technology
Journal of College Writing
Journal of Computer Science
Journal of Emergency Medical Services
Journal of Human Services
Journal of Mathematics and Statistics
Journal of Real Estate Research

(continued)

Journal of Social Sciences
Law and Order Magazine
Research Journalism
The Journal of American History
The Welding Journal

A query letter asks if the publisher is interested in an article on a particular topic.

1. What is the mission of the journal?
2. On what types of articles does the journal focus?
3. For what organization, association, or club is this an official publication?
4. Is an author required to first submit a query letter, or may he or she send the manuscript without one?
5. Does each issue of the journal have a designated theme?
6. Must an author include biographical information?
7. Does the journal require other information? If so, what?
8. Cite the name, title, and address of the person to whom the author must submit the article.
9. What are the minimum and maximum lengths for manuscript submissions?
10. Does the maximum length include references, charts, figures, and tables, or may they be counted separately?
11. Must the manuscript be double-spaced?
12. What margins must be used?
13. Must graphic aids (e.g., tables, charts, figures, appendices, and illustrations) be placed inside the manuscript or at the end?
14. If the submission must have a cover page, what must be included on the cover page?
15. If the submission must have an abstract, what are the minimum and maximum word lengths required for the abstract?
16. If the submission must have an abstract, what must be addressed in the abstract?
17. Should the manuscript be submitted in paper form, electronic form, or both?
18. If electronic form is used for a submission, is a particular word-processing program required?
19. If the manuscript must be submitted in paper form, how many copies are required?
20. Does the journal return copies of manuscripts that are submitted to it?

TECHNO TIP

Students who speak a foreign language may be interested in professional journals published in other countries. Use an online search engine to access articles from one of the journals below. Alternately, use an online search engine to find a professional journal published in another country.

Alshark Alawsat (Lebanon)
Farhangistan (Afghanistan)
Française Nouvelles (France)
Periodistas–es (Spain)
Saansadji (India)
Factor K-1 (Dominican Republic)
FilBalad (Egypt)
In Time News Network (Pakistan)
Srm Psicologia (Italy)
SBNotícias (Brazil)

TECHNO TIP

For more on professional journals, search the Internet for these videos:
Research Minutes: How to Identify Scholarly Journal Articles
Scholarly vs. Popular Periodicals
Magazines vs. Scholarly Journals
Magazines and Scholarly Journals (University of Arkansas Libraries)

Literary Journals

LEARNING GOAL
❸ Discover and evaluate works in literary journals

Literary journals are periodicals devoted to showcasing finely crafted, insightful literature—most often fiction, poetry, and creative nonfiction—that might not be published elsewhere. While some journals highlight established writers, most feature new or emerging writers and are therefore the places that many previously unpublished writers first submit their work.

A few literary journals, like *Glimmer Train Stories*, focus on little other than short stories; others, like *Poet Lore*, feature poetry. Still others broaden their offerings. In addition to fiction, poetry, and creative nonfiction, you might also find

Literary journals are also called literary magazines, lit mags, or little magazines.

art	essays	memoirs	reviews
biographies of authors	illustrations	music	screenwriting
children's writing	interviews	novellas	translations
columns	letters	photography	
comics	literary criticism	plays	

Literary criticism analyzes and evaluates literary works.

Memoirs are autobiographical accounts of part of the author's life.

Novellas are short novels.

Commercial magazines generate a great deal of revenue through advertising. On the other hand, literary journals carry little or no advertising, so they operate on a tight budget. Their funding comes from subscriptions and, for some, from donors, grants, and other subsidies supporting the arts. Because literary journals are not commercial ventures, most offer little or no monetary payment for contributions; instead, as payment they offer free copies of the magazine featuring the author's work.

Some literary journals are published by independent companies, but many originate from the English department in colleges or universities. Ask your writing instructor for details about a writing magazine published by your school, including

The magazine may accept submissions from all over the world, or it may accept submissions from only those in a certain geographical area.

- categories accepted for submission
- who is eligible to submit
- monetary or other prizes awarded in categories
- submission deadlines
- submission word count and format requirements
- if previously published work is eligible for submission

Hundreds of literary journals are published each year. EBSCO's Literary Reference Center is an online database that includes more than 460 literary journals. Check to see if it is offered in your college library.

19.11 Writing a Criticism of an Entry in a Literary Journal

Directions: Many literary journals offer some or all of their articles online. You can find articles from *2River View* at **2river.org/**; you can find articles from *Slope* at **slope.org/**; and you can find articles from *Word Riot* at **wordriot.org/**. Access a selection in one of those journals or use an online search engine to find one of the journals listed below. Use links to access a selection. Then write a criticism of the selection you read, supporting the reasons for your criticism.

Heads Up!
Remember that in a criticism you present positive and negative points about a piece.

3:AM Magazine
Apple Valley Review
Barcelona Review
Big Bridge
Blackbird
Carve Magazine
Cerise Press
CrossConnect
Evergreen Review
Exquisite Corpse
Mudlark
Narrative Magazine
Per Contra
Pi
Shampoo
Slate
Splash of Red
The Cortland Review
The Mad Hatters' Review
Timothy McSweeney's Internet Tendency
Toasted Cheese

Guidelines

As a writing student, you probably must adhere to guidelines your instructor gives, including topics you may address and formats you must use when presenting your paper in its final draft (that is, when you *publish* it in your class).

Material submitted to literary journals must follow specific guidelines as well, and often these guidelines are far more rigorous than those in classrooms. Just as each writing instructor requires different guidelines, so does each literary journal. Examining guidelines for submitting to a literary journal will give you a greater understanding of the meticulous process a writer must undertake, as well as the requirements the author and the material must meet.

19.12 Examining Literary Journal Guidelines

Directions: Many literary journals have online pages that note their writer's guidelines (sometimes called submission guidelines). Guidelines for the following journals are available at the listed URLs:

2River View	**2river.org/office/submit.html**
Ploughshares	**pshares.org/submit/guidelines.cfm**
The Paris Review	**theparisreview.org/about/submissions**
TriQuarterly	**triquarterly.org/submissions**
Zyzzyva	**zyzzyva.org/contact/**

Access one of those journals or use an online search engine and find one of the journals listed below. Use links from the journal's homepage to answer the questions below about the journal you selected. (Write n/a if the information is not available.)

3:AM Magazine
African American Review
Alaska Quarterly Review
Blackbird
Carve Magazine
The Chicago Review
The Mad Hatters' Review
Prairie Schooner
Slate
Timothy McSweeney's Internet Tendency
Zoetrope

1. What types of submissions does the magazine accept?
2. List any famous writers whose works the magazine has published.

(continued)

3. Should the manuscript be submitted in paper form, electronic form, or both?
4. Are submissions accepted year-round or only during specific periods?
5. Must the submission be double-spaced?
6. What margins are required for submissions?
7. Is a self-addressed, stamped envelope required?
8. Does the magazine allow simultaneous submissions?
9. Does the magazine allow previously published work?
10. Does the magazine specify how many submissions it allows at a single time?
11. Does the magazine specify minimum and maximum word lengths?
12. Does the magazine sponsor any writer's contests?
13. To whom should submissions be sent?
14. What type of payment is given for work that is published by the magazine?
15. Does the magazine accept unsolicited submissions?
16. How often is the magazine published?

A self-addressed, stamped envelope is sometimes abbreviated SASE.

TECHNO TIP

For more on literary journals, search the Internet for this video:
NSU's Digressions Student Literary Magazine

RUN THAT BY ME AGAIN

LEARNING GOAL
1. Recognize and write various types of newspaper articles

- **Newspapers include . . .** three main types of articles: news articles, editorials, and feature articles.
- **Writers of news articles focus solely on . . .** factual information.
- **In an inverted pyramid, the writer presents . . .** the most important points of the story first.
- **The lead gives readers . . .** short answers to reporter's questions.
- **Editorials are . . .** usually unsigned, often state the perspective of those who run the newspaper, and run on the same page and same section in each issue of a newspaper.
- **The purpose of an editorial is . . .** to convince readers the writer is correct on a certain issue.
- **A feature article is often found in . . .** inside sections of newspapers or in newspaper supplements.

- **A feature article . . .** is not centered only on "breaking news"; is not written in an inverted pyramid format; is usually longer than a news article; does not answer reporter's questions in its lead.
- **Articles in professional journals are . . .** authored by leaders in their field.
- **Articles in professional journals often include . . .** an abstract, descriptions of methods authors used in their research, conclusions authors reach as a result of their research, and a bibliography.
- **In a peer-review process for a professional journal, peers judge the article to see if . . .** its content is relevant and significant to the field, its data is well researched and well documented, and it is written clearly and logically.

LEARNING GOAL

❷ Find and examine articles in professional journals

- **Most often, literary journals include . . .** fiction, poetry, and creative nonfiction.
- **Many literary journals originate from . . .** the English department in colleges or universities.

LEARNING GOAL

❸ Discover and evaluate works in literary journals

NEWSPAPERS AND JOURNALS LEARNING LOG MyWritingLab™

MyWritingLab™

Answer the questions below to review your mastery of newspapers and journals.

1. Where are news articles found?
   News articles are found on the front pages of papers, as well as in many inside pages.
2. In what format are news articles written?
   News articles are written in an inverted pyramid format.
3. What are editorials?
   Editorials are articles that state writers' opinions on current topics.
4. What are feature articles?
   Feature articles educate or entertain readers by providing well-researched, in-depth information on events, issues, trends, or people.
5. What do articles in professional journals include?
   Articles in professional journals include authoritative information devoted to the latest trends or research in a specific occupation or industry.
6. What is an abstract?
   An abstract is a short summary of important points of the text.
7. Where in your college library can you find professional journals?
   Hard copies of professional journals are in the reference section; articles are available in online databases; copies of articles can be obtained via interlibrary loan.
8. What is the process for peer review for a professional journal?
   The finished article is sent to a group of scholars in the field, and that group decides if the material in the article is worthy of publication.
9. What are literary journals?
   Literary journals are periodicals devoted to showcasing finely crafted, insightful literature that might not be published elsewhere.

Grammar and Mechanics

PART 4

For readers to understand and appreciate what you write, you must express yourself clearly and effectively. Mistakes in grammar, punctuation, mechanics, usage, and spelling can detract from your finished product, and they may even convey ideas that are different from what you intend. Part 4 helps you identify and correct problem areas.

"People who cannot distinguish between good and bad language, or who regard the distinction as unimportant, are unlikely to think carefully about anything else."

—B. R. Myers

Sentence Fragments and Run-Ons

LEARNING GOALS

In this chapter, you'll learn and practice how to

1. Recognize sentence fragments
2. Repair sentence fragments
3. Recognize run-on sentences
4. Repair run-on sentences

GETTING THERE

- A sentence fragment is an incomplete thought.
- A run-on sentence has two or more complete thoughts that should be separated with some form of punctuation.
- In academic writing, fragments and run-ons are generally prohibited because they can prevent you from clearly conveying your point.

What Are Sentence Fragments?

After I lost my iPod.	Until Rich starts his new job.
While watching the Super Bowl.	To find a cure for the dreaded disease.
Excluding the rest of the day.	Wanted to see the rest of the film.

LEARNING GOAL
1. Recognize sentence fragments

When you read the examples above, you're confused because information is missing. All of the examples are **sentence fragments**; they are not complete sentences. To repair a sentence fragment, you have to revise it so it becomes a complete sentence.

Complete Sentences versus Sentence Fragments

If a sentence is **complete**, it must have an independent clause; it may also have one or more subordinate clauses. An independent clause

- has a **verb**
- has a **subject**
- **makes sense**

The third bullet means that if you were to write only the words in the clause—nothing more—those words would make sense (that is, they would express a complete thought).

Here's an example of an independent clause:

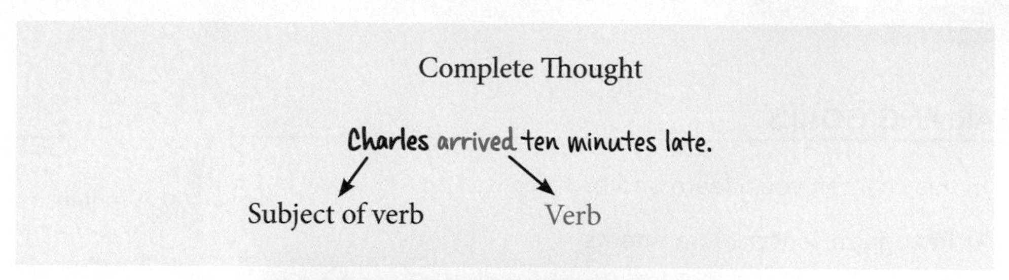

This group of words has a verb (**arrived**) and its subject (**Charles**), and if you read it alone, the words make sense. That makes it an independent clause (a complete sentence).

Fragments may also contain a verb and its subject, but the words don't make sense if you read them alone. Look at this example:

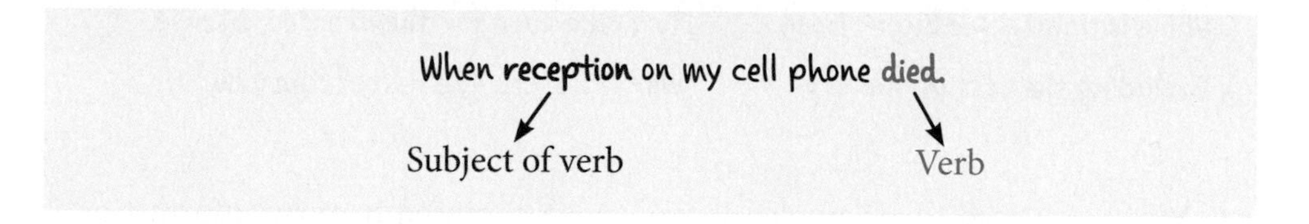

These words have a verb (**died**) and its subject (**reception**). The problem is that when you read the words, you'd be scratching your head and wondering what happened when the phone's reception died. The writer did not express a complete thought. That's what makes these words a **sentence fragment**.

Recognizing Fragments

You need to be able to recognize sentence fragments before you can repair them. You can decide if a group of words is a sentence or a fragment by using one of four methods:

Four Methods for Recognizing Sentence Fragments	
the **"makes sense"** method	the **"yes-or-no question"** method
the **"Is it true that . . ."** method	the **"I heard that . . ."** method

"Makes Sense" Method

Read *just* the words in question and then ask yourself if they make sense by themselves. If they do, fine. If they don't, you have a fragment you need to repair.

For example, when you read But waited too long.

you see that the words don't make sense. You need something extra.

Complete Thought

Stephen really wanted another soda but waited too long.

Subject of verb (Stephen) — Verb (wanted) — Verb (waited)

"Yes-or-No Question" Method

Turn the words into a yes-or-no question. If you answer *yes* to the question, then your words form a sentence. However, if you answer *no* or if your question makes no sense, you have a fragment.

Use the yes-or-no question method on these examples:

When the basketball goes through the hoop.

Sarah stands and cheers.

If you change the first example into a yes-or-no question, you have

Does when the basketball goes through the hoop?

Because that makes no sense, you know "when the basketball goes through the hoop" is a fragment. Using the same method with the second example, you have

Does Sarah stand and cheer?

That makes sense, so you know it's not a fragment.

"Is it true that . . ." Method

Before the group of words you're uncertain about, place the question "Is it true that. . . ." If the answer is *yes*, you have a complete sentence. If the answer is *no* (or if the answer makes no sense), you have a fragment.

For example, maybe you're wondering if this is a complete sentence.

> Because he needed a new pair of jeans.

If you put "Is it true that . . ." in front of it, you have

> Is it true that because he needed a new pair of jeans?

Because this doesn't make sense, you know you have a fragment and you should repair it. Now you decide to try

> He needed a new pair of jeans.

If you put "Is it true that . . ." in front of it, you have

> Is it true that he needed a new pair of jeans?

The question makes sense, and you can answer it with *yes*, so you have a complete sentence.

"I heard that . . ." Method

Before the group of words you're uncertain about, place "I heard that. . . ." If the statement makes sense, you have a complete sentence; if it doesn't, you have a fragment.

> Since my cell phone lost its connection.

If you put "I heard that . . ." in front of it, you have

> I heard that since my cell phone lost its connection.

Because that doesn't make sense, you need to change your wording. If you put "I heard that . . ." in front of it, you have

> I heard that my cell phone lost its connection.

That makes sense; it's a complete sentence.

20.1 Identifying Sentences and Fragments

Directions: Read the paragraph below. Working individually or in groups, use the "**makes sense,**" "**yes-or-no question,**" "**Is it true that . . .,**" or "**I heard that . . .**" method to determine if each group of words is a sentence or a fragment. Then underline ***S*** if the words are a sentence and ***F*** if the words are a fragment.

Even though Thanksgiving was just a week away. (S F) Megan thought the holiday would never arrive. (S F) Because she felt she desperately needed a break from classes. (S F) After calling several members of her family. (S F) She felt certain that she would be welcome at three separate Thanksgiving celebrations. (S F) Megan knew that she would need to work on a paper. (S F) Concerning the book that her class had read. (S F) She told herself that she would worry about the paper on the holiday weekend. (S F) All she could think of at the present. (S F) Was getting a little rest and a lot of good food. (S F)

Repairing Fragments

LEARNING GOAL
❷ Repair sentence fragments

Problems with sentence fragments come in five common categories:

- fragments introduced by a subordinating conjunction
- fragments introduced with a participial phrase with *-ing* words at or near the beginning of a sentence
- fragments introduced by an infinitive phrase
- fragments caused by added details
- fragments caused by missing subjects

Each problem fragment can be easily repaired.

THE PROBLEM

A fragment introduced by a subordinating conjunction

Subordinating conjunctions (you may know these as **dependent words**) introduce words that depend on other words to make sense. In the example below, *unless* is a subordinating conjunction:

Unless David comes home soon.

Subject of verb — David; Verb — comes

The word *unless* makes the rest of the thought confusing because readers don't know what will happen if David doesn't come home soon.

To decide if these words are a fragment, try the four tests.

- **Using the "makes sense" method, read only the words**

  Unless David comes home soon.

  and ask yourself if you would understand them. Because the words don't make sense as they are written, this is a fragment.

- **Using the "yes-or-no question" method, would you understand**

  Did David unless comes home soon?

  The question makes no sense, so this is a fragment.

- **Using the "Is it true that . . ." method, you would have**

  Is it true that did David unless comes home soon?

  The question makes no sense, so this is a fragment.

- **Using the "I heard that . . ." method, you would have**

  I heard that did David unless comes home soon.

  The statement makes no sense, so this is a fragment.

THE FIX

Join the fragment to an adjacent sentence

To repair a fragment introduced by a subordinating conjunction, join the fragment to an adjacent sentence.

Now look at this example

Incomplete Thought Complete Thought

Unless David comes home soon, **we'll have to leave without him.**

the words Unless David comes home soon

are dependent on **we'll have to leave without him.**

Once you read the extra part of the sentence, you understand what the writer is saying.

Here is a list of thirty common subordinating conjunctions:

Common Subordinating Conjunctions					
after	by	into	so long as	unless	wherever
although	concerning	like	than	until	whether
as much as	even though	of	that	when	while
because	how	once	though	whenever	with
before	if	since	through	where	without

In this example Whenever court is in session

all you need to do is add what happens at that time. For instance, you might have

Incomplete Thought Complete Thought

Whenever court is in session, **finding a parking place becomes a problem.**

Or in this example If I don't turn in my paper

you might add what will happen if you don't turn in your paper, as in

Incomplete Thought Complete Thought

If I don't turn in my paper, **my grade will drop.**

20.2 Determine Sentence or Fragment

Directions: In the paragraph below are many subordinating conjunctions. Working individually or in groups, underline ***S*** if the word group is a sentence. If it is a fragment, underline ***F*** and then revise it, adding or deleting words as needed.

Congress is convening on January 25. (S F) After its winter break. (S F) One of the topics on the minds of most Americans is the state of the economy. (S F) Most members of Congress have been bombarded with questions. (S F) Concerning what they will do or not do to get the country on a more stable economic footing. (S F) Unless something is done soon. (S F) Some fear a depression may be in the country's future. (S F) Without bailouts for certain industries that employ thousands of people. (S F) The economic outlook is bleak. (S F) So long as so many people are out of work. (S F)

THE PROBLEM

A fragment introduced by a participial phrase with *–ing* words at or near the beginning of a sentence

Heads Up!
A **participle** is a verb form that ends in *-ing* or *-ed* and that functions as an adjective (it describes a noun or pronoun). A **participial phrase** is the group of words that accompanies a participle.

Participial phrases with ***–ing* words** at or near the beginning of the sentence can cause problems because readers are unclear about to whom or what the phrases refer, as in this example:

By working three part-time jobs.

To decide if these words are a fragment, try the four tests.

- **Using the "makes sense" method, read only the words**

  By working three part-time jobs.

  and ask yourself if you'd understand them. Because you don't know what would happen as a result of working, the words don't make sense. They are a fragment.

- **Using the "yes-or-no question" method, would you understand**

  Did by working three part-time jobs?

  The question makes no sense, so you know the example is a fragment.

- **Using the "Is it true that . . ." method, would you understand**
  Is it true that by working three part-time jobs?
  The question is incomplete, so you know the example is a fragment.
- **Using the "I heard that . . ." method, would you understand**
  I heard that by working three part-time jobs.
  The statement is incomplete, so you know the example is a fragment.

Explain to whom or what the fragment refers

To repair a fragment introduced by a participial phrase with *–ing* words at or near the beginning of a sentence, explain to whom or what the fragment refers.

You might correct the fragment in this way — **By working three part-time jobs.**

Incomplete Thought	Complete Thought
By working three part-time jobs,	**Chris could afford a new car.**

20.3 Determine Sentence or Fragment

Directions: The paragraph below contains several participial phrases. Working individually or in groups, underline ***S*** if the word group is a sentence. If it is a fragment, underline **F** and then revise, adding or deleting words as needed.

After alighting from the small boat that had carried her downstream. (S F) Teacher Elizabeth Blackwell looked around in the village that would be her new home. (S F) The March day was dismal, with rain and strong winds. (S F) By settling in with a local family. (S F) Elizabeth soon came to know a number of the citizens. (S F) In readying the schoolhouse. (S F) She realized that the building would need to be painted before the students arrived. (S F) After asking several people for help. (S F) That job was finished in two weeks. (S F)

A fragment introduced by an infinitive phrase

Heads Up!
In infinitive phrases, the word *to* is either written or implied.

Fragments caused by infinitive phrases are problems because readers can't see the correlation between the action and the outcome. **Infinitive phrases** usually begin with ***to*** followed by a **verb**, as in this example:

To sell peaches to her coworkers.

To decide if these words are a fragment, try the four tests.

- **Using the "makes sense" method, read only the words**

  To sell peaches to her coworkers.

  and ask yourself if you understand just those words. Because you don't know what happened before the peaches were to be sold or why they were being sold, the words don't make sense. They are a fragment.

- **Using the "yes-or-no question" method, would you understand**

  Did to sell peaches to her coworkers?

  The question makes no sense, so you know the example is a fragment.

- **Using the "Is it true that . . ." method, would you understand**

  Is it true that to sell peaches to her coworkers?

  The question is incomplete, so you know the example is a fragment.

- **Using the "I heard that . . . " method, would you understand**

  I heard that to sell peaches to her coworkers.

  The statement is incomplete, so you know the example is a fragment.

Explain to whom or what the fragment refers

To repair a fragment caused by an introductory infinitive phrase, add words to explain to whom or what the fragment refers.

In the fragment	To sell peaches to her coworkers.
you might add the words	Arlene set up a tent
and have the sentence	Arlene set up a tent to sell peaches to her coworkers.
	Complete Thought

20.4 Determine Sentence or Fragment

Directions: The paragraph below contains several infinitive phrases. Working alone or in groups, underline ***S*** if the word group is a sentence. If it is a fragment, underline ***F*** and then revise it, adding or deleting words as needed.

Marissa thought she would find it difficult. (S F) To learn human anatomy in just one semester. (S F) She had read that the human body has about two hundred bones. (S F) And to memorize all the names of those seemed impossible. (S F) She knew that medical mnemonics have been developed. (S F) To help students memorize the many names. (S F) To find mnemonics. (S F) She went online. (S F) There she found "Remember to deliver campus bulletin" as a mnemonic for the nerve fibers of the brachial plexus (roots, trunks, divisions, cords, branches). (S F)

THE PROBLEM

A fragment caused by added details

Sometimes writers **add details** *that should be part of a preceding sentence*. In this example, the second "sentence" is incomplete and should be a part of the first sentence:

Complete Thought	Incomplete Thought
Barry always eats in the cafeteria.	Except on Wednesdays.

To decide if these words are a fragment, try the four tests.

- **Using the "makes sense" method, read only the words**

  Except on Wednesdays.

  and ask yourself if they make sense. Because you don't know what happens on other days, the words are a fragment (the words don't make sense).

- Using the **"yes-or-no question" method, would you understand**

  Did except on Wednesdays?

  Because that makes no sense, you know you have a fragment.

- **Using the "Is it true that . . ." method, would you understand**

  Is it true that except on Wednesdays?

  Because the question is incomplete, you know you have a fragment.

- **Using the "I heard that . . ." method, would you understand**

  I heard that except on Wednesdays?

  Because the statement is incomplete, you know you have a fragment.

Attach the fragment onto the sentence that precedes it

To repair a fragment caused by added details, attach the fragment onto the sentence that precedes it. The repaired fragment could be worded this way:

Barry always eats in the cafeteria, except on Wednesdays.

20.5 Determine Sentence or Fragment

Directions: The paragraph below contains several infinitive phrases. Working individually or in groups, underline ***S*** if the word group is a sentence. If it is a fragment, underline ***F*** and then revise it, adding or deleting words as needed.

Stephanie and Derrick were both interested in becoming social workers. (S F) When they graduated from college. (S F) In at least four years. (S F) The two especially wanted to assist families. (S F) That have domestic conflicts. (S F) They studied the field of social work and learned that employment was expected to grow. (S F) Much faster than average. (S F) They also read that jobs might be competitive. (S F) Except in rural areas. (S F)

A fragment caused by a missing subject

Sometimes writers **fail to identify the subject(s)** of a sentence. This results in fragments, such as in the second of part of the following "sentence":

Complete Thought	Incomplete Thought
Our History Club met on Monday.	And welcomed four new members.

To decide if these words are a fragment, try the four tests.

- **Using the "makes sense" method, read only the words**

  And welcomed four new members.

and ask yourself if they make sense. Because you don't know who welcomed the members, the words are a fragment

- **Using the "yes-or-no question" method, would you understand**

  Did and welcomed four new members?

  Because that makes no sense, you know you have a fragment.

- **Using the "Is it true that . . ." method, would you understand**

  Is it true that and welcomed four new members?

  The question makes no sense, so you know you have a fragment.

- **Using the "I heard that . . ." method, would you understand**

  I heard that and welcomed four new members?

  The statement makes no sense, so you know you have a fragment.

THE FIX

Attach the fragment to the preceding sentence

To repair a fragment caused by a missing subject, attach the fragment to the preceding sentence. The repaired sentence could read like this:

Our History Club met on Monday and welcomed four new members.

20.6 Determine Sentence or Fragment

Directions: The paragraph below contains several infinitive phrases. Working individually or in groups, underline ***S*** if the word group is a sentence. If it is a fragment, underline ***F*** and then revise it, adding or deleting words as needed.

The assignment was driving me crazy. (S F) And should've been started earlier. (S F) I was supposed to interview someone from an older generation. (S F) But not a member of my family. (S F) I knew I could talk with my next-door neighbor. (S F) Or someone at the Senior Citizens Center. (S F) Finally, I decided that my neighbor would be a better choice. (S F) I called her to set up a time to visit. (S F) She seemed happy to be interviewed. (S F) Yet wondered how long the interview would take. (S F)

20.7 Repair Sentence Fragments

Directions: Read the paragraph below. Work individually or in groups to find and repair all the fragments, and then write the corrected paragraph. Answers may vary.

I know the saying is true—truth is stranger than fiction. I was visiting my great aunt and uncle on St. George Island, Florida. Where they've lived since they retired ten years ago. At the age of seventy. Every time I come for a visit. Uncle Lee is always fishing. While Aunt Lois is inside crocheting. Though this may not be strange, what is strange is the company they keep. On my last visit. I was blown away to see my Aunt Lois having coffee and cracking jokes with another island homeowner. Hank Williams, Jr. I couldn't believe my eyes. Before I could speak. Uncle Lee walked in from the beach with another out-of-place character. Who had long stringy hair and quite a few tattoos. They were carrying their nets and a bucket of fish. Uncle Lee said, "Aaron, meet my friend Bobby Ritchie. Some call him 'Kid Rock,' but I just can't bring myself to." I couldn't move. Tried to speak but couldn't. "I guess he's never seen that much fish before," Hank said. Laughing with Kid, Uncle Lee, and Aunt Lois.

I know the saying is true—truth is stranger than fiction. I was visiting my great aunt and uncle on St. George Island, Florida, where they've lived since they retired ten years ago at the age of seventy. Every time I come for a visit, Uncle Lee is always fishing while Aunt Lois is inside crocheting. Though this may not be strange, what is strange is the company they keep. On my last visit, I was blown away to see my Aunt Lois having coffee and cracking jokes with another island homeowner, Hank Williams, Jr. I couldn't believe my eyes. Before I could speak, Uncle Lee walked in from the beach with another out-of-place character who had long stringy hair and quite a few tattoos. They were carrying their nets and a bucket of fish. Uncle Lee said, "Aaron, meet my friend Bobby Ritchie. Some call him 'Kid Rock,' but I just can't bring myself to." I couldn't move. I tried to speak but couldn't. "I guess he's never seen that much fish before," Hank said, laughing with Kid, Uncle Lee, and Aunt Lois.

MyWritingLab™ Visit *MyWritingLab.com* and complete the exercises and activities in the **Fragments** topic area.

RUN THAT BY ME AGAIN

- **Sentence fragments are . . .** groups of words that are not complete sentences. They may lack either a subject or a verb, or they may have both a subject and a verb but don't make sense.
- **Sentence fragments might be introduced by . . .** a subordinating conjunction, a participial phrase with *–ing* words at or near the beginning of a sentence, or an infinitive phrase; or **they might be caused by . . .** added details or a missing subject.
- **Sentence fragments can be recognized by the four methods called . . .** "makes sense," "yes-or-no question," "Is it true that . . .," and "I heard that . . . ."
- **Sentence fragments can be repaired by . . .** (1) using a subordinating conjunction to join the fragment to an adjacent sentence; (2) explaining to whom or what each of the fragments refers; and (3) attaching the fragment onto the sentence that precedes it.

LEARNING GOAL
❶ Recognize sentence fragments

LEARNING GOAL
❷ Repair sentence fragments

SENTENCE FRAGMENTS LEARNING LOG MyWritingLab™

Complete this Exercise

MyWritingLab™

Answer the questions below to review your mastery of repairing sentence fragments.

1. What is a sentence fragment?

   A sentence fragment is a group of words that may contain a subject or a verb or both, but the words do not make sense when read alone. Sentence fragments may also be defined as words that are not independent clauses.

2. What are the three parts of an independent clause (a complete sentence)?

   An independent clause (a complete sentence) must have a verb and its subject and must make sense if read alone.

3. What are the four methods for determining if a group of words is a fragment?

   The four methods for recognizing fragments are the "makes sense" method, "yes-or-no question" method, "Is it true that . . ." method, and "I heard that . . ." method.

4. To repair a sentence fragment caused by words beginning with a subordinating conjunction, what should you do?

   To repair a fragment introduced by a subordinating conjunction, join the fragment to an adjacent sentence.

5. To repair a sentence fragment introduced by a participial phrase with *–ing* words at or near the beginning of a sentence, what should you do?

   To repair a fragment introduced by a participial phrase with *–ing* words at or near the beginning of a sentence, explain to whom or what the fragment refers.

6. To repair a sentence fragment introduced by an infinitive phrase, what should you do?

   To repair a fragment caused by an introductory infinitive phrase, add words to explain to whom or what the fragment refers.

7. To repair a sentence fragment caused by added details, what should you do?

   To repair a fragment caused by added details, hook the fragment onto the sentence that preceded it.

8. To repair a sentence fragment caused by missing subjects, what should you do?

   To repair a fragment caused by a missing subject, attach the fragment to the preceding sentence.

What Are Run-On Sentences?

The term **run-on sentence** refers to words that are fused into one sentence—they "run on"—but should be separated into more than one sentence. Read these examples:

> LEARNING GOAL
> ❸ Recognize run-on sentences

The new action movie opened Friday night it took in lots of money. John and I went to see the movie, we were not impressed.

These sentences are confusing. Both contain two complete thoughts (that is, two independent clauses), but they lack the punctuation necessary for clarity. In each example, the writer's meaning is unclear. The first example is called a **fused sentence**. It is confusing because readers don't know if the movie took in lots of money, took lots of money to make, or was very expensive to see. The second sentence is a **comma splice**. In this type of run-on, a comma alone is not enough to join the two thoughts. Something else—a different form of punctuation or an extra word—is necessary.

Recognizing Run-On Sentences

Run-on sentences are divided into two categories:

- **fused sentences**
- **comma splices**

> **Heads Up!**
> If you're correcting a run-on sentence, be sure you have two (or more) complete thoughts. Ask yourself if each group of words could be a sentence by itself. If one group of words isn't a sentence, you don't have a complete thought and you need to repair that grouping.

THE PROBLEM

Run-on sentences caused by fused sentences

Fused sentences are two or more sentences (complete thoughts) strung together without the necessary punctuation to show readers where one thought stops and another begins, as in

The new action movie opened Friday night it took in lots of money.

Because these two thoughts run together, they baffle readers and slow them down.

THE PROBLEM

Run-on sentences caused by comma splices

Comma splices are also confusing to readers. Sentences with comma splices have two or more sentences (complete thoughts) separated by a comma, but a comma is the wrong type of punctuation. This example has two complete thoughts:

John and I went to see the movie, we were not impressed.

The problem is that the comma is the wrong type of punctuation to separate these two thoughts. Readers are confused because the sentence needs a stronger punctuation mark than a comma, or it needs to be separated into two sentences.

Repairing Run-On Sentences

LEARNING GOAL
4 Repair run-on sentences

Repairing run-on sentences—whether they're fused sentences or comma splices—isn't hard. All you need to do is insert the correct punctuation. Several punctuation choices repair run-on sentences. Just remember that you must show readers where one thought ends and another begins.

THE FIX

Separate complete thoughts with appropriate punctuation

Insert punctuation only between complete thoughts.

You can do this in one of three ways:

- create two separate sentences by inserting a period

John and I went to see the movie. We were not impressed.

Depending on the wording, you might also use a colon.

- insert a semicolon

John and I went to see the movie; we were not impressed.

Remember these conjunctions with the mnemonic **boysfan**: but, or, yet, so, for, and, nor.

- insert a comma followed by one of these seven conjunctions: *but, or, yet, so, for, and, nor*

John and I went to see the movie, and we were not impressed.

THE FIX

Make one of the clauses dependent by adding a subordinating word and appropriate punctuation

Although John and I went to see the movie, we were not impressed.

John and I went to see the movie; however, we were not impressed.

TICKET to WRITE

20.8 Repair Run-On Sentences

Directions: Read the paragraph below. Then work individually or in groups to find and repair all the run-ons. Answers will vary.

I tried to sleep but I just tossed and turned I finally turned on the television all that was on was infomercials, some of them assured me that I'd be a richer person if I just ordered some CDs and followed their instructions, others tried to sell me products that would make me more appealing for instance, some would help me get rid of wrinkles others would show me how to grow more hair, several advertised ways to help me lose weight, the third category centered on how my home could be more attractive, one said all I had to do was buy a new vacuum cleaner, another hyped some furniture it claimed is more comfortable than what I already have.

I tried to sleep, but I just tossed and turned. I finally turned on the television. All that was on was infomercials. Some of them assured me that I'd be a richer person if I just ordered some CDs and followed their instructions, but others tried to sell me products that would make me more appealing. For instance, some would help me get rid of wrinkles; others would show me how to grow more hair; several advertised ways to help me lose weight. The third category centered on how my home could be more attractive. One said all I had to do was buy a new vacuum cleaner, and another hyped some furniture it claimed is more comfortable than what I already have.

TECHNO TIP

For more review of run-on sentences, search the Internet for this video:

English Corner: Sentence Errors—Run-ons/Comma Splices

TECHNO TIP

For more review of fragments and run-ons, search the Internet for this video:
How to Edit Your Writing: What Are Run-ons and Sentence Fragments?

MyWritingLab™ Visit *MyWritingLab.com* and complete the exercises and activities in the **Run-Ons** topic area.

RUN THAT BY ME AGAIN

LEARNING GOAL
❸ Recognize run-on sentences

- **Run-on sentences are . . .** words that are fused into one sentence but should be separated into more than one sentence.
- **The two types of run-on sentences are . . .** fused sentences and comma splices.
- **Fused sentences can be recognized by looking for . . .** two or more sentences strung together without the necessary punctuation to show readers where one thought stops and another begins.
- **Sentences with comma splices can be recognized by looking for . . .** two or more sentences separated by a comma, but a comma is the wrong type of punctuation.

LEARNING GOAL
❹ Repair run-on sentences

- **Either type of run-on sentence can be repaired by . . .** separating the complete thoughts with appropriate punctuation.

RUN-ONS LEARNING LOG MyWritingLab™

Complete this Exercise MyWritingLab™

Answer the questions below to review your mastery of repairing run-on sentences.

1. What is the problem with a run-on sentence?
   A run-on sentence is confusing to readers.
2. What is a fused sentence?
   A fused sentence has two or more sentences (complete thoughts) strung together without the necessary punctuation to show readers where one thought stops and another begins.
3. What does a fused sentence lack?
   A fused sentence lacks necessary punctuation.
4. Comma splices contain at least how many complete thoughts?
   Comma splices contain at least two complete thoughts.
5. What are three ways to correct sentences that contain comma splices?
   Three ways to correct comma splices are (1) create two separate sentences, (2) insert a semicolon, and (3) insert a comma followed by an appropriate conjunction.
6. What seven conjunctions can you use to join complete thoughts?
   Complete thoughts can be joined by the conjunctions *but, or, yet, so, for, and,* and *nor.*
7. What mnemonic helps you remember the seven conjunctions you may use to join complete thoughts?
   The mnemonic **boysfan** aids in remembering these seven conjunctions: *but, or, yet, so, for, and, nor.*

CHAPTER

21

Consistency

LEARNING GOALS

In this chapter, you'll learn and practice how to

- Identify and correct problems in verb consistency
- Identify and correct problems in pronoun consistency

GETTING THERE

- Maintaining consistency in verbs and pronouns ensures clarity for readers.
- Problems with verb consistency arise from verbs with incorrect tenses.
- Problems with pronoun consistency arise from pronouns with points of view that do not correspond with the pronouns' antecedents.

What Is Consistency?

In writing, **consistency** is concerned with verbs and pronouns. To ensure that your meaning is clear, use the form of a verb that agrees with (is *consistent* with) another verb, and use a pronoun that agrees with (is *consistent* with) its antecedent.

Verb Consistency

LEARNING GOAL

1. Identify and correct problems in verb consistency

Verb tenses show readers when various events occurred in relation to each other. Problems in **verb consistency** occur when tenses are used incorrectly. When this happens, readers are confused because the time frames described are incorrect or unclear.

Verb Tenses

A **verb** is a word that shows either the *existence* or *action* of a person, place, or thing:

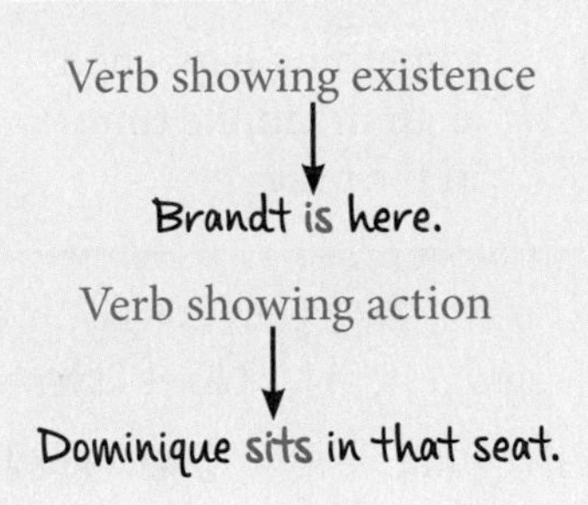

Six Main Tenses

Through various forms (called **tenses**), a verb also shows readers the *time* of the action or existence. English verbs have six main tenses.

Tense	Tells readers . . .	Formed by . . .
Present	something is happening now or is a habit	the basic form of the verb

Today I listen to the song.

Today Jamal begins his paper.

Past	something happened but is no longer happening	adding *-d* or *-ed* to the present tense of regular verb; irregular verbs are formed in other ways

Past tense of regular verb — Yesterday I listened to the song.

Past tense of irregular verb — Yesterday Jamal began his paper.

Future	something will happen at some time to come	adding *will* or *shall* before the present tense of the verb

Tomorrow I will listen to the song.

Tomorrow Jamal will begin his paper.

Present Perfect	something happened at an indefinite time in the past	adding *have* or *has* before the past participle of the verb

Past participle of regular verb — I **have listened** to the song many times.

Past participle of irregular verb — Jamal **has begun** his paper.

Past Perfect	something happened in the past prior to another event that happened in the past	adding *had* before the past participle of the verb

Past participle of regular verb — I **had listened** to the song many times before I bought the CD.

Past participle of irregular verb — Jamal **had begun** his paper before class met last Monday.

Future Perfect	something will be completed in a time to come before some other event that will be completed in a time to come	adding *will have* or *shall have* before the past participle of the verb

Past participle of regular verb — By the time I buy the CD, I **will have listened** to the song many times.

Past participle of irregular verb — Jamal **will have begun** his paper before class meets on Monday.

21.1 Identifying Verb Tenses

Directions: In each sentence below, identify the tense of the verb in italics.

1. The deejay *will mix* new songs for the party. future
2. The quarterback *has passed* the ball three times. present perfect

3. Emmett and Kia *had hoped* that the news from the hospital would be good. past perfect
4. Before the test, everyone *compared* notes. past
5. This seat belt *pinches* my shoulder. present
6. Kate and Carl *had agreed* to meet at the fountain. past perfect
7. The snow *will have melted* by the time spring arrives. future perfect
8. My brother joked that my new haircut *will frighten* his daughter. future
9. Our mayor *has welcomed* the visitors from Japan. present perfect
10. Susan's brother *disapproved* of her new boyfriend. past

Problems with Verb Consistency

"There must be consistency in direction."
—W. Edwards Deming

Problems with **verb consistency** arise when writers use an incorrect tense. This often occurs when writers change tenses unnecessarily, which is confusing to readers. To maintain verb consistency, the rule to remember is this: All verbs should be in the same tense as the first verb, unless the time frame changes.

To determine the correct verb form, note the time frame and tense of the first verb. Then compare the time frame of following verbs to the first one:

- If the time frame is the *same* as that of the first verb, the tense of following verbs is the same.
- If the time frame of a following verb *changes* from the first verb, the tense of that verb also changes.

In the passage below, the action of the first sentence occurred in the past (its time frame), so the writer uses past tense. The second sentence is a continuation of what the writer is relating (it also happened in the past), but its verbs are in present tense. Because of this, the verbs are inconsistent.

Incorrect verb consistency

Past tense verb (took) · Past tense verb (decided)

Cheyenne **took** her time before she **decided** what to order. Then she **says** she **wants** hamburgers.

Present tense verb (says) · Present tense verb (wants)

For the sentences to read correctly, their verb tenses should be consistent (the same). The first verb this writer uses is in the past tense. The time frame of the second sentence is the same as the first, so the sentence would be correct if verbs in both sentences were in the past tense.

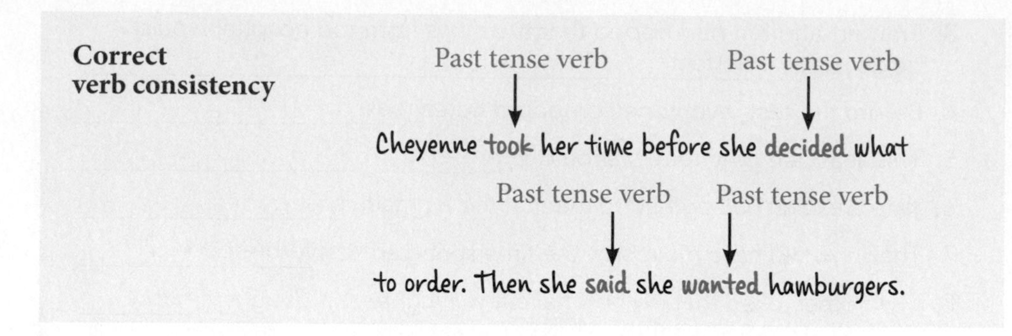

The verbs are also consistent if the writer's first verb is in the present tense and the verbs that follow are also in the present tense.

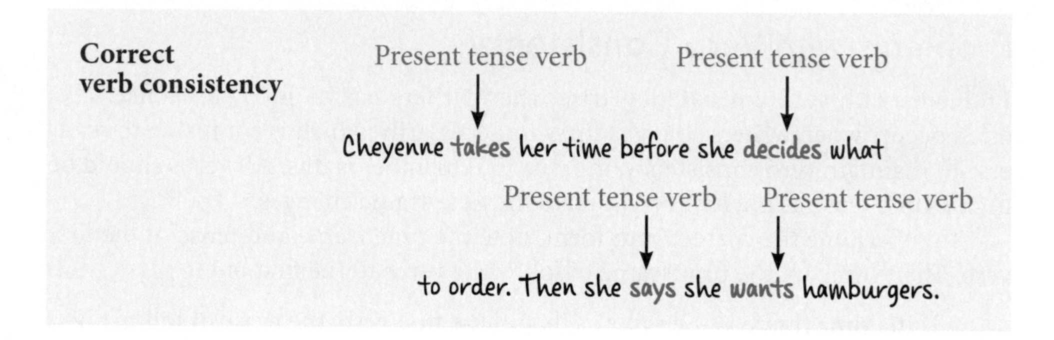

TECHNO TIP

For online quizzes on verb tense consistency, search the Internet for these sites:

Grammar: Verb Tense Consistency Tense Consistency Exercises

21.2 Identifying and Correcting Incorrect Verb Consistency

Directions: Mark the sentences that have correct verb consistency with a *C*. Mark the sentences that have incorrect verb consistency with an *I*, underline the incorrect verb tense, and write the correct form. Answers will vary.

C 1. Nico rode the bus for two weeks and saved ten dollars on gas. ______

I 2. In his latest movie, Daniel Radcliffe plays a Martian villain who visits earth but was disappointed in what he sees. is

I 3. Shayla and Fernando wanted to go home for the weekend, but then they decide to stay for the game. decided

C 4. Shannon called last night and later dropped off my netbook. ______

I 5. Nursing students practiced getting blood pressure readings, and Rhonda allows them to test her first. allowed

C 6. When the power outage occurred, we lost all data in the computer file. ______

I 7. Monica came into Lab 3, and she immediately realizes she is in the wrong room. realized, was

I 8. Yesterday we had given our group presentation and then went out to celebrate. gave

I 9. Most people who took Dr. Will's class like it. take

C 10. Some students mentioned they thought the second test was harder than the first. ______

What Is Pronoun Consistency?

Pronoun consistency deals with using pronouns in the same point of view (that is, making them refer to the same person, place, or thing). When an incorrect pronoun is used, readers are not sure about who or what is being described in the sentence.

LEARNING GOAL

❷ Identify and correct problems in pronoun consistency

Point of View

Point of view refers to the relationship between a writer and a reader; it tells the reader who is performing the action of a sentence. Pronouns are divided into three points of view: *first person, second person,* and *third person.*

Heads Up!
For other problems with pronouns, see Chapter 23.

Pronouns			
Point of View	**Refers to . . .**	**Singular Form**	**Plural Form**
First Person	the person writing I ordered a pizza. (the **writer** ordered)	I, me, my, mine, myself, ourselves	we, us, our, ours
Second Person	the person reading You ordered a pizza. (the **reader** ordered)	you, your, yourself	you, your, yours, yourselves
Third Person	a person or thing written about She ordered a pizza. (**someone other than the writer or reader** ordered)	he, she, it, him, his, her, hers, its, himself, herself, itself	they, them, their, theirs, themselves

For writing to be clear, maintain the same point of view throughout your piece, unless you have a logical reason for a change.

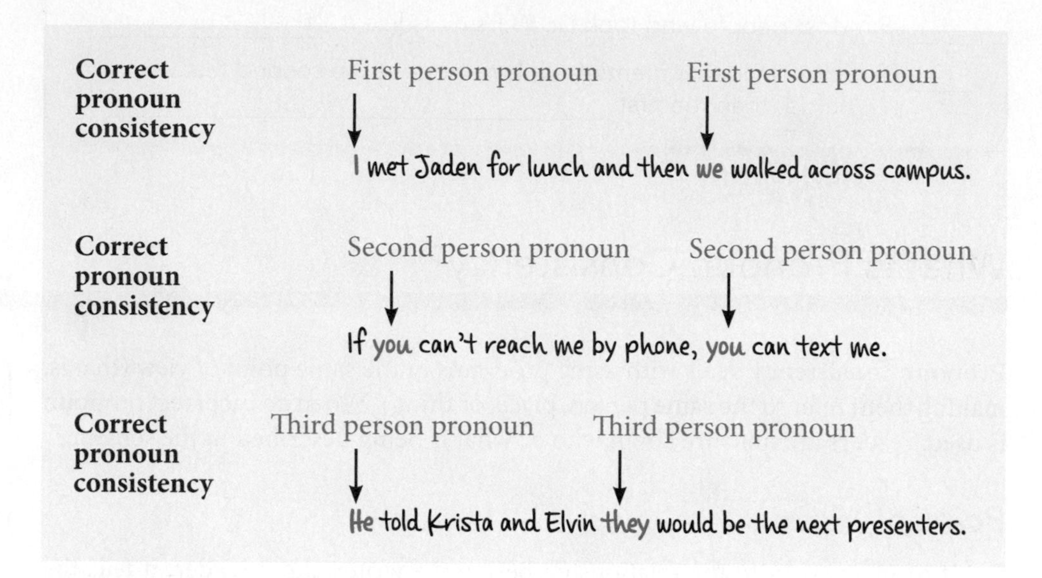

21.3 Identifying Point of View in Pronouns

Directions: Determine the point of view of the italicized pronouns. In the space, write *1S* for a first person singular pronoun; *1P* for a first person plural pronoun; *2S* for a second person singular pronoun; *2P* for a second person plural pronoun; *3S* for a third person singular pronoun; or *3P* for a third person plural pronoun.

3P 1. After *they* left for the day, Jamal finally felt peaceful.
1S 2. *I* will eat lunch outside today.
3S 3. That book began to bore me in the first chapter, and *it* hasn't changed yet.
1P 4. If *we* arrive at the lecture early, we can get a better seat.
2S 5. If *you* sit next to An-ming, tell her I said hello.
1P 6. Give *us* the ball!
3S 7. Abdullah saw *him* at the science club meeting at seven tonight.
2P 8. Willie looks forward to the time the three of *you* are together.
3S 9. Is *it* true that class is cancelled?
3P 10. Without question, the books are *theirs*.

Problems with Pronoun Consistency

Heads Up!
This is also called *pronoun shift*.

A frequent problem with pronoun consistency comes when writers needlessly change person (point of view) in a sentence or paragraph. The pronoun does not agree with the person (point of view) of its antecedent.

This usually occurs with a shift to second person (using *you* or *yours*), which means the reader is addressed directly. When this happens, readers can be confused about whom or what the writer is indicating or noting.

To keep from making this error, keep pronouns in the same person.

Incorrect pronoun consistency

First person pronoun → we; Second person pronoun → you

When we take notes in class, you should skip lines.

Correct pronoun consistency

First person pronoun → we; First person pronoun → we

When we take notes in class, we should skip lines.

Incorrect pronoun consistency

Third person pronoun → they

Professor Chan's students know if they take good notes

Second person pronoun → you

and review frequently, you will probably pass.

Correct pronoun consistency

Third person pronoun → they

Professor Chan's students know if they take good notes

Third person pronoun → they

and review frequently, they will probably pass.

In academic writing, the easiest way to avoid the problem of pronoun inconsistency in second person is to refrain from using *you* or *yours*, unless you are

Heads Up! Many instructors require only third person be used in academic material.

- quoting someone — The protagonist said, "You should arrive at eight."
- providing instructions — You will need eight 2 x 4s for this project.
- giving directions — If you drive past the gas station, you've gone too far.

TECHNO TIP

For an online quiz on verb and pronoun consistency, search the Internet for this site: Quiz on Consistency in Tense and Pronouns

21.4 Correcting Point of View in Pronouns

Directions: Delete incorrect pronouns; write the correct ones above them.

1. People should never be unkind, especially when ~~you~~ they are talking about those less fortunate.
2. Dino and Leslie were talking, and before ~~you~~ they knew it, they discovered they were from the same hometown.
3. Math students should come to the tutoring session, especially if ~~you~~ they have grades that need to be improved.
4. Kavon wanted to drive through the night, but he found there's a limit to how long ~~you~~ he can drive without a break.
5. Reggie and Lori want to stop using credit cards because ~~you~~ they can quickly pile up high debt.
6. Although they were hungry, Yasmin and Melanie didn't go to the cafeteria because the day's menu didn't appeal to ~~you~~ them.
7. When a student is going through a rough time, ~~you~~ he or she can always talk to a college counselor.
8. Lily knows that working too many hours might hinder ~~your~~ her success in class.
9. Payton's boss said that if employees are diligent, ~~you~~ they can quickly earn a raise.
10. The returning vet said that before he was deployed, ~~you~~ he could not enroll in more than four classes.

MyWritingLab™ Visit *MyWritingLab.com* and complete the exercises and activities in the **Consistent Verb Tense and Active Voice, Tense, Pronoun References and Point of View,** and **Pronoun Case** topic areas.

RUN THAT BY ME AGAIN

- **Consistency in writing is concerned with using . . .** verbs and pronouns in their correct forms.
- **Verb consistency deals with using . . .** verb tenses that are identical in their time frame.
- **A verb shows either . . .** the existence or action of a person, place, or thing.
- **English verbs have . . .** six main tenses.
- **Problems with verb consistency often occur when . . .** writers change tenses unnecessarily.
- **To maintain verb consistency . . .** keep verbs in the same tense as the first verb, unless the time frame changes.
- **Pronoun consistency deals with . . .** using pronouns in the same point of view.
- **Point of view tells the reader . . .** who is performing the action of the sentence.
- **A frequent problem with pronoun consistency comes when . . .** writers needlessly change person (point of view) in a sentence or paragraph.

LEARNING GOAL
1. Identify and correct problems in verb consistency

LEARNING GOAL
2. Identify and correct problems in pronoun consistency

CONSISTENCY LEARNING LOG MyWritingLab™

MyWritingLab™

Answer the questions below to review your mastery of consistency.

1. With what is consistency in writing concerned?

   Consistency in writing is concerned with using verbs and pronouns in their correct form to ensure that meaning is clear.

2. With what does verb consistency deal?

   Verb consistency deals with using the correct verb tenses.

3. What is the effect on readers when verb tenses are used incorrectly?

   Readers are confused because the time frames described are incorrect or unclear.

4. What do verb tenses show readers?

   Verb tenses show readers the time of the action or existence.

5. To maintain verb consistency, what is the rule to remember?

   All verbs should be in the same tense as the first verb, unless the time frame changes.

6. With what does pronoun consistency deal?

   Pronoun consistency deals with using pronouns in the same point of view.

7. What does point of view tell the reader?

   Point of view tells the reader who is performing the action of the sentence.

8. A frequent problem with pronoun consistency comes when writers needlessly do what?

   A frequent problem with pronoun consistency comes when writers needlessly change person (point of view) in a sentence or paragraph.

9. In second person pronouns (*you* or *yours*), who is being addressed?

   In second person, the reader is being addressed directly.

Subject-Verb Agreement

LEARNING GOALS

In this chapter, you'll learn and practice how to

1. Identify subject-verb agreement errors
2. Correct common problems that prepositional phrases and clauses create with verbs
3. Correct common problems that pronouns and nouns create with verbs

GETTING THERE

- Subjects and verbs must agree in number.
- Prepositional phrases that come between subjects and verbs often create problems in determining the correct verb.
- Nouns and pronouns used as subjects often present problems in determining the correct verb.

What Is Subject-Verb Agreement?

LEARNING GOAL

1. Identify subject-verb agreement errors

The **subject** of a sentence is the person, place, or thing the sentence is about; a subject is either a noun or pronoun. The **verb** of a sentence tells about either the action or existence of the subject. Subjects and verbs must agree in number (that is, whether they're *singular* or *plural*). To make them agree,

- use a singular verb with a singular subject
- use a plural verb with a plural subject

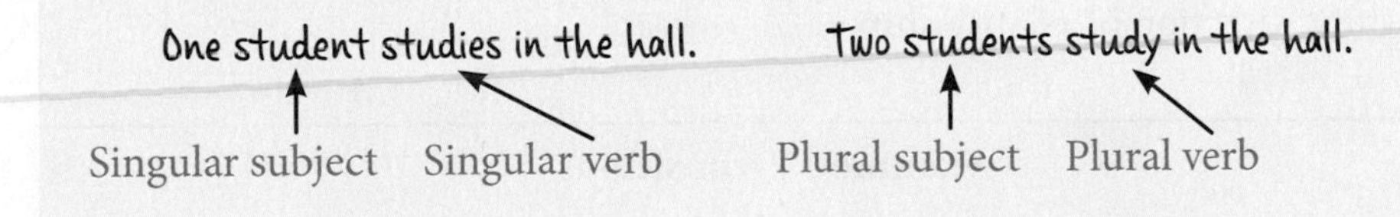

Correcting Subject-Verb Agreement Errors

Problems with subject-verb agreement usually come from incorrectly identifying the subject. When correcting subject-verb agreement errors, begin by identifying the subject of your sentence. To find the subject, find the main verb of the sentence and ask yourself *who* or *what* is performing the action (doing the verb).

The hamster in the cage runs in its wheel.

In this sentence, the verb is *runs*. When you answer the question "Who or what runs?" you find the subject, *hamster*.

Two lamps by the window showed me the way.

In this sentence, the verb is *showed*. Answering the question "Who or what showed?" gives you the subject, *lamps*.

The nearest bridge over the river has terrible traffic congestion.

The verb is *has*. Answering the question "Who or what has?" gives you the subject, *bridge*.

Other subject-verb agreement errors relate to (1) words that come between subjects and verbs, like prepositional phrases and clauses; (2) indefinite pronouns, (3) collective nouns, (4) expressions of amount, (5) unusual nouns, (6) compound subjects, and (7) inverted word order.

Correcting Errors Caused by Prepositional Phrases and Clauses

LEARNING GOAL

❷ Correct common problems that prepositional phrases and clauses create with verbs

Prepositions are words or phrases that

- link a noun or pronoun to another word in a sentence
- show direction or relationship

Over 150 English words and phrases can be used as prepositions.

A preposition often explains the position of something.

Common Prepositions

about	amid	behind	by way of
above	among	below	considering
according to	around	beside(s)	despite
across	as	between	down
after	at	beyond	during
against	because of	but	except
along	before	by	excluding

following	into	over	toward(s)
for	like	past	under
from	minus	plus	underneath
in	near	regarding	unlike
in addition to	of	since	until
in place of	off	than	up
in spite of	onto	through	with
inside	out (of)	throughout	within
instead of	outsie	to	without

Prepositional Phrases

A **prepositional phrase** includes a preposition, the object of that preposition (the noun or pronoun it refers to), and any words that describe the noun or pronoun.

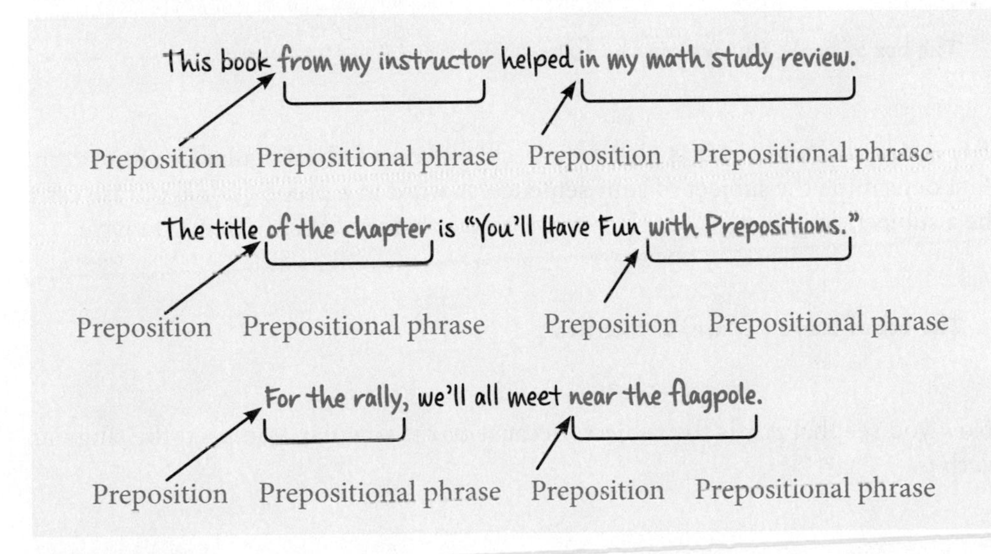

TECHNO TIP

For a quiz to review prepositional phrases, search the Internet for this site:

Big Dog's Exercises: Prepositional Phrases

22.1 Identifying Prepositional Phrases

Directions: Underline the prepositional phrases in each sentence.

1. For the following analysis, click on the button you see at the top of the page.
2. Are you going with me to the game on Saturday afternoon?
3. After math class, come to the cafeteria and we'll eat lunch between classes.
4. In regard to your grade, consult with your instructor for your current standing.
5. The timing of my next test is terrible on account of its proximity to the due date of my paper.

A prepositional phrase or phrases between the subject and the verb

Prepositional phrases that appear between a subject and a verb may prevent you from choosing the correct form of the verb.

Disregard prepositional phrases between subjects and verbs

To decide if a verb should be singular or plural, cross out any prepositional phrases between the subject and the verb in a sentence. For example, in this sentence

The box of books (is, are) on the floor.

box is singular, and *books* is plural. To decide if you need a singular or plural verb, first determine the subject of your sentence. A word in a prepositional phrase can't be a subject, so you cross out the prepositional phrase *of books*. Then you have

The box ~~of books~~ (is, are) on the floor.

Now you see that *box* is the subject. Because *box* is singular, you need the singular verb *is*.

The box of books is on the floor.

You should eliminate prepositional phrases to find your subject no matter

- how many prepositional phrases you have between your subject and verb
- how long the prepositional phrases are

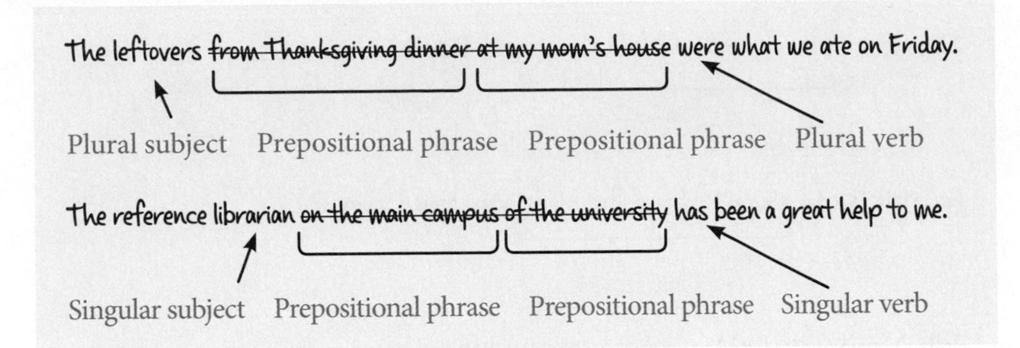

TICKET to WRITE

22.2 Identifying the Correct Verb (Prepositional Phrases)

Directions: In the sentences below, cross out any prepositional phrases between the subject and the verb, underline the subject, and then underline the correct verb.

1. Stephanie's interview ~~with the job recruiters~~ (is, are) scheduled for tomorrow.
2. One ~~of Carmen's friends~~ (visit, visits) here often.
3. Yesenia's resume, ~~including her jobs at the army bases~~, (is, are) impressive.
4. Marisol's present job ~~with an industry in her major~~ (pays, pay) just minimum wage.
5. The telephone book ~~in the lounge~~ close ~~to the classes~~ does not (has, have) campus numbers.
6. Most apartments ~~at the time of Evelyn's college visit~~ (was, were) located close to campus.
7. Noah and Hayden, ~~like Roberto~~, (tries, try) to get to class as early as possible.
8. Three ~~of the students in the room by the side door~~ (was, were) witnesses to the accident.
9. The onset ~~of the flu, in addition to problems with flooding,~~ (has, have) contributed to decreased enrollment.
10. The building ~~between the two parking lots~~ (is, are) where class will be next Tuesday.

Clauses

Clauses are groups of words that have a subject, a verb, and words relating to that subject and verb.

The tutors who are working today have been helpful to me.

Clause

Gabrielle's car, which had two flat tires, was towed away.

Clause

Colby's paper that is due in two weeks is what he's working on now.

Clause

THE PROBLEM

A clause or clauses between the subject and the verb

Clauses that begin with *who*, *which*, or *that* can create subject-verb agreement problems when they come between a subject and a verb.

THE FIX

Disregard clauses that begin with *who, which,* or *that* between subjects and verbs

To determine the correct verb to use, cross out any clause that

- begins with *who*, *which*, or *that*
- comes between a subject and a verb

In the sentence

The designer who is creating Trenton's Web pages (has, have) just called.

designer is singular and *Web pages* is plural. To decide if you need a singular or plural verb, determine which word is the subject of your sentence. If you cross out the clause *who is creating Trenton's Web pages*, you have

The designer ~~who is creating Trenton's Web pages~~ (has, have) just called.

Now you see the subject of your sentence: *designer*. Because *designer* is singular, you need the singular verb *has*.

22.3 Identifying the Correct Verb (*Who, Which, That* Clauses)

Directions: In the sentences below, cross out any *who*, *which*, or *that* clauses between the subject and the verb, underline the subject, and then underline the correct verb.

1. Tyrone~~, who enrolled in two night classes,~~ (is, are) adjusting his work schedule.
2. The cat ~~that belongs to Angel and Tia~~ (is, are) not too friendly.
3. Sadie's coffee~~, which has been sitting on the table for an hour,~~ (is, are) surely cold by now.
4. Colas ~~that were left in the student lounge~~ (was, were) put in the trash.
5. Coach Liles, ~~who keeps tabs on her players,~~ (is, are) asking about Elisabeth's grades.
6. The car ~~that crashed into the railings~~ (was, were) totaled.
7. The movies ~~that are playing tonight~~ (has, have) not gotten good reviews.
8. Arthur and Daniela~~, who live in a downtown apartment,~~ (has, have) become friends of mine.
9. The back fence~~, which was supposed to keep out stray dogs,~~ (has, have) failed.
10. The energy costs ~~that will be saved by shutting down the building between Christmas and New Year's~~ (amount, amounts) to over $4,000.

Correcting Errors Caused by Indefinite Pronouns and Nouns

Pronouns take the place of nouns, which name persons, places, or things. **Indefinite pronouns** take the place of nouns that are not named (the nouns are not definite).

LEARNING GOAL

❸ Correct common problems that pronouns and nouns create with verbs

Indefinite Pronouns

all	anything	everyone	most	none	several
another	both	everything	much	nothing	some
any	each	few	neither	one	somebody
anybody	either	little	no one	other	someone
anyone	everybody	many	nobody	others	something

In these examples, the indefinite pronouns pose no problem. *Others* is plural, so it takes the plural verb *are*; *somebody* is singular, so it takes the singular verb *texts*.

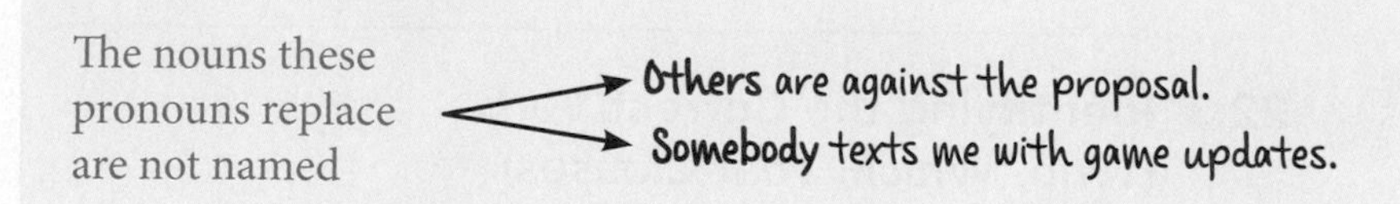

THE PROBLEM

Indefinite pronouns that look plural but are singular

Some indefinite pronouns (*everybody*, *each*, *everyone*, and *everything*) look as if they are plural but are considered singular.

THE FIX

Use a *singular* verb with *everybody, each, everyone,* and *everything*

Use a singular verb with the indefinite pronouns *everybody*, *each*, *everyone*, and *everything*, as in the following examples:

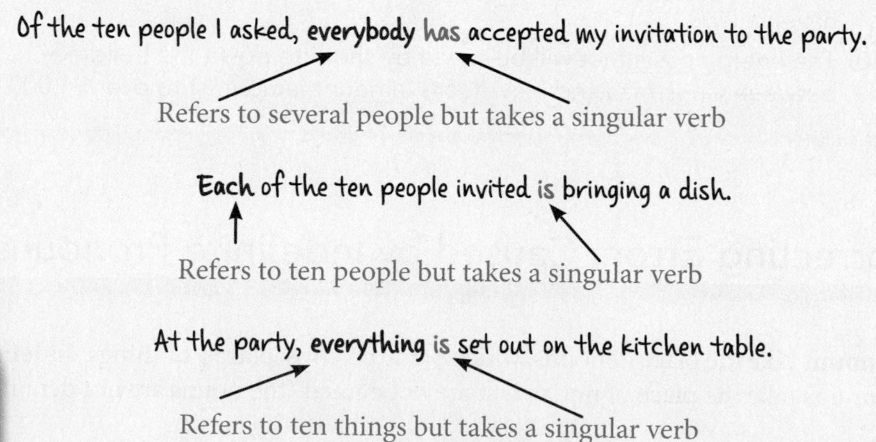

In these examples, *each, everybody, everything*, and *everyone* refer to plural people (ten of them) or things (ten dishes), but each of these takes a singular verb.

22.4 Identifying the Correct Verb (Indefinite Pronouns #1)

Directions: Determine the subject and underline the correct verb in each sentence.

1. Each of the computers (is, are) booting up now.
2. My four-year-old niece thinks everything in the mall (is, are) interesting.
3. Everyone in the library (was, were) talking and I couldn't concentrate.
4. Everything the fans do (seems, seem) to get on the star's nerves.
5. Kendra thinks everybody (was, were) working on the paper over the weekend.

TECHNO TIP

To review indefinite pronouns and agreement, search the Internet for this video: Mrs. Wood's Pronouns 2

THE PROBLEM

Indefinite pronouns that can be singular or plural

These five indefinite pronouns can be either singular or plural:

- all
- any
- most
- none
- some

THE FIX

Look at the object of the preposition to decide whether to use a singular or plural verb

If the indefinite pronoun *all*, *any*, *most*, *none*, or *some* is followed by a prepositional phrase, look at the object of the preposition. If the object is *singular*, use a *singular* verb; if the object is *plural*, use a *plural* verb.

Both of the following sentences have the indefinite pronoun *none* as their subject:

Heads Up! This is the only time you consider the prepositional phrase.

None of the money was recovered.

Money is singular, so the verb is singular

None of the suspects were captured.

Suspects is plural, so the verb is plural

22.5 Identifying the Correct Verb (Indefinite Pronouns #2)

Directions: Determine the subject and then underline the correct verb in each sentence.

1. Some of the students (was, were) still deciding their topics.
2. Any of the phone numbers I gave you today (is, are) okay to use to call me.
3. None of my phone numbers from last year (is, are) still in service.
4. All of the people (is, are) gathering for the parade.
5. Because Hayley didn't save it, most of her document (is, are) lost.
6. I was amazed that some of the pizza (was, were) not eaten.
7. (Is, Are) any of the resumes we received impressive?
8. None of the professors I have this semester (is, are) teaching night classes.
9. All of the town (was, were) quiet after the storm passed.
10. Most of the soft drinks we brought to class (is, are) still cold.

Collective Nouns

Sometimes these are called *group nouns.*

A **noun** names a person, place, or thing. One type of noun is a *collective noun*, which refers to a group of individuals (a *collection* of them), like *family*, *department*, and *audience*. Each of those nouns is one group, but that group has several members.

Common Collective Nouns

army	cabinet	committee	crowd	gang	navy
audience	cast	company	department	group	party
band	choir	congregation	faculty	jury	public
board	chorus	corporation	family	majority	staff
bunch	class	council	firm	minority	team

THE PROBLEM

Collective nouns take different verb forms

Sometimes collective nouns take a singular verb; sometimes they take a plural verb.

You probably know some collective nouns associated with animals, like a *flock of geese*, a *bed of oysters*, or a *gaggle of geese*.

Other collective nouns for animals are uncommon, like an *aerie* of eagles, a *bale* of turtles, a *float* of crocodiles, a *kettle* of hawks, a *mischief* of rats, a *pod* of dolphins, a *prickle* of porcupines, a *rhumba* of rattlesnakes, a *streak* of tigers, a *tribe* of baboons, or an *unkindness* of ravens.

To see more complete lists, search the Internet for these sites:

Animal Groups
Animal Group Terminology

THE BEST FIX

Look at how the group acts or thinks

The actions or thoughts of the group (those that make up the collective noun) determine if you use a singular or plural verb.

- If the parts of the group act or think together (as one), use a singular verb.
- If the parts of the group act or think separately, use a plural verb.

The family **is** donating $5,000 in memory of their late grandfather.

↑ Singular verb = individual members acting as one unit

The family **are** divided over how to allocate the rest of the money.

↑ Plural verb = individual members acting separately

TWO MORE FIXES

Add a word or substitute a word

If you're not sure about which verb to use, try one of these solutions:

- Add the word *members* after the problem collective noun. That turns *members* (a plural noun) into the subject, so you'll always use a plural verb.

The family **members** are divided over how to allocate the rest of the money.

TECHNO TIP

For more on subject-verb agreement using collective (group) nouns, search the Internet for this title

Language Portal: Subject-Verb Agreement with Collective Nouns

- Substitute a different word altogether.

The heirs are divided over how to allocate the rest of the money.

22.6 Identifying the Correct Verb (Collective Nouns)

Directions: Underline the correct verb in each sentence.

1. The basketball team (run, runs) out of the tunnel before the game begins.
2. After the game, the team (go, goes) their separate ways before the next day's practice.
3. Brady's class (is, are) trying to decide how to celebrate Professor Brand's fortieth birthday.
4. The jury (has, have) reached a verdict.
5. After his speech in communications class, Frank's audience (was, were) so impressed they gave him a standing ovation.

Expressions of Amount

Measurements or quantities are sometimes singular and sometimes plural.

THE PROBLEM

Expressions of amount can be singular or plural

THE FIX

Decide if the amount is one group or separate units of groups

If the subject expresses an amount (a measurement or quantity of money, time, weight, volume, or fractions), use

- a *singular* verb if the amount is considered *one unit or group*
- a *plural* verb if the amount is considered as a *group of separate units or groups*

Money as a single unit

Eighty dollars is a steep price to pay for lab equipment.

Money as separate units

The three shiny nickels on the ground attract the little girl.

Time as a single unit

Two weeks is a long time to wait to see what grade we made.

Time as a separate unit

Our first two days in class were homework-free.

22.7 Identifying the Correct Verb (Amounts, Measurements, Quantities)

Directions: Underline the correct verb in each sentence.

1. Fifteen minutes (was, were) all Adriana needed to finish the test.
2. Five pounds of illegal drugs (was, were) confiscated in the raid.
3. One-half of the class (sits, sit) in the first three rows.
4. Over thirteen miles (is, are), the distance of a half-marathon, seemed a short distance to run.
5. Sixty-four ounces of water (seem, seems) like a lot to drink.
6. Two gallons of milk (is, are) needed in the recipe.
7. "This six months with you (have, has) passed too quickly," Dustin said to Cassandra.
8. Four dollars (is, are) too much to pay for that hamburger at the ball game.
9. Two weeks (is, are) all the time left until spring break.
10. Fifteen pounds of groceries (is, are) quite a lot to carry up three flights of stairs.

Unusual Nouns

Most nouns that end in *–s* are plural, like *books*, *trucks*, and *baseballs*.

THE PROBLEM

Singular nouns that look like plural nouns

THE FIX

Determine if the noun is singular or plural

A few tricky nouns end in *–s* but name just one person, place, or thing, so they are singular and take a singular verb.

Economics **was** Jacob's favorite subject in high school.

Looks plural but takes a singular verb

Mumps **is** often a childhood disease.

Looks plural but takes a singular verb

Heads Up! Subjects that end in *–ics* are usually singular

Singular Nouns That Look Plural		
Diseases	**Subjects**	**Others**
AIDS, diabetes, herpes, measles, mumps, shingles, tuberculosis	civics, economics, fine arts, linguistics, physics, mathematics, statistics	aerobics, aesthetics, athletics, ethics, gymnastics, molasses, news, politics

THE PROBLEM

Singular nouns that take a plural form

Some nouns represent one thing (they are singular), but they

- end in *–s*
- take a plural verb

These nouns are things that have two equal parts, like *scissors* or *tights*.

My favorite shorts are in the hamper.

One item of clothing but takes a plural verb

The pliers were not on the workbench.

One tool but takes a plural verb

Determine if the noun refers to something that has two equal parts

When a noun refers to one thing that has two equal parts, it takes a plural verb. Here are some common examples:

Singular Nouns That Take Plural Verbs

clippers	pants	slacks	trousers
glasses (eyeglasses)	pliers	spectacles (eyeglasses)	trunks (swimming)
leggings	scales		tweezers
jeans	scissors	suspenders	
overalls	shears	tights	
pajamas	shorts	tongs	

22.8 Identifying the Correct Verb (Unusual Nouns)

Directions: Underline the correct verb in each sentence.

1. Fine arts (include, includes) drawing, painting, and architecture.
2. (Is, Are) Amelia's glasses on the table?
3. AIDS (has, have) become a disease of epidemic proportions.
4. Alexander's scissors (is, are) sticky from the glue he used in his project.
5. At this point in the semester, mathematics (is, are) Colton's favorite class.
6. The news that Ronald and Dominique brought to class (was, were) not good.

(continued)

7. Molasses (is, are) made from boiling down vegetable or fruit juice.
8. Where (were, was) the pliers?
9. The tweezers (is, are) in the drawer; use them to pick up the rare stamp.
10. Diabetes (is, are) a disease that results from lack of insulin.

Compound Subjects

Compound subjects are two or more subjects in a sentence. Compound subjects may be

- nouns, pronouns, or a combination of nouns and pronouns
- joined by one of three words: *and*, *or*, or *nor*

Compound subjects joined by *and*

Two or more subjects joined by *and* take a plural verb

Two or more subjects (either singular or plural) joined by *and* take a plural verb.

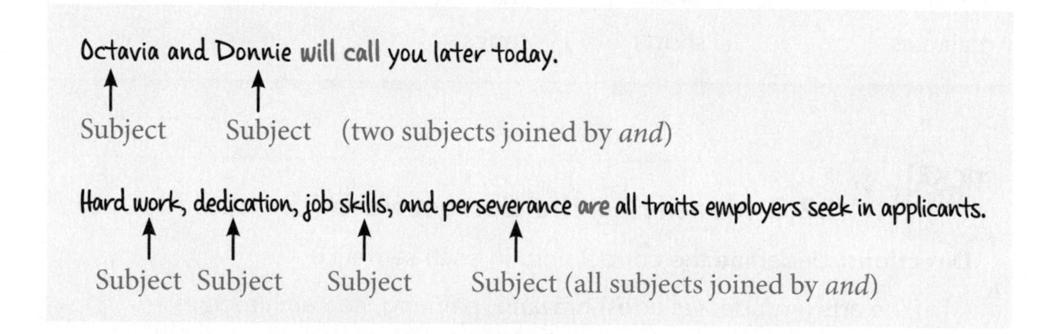

To check that you have the correct verb when all the subjects are joined by *and*, replace all the subjects with the word *they*. Then ask yourself what verb would be correct.

The page numbers and the glossary (is, are) in the back of the book.

They are in the back of the book.

CAREFUL! An exception to this rule comes when two or more subjects joined by *and* are thought of as one unit. In that case, the subjects take a singular verb.

Peanut butter and jelly is often found on kids' plates.

Subject + Subject Thought of as one unit

THE PROBLEM

Singular compound subjects joined by *or* or *nor*

THE FIX

Use a singular verb

Singular compound subjects joined by *or* or *nor* take a singular verb.

The air conditioner or the heater is always too loud in the classroom.

Singular subject Singular subject Singular verb

THE PROBLEM

Plural compound subjects joined by *or* or *nor*

THE FIX

Use a plural verb

Plural compound subjects joined by *or* or *nor* take a plural verb.

The air conditioners or the heaters are always too loud in both classrooms.

Plural subject Plural subject Plural verb

THE PROBLEM

Singular and plural compound subjects joined by *or* or *nor*

THE FIX

Look at the subject closest to the verb and use a singular verb if it is singular and plural if it is plural

If you have a singular subject and a plural subject joined by *or* or *nor*, look at the subject closer to the verb. If the subject closer to the verb is

- *singular*, use a *singular* verb
- *plural*, use a *plural* verb

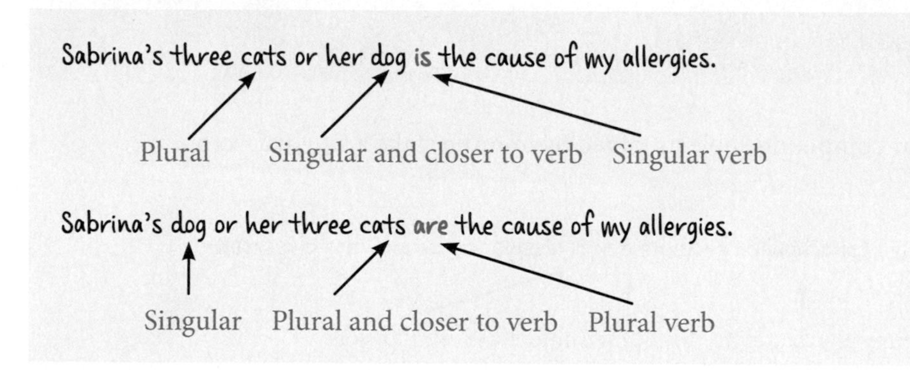

TECHNO TIP

For exercises on subject-verb agreement with compound subjects, search the Internet for these sites:

Towson University: Subject-Verb Agreement-Exercise 2

Commnet Compound Subjects 12.9

TICKET to WRITE

22.9 Identifying the Correct Verb (Compound Subjects)

Directions: Underline the correct verb in each sentence.

1. Plastic bottles or aluminum cans (is, are) supposed to go in this box for recycling.
2. Alan and Michaela (has, have) been dating for about six weeks.
3. Gavin's books or his backpack (is, are) always in his car.
4. Florida and Texas (is, are) both sending teams to the convention.
5. Spaghetti and meatballs (is, are) the specialty at Rocco's Italian Restaurant.
6. The transmission or the tires (was, were) replaced in the last year.
7. Jeremiah or Dana (is, are) picking me up for class.
8. A black ink pen or colored markers (is, are) acceptable for this assignment.

9. Rebecca Brady or Jack and Jill Jones (is, are) behind the anonymous donation.
10. (Is, Are) macaroni and cheese on the menu in the cafeteria tonight?

Inverted Order: When the Subject Follows the Verb

In most sentences, the verb follows the subject. Sentences written in **inverted order** (with the subject following the verb) sometimes pose problems in subject-verb agreement.

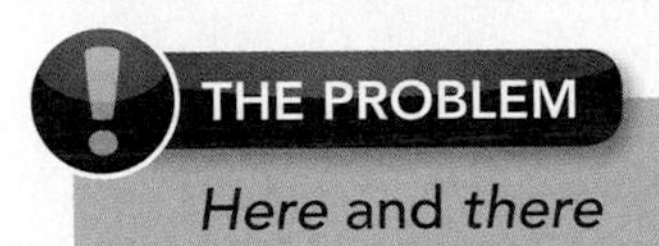

Here and there

When they are the first word of a sentence, *here* and *there* are never subjects of that sentence. Sentences that begin with *here* or *there* sometimes pose problems because the subject follows the verb.

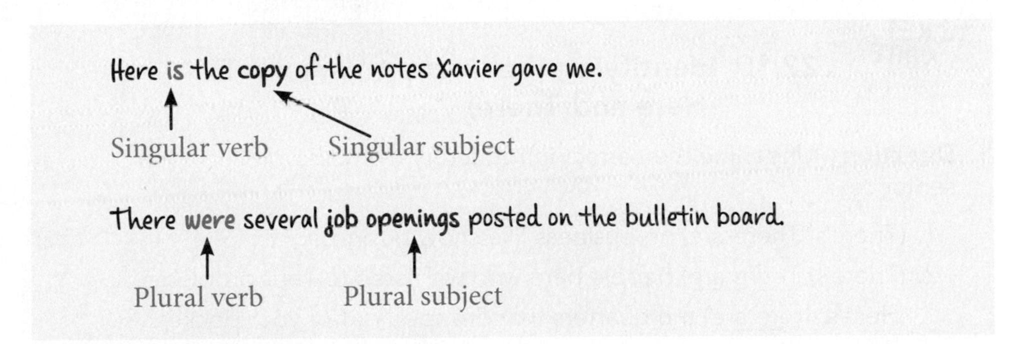

A common mistake is using the contraction *here's* or *there's* with a plural subject, as in these sentences:

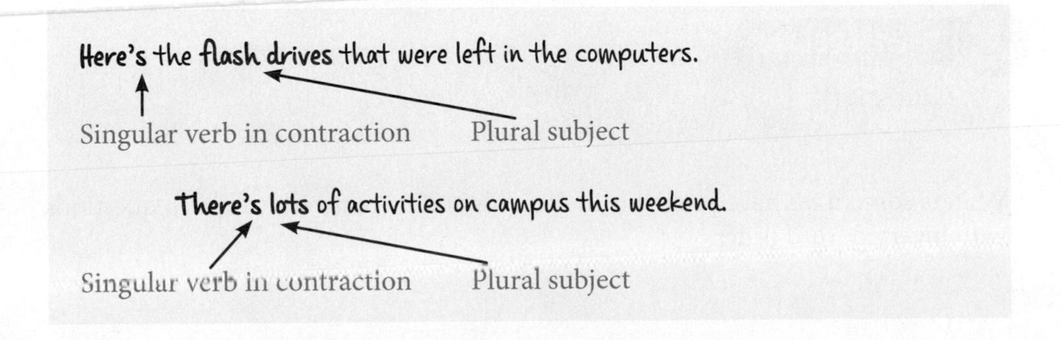

Use *here's* and *there's* only with singular verbs

The *'s* in the contractions stands for *is* (a singular verb), so these sentences are saying "Here is the flash drives . . ." and "There is lots of activities. . . ." Both subjects, *flash drives* and *activities*, are plural, so the sentences should read

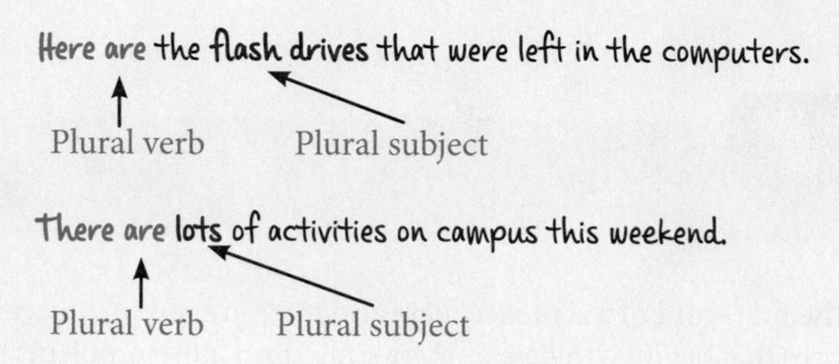

22.10 Identifying the Correct Verb (*Here* and *There*)

Directions: Underline the correct introductory word or words in each sentence.

1. (There's, There are) no business like show business.
2. (There's, There are) people here who will listen to your problems.
3. (Here's, Here are) the quarters that dropped out of your pocket.
4. (Here's, Here are) my assignments that should be in my portfolio.
5. (Here's, Here are) the syllabus for the class.

Questions

Writers sometimes have problems in deciding the correct verb to use in questions with inverted word order.

What is the name of the person in charge?

Singular verb Singular subject

When are they supposed to arrive?

Plural verb Plural subject

CAREFUL! A common problem comes when question words combine with verbs to form contractions.

	Singular	**Plural**
Who	Who's = *Who is*	Who're = *Who are*
What	What's = *What is*	What're = *What are*
When	When's = *When is*	When're = *When are*
Where	Where's = *Where is*	Where're = *Where are*
Why	Why's = *Why is*	Why're = *Why are*
How	How's = *How is*	How're = *How are*

Place the question word at the end of the sentence

To determine the subject, rearrange the words so that the question word (*Who? What? When? Where? Why?* or *How?*) comes at the end of the sentence.

The name of the person in charge is what?

They are supposed to arrive when?

22.11 Identifying the Correct Verb (Questions)

Directions: Underline the correct contraction in each sentence.

1. (Who's, Who're) the people going to the party?
2. (What's, What're) you taking to the party?

(continued)

3. (When's, When're) Tiffany leaving?
4. (Why's, Why're) they going so early?
5. (How's, How're) your date getting to the party?

THE PROBLEM

Don't and *doesn't*

Don't and *doesn't* pose particular problems in both spoken and written English.

TECHNO TIP

For a quiz to review using don't and doesn't correctly, search the Internet for this site:

Using English Quiz: Don't and Doesn't

THE FIX

Remember that *don't* contains the plural verb *do* and *doesn't* contains the singular verb *does.*

Don't is the contraction that stands for *do not.* Because it contains the plural verb *do,* use *don't* with a plural subject. *Doesn't* is the contraction that stands for *does not*; because it contains the singular verb *does,* use *doesn't* with a singular subject.

TECHNO TIP

To take a quiz on subject-verb agreement, search the Internet for this site

Subject-Verb Agreement: Additional Practice 2

doesn't
Lori ~~don't~~ like working with the new software.

doesn't
Neil said he ~~don't~~ know if he wants to stay here for the weekend or not.

TECHNO TIP

To review subject-verb agreement, search the Internet for these videos:

English-Subject/Verb Agreement

Subject Verb Agreement 1

Subject Verb Agreement 2

TICKET to WRITE

22.12 Identifying the Correct Verb (*Don't* and *Doesn't*)

Directions: Underline the correct verb in each sentence.

1. Jakob and Mckenna (don't, doesn't) live here.
2. Jerome (don't, doesn't) want to get caught in the snowstorm.
3. He (don't, doesn't) know what to think about the situation.
4. If Kristi (don't, doesn't) call me soon, I'll give her a buzz.
5. Donte (don't, doesn't) think his new haircut suits him.

MyWritingLab™ Visit *MyWritingLab.com* and complete the exercises and activities in the **Subjects and Verbs** and **Subject-Verb Agreement** topic areas.

RUN THAT BY ME AGAIN

- **Subjects and verbs must agree . . .** in number.
- **With a singular subject, use . . .** a singular verb.
- **With a plural subject, use . . .** a plural verb.
- **A prepositional phrase includes . . .** a preposition, the object of that preposition (the noun or pronoun it refers to), and any words that describe the noun or pronoun.

LEARNING GOAL

❶ Identify subject-verb agreement errors

LEARNING GOAL

❷ Correct common problems that prepositional phrases and clauses create with verbs

- **To find a subject, eliminate . . .** prepositional phrases, no matter how many prepositional phrases you have between your subject and verb and how long the prepositional phrases are.
- **Clauses are groups of words that have . . .** a subject, a verb, and words relating to the subject and verb.
- **To determine the correct verb with sentences that have clauses (1) beginning with *who, which,* or *that* and (2) coming between a subject and a verb . . .** cross out the clauses.
- **Indefinite pronouns take the place of . . .** nouns that are not named (the nouns are not definite).

LEARNING GOAL

❸ Correct common problems that pronouns and nouns create with verbs

- **Indefinite pronouns *everybody, each, everyone,* and *everything* . . .** look as if they are plural but are considered singular.
- **Indefinite pronouns *all, any, most, none,* and *some* . . .** can be either singular or plural.
- **When one of the pronouns *all, any, most, none,* or *some* is followed by a prepositional phrase . . .** look at the object of the preposition. If that object is *singular*, use a *singular* verb; if the object is *plural*, use a *plural* verb.
- **Collective nouns name . . .** groups of individuals (a collection of them).
- **With collective nouns, use a singular verb if . . .** the parts of the group act or think together (as one).
- **With collective nouns, use a plural verb if . . .** the parts of the group act or think separately.
- **With a problem collective noun, adding the word *members* after the noun . . .** turns *members* (a plural noun) into the subject.
- **Expressions of amount are singular if . . .** the amount is considered one unit or group.

- **Expressions of amount are plural if . . .** the amount is considered as a group of separate units or groups.
- **Some nouns are singular but . . .** they end in –*s* and take a plural verb.
- **Compound subjects are . . .** two or more subjects in a sentence.
- **Compound subjects joined by *and* . . .** take a plural verb.
- **When two or more subjects joined by *and* are thought of as one unit . . .** they take a singular verb.
- **Singular subjects joined by *or* (or *nor*) . . .** take a singular verb.
- **Plural subjects joined by *or* (or *nor*) . . .** take a plural verb.
- **When a singular subject and a plural subject are joined by *or* (or *nor*) . . .** use a singular verb if the subject closer to the verb is singular.
- **When a singular subject and a plural subject are joined by *or* (or *nor*) . . .** use a plural verb if the subject closer to the verb is plural.

SUBJECT-VERB AGREEMENT LEARNING LOG MyWritingLab™

Answer the questions below to review your mastery of subject-verb agreement.
Answers will vary.

1. **Subjects and verbs must agree in what?**

   Subjects and verbs must agree in number.

2. **To make subjects and verbs agree, begin by doing what?**

   Begin by identifying the subject of the sentence.

3. **What are two functions of prepositions?**

   Prepositions (1) link a noun or pronoun to another word in a sentence and (2) show direction or relationship.

4. **What does a prepositional phrase include?**

   A prepositional phrase includes a preposition, the object of that preposition (the noun or pronoun it refers to), and any words that describe the noun or pronoun.

5. **What should you do to determine the correct verb in a sentence that has one or more prepositional phrases between the subject and verb?**

   Cross out all prepositional phrases between the subject and verb.

6. **What are clauses?**

   Clauses are groups of words that have a subject, a verb, and words relating to that subject and verb.

7. **Which clauses often cause problems in subject-verb agreement?**

   Clauses that begin with *who*, *which*, or *that* often create subject-verb agreement problems when they come between a subject and a verb.

8. **To determine the correct verb in sentences that have clauses that begin with *who*, *which*, or *that*, what should you do?**

   Cross out clauses that begin with *who*, *which*, or *that* when they come between a subject and a verb.

9. **Why do indefinite pronouns have that name?**

   Indefinite pronouns have that name because the nouns they take the place of are not named (the nouns are not definite).

10. What four indefinite pronouns look as if they are plural but are considered singular?

    Everybody, *each*, *everyone*, and *everything* look as if they are plural but are considered singular.

11. What five indefinite pronouns can be either singular or plural?

    All, *any*, *most*, *none*, or *some* can be either singular or plural.

12. How do you decide whether to use a singular or plural verb with those five indefinite pronouns?

    Look at the object of the prepositional phrase. If it is singular, use a singular verb; if the object is plural, use a plural verb.

13. To what does a *collective noun* refer?

    A *collective noun* refers to a group of individuals.

14. With a collective noun, how do you determine whether to use a singular or plural verb?

    If the parts of the group act or think together (as one), use a singular verb.

    If the parts of the group act or think separately, use a plural verb.

15. In expressions of amount, how do you decide whether to use a singular or plural verb form?

    Use a singular verb if the amount is considered one unit or group; use a plural verb if the amount is considered as a group of separate units or groups.

16. Cite four examples of singular nouns that end in –*s* and therefore look as if they are plural but take a singular verb.

    Answers will vary.

17. Cite four examples of singular nouns that end in -*s* and take a plural verb.

    Answers will vary.

18. Compound subjects joined by *and* take what verb form?

    Compound subjects joined by *and* take a plural verb.

19. If two or more subjects joined by *and* are thought of as one unit, what verb form do they take?

    If two or more subjects joined by *and* are thought of as one unit, they take a singular verb.

20. If singular subjects are joined by *or* (or *nor*), what verb form is used?

    Singular subjects joined by *or* or *nor* take a singular verb.

21. If plural subjects are joined by *or* (or *nor*), what verb form is used?

    Plural subjects joined by *or* or *nor* take a plural verb.

22. If a singular subject and a plural subject are joined by *or* or *nor*, what verb is used?

    If the subject closer to the verb is singular, a singular verb is used; if the subject closer to the verb is plural, a plural verb is used.

23. How do you determine the correct verb in questions?

    Rearrange the words so that the question word comes at the end of the sentence.

24. *Don't* and *doesn't* should be used with what types of subjects?

    Use *don't* with a plural subject; use *doesn't* with a singular subject.

CHAPTER

23 Pronouns

LEARNING GOALS

In this chapter, you'll learn and practice how to

1. Identify pronouns and antecedents
2. Determine the correct pronoun to use, according to its antecedent
3. Ensure pronouns clearly identify their antecedents
4. Ensure pronouns agree in number with their antecedents

GETTING THERE

- Pronouns refer back to antecedents, which are nouns.
- *Who* is a subject; *whom* is an object.
- Possessive pronouns do not take apostrophes.
- Compound pronouns can cause problems.
- Reflexive pronouns must have antecedents.
- Singular indefinite pronouns require singular verbs; plural indefinite pronouns require plural verbs.

What Are Pronouns and Antecedents?

LEARNING GOAL

1. Identify pronouns and antecedents

Pronouns are words that take the place of nouns or other pronouns. **Antecedents** are the nouns (persons, places, things, ideas) the pronouns stand for.

Antecedent ← Pronoun

Raul isn't in class because he had to go to the hospital.

Sometimes a pronoun refers to an antecedent in a preceding sentence.

Antecedent ← Pronoun

Olivia hopes to finish her paper today. She is working on her final draft.

These words can be used as pronouns:

all	hers	much	somebody	which
another	herself	myself	someone	whichever
any	him	neither	something	who
anybody	himself	nobody	that	whoever
anyone	his	none	theirs	whom
anything	I	no one	them	whomever
both	it	nothing	themselves	whose
each	its	one	these	you
either	itself	other	they	yours
everybody	little	others	this	yourself
everyone	many	ours	those	yourselves
everything	me	ourselves	us	
few	mine	several	we	
he	more	she	what	
her	most	some	whatever	

You may hear or read these archaic pronouns:

archaic	**modern**
thy	your
thine	your, yours
ye	you
thou	you
thee	yourself

Heads Up!
Some of these also function as other parts of speech.

Types of Pronouns

Pronouns are divided into six categories:

demonstrative	interrogative	reflexive and intensive
relative	personal	indefinite

LEARNING GOAL
❷ Determine the correct pronoun to use, according to its antecedent

Heads Up!
Some pronouns can be used in more than one category.

Demonstrative Pronouns

Demonstrative pronouns point out (demonstrate) a definite person, place, or thing.

Refers to a singular noun that is nearby	this	that	Refers to a singular noun that is farther away
Refers to a plural noun that is nearby	these	those	Refers to a plural noun that is farther away

CAREFUL! Using the words *here* or *there* immediately after demonstrative pronouns is nonstandard and not acceptable in academic writing.

Incorrect	This **here** computer is new.
Correct	This computer is new.
Incorrect	I've had problems with that **there** vending machine.
Correct	I've had problems with that vending machine.

TECHNO TIP

For an online quiz on demonstrative pronouns, search the Internet for this site: Demonstrative Pronoun Quiz

TICKET to WRITE

23.1 Identifying Demonstrative Pronouns

Directions: Read the following sentences and underline the demonstrative pronoun and cross out any incorrect words in each.

1. That is the information I've tried to find.
2. Are those your papers?
3. This ~~here~~ was where the house stood.
4. I'll give these to the officials.
5. That ~~there~~ car you're driving is new, isn't it?
6. Please return this to the division secretary.
7. Is that the paper you needed to copy?
8. This ~~here~~ paper is going to get an A!
9. Louisa didn't realize she had left those in the car.
10. I'll never go to that ~~there~~ gas station again.

Relative Pronouns

Relative pronouns connect (relate) a subordinate clause to the rest of the sentence.

who whom that which whoever whomever whichever

Relative pronoun ↓

Correct That man, **who** is my cousin, will give me a ride back to campus.

Who is my cousin is a subordinate clause (it has a subject and verb but cannot stand alone as a sentence). The pronoun *who* relates the clause back to *man*.

THE PROBLEM

Deciding when to use *who* and *whom*

Many writers have a problem determining when to use the relative pronouns

who or *whom* *whoever* or *whomever*

THE FIX #1

Determine if the pronoun is used as the subject or object of the clause

Isolate the clause *who* or *whom* is in. If the pronoun is used as a subject, use *who*; if it is used as an object, use *whom*.

Makayla wondered (who, whom) had called.

Isolate the clause (who, whom) had called.

The word you need is the subject of the clause, so use *who*.

Brandon had called (who, whom)?

This sentence has only one clause, so isolating it is easy:

Brandon had called (who, whom).

The word you need is the object of the clause, so use *whom*.

The same rule holds for *whoever* or *whomever*. If you need a subject, use *whoever*; if you need an object, use *whomever*.

THE FIX #2

Substitute *he* or *him*

Isolate the clause *who* or *whom* is in. Rearrange the words to form a declarative sentence (a sentence that states a fact). Then substitute *he* or *him* for *who* or *whom*. If your ear tells you that *he* is correct, use *who*; if your ear tells you that *him* is correct, use *whom*.

Heads Up!
Use this mnemonic:
he = who
hiM = whoM

Read the sentence

Raul asked (who, whom) he would be paired with for the presentation.

and isolate the clause.

(who, whom) he would be paired with for the presentation.

Rearrange the wording into a declarative sentence.

He would be paired with (who, whom) for the presentation.

Substitute *he* and *him* and let your ear determine which sounds correct.

Incorrect He would be paired with **he** for the presentation.

Correct He would be paired with **him** for the presentation.

Because *him* is correct, you would use *whom* in the sentence:

Raul asked **whom** he would be paired with for the presentation.

TECHNO TIP

For more about *who* and *whom*, search the Internet for this video: Halt! Whom Goes There? Pronoun Case

For a quiz on relative pronouns, search the Internet for this site: Online Relative Pronoun Quiz

The same rule holds for *whoever* and *whomever*. If you can substitute *he*, use *whoever*; if you can substitute *him*, use *whomever*.

23.2 Using Correct Relative Pronouns: *Who* or *Whom*

Directions: Read the following sentences and underline the correct relative pronoun in each.

1. Professor Randall, who/whom I respect, is speaking at an assembly in Strawn Auditorium.
2. My roommate, who/whom is from Haiti, will not go home for Thanksgiving.
3. The woman who/whom sits beside me in class has three children.
4. The woman's children, who/whom she speaks of often, are proud that she's in college.
5. I'm conducting a survey with Heather, who/whom I just met.
6. The student who/whom is in line has already called in an order.
7. I've often wondered who/whom was the architect for the football field.
8. For this assignment, we can pair with whoever/whomever we want.
9. Chef John Abbott, who/whom is an adjunct professor here, recently appeared on television.
10. Our college's Quick Start program provides college courses for high school students who/whom want to begin classes early.

Interrogative Pronouns

In questions, **interrogative pronouns** take the place of unnamed nouns.

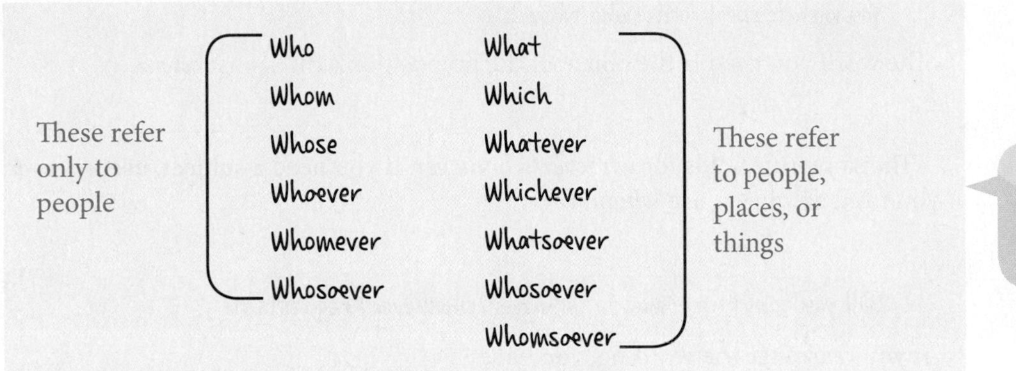

Whosoever and *whomsoever* are archaic words not used often today.

Whom did Desiree call yesterday afternoon?

Whom takes the place of an unnamed person.

What will Rafael bring to help repair the computer?

What takes the place of an unnamed thing.

THE PROBLEM

When to use *who* or *whom*

When composing questions, many writers have a problem determining whether to use

who or *whom* *whoever* or *whomever*

Determine if the pronoun is used as the subject or object

Isolate the clause *who* or *whom* is in. Rearrange the wording to create a declarative sentence (a sentence that states a fact) rather than a question. If the pronoun is used as a subject, use *who*; if it is used as an object, use *whom*.

(Who/Whom) did you wish to speak with?

If you rearrange the wording, you have

You wish to speak with (who/whom).

The word you need is the object of the preposition with, so use **whom**.

The same rule holds for *whoever/whomever*. If you need a subject, use *whoever*; if you need an object, use *whomever*.

Will you copy the e-mail to (whoever/whomever) requests it?

If you rearrange the wording, you have

You will copy the e-mail to (whoever/whomever) requests it.

The word you need is the subject of the clause (whoever/whomever) *requests it*, so use **whoever**.

Substitute *he* or *him*

Heads Up!
Use this mnemonic:
he = who
hiM = whoM

Isolate the clause *who* or *whom* is in. Rearrange the wording to create a declarative sentence (a sentence that states a fact) rather than a question. Then substitute *he* or *him* for *who* or *whom*. If your ear tells you that *he* is correct, use *who*; if your ear tells you that *him* is correct, use *whom*.

TECHNO TIP

For an online quiz on *who* and *whom*, search the Internet for this site:

Grammar: Quiz on Forms of Who

For an online quiz on interrogative pronouns, search the Internet for this site:

Interrogative Pronoun Quiz

In the sentence

(Who/Whom) will watch the DVD with me?

substitute *he* and *him* and let your ear determine which sounds correct.

Correct **He** will watch the DVD with me.

Incorrect **Him** will watch the DVD with me.

Because *he* sounds correct, use *who* in the question:

Who will watch the DVD with me?

23.3 Identifying Interrogative Pronouns

Directions: Read the following sentences and underline the interrogative pronoun in each.

1. Which shirt did you decide to buy?
2. Whatever happened to your "true love" from junior high?
3. Whose cell phone interrupted class?
4. When I started singing "Who Let the Dogs Out?" several others joined in.
5. What did I do with my car keys?

23.4 Using Correct Interrogative Pronouns: *Who or Whom*

Directions: Read the following sentences and underline the correct interrogative pronoun in each.

1. Who/Whom did you see at the Tips for Success study session?
2. Are we supposed to give the papers to whoever/whomever asks for them?
3. Who/Whom has the best sense of humor in class?
4. From who/whom did you get your academic grant?
5. Is Sarah Bruner the woman who/whom is our guest lecturer today?
6. Who/Whom was eliminated last night on *The Bachelor*?
7. Whoever/Whomever runs for student government will work with Dr. Tweddell.
8. Who/Whom was Ms. Jennings referring to when she said "outstanding student"?
9. From who/whom did you get the idea that class will be cancelled tomorrow?
10. With who/whom are you riding home?

Personal Pronouns

Personal pronouns refer to specific persons, places, or things. They change their form depending on *person*, *number*, *case*, or *gender*.

1. **Person** refers to *point of view*, the relationship between a writer and a reader.

First person refers to the person writing.
Second person refers to the person written to.
Third person refers to a person or thing written about.

2. **Number** refers to **singular** and **plural** words.

The writer	first person singular	I, me, my, mine
The writers	first person plural	we, us, our, ours
The person addressed	second person singular	you, yours
The persons addressed	second person plural	you, yours
Someone, something else	third person singular	he, she, it, her, him, hers, his, its
Other people or things	third person plural	they, them, their, theirs

TECHNO TIP

To review singular and plural personal pronouns, search the Internet for these videos:

Personal Pronouns–Singular Subject & Object Form

Personal Pronouns–Plural Subject Form

3. **Case** refers to whether a pronoun is used as a subject, an object, or a possessive.

Subject:	I, my, mine, we, our, ours
Object:	me, my, you, yours, he, him, his, she, her, hers, it, its
Possessive:	my, mine, our, ours, yours, their, theirs, her, hers, his, its

4. **Gender** refers to whether the pronoun stands for a male or female person or a neutral object.

Male:	he, him, his
Female:	she, her, hers
Neutral:	it, its

Problems with Personal Pronouns

Personal pronouns present a number of problems.

Using apostrophes with possessive pronouns

Writers sometimes use apostrophes with possessive pronouns.

Incorrect	The books my study partners left on the desk are their's.
Incorrect	The dog ate it's bone.

Heads Up!
Remember that *it's* means *it is; its* means belonging *to it.*

Never use an apostrophe with a possessive pronoun

No possessive pronoun uses an apostrophe.

Correct	The books my study partners left on the desk are **theirs.**
Correct	The dog ate **its** bone.

23.5 Using Correct Possessive Pronouns

Directions: Read the following sentences and underline the correct possessive pronoun in each.

1. The police officer asked if the car parked on the grass was (their's, theirs).
2. I study in the library because (it's, its) computers have the best programs.
3. Julio and (his', his) family sent a card when I was sick.
4. This billfold is (yours, your's), isn't it?
5. The moon lost (its, it's) glow after midnight.

Which pronoun to use with the compound subject of a verb

Writers sometimes are unsure of the correct pronoun to use as part of the compound subject of a sentence.

Gustavo and (I, me) can't wait for the game to start.

Rosa and (she, her) are having car troubles.

Change the verb; say only one of the pronouns and the verb

Change the verb to a singular form; then say the verb and only one of the pronouns. Your ear will tell you if the pronoun is correct. In the examples, you would never write

Incorrect	**Me** can't wait for the game to start.
Incorrect	**Her** is having car troubles.

Instead, your ear would tell you to write

Correct	**I** can't wait for the game to start.
Correct	**She** is having car troubles.

So you would write the sentences

Gustavo and I can't wait for the game to start.

Rosa and she are having car troubles.

23.6 Using Correct Pronouns as Subjects

Directions: Read the following sentences and underline the correct subject in each.

1. Enrique and (I, me) will bring the food for the party.
2. Maddie and (she, her) volunteered to donate at the blood drive.
3. Why don't you and (me, I) walk through campus?
4. The guys and (us, we) were trying to sing along to the new CD.
5. Kelsey and (we, us) are headed to math class.
6. The Carlsons and (them, they) live in second-floor apartments.
7. (Her, She) and her roommate are grilling tonight.
8. Grant and (me, I) are on our way to the mall.
9. Why do you and (him, he) always disagree?
10. Shaquille and (him, he) wore the same number on their jerseys.

THE PROBLEM

Which pronoun to use with the compound object of a verb

Some writers have a problem deciding the correct pronoun to use when writing compound objects of a verb.

Our instructor gave Cameron and (me, I) good grades.

The letter was addressed to Chelsea and (he, him).

THE FIX

Use only the pronouns

Read or say the sentence with only one of the pronouns (eliminate any nouns in the compound). Your ear will tell you which is correct. In the examples, you would never write

Incorrect	*Our instructor gave I good grades.*
Incorrect	*The letter was addressed to he.*

Instead, your ear would tell you to write

Correct Our instructor gave **me** good grades.

Correct The letter was addressed to **him**.

So you would write the sentences

Our instructor gave Cameron and me good grades.

The letter was addressed to Chelsea and him.

TECHNO TIP

For an online quiz on personal pronouns, search the Internet for this site:
Personal Pronouns Quiz

23.7 Using Correct Pronouns as Objects

Directions: Read the following sentences and underline the correct object in each.

1. You called Eliza and (I, me) last night.
2. Brock's night class bores Kasey and (he, him).
3. Mr. Franklin teaches Aidan and (me, I).
4. Did you want Chaz and (her, she) to wait for you?
5. Grandmother often texts Ella and (he, him).
6. Grandfather sends (us, we) cards through the mail.
7. Uncle Albert called Kim and told (her, she) he'd take her out to lunch.
8. Will you remind Kyla and (he, him) of the assignment?
9. The constant ringing of cell phones is driving Madelyn and (I, me) both up the wall.
10. Please give (us, we) some help with our math problems.

Reflexive and Intensive Pronouns

The same eight words act as reflexive and intensive pronouns.

myself yourself himself herself itself ourselves yourselves themselves

Reflexive pronouns are always objects and always refer action back to the subject.

Reflexive pronoun (object)

I surprised **myself** with the grade on my paper.

Myself refers to the subject of the sentence, I.

THE PROBLEM

Reflexive pronoun with no antecedent

Some writers use a reflexive pronoun without having named whom or what the pronoun refers to (the reflexive pronoun does not have an antecedent).

Incorrect Give the papers to Noreen and **myself** before you leave.

(The antecedent of myself is not named in the sentence.)

THE FIX

Use an antecedent with a reflexive pronoun

Use a reflexive pronoun only if its antecedent is named in the same sentence.

Correct Give the papers to Noreen and **me** before you leave.

Intensive pronouns, which come immediately after their antecedents, help to emphasize their antecedents.

Intensive pronoun

I **myself** was surprised with the grade on my paper.

Myself emphasizes its antecedent, I.

Intensive pronouns may be omitted without altering the meaning of the sentence.

THE PROBLEM

Nonstandard reflexive and intensive pronouns

These words are nonstandard and should not be used in academic or other formal writing:

hisself theyself theyselves theirself theirselves

These words are considered colloquial. Although they are sometimes heard in speech, they should not appear in academic or other formal writing.

Use the correct form of the word

~~hissself~~ should be **himself**
~~ourself~~ should be **ourselves**
~~theyself~~ should be **themselves**
~~theyselves~~ should be **themselves**
~~theirself~~ should be **themselves**
~~theirselves~~ should be **themselves**

TECHNO TIP

For more on reflexive and intensive pronouns, search the Internet for this video:

Reflexive and Intensive Pronouns

To find online quizzes on reflexive and intensive pronouns, search the Internet for these sites:

Reflexive and Intensive Pronoun Quizzes

23.8 Using Reflexive and Intensive Pronouns Correctly

Directions: Read the following sentences, identify the error(s) in each, and then write the corrected version on the line below.

1. Shawn helped hisself to the candy on the desk.
   Shawn helped himself to the candy on the desk.
2. Claudia was surprised herself with how hard the test was.
   Claudia herself was surprised with how hard the test was.
3. Papers were handed out by herself.
   Papers were handed out by her.
4. Those toddlers can't stop theirselves after they start giggling.
   Those toddlers can't stop themselves after they start giggling.
5. Austin knew himself he'd get a call from the dean.
   Austin himself knew he'd get a call from the dean.
6. All the class members have introduced theyselves to each other.
   All the class members have introduced themselves to each other.
7. Leave your paper with either Tori or myself and we'll turn it in.
   Leave your paper with either Tori or me and we'll turn it in.

8. We ourself are proud of the money the club raised to support underprivileged children.

   We ourselves are proud of the money the club raised to support underprivileged children.

9. If Henry and Kaylee want copies of my notes, they can help theirself to them.

   If Henry and Kaylee want copies of my notes, they can help themselves to them.

10. How can class go on without Bryce or myself?

    How can class go on without Bryce or me?

Indefinite Pronouns

Indefinite pronouns refer to nouns that are not identified.

all	anything	everyone	most	none	several
another	both	everything	much	nothing	some
any	each	few	neither	one	somebody
anybody	either	little	no one	other	someone
anyone	everybody	many	nobody	others	something

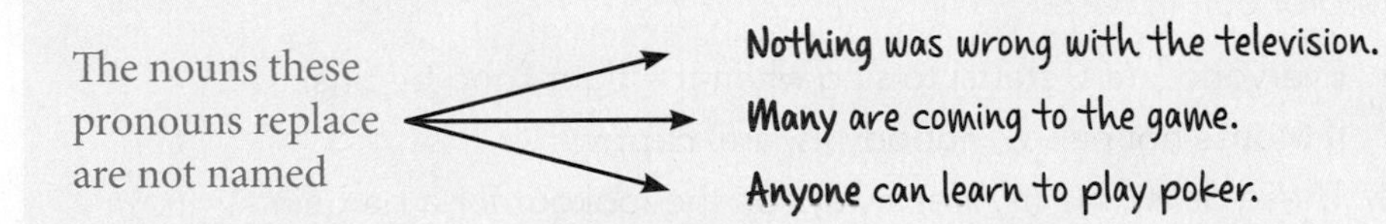

THE PROBLEM

Indefinite pronouns that look plural but are singular

Some indefinite pronouns look as if they are plural, but they are singular.

THE FIX

Always use a singular verb with certain indefinite pronouns

These indefinite pronouns may look plural, but each is singular and takes a singular pronoun and a singular verb.

anybody	each	everybody	everything	nobody	somebody
anyone	either	everyone	neither	one	someone

Some instructors classify *my, our, your, his, her, their,* and *its* as possessive adjectives.

Correct Everyone is waiting for his or her grade.

Singular verb (is) Singular pronoun (his or her)

Even though *everyone* means *many people*, it is singular.

Correct Each of the books has had its spine repaired.

Singular pronoun (Each) Singular verb (has)

Even though *each* means *many things*, it is singular.

TICKET to WRITE

23.9 Using Correct Singular Indefinite Pronouns

Directions: Read the following sentences and underline the correct indefinite pronoun in each.

1. Everyone (start, starts) to sing when the fight song begins.
2. If Mom's not happy, nobody (is, are) happy.
3. These days, one (is, are) always on the lookout for a bargain.
4. Either (is, are) an acceptable answer.
5. Everybody (stop, stops) talking when the lights go down.
6. Anyone in class (is, are) eligible to receive free tutoring.
7. For the survey, each of the students in the third row (get, gets) a #2 pencil.
8. Somebody (is, are) hosting a surprise party in the Student Center.
9. Everything (was, were) in place when we arrived.
10. Someone (sit, sits) in that seat.

Indefinite pronouns that can be singular or plural

Five indefinite pronouns can be either singular or plural:

all any most none some

Heads Up!
Use the mnemonic **A MANS** to remember these:
All
Most
Any
None
Some

THE FIX

Look at the object of the preposition to decide whether to use a singular or plural verb

If one of the indefinite pronouns *all*, *any*, *most*, *none*, or *some* is followed by a prepositional phrase, look at the object of the preposition. If that object is singular, use a singular verb; if the object is plural, use a plural verb.

In both of these sentences, the indefinite pronoun *none* is the subject:

Money is singular, so the verb is singular.

None of the money was recovered.

Suspects is plural, so the verb is plural.

None of the suspects were captured.

23.10 Using Correct Verbs with Pronouns That Can Be Singular or Plural

Directions: Read the following sentences and underline the correct indefinite pronoun in each.

1. Any of the students (is, are) entitled to get extra tickets to the game.
2. All of the equipment (was, were) ready when we arrived.
3. None of the people (is, are) enrolling in the early session.
4. Most of the food (was, were) donated by a local restaurant.
5. All of the papers (was, were) accepted in the competition.
6. Some of the juice (has, have) lost its flavor.
7. Most of the sandwiches (was, were) eaten by the hungry staff.
8. None of the instructors (has, have) signed up to help on Friday.
9. Some of the visitors (is, are) people I recognize.
10. (Is, Are) any of these seats taken?

Pronouns Must Clearly Identify Their Antecedents

LEARNING GOAL

3 Ensure pronouns clearly identify their antecedents

For sentences to be understood correctly, antecedents of pronouns must be clear.

Unclear antecedents

To avoid confusion, be sure a pronoun has a clear antecedent.

"Every pronoun necessarily has an antecedent. Which person or thing in the sentence that antecedent is must be immediately clear to the reader."
—Jacques Barzun

Incorrect Allison hoped Kelsey would join her study group because she was shy.
Allison Kelsey Who is she?

The meaning is unclear; the reader doesn't know if Allison or Kelsey is the person who is shy.

Incorrect Kyle took the doors off the jambs and painted them.
doors jambs What is them?

The meaning is unclear because the reader doesn't know if Kyle painted the jambs or the doors.

Incorrect They said class will be in the library today.

The meaning is unclear because the reader doesn't know who *they* is; the pronoun has no antecedent.

THE FIX

Reword the sentence to eliminate the vague or ambiguous pronoun

Rewording the sentence lets the reader know which meaning was intended:

Correct Allison hoped Kelsey would join her study group because **Allison** was shy.

or Now the reader understands who was shy.

Correct Allison hoped Kelsey would join her study group because **Kelsey** was shy.

Correct Kyle painted the **jambs** after taking the doors off them.

or Now the reader understands what Kyle painted.

Correct Kyle painted the **doors** after taking them off the jambs.

Correct **Three people in English 085** said class will be in the library today.

23.11 Correcting Unclear Pronoun References

Directions: Read the sentences below and rewrite them so they are no longer unclear. Answers will vary.

1. Megan told Stephanie that her instructor had arrived.
2. They bought Ciarra a new motorcycle.
3. Justin's assignment was in his backpack, but he can't find it.
4. The temperature reached ninety, and the noise from the paving in the parking lot was loud. It made concentrating on the test difficult.
5. After she moved, Mariah took her books out of the boxes and stacked them in the spare room.
6. In his speech on campus, Governor Washburn said we'd get a new student center, which sounded good.
7. They told me to pick up my car this afternoon.
8. Vincent watched the DVD for five minutes, which was not the right thing to do.
9. Angelica asked her landlady if she could cook a turkey.
10. The flood had caused the road to be covered with water; it was a big problem.

Pronouns Must Agree in Number with Their Antecedents

LEARNING GOAL

4 Ensure pronouns agree in number with their antecedents

Agreement in *number* refers to pronouns and antecedents being *singular* or *plural.*

Because pronouns refer back to nouns (their antecedents), the rule that pronouns must agree in number means that

- if the antecedent of a pronoun is singular, the pronoun must be singular
- if the antecedent of a pronoun is plural, the pronoun must be plural

Singular antecedent → essay; Plural pronoun → them

Incorrect After he read my essay, Julio lavishly praised them.

This example is incorrect because the pronoun *them* is plural and its antecedent *essay* is singular.

Singular antecedent → essay; Singular pronoun → it

Correct After he read my essay, Julio lavishly praised it.

This example is correct because both the pronoun *it* and its antecedent *essay* are singular.

THE PROBLEM #1

Using *they, them,* or *their* incorrectly

Some instructors classify *their* as a possessive adjective.

Some writers incorrectly use *they*, *them*, or *their* to refer to an antecedent that is singular and can be either masculine or feminine.

Incorrect When a professor is out of town, they might cancel class.

Professor is singular and can be masculine or feminine, so they is incorrect.

THE FIX

Change to gender-neutral constructions that are singular or change construction so both pronoun and antecedent are plural

Use "he or she," "him or her," or "his or her." These are called *gender-neutral* constructions because they are free of reference to a specific sex.

Some instructors classify *his* and *her* as possessive adjectives.

Correct When a professor is out of town, **he or she** may cancel class.

Professor is singular and can be masculine or feminine, so he or she is correct.

Correct When professors are out of town, **they** may cancel class.

Professors is plural and can be masculine or feminine, so they is correct.

THE PROBLEM #2

Overusing gender-neutral pronouns

Overusing "he or she" and other gender-neutral constructions is repetitive.

Incorrect When a professor is out of town, **he or she** may cancel class. If **he or she** must do so, **he or she** will tell students.

THE FIX

Change to plural pronouns and antecedents

If possible, reword with plurals.

Correct When professors are out of town, **they** may cancel class. If **they** must do so, **they** will tell students.

TECHNO TIP

For more on pronoun-antecedent agreement, search the Internet for this video:

English Grammar & Punctuation: What Is a Pronoun-Antecedent Agreement?

23.12 Correcting Pronoun Problems in Number

Directions: Read the sentences below and rewrite them to correct problems with pronoun number. Answers will vary.

1. Everyone should do their part to help the environment.
2. Each student should have their ID to get in free.

(continued)

3. If a caller asks for personal information, don't give it to them.
4. When the Phi Delta Kappa representative contacts you, tell them you're interested in joining.
5. Reading essays written by another student will help you understand their point of view.
6. Do you always get up with your child if they cry during the night?
7. Each member of the US Senate may e-mail their constituents an update on the bill.
8. The attendant at the gas station knows the way; ask them for directions.
9. If that fan keeps jumping in the stands, they're going to get hurt.
10. Every nursing student will take their finals during the same week.

MyWritingLab™ Visit *MyWritingLab.com* and complete the exercises and activities in the **Pronouns** and **Pronoun-Antecedent Agreement** topic areas.

RUN THAT BY ME AGAIN

LEARNING GOAL
❶ Identify pronouns and antecedents

- **Pronouns take the place of . . .** nouns or other pronouns; **antecedents** are the nouns the pronouns stand for.
- **Demonstrative pronouns point out . . .** a definite person, place, or thing.
- **Relative pronouns connect . . .** a subordinate clause to the rest of the sentence.
- **Personal pronouns refer to . . .** specific persons, places, or things.
- **Reflexive pronouns are always . . .** objects and always refer action back to the subject.
- **Intensive pronouns help to emphasize . . .** their antecedents.
- **Indefinite pronouns refer to . . .** nouns that are not identified.

LEARNING GOAL
❷ Determine the correct pronoun to use, according to its antecedent

LEARNING GOAL
❸ Ensure pronouns clearly identify their antecedents

- **A pronoun must have . . .** a clear antecedent.

LEARNING GOAL
❹ Ensure pronouns agree in number with their antecedents

- **A singular pronoun must refer to . . .** a singular noun; **a plural pronoun** must refer to a plural noun.

PRONOUN LEARNING LOG MyWritingLab™

Answer the questions below to review your mastery of pronouns.

1. What is a pronoun?

   A pronoun is a word that takes the place of a noun or other pronoun.

2. What is an antecedent?

   An antecedent is the noun a pronoun stands for.

3. What are the four demonstrative pronouns?

   Demonstrative pronouns are *this, that, these,* and *those.*

4. What does a relative pronoun do?

   A relative pronoun connects a subordinate clause to the rest of the sentence.

5. How does a writer determine when to use *who* and when to use *whom*?

   Isolate the clause the pronoun is in and decide if the pronoun is used as a subject or an object. Use *who* if it is a subject; use *whom* if it is an object.

6. Interrogative pronouns are used in what type of sentence?

   Interrogative pronouns are used in questions.

7. Personal pronouns change form, depending on what?

   Personal pronouns change their form depending on *person, number, case,* or *gender.*

8. When should a personal pronoun be written with an apostrophe?

   A personal pronoun should never be written with an apostrophe.

9. What do reflexive pronouns always refer?

   Reflexive pronouns always refer action back to the subject.

10. What is the function of an intensive pronoun?

    An intensive pronoun helps to emphasize its antecedent.

11. What is an indefinite pronoun?

    An indefinite pronoun is a word that refers to a noun that is not identified.

12. If a pronoun has an unclear antecedent, what should a writer do?

    A writer should reword a sentence to eliminate an unclear antecedent.

13. What does the rule that pronouns must agree in number mean?

    The rule that pronouns must agree in number means that if the antecedent of a pronoun is singular, the pronoun must be singular; if the antecedent of a pronoun is plural, the pronoun must be plural.

Parallel Structure

LEARNING GOALS

In this chapter, you'll learn and practice how to

1. Recognize effective parallel structure
2. Correct faulty parallelism

GETTING THERE

- Using parallel structure gives balance and rhythm to writing.
- Mistakes in parallel structure are called *faulty parallelism.*
- Faulty parallelism can be repaired by changing words or phrases so all items are in the same grammatical form.

What Is Parallel Structure?

LEARNING GOAL

1. Recognize effective parallel structure

Parallel structure (parallelism) refers to consistency in the way writers present similar ideas in a sentence. These ideas might be words, phrases, or clauses that have the same—or nearly the same—grammatical structure. When writers use parallel structure, their sentences have a rhythm that emphasizes what they say.

Here are examples of famous sentences that use parallel structure:

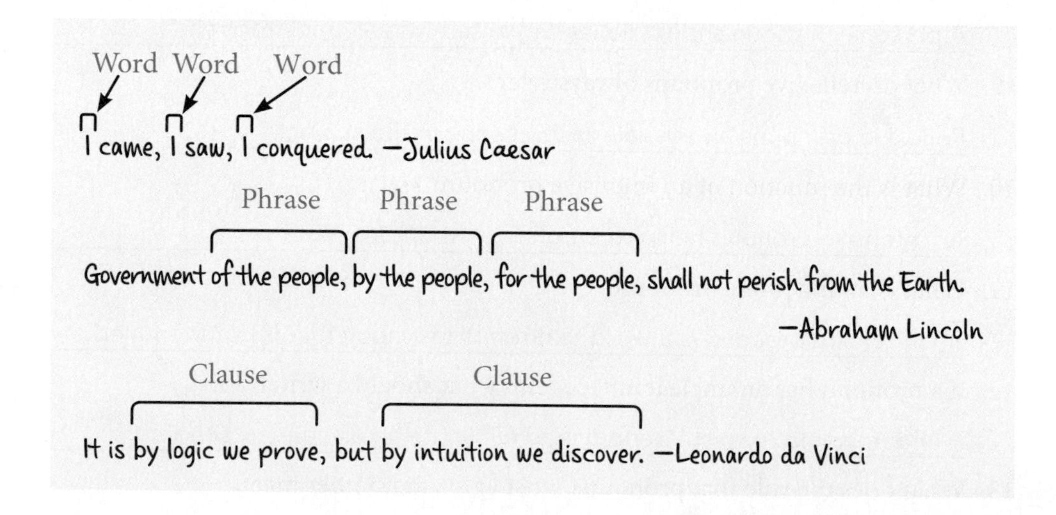

Correcting Faulty Parallelism

LEARNING GOAL
❷ Correct faulty parallelism

Mistakes in parallel structure are called *faulty parallelism*; these mistakes occur when writers are inconsistent in the way they present items in a series. With faulty parallelism, writing is unbalanced and can confuse readers.

Faulty parallelism can be found in almost any part of speech, like verbs, nouns, adjectives, or adverbs.

THE PROBLEM #1

Items in pairs or in a series

Problems with Verbs

When you write verbs in pairs or in a series, list them all in the same grammatical form. For example, if you begin with a verb in the past tense, put all other verbs in the past tense. This example of verbs in a series breaks that rule:

Verb in past tense → walked; Verb in past perfect tense → had fed

Incorrect Last night we walked the dog and had fed the cat.

THE FIX

Change all items in pairs or a series to the same grammatical form

The grammatical form of the first item is usually the one that works best for an entire list. You could think through fixing the preceding example this way:

we	walked	(past tense verb)
[we]	had fed	(past perfect tense verb)

Because *had fed* has a different grammatical form (a different tense) from *walked*, it should be changed.

[we]	fed	(past tense verb)

Because *fed* is the same tense as *walked*, the sentence is now parallel.

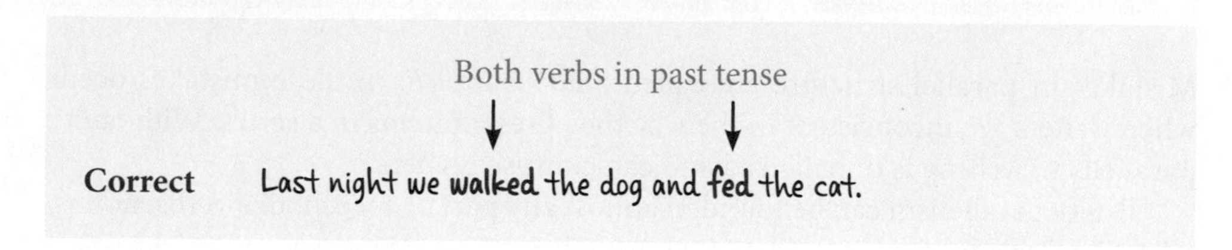

Problems with Nouns or Pronouns

A pair or series that begins with a noun or pronoun should contain only nouns or pronouns. This example of nouns in a series breaks that rule:

In a pair or series of nouns or pronouns, be sure all the items in the pair or series are nouns or pronouns

The grammatical form of the first item in the series above is a noun, so all the items should be nouns or pronouns. You could think through fixing the preceding example this way:

was	a traveler
[was]	an entrepreneur
[was]	entertaining

Because *entertaining* (an adjective) has a different grammatical form from *traveler* and *entrepreneur* (both nouns), *entertaining* needs to be changed.

	Change to noun
[was]	an entertainer

Because *an entertainer* has the same grammatical form as *a traveler* and *an entrepreneur*, the sentence is now parallel. The repaired sentence could read:

Correct	The man interviewed by Piers Morgan was a **traveler**, an **entrepreneur**, and an **entertainer**.

Problems with Adjectives

A pair or series that begins with an adjective should contain only adjectives. This example of adjectives in a series breaks that rule:

Incorrect	Wednesday's guest lecturer was **disorganized**, **rude**, and **showed up unprepared**.

THE FIX

In a pair or series of adjectives, be sure all the items in the pair or series are adjectives

The grammatical form of the first item in the series above is an adjective, so all the items should be adjectives. You could think through fixing the preceding example this way:

was	disorganized
[was]	rude
[was]	showed up unprepared

Because *showed up* has a different grammatical form from *disorganized* and *rude*, it needs to be changed.

	Change to adjective
[was]	unprepared

Because *unprepared* has the same grammatical form as *disorganized* and *rude*, the sentence is now parallel. The repaired sentence could read:

Correct	Wednesday's guest lecturer was **disorganized**, **rude**, and **unprepared**.

24.1 Repairing Faulty Parallelism with Items in Pairs or in a Series: Verbs, Nouns, and Adjectives

Directions: Rewrite and repair each sentence below to make its parts parallel. Answers will vary.

1. Miguel ate popcorn, drank water, and was watching *Survivor.*
2. I need to find a parking space with shade and having an extra-wide slot.
3. During the movie, I felt bewildered, stunned, and I was scared.
4. Jean-Claude prefers reggae to listening to hip-hop.
5. Professor Benson said he could walk, talk, and be chewing gum at the same time.
6. Arturo was athletic, rich, and a snob.
7. Professor Groves said my thesis statement was coherent and it was written concisely.
8. For the project, I needed paper, glue, and to get a pair of scissors.
9. The defendant stood up, screamed at the judge, and was ranting at his lawyer.
10. I didn't recognize Marty because he now has a goatee, glasses, and wears his hair long.

Problems with Adverbs

A pair or series that begins with an adverb should contain only adverbs. This example of adverbs in a series breaks that rule:

Incorrect Friday's guest lecturer spoke **accurately**, **thoroughly**, and in a **knowledgeable manner.**

THE FIX

In a pair or series of adverbs, be sure all the items in the pair or series are adverbs

The grammatical form of the first item in the series above is an adverb, so all the items should be adverbs. You could think through fixing the preceding example this way:

spoke	accurately
[spoke]	thoroughly
[spoke]	in a knowledgeable manner

Because *in a knowledgeable manner* has a different grammatical form from *accurately* and *thoroughly*, it needs to be changed.

	Change to adverb
[spoke]	knowledgeably

Because *knowledgeably* has the same grammatical form as *accurately* and *thoroughly*, the sentence is now parallel. The repaired sentence could read:

Correct	Friday's guest lecturer spoke **accurately, thoroughly,** and **knowledgeably.**

Problems with Phrases

A pair or series that begins with a phrase should contain only phrases. This example of phrases in a series breaks that rule:

Incorrect	Jon has class **in the Henkel Building** and **Sullivan Center.**

THE FIX

In a pair or series of phrases, be sure all the items in the pair or series are phrases

The grammatical form of the first item in the series above is a phrase, so all the items should be phrases. You could think through fixing the preceding example this way:

"One must be drenched in words, literally soaked in them, to have the right ones form themselves into the proper patterns at the right moment."
—Hart Crane

has class	in the Henkel Building
[has class]	Sullivan Center

Because *Sullivan Center* has a different grammatical form from *in the Henkel Building*, something needs to be added.

	Change to phrase
[has class]	in the Sullivan Center

Because *in the Sullivan Center* has the same grammatical form as *in the Henkel Building*, the sentence is now parallel. The repaired sentence could read:

Correct	Jon has class **in the Henkel Building** and **in the Sullivan Center.**

Problems with Clauses

A pair or series that begins with a clause should contain only clauses. This example of clauses in a series breaks that rule:

Incorrect	Katy's son promised he **would make his bed** and **not to leave empty glasses in his room.**

THE FIX

In a pair or series of clauses, be sure all the items in the pair or series are clauses

The grammatical form of the first item in the series above is a clause, so all the items should be clauses. You could think through fixing the preceding example this way:

promised he	would make his bed
[promised he]	not to leave empty glasses in his room

Because *not to leave empty glasses in his room* has a different grammatical form from *would make his bed*, it needs to be changed.

	Change to clause
[promised he]	would not leave empty glasses in his room

Because *would not leave empty glasses in his room* has the same grammatical form as *would make his bed*, the sentence is now parallel. The repaired sentence could read:

Correct Katy's son promised he **would make his bed** and **would not leave empty glasses in his room.**

24.2 Repairing Faulty Parallelism with Items in Pairs or in a Series: Adverbs, Phrases, and Clauses

Directions: Rewrite and repair each sentence below to make its parts parallel. Answers will vary.

1. Eating an occasional candy bar, watching reality television, and graphic novels are the ways Carlos rewards himself.
2. Matt accepted the promotion cheerfully and with gratitude.
3. Kevin hoped he could get time off from work, borrow the tools he needed, and to get his car running again.
4. I'm in class today because I need to turn in my paper, because I want to get notes, and to see the PowerPoint presentation.
5. Driving on the black ice, I approached the stoplight slowly and cautious.
6. Professor Busby told the class that we should study the first three chapters and to answer all the review questions.
7. Freshmen are in their first year of college, while sophomores are those who already have college hours.
8. Zoey sang confidently, with boldness, and as if she were putting out no effort.
9. Trinity doesn't know where Troy is or will he be back soon.
10. At the soccer game, Giovanni cheered loudly and with enthusiasm.

PROBLEM #2

Chronology or degree of importance

Problems with Chronology (Time Order)

Items that are not listed in chronological (time) order are confusing to readers.

Incorrect Henry laughed at his pictures from **kindergarten, high school,** and **sixth grade.**

Henry attended sixth grade before he went to high school, so those two items should be reversed.

Place a series of items in chronological (time) order, from earliest to most recent or most recent to earliest

You could think through fixing the preceding example this way:

pictures from	kindergarten
[pictures from]	high school
[pictures from]	sixth grade

Because *high school* comes last chronologically, it should be last in the list.

[pictures from]	kindergarten
[pictures from]	sixth grade
[pictures from]	high school

The repaired sentence could read:

Correct Henry laughed at his pictures from **kindergarten, sixth grade,** and **high school.**

Problems with Order of Importance

Items that are not listed in order of their degree of importance are confusing to readers.

Incorrect Taking an incorrect dose of the drug can result in **death** or **anxiety.**

Because death is the more drastic result, it should be listed last.

THE FIX

Place the most important, most drastic, most emphatic, or strongest item in a series in the last position

This is how you could think through fixing the preceding example:

can result in	death → a more drastic outcome, so it should be last
[can result in]	anxiety → a less drastic outcome, so it should be first

The repaired sentence could read:

Correct Taking an incorrect dose of the drug can result in **anxiety** or **death**.

PROBLEM #3

Correlative conjunctions

Correlative conjunctions are words or phrases that need other words or phrases to be complete and to make sense. When only one part of a correlative conjunction is used, the sentence is incomplete and readers are confused.

either . . . or neither . . . nor both . . . and
not only . . . but also whether . . . or

Incorrect The latest blockbuster movie is **not only** dull.

Correct The latest blockbuster movie is **not only** dull **but also** poorly edited.

THE FIX

Write the first correlative conjunction and the word or phrase that accompanies it; do the same with the second. Match the grammatical form of the second word or phrase to the form of the first word or phrase.

Incorrect The latest blockbuster movie is **neither exciting nor is it amusing.**

This is how you could think through fixing the preceding example:

Correlative conjunction	Word or phrase that accompanies it
neither	exciting
nor	is it amusing

Because *is it amusing* has a different grammatical form from *exciting*, it needs to be changed.

nor	amusing

Because *amusing* is the same grammatical form as *exciting*, the sentence is now parallel. The repaired sentence could read:

Correct The latest blockbuster movie is **neither exciting nor amusing.**

24.3 Repairing Faulty Parallelism: Chronology, Importance, and Correlative Conjunctions

Directions: Rewrite and repair each sentence below to make its parts parallel. Answers will vary.

1. Chanika has both a witty personality, and she has good study habits.
2. Being accepted into college, being awarded a diploma, and completing a certain number of hours of classes are all part of a college graduate's life.
3. Students who plagiarize may be expelled, suspended, or fail the class.
4. Either Kelly is going to shadow his mentor tonight.
5. DeJuan is not only president of student government.
6. The baseball team got a standing ovation when they scored the winning run in the bottom of the ninth, were introduced, and tied the game 4-4.

7. Dante's memoir focused on recovering from bites by a mosquito, a shark, and a dog.
8. After hearing our test was cancelled, I didn't know whether to be glad.
9. Because of the employee's outburst, he may have his pay docked, be fired, or be demoted.
10. To secure your apartment, sign the contract, contact the landlord, and give a deposit.

Listen for Parallelism

In addition to *looking* for parallel structure, good writers also *listen* for it. Reading your material aloud will sometimes help you hear mistakes that may slip by if you read only silently. Reading aloud also enhances the beauty of parallelism. Read these two examples aloud to further appreciate their impact:

> With this faith we will be able to work together, to pray together, to struggle together, to go to jail together, to stand up for freedom together. . . .
>
> —Rev. Martin Luther King, Jr.

> I hear and I forget. I see and I remember. I do and I understand.
>
> —Confucius

TECHNO TIP

For online quizzes on parallelism, search the Internet for these sites:

Commnet Quiz on Parallel Structures

Second Exercise in Parallelism

Recognizing Parallel Structure

Search the Internet for this video:

Parallel Structure

MyWritingLab™ Visit *MyWritingLab.com* and complete the exercises and activities in the **Parallelism** topic area.

RUN THAT BY ME AGAIN

- **Parallel structure (parallelism) refers to . . .** consistency in the way writers present similar ideas in a sentence.
- **Mistakes in parallel structure are called . . .** faulty parallelism.
- **When you write verbs in pairs or in a series, list them all . . .** in the same grammatical form.
- **A pair or series that begins with a noun or pronoun should contain . . .** only nouns or pronouns.
- **A pair or series that begins with an adjective should contain . . .** only adjectives.

LEARNING GOAL

1 Recognize effective parallel structure

LEARNING GOAL

2 Correct faulty parallelism

- **A pair or series that begins with an adverb should contain . . .** only adverbs.
- **A pair or series that begins with a phrase should contain . . .** only phrases.
- **A pair or series that begins with a clause should contain . . .** only clauses.
- **A series of items in chronological (time) order should be presented . . .** from earliest to most recent or most recent to earliest.
- **Place the most important, most drastic, most emphatic, or strongest item in a series . . .** in the last position.
- **With words or phrases that accompany correlative conjunctions . . .** match the grammatical form of the second word or phrase to the grammatical form of the first word or phrase.

PARALLEL STRUCTURE LEARNING LOG MyWritingLab™

Answer the questions below to review your mastery of parallelism.

1. What is parallel structure?
   Parallel structure (parallelism) is consistency in the way similar ideas are presented in a sentence.
2. What are mistakes in parallel structure called?
   Mistakes in parallel structure are called *faulty parallelism.*
3. To avoid faulty parallelism, what should you do when listing verbs in pairs or in a series?
   When you write verbs in pairs or in a series, list them all in the same grammatical format.
4. To avoid faulty parallelism, what should you do when listing nouns or pronouns in pairs or in a series?
   When you write nouns or pronouns in pairs or in a series, list only nouns or pronouns.
5. To avoid faulty parallelism, what should you do when listing adjectives in pairs or in a series?
   When you write adjectives in pairs or in a series, list only adjectives.
6. To avoid faulty parallelism, what should you do when listing adverbs in pairs or in a series?
   When you write adverbs in pairs or in a series, list only adverbs.
7. To avoid faulty parallelism, what should you do when listing phrases in pairs or in a series?
   When you write phrases in pairs or in a series, list only phrases.
8. To avoid faulty parallelism, what should you do when listing clauses in pairs or in a series?
   When you write clauses in pairs or in a series, list only clauses.
9. To avoid faulty parallelism, what should you do with a series of items in chronological (time) order?
   When you write a series of items in chronological (time) order, place the items from earliest to most recent or most recent to earliest.
10. To avoid faulty parallelism, what should you do with a series of items in order of importance?
    When you write a series of items in order of importance, place the most important, most drastic, most emphatic, or strongest item in a series in the last position.

11. To avoid faulty parallelism, what should you do with a series of items that contain correlative conjunctions?

    When you write a series of items that have correlative conjunctions, write the first correlative conjunction and the word or phrase that accompanies it; do the same with the second. Match the grammatical form of the second word or phrase to the form of the first word or phrase.

Sentence Variety

LEARNING GOALS

In this chapter, you'll learn and practice how to

1. Identify primary sentence functions
2. Use four sentence types to vary sentence construction
3. Vary sentence length
4. Vary sentence beginnings

GETTING THERE

- Sentences can be constructed as declarative, interrogative, imperative, or exclamatory.
- Using different simple, compound, complex, or compound-complex sentences aids sentence variety.
- Varying sentence length and sentence beginnings helps keep writing fresh and interesting.

What Is Sentence Variety?

Sentence variety involves adjusting the way sentences are structured to keep writing from being monotonous. Mature readers expect variety. In children's books, sentences are often structured the same way:

> The dog chased the cat. The cat went over the fence. The dog went over the fence. The cat ran to the yard. The ground was cold.

While that construction is fine for children as they learn to read, it is boring for advanced readers because all the sentences are the same type (all have one subject and one verb), they all have the same beginning (all start with *the*, followed by a subject), and they all are about the same length (all are four or five words).

To keep from being boring and to add emphasis to points, advanced writers practice sentence variety; they vary the length, format, and rhythm of their sentences.

When talking, we often vary the length of our sentences.

You can achieve sentence variety in several ways, including

- using different types of sentences
- changing the length of sentences
- altering the beginnings of sentences

"Variety's the very spice of life, That gives it all its flavor." —William Cowper

If you examine sentences in a book or magazine, you'll probably find that the author used different sentence types. This keeps the material from being monotonous and also presents ideas in a fresh way. Understanding the types and subdivisions of sentences can help you vary the length of sentences and alter their beginnings.

Sentence types are classified in two ways: according to their *function* and according to their *construction*.

Sentence Functions

LEARNING GOAL
1. Identify primary sentence functions

Sentences are written so readers can determine if the information in them states a fact, asks a question, gives a command, or expresses an emotion.

In essay and paragraph writing, most sentences are declarative; the author states facts.

Sentence Type	Sentence Function
Declarative Bottled water can be expensive.	states a fact
Interrogative How much does that bottle of water cost?	asks a question
Imperative Give me only bottled water.	gives a command
Exclamatory That water smells awful!	expresses an emotion

TICKET to WRITE

25.1 Identifying Sentence Types According to Function

Directions: In the space before each sentence, mark *D* if the sentence is declarative, *IN* if it is interrogative, *IM* if it is imperative, and *E* if it is exclamatory.

E 1. Let freedom ring!

D 2. The only thing we have to fear is fear itself.

IN 3. How much is that new car?

IM 4. Bring me that memo now.

D 5. My cell phone must be hiding from me.

IN 6. Have you seen my cell phone anywhere in the apartment?

E 7. "Turn off your cell phone when you're in my class!"

IM 8. "Turn that darn cell phone off right now," said Professor Shaffer.

IN 9. May I borrow your cell phone?

D 10. This is the last sentence about cell phones.

Sentence Constructions and How to Vary Them

Sentences are classified by the type and number of clauses they contain.

An **independent clause** contains a subject and its verb, and it can stand alone as a complete thought.

LEARNING GOAL

❷ Use four sentence types to vary sentence construction

Complete Thought

Serafina wants bottled water.

Subject of verb Verb

A **subordinate (dependent) clause** has a subject and its verb but cannot stand alone as a complete thought.

Heads Up!
A subordinate clause is a sentence fragment.

Incomplete Thought

If Serafina wants bottled water

Subject of verb Verb

Sentences can be constructed in four ways: as a *simple* sentence, *compound* sentence, *complex* sentence, or *compound-complex* sentence.

Sentence Type	Sentence Construction
Simple Bruno called the radio show.	one independent clause, no subordinate clauses
Compound Bruno called the radio show, and he expressed outrage about the law.	at least two independent clauses, no subordinate clauses
Complex Although he needed to get to work, Bruno called the radio show to express outrage about the law.	one independent clause, one or more subordinate clauses
Compound-Complex Although he needed to get to work, Bruno called the radio show to express outrage about the law; however, he was nervous once he was on the air.	at least two independent clauses, one or more subordinate clauses

Simple Sentences

No matter how many words they have, **simple sentences** have only one clause.

Simple sentence (only one clause) You're here!

Simple sentence (only one clause) During the ceremony on Veterans Day, the middle-school students respectfully stood in honor of the numerous military men and women in the crowded gymnasium.

"My philosophy? Simplicity plus variety."
—Hank Stram

One way to create sentence variety is to combine two or more simple sentences into a single sentence or a compound sentence, using the correct punctuation.

Combine Two or More Simple Sentences

You can combine two or more simple sentences into one sentence:

Simple sentence	Simple sentence
Gerardo is waiting for Antoine.	Darius is waiting for Antoine.

Simple sentences combined into one simple sentence

Correct Gerardo and Darius are waiting for Antoine.

However, you may combine sentences only if the thoughts are logically connected.

Simple sentence	Simple sentence
Gerardo is waiting for Antoine.	Gerardo wants spaghetti for dinner.

Simple sentences combined into one simple sentence

Incorrect Gerardo is waiting for Antoine and wants spaghetti for dinner.

The fact that Gerardo is waiting for Antoine has nothing to do with what Gerardo wants for dinner, so combining these sentences is incorrect.

Combine Two or More Simple Sentences into a Compound Sentence

You can also combine two or more simple sentences into one compound sentence:

Simple sentences combined into compound sentence

Correct Gerardo is waiting for Antoine, and so is Darius.

However, you may combine sentences only if the thoughts are logically connected.

Simple sentence	Simple sentence
Gerardo is waiting for Antoine.	The cable went out.

Simple sentences combined into compound sentence

Incorrect Gerardo is waiting for Antoine, and the cable went out.

The writer made no connection between Gerardo, Antoine, and the cable, so combining the two sentences is incorrect.

Punctuation of Compound Sentences

Use correct punctuation when you create a compound sentence. If you join sentences with a *FANBOYS* word (*for*, *and*, *nor*, *but*, *or*, *yet*, *so*), use a comma between the sentences.

Simple sentence	Simple sentence
Gerardo is waiting for Antoine.	Molly is waiting for Marcella.

Here *and* plus a comma joins the two simple sentences into a compound sentence:

Correct Gerardo is waiting for Antoine, and Molly is waiting for Marcella.

Comma between complete thoughts

If you do not join the sentences with a *FANBOYS* word (*for, and, nor, but, or, yet, so*), use a semicolon between the sentences.

Simple sentence	Simple sentence
Gerardo is waiting for Antoine.	Molly is waiting for Marcella.

Here the two simple sentences are combined into a compound sentence using a semicolon:

Correct Gerardo is waiting for Antoine; Molly is waiting for Marcella.

Semicolon between complete thoughts

25.2 Creating Compound Sentences

Directions: Combine the simple sentences below into compound sentences. Use correct punctuation. Answers will vary.

1. Investigators examined bank records. They found irregularities.
2. Mohammad and Monique ate at the Thai restaurant at lunch. They went to a hamburger joint for dinner.
3. The telephone solicitor called three times. I hung up every time.
4. Lincoln wasn't in class today. Monique wasn't in class.
5. The book I needed was checked out. I asked for the librarian to put a hold on it.
6. The pitcher looked as if he was too tired to complete the game. He stayed in.

7. Most people liked the new movie. A few thought it was terrible.
8. Lynn is a vegetarian. Her roommate Kiana is, too.
9. Tracy brought the notes to class. I photocopied them.
10. The veteran was decorated for his actions during the war. His family was proud of him.

Complex Sentences

Another way to increase sentence variety is to create **complex sentences**, which are formed by combining one independent clause and one or more subordinate (dependent) clauses.

Subordinate clauses are introduced by either a **subordinating conjunction** (which links a subordinate clause to an independent clause) or a **relative pronoun** (which relates a group of words to a noun or pronoun).

Commonly Used Subordinating Conjunctions

after	even	since	unless	whereas
although	if	so that	until	whether
as	now	than	when	which
because	once	that	whenever	while
before	provided	though	where	why

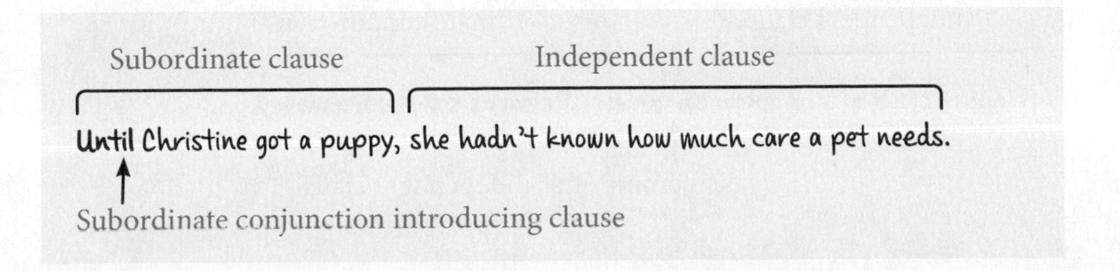

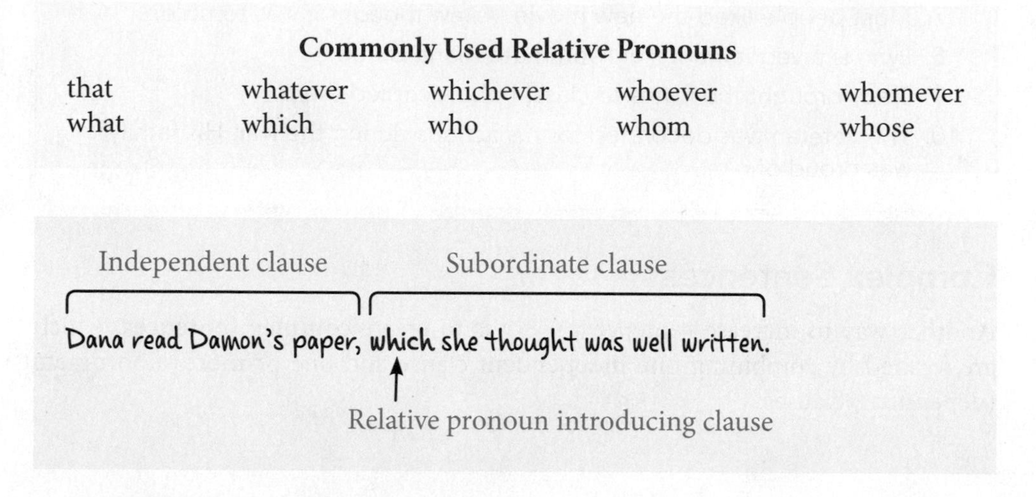

Commonly Used Relative Pronouns

that	whatever	whichever	whoever	whomever
what	which	who	whom	whose

Punctuation with Complex Sentences

When you create a complex sentence, use the correct punctuation:

1. **With subordinate conjunctions:** If you introduce a sentence with a subordinate clause that begins with a subordinating conjunction, use a comma between the subordinate clause and the independent clause.

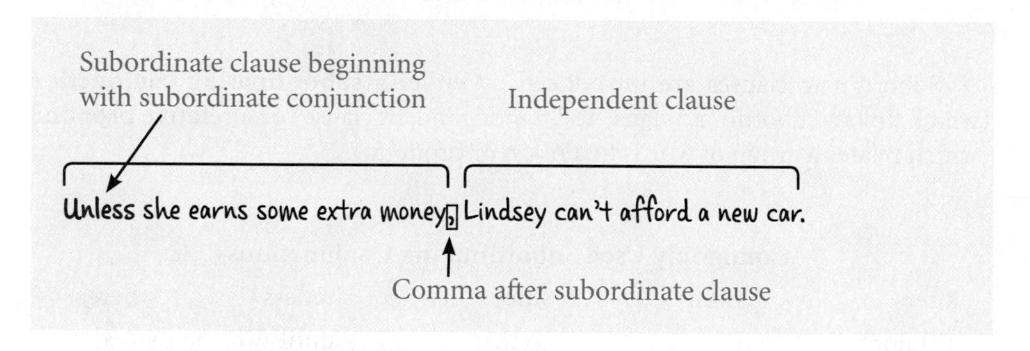

If you end a sentence with a subordinate clause that begins with a subordinate conjunction, use no comma after the independent clause.

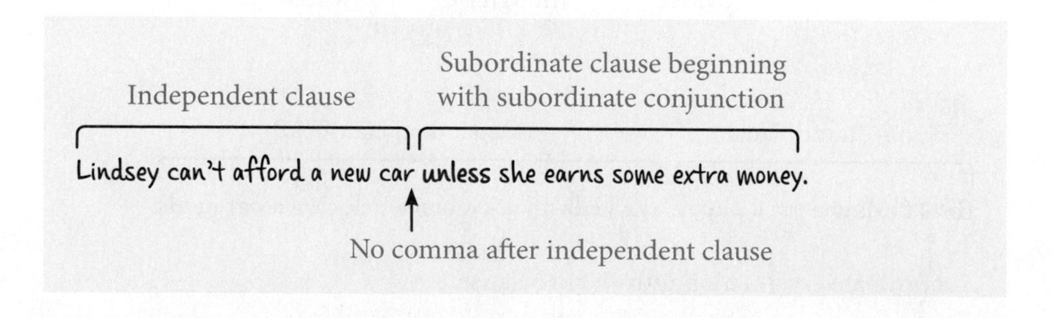

2. **With relative pronouns:** If the subordinate clause is not necessary for the meaning of the sentence, use commas around it.

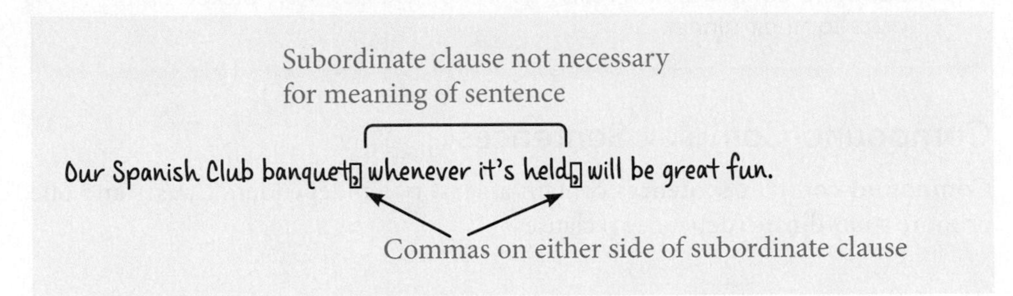

If the subordinate clause is necessary for the meaning of the sentence, use no commas around it.

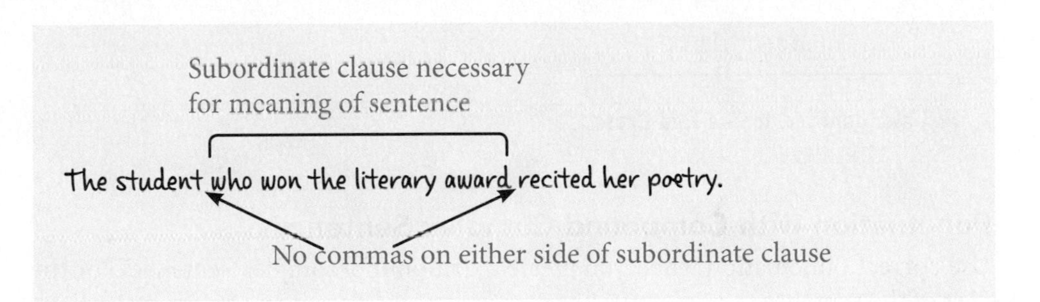

25.3 Creating Complex Sentences

Directions: Use either a subordinating conjunction or relative pronoun to combine the sentences below to create complex sentences. Use correct punctuation. Answers will vary.

1. Anastasia recently returned to campus. She spent last semester studying in Japan.
2. Demetrius prefers outdoor sports. He enrolled in a golf class.
3. Max bought Robyn a goldfish. She named it Traveler.
4. Maurice lost his laptop. He filed a claim with his insurance company.
5. Skylar noticed the e-mail from the company she had interviewed with. She opened the e-mail immediately.
6. Marcel revised his paper three times. His instructor asked for additional changes.
7. I watched a new movie last night. It was hard to follow.
8. Susannah had worked for the company two years. She did not get a raise.

(continued)

9. Rex and Rihanna traveled to the beach. They had not appreciated sunsets.
10. Salvatore completed his reading for the next day. Kurt picked up a pizza for their dinner.

Compound-Complex Sentences

Compound-complex sentences contain at least two independent clauses and one or more subordinate (dependent) clauses.

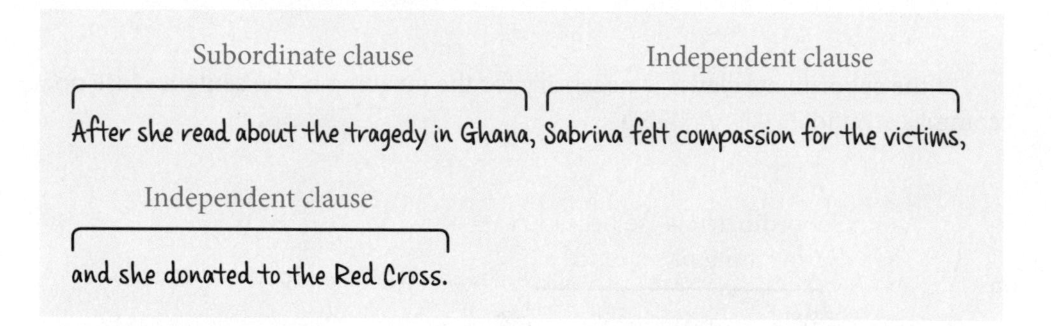

Punctuation with Compound-Complex Sentences

Use correct punctuation when you create a compound-complex sentence. For the part of the sentence that is

complex, use punctuation rules that apply to complex sentences

compound, use punctuation rules that apply to compound sentences

25.4 Creating Compound-Complex Sentences

Directions: Combine the sentences below to create compound-complex sentences. Use correct punctuation. Answers will vary.

1. The rain began. Paige and Pablo had begun to walk across campus. They ran for shelter.
2. Quentin wanted to stay for the rest of the game. His roommates wanted to leave. He needed to ride with them.
3. Tyrone was thirsty. He went to the vending machine. He found five quarters.
4. Savannah and Sawyer stopped at the grocery. They needed a pie. Then they took it to the party.
5. I study flash cards. I won't remember the vocabulary. I won't do well on the test.
6. Marcus went home for the weekend. Tatum stayed on campus. She wanted to go to the rally on Saturday.

7. Here are the notes. I borrowed them from you. Here is your flash drive.
8. Professor Gunderson schedules a test. We should get together to study. We'll have to wait to schedule a time.
9. Today is sunny. We should plan a picnic. The weather is still good.
10. The president of our college was Dr. Lake. He was here for twenty-five years. He was succeeded by Dr. Williams.

Varying Sentence Length

LEARNING GOAL
3 Vary sentence length

Take care to vary the length of your sentences. One short sentence can emphasize a point, but a series of short sentences will sound as if your ideas lack depth, as in this paragraph:

> Phoebe has an important decision to make today. Phoebe had joined the local Big Sisters program. She had been assigned a little sister, Stephanie. Phoebe has grown to be fond of Stephanie. Stephanie wants to go to an afternoon performance. Colonel Kazoo is performing at the arts center. Phoebe wants to take Stephanie to the performance. Phoebe also wants to go to the mall. Phoebe's current roommate at college is named Rochelle. Rochelle has a birthday on Sunday next week. Phoebe wants to get Rochelle a new necklace. Phoebe and Rochelle had seen the necklace Tuesday. They had been together at the mall then.

"Sameness is the mother of disgust, variety the cure."
—Petrarch

Paragraphs with sentences like this can seem as if they are written for a children's book.

This paragraph is monotonous and lacks any pizzazz. One problem is that each of the sentences is just eight words long. If, however, the writer revised the paragraph to vary the length of the sentences, then the paragraph would become stronger and far more interesting. The writer can do this by combining sentences to create a mixture of simple, compound, complex, and compound-complex sentences as shown here:

> Phoebe has an important decision to make. Through the local Big Sisters program, Phoebe has a little sister, Stephanie, of whom she is very fond. Stephanie wants to go to the local arts center for an afternoon performance by Colonel Kazoo. While Phoebe wants to take her, she also wants to go to the mall because her roommate, Rochelle, has a birthday next week. Phoebe wants to get Rochelle a necklace that they had seen when they were at the mall on Tuesday.

The writer combines simple sentences into compound, complex, and compound-complex sentences.

The revised paragraph is much stronger because the length of the sentences varies. The writer begins with a short, simple sentence and then combines other sentences so they no longer have the same number of words.

25.5 Varying Sentence Length

Directions: Combine the sentences below to create a paragraph with varying sentence length. Answers will vary.

My friend Kelsey worked at Lion's. That is a local restaurant. It specializes in American food, like burgers and steaks. Kelsey was part of the wait staff at Lion's. Kelsey's boss is named Frank Galloway. Kelsey was waiting on a group of people. They were eating at Lion's. This happened last Thursday. One man in the group didn't like his food. He told Kelsey his steak had a funny taste. Kelsey told him he was drunk. She wouldn't take his food back to the kitchen. The man said he hadn't had anything to drink. The man complained to Mr. Galloway. Mr. Galloway told Kelsey she was wrong. Kelsey thought about it all the next day. Then she went on Facebook. She wrote about the customer. She told what Mr. Galloway had done. She called Mr. Galloway and the customer several vulgar names. Mr. Galloway heard about this. He fired Kelsey.

Varying Sentence Beginnings

LEARNING GOAL
❹ Vary sentence beginnings

Novice essay writers often fall into the habit of beginning their sentences the same way, as in the paragraph below:

The two soap operas had characters with similar traits. The characters wore only designer clothes. The characters all were good-looking. The main male character was attracted to two different women. The main female character had her eyes on another man. The new characters had some unexplained secrets. The other characters had been married at least five times.

Heads Up!
As you revise, read your sentences out loud and listen for repetition.

All these sentences begin with *the*, followed closely by the subject of the sentence and then the verb. This makes the sentences choppy and flavorless. By varying the way the sentences begin, the writer could improve the problem paragraph, perhaps with this wording:

The two soap operas were similar. Both were set in a small town near a large city. All their characters were good-looking and wore only designer clothes. While the main male character was attracted to two different women, the main female character had her eyes on a different man. Characters who were new to the show had some unexplained secrets, and characters who had been around many years had been married at least five times.

Rearranging words, phrases, or clauses so that you change the subject-verb construction will make your writing more creative and will give you ways to better emphasize certain ideas. You can revise sentence beginnings in a variety of ways.

Rearrange Adjectives, Adverbs, and Prepositional Phrases

One way to vary sentence beginnings is to move adjectives, adverbs, and prepositional phrases to the beginnings of sentences.

1. **Adjectives:** Rearrange sentences to begin with an **adjective**, a word that describes a noun or pronoun; it answers the question *Which one? What kind?* or *How many?*

Describes a noun; answers the question "Which one?"
↓
The impatient candidate fidgeted until it was her turn to speak.

Impatient, the candidate fidgeted until it was her turn to speak.

2. **Adverbs:** Rearrange sentences to begin with an **adverb**, a word that describes a verb, adjective, or another adverb; it answers the question *How? When? Where? Why? Under what circumstances? How much? How often?* or *To what extent?*

Describes a verb; answers the question "How?"
↓
Darnell and Alyssa will nervously watch the finals of the tournament.

Nervously, Darnell and Alyssa will watch the finals of the tournament.

3. **Prepositional phrases:** Rearrange sentences to begin with a **prepositional phrase**, a group of words that begins with a preposition and ends with a noun or pronoun.

Begins with a preposition; ends with a noun

Donte left his apartment in a hurry and he forgot to blow out the candles.

In a hurry, Donte left his apartment and he forgot to blow out the candles.

25.6 Varying Sentence Beginnings: Adjectives, Adverbs, and Prepositional Phrases

Directions: Rewrite the sentences below to begin with an adjective, an adverb, or a prepositional phrase. Answers will vary.

1. The star thought he was entitled to extra privileges because he was so well known.
   Because he was so well known, the star thought he was entitled to extra privileges.
2. The campers, tired and confused, arrived home after midnight.
   Tired and confused, the campers arrived home after midnight.
3. Chantal and Kendrick left yesterday to spend the weekend at home.
   To spend the weekend at home, Chantal and Kendrick left yesterday.
4. We bought extra food in anticipation of the oncoming snowstorm.
   In anticipation of the oncoming snowstorm, we bought extra food.
5. Four earthquakes, which were sudden and frightening, rocked our area.
   Sudden and frightening, four earthquakes rocked our area.
6. The police investigated the crime scene quickly and thoroughly.
   Quickly and thoroughly, the police investigated the crime scene.
7. This is Brook's best paper by far.
   By far, this is Brook's best paper.
8. Breck's dog Bowser, frisky and hungry, ran to greet his owner.
   Frisky and hungry, Breck's dog Bowser ran to greet his owner.
9. Area restaurants were full on the eve of classes starting.
   On the eve of classes starting, area restaurants were full.
10. Jonah clearly put no time into his peer review.
    Clearly, Jonah put no time into his peer review.

Rearrange Adverb Clauses, Noun Clauses, and Infinitive Phrases

Another way to vary sentence beginnings is to rearrange sentences so they begin with an adverb clause, noun clause, or infinitive phrase.

1. **Adverb clause:** Rearrange sentences to begin with an **adverb clause**, a group of words that has a subject and verb and cannot stand alone as a sentence; it answers the question *When? Where? How? Why? To what extent? With what goal or result?* or *Under what condition or circumstances*?

Adverb clause answering the question "When?"

I realized I'd studied the wrong chapter **when I heard Professor Green review for the test.**

When I heard Professor Green review for the test, I realized I'd studied the wrong chapter.

2. **Noun clause:** Rearrange sentences to begin with a **noun clause**, a group of words that has a subject and predicate that cannot stand alone as a sentence and that acts as a noun; it answers the question *Who? Whom?* or *What?*

Noun clause answering the question "What?"

A real eye-opener for me was **what we learned in biology class today.**

What we learned in biology class today was a real eye-opener for me.

3. **Infinitive phrase:** Rearrange sentences to begin with an **infinitive phrase**, a group of words that consists of *to* plus a verb and all words that modify it; it usually acts as a noun but can sometimes act as an adjective or adverb.

Infinitive phrase ("to" plus the verb "return," and words that modify it)

Press the pound sign **to return this call.**

To return this call, press the pound sign.

25.7 Varying Sentence Beginnings: Adverb Clauses, Noun Clauses, and Infinitive Phrases

Directions: Rewrite the sentences below to begin with an adverb clause, a noun clause, or an infinitive phrase. Answers may vary.

(continued)

1. You'll find Serina wherever Jason is.

   Wherever Jason is, you'll find Serina.

2. Tia and Matthew drove to the bookstore to finish buying their supplies for the semester.

   To finish buying their supplies for the semester, Tia and Matthew drove to the bookstore.

3. One nice thing about my roommate and me is that we get along so well.

   That we get along so well is one nice thing about my roommate and me.

4. Lily and Oliver signed up for a night class so they could advance their careers.

   So they could advance their careers, Lily and Oliver signed up for a night class.

5. Jorge took several deep breaths to calm himself before he shot the free throw.

   To calm himself before he shot the free throw, Jorge took several deep breaths.

6. Miguel's main point is that certain brands of coffee are overpriced.

   That certain brands of coffee are overpriced is Miguel's main point.

7. Lexie made an appointment with her advisor to plan next semester's schedule.

   To plan next semester's schedule, Lexie made an appointment with her advisor.

8. Tianna found many hits for her research topic when she went online.

   When she went online, Tianna found many hits for her research topic.

9. The reference librarian can help whoever has questions.

   Whoever has questions can be helped by the reference librarian.

10. Tyrell realized he needed to revise his paper, although he had hoped to join friends for pizza.

    Although he had hoped to join friends for pizza, Tyrell realized he needed to revise his paper.

Rearrange Present Participial Phrases, Past Participial Phrases, and Gerund Phrases

Sometimes you may need to reword slightly.

A third way to vary sentence beginnings is to rearrange sentences so they begin with a present participial phrase, a past participial phrase, or a gerund phrase.

1. **Present participial phrase:** Rearrange sentences to begin with a present participial phrase. A **present participle** is a verb plus *–ing*; it functions as an adjective, so it modifies a noun or pronoun and answers the question *Which one? What kind?* or *How many?* A **present participial phrase** is a present participle and any words that modify it.

Present participial phrase describing "student"

The student felt nervous **asking for an extension on the deadline.**

Asking for an extension on the deadline, the student felt nervous.

2. **Past participial phrase:** Rearrange sentences to begin with a past participial phrase. A **past participle** is a verb plus an ending like *-ed, -d, -t, -en,* or *-n*; it functions as an adjective and answers the question *Which one? What kind?* or *How many?* A **past participial phrase** is a past participle and any words that modify it.

Both present and past participial phrases act only as adjectives

Past participial phrase describing "newspaper"

The school newspaper, **written by a team of twenty journalists,** won first place.

Written by a team of twenty journalists, the school newspaper won first place.

3. **Gerund phrase:** Rearrange sentences to begin with a gerund phrase. A **gerund** is a verb form ending in *–ing* that functions as a noun. A **gerund phrase** consists of a gerund and any words that modify it.

Gerund phrase

One of Eli's hobbies is **working Sudoku puzzles.**

Working Sudoku puzzles is one of Eli's hobbies.

25.8 Varying Sentence Beginnings: Present Participial Phrases, Past Participial Phrases, and Gerund Phrases

Directions: Rewrite the sentences below to begin with a present participial phrase, a past participial phrase, or a gerund phrase. You may need to reword slightly. Answers may vary.

(continued)

1. *To Kill a Mockingbird* remains a bestseller more than fifty years after it was published. It is studied by many.

   Studied by many, *To Kill a Mockingbird* remains a bestseller more than fifty years after it was published.

2. Elizabeth Cady Stanton's desires included abolishing slavery and gaining suffrage for women.

   Abolishing slavery and gaining suffrage for women were two of Elizabeth Cady Stanton's desires.

3. The coffee spread all over Diego's newly printed paper. The coffee was leaking from the cup.

   Leaking from the cup, the coffee spread all over Diego's newly printed paper.

4. The rescued miners rushed to the dinner tent. They were crazed with hunger.

   Crazed with hunger, the rescued miners rushed to the dinner tent.

5. I enjoy walking in the rain.

   Walking in the rain is something I enjoy.

6. Carson learned that finding time for studying and for working left little time for socializing. He is a college student.

   Being a college student, Carson learned that finding time for studying and for working left little time for socializing.

7. Take extra time when you're driving on ice.

   Driving on ice means you should take extra time.

8. Magdalena was exhausted from working at the daycare center. She took a nap.

   Exhausted from working at the daycare center, Magdalena took a nap.

9. Sherry ran out into the rain. She dared Noelle to join her.

   Daring Noelle to join her, Sherry ran out into the rain.

10. You shouldn't read the second book in the series before you read the first. It will confuse you.

    Reading the second book in the series before you read the first will confuse you.

Sentences That Begin with *There Are* or *There Is*

Sentences that begin with *There are* or *There is* are often weak. Rewording these sentences makes them more concise and more powerful. To reword sentences, first eliminate *There are* or *There is*. Then find the subject and verb of the sentence and rearrange the remaining words so that the construction becomes stronger.

Weak, wordy	~~There is~~ evidence that shows that the earth's climate is changing.
Concise, powerful	Evidence shows that the earth's climate is changing.
Weak, wordy	~~There are~~ many reasons I can think of for revising my paper.
Concise, powerful	I can think of many reasons for revising my paper.

25.9 Varying Sentence Beginnings: There Are, There Is

Directions: Rewrite the sentences below to eliminate beginning with *There are* or *There is*. You may need to reword slightly. Answers may vary.

1. There is one person whom Mason feels he can speak privately with.
   Mason feels he can speak privately with one person.
2. There are many ways you can examine this problem.
   You can examine this problem in many ways.
3. There are three articles in the book that I may quote.
   I may quote three articles in the book.
4. There is only one brand of car that Rosa is interested in buying.
   Rosa is interested in buying only one brand of car.
5. There are many students milling about in the back parking lot.
   Many students are milling about in the back parking lot.
6. There is something about Lorenzo's accent that I find fascinating.
   I find something about Lorenzo's accent fascinating.
7. There are many excuses Chantal uses for continually being late for work.
   Chantal uses many excuses for continually being late for work.
8. There is one main lesson I learned in math class today.
   I learned one main lesson in math class today.
9. There are two television shows I never miss.
   I never miss two television shows.
10. There is a mysterious package waiting for Terrence.
    A mysterious package is waiting for Terrence.

25.10 Varying Sentence Beginnings

Directions: Rewrite the sentences below to create a paragraph with varying sentence beginnings. Use correct punctuation. Answers will vary.

George Anderson's nickname was "Sparky." Anderson got that nickname when he played minor league baseball. Anderson played one season of major league baseball. Anderson played with the Philadelphia Phillies in 1959. Anderson played second base. Anderson hit .218. Anderson began managing minor league teams. Anderson was so good he became manager of the Cincinnati Reds in 1970. Anderson led the Reds to two consecutive World Series titles. Anderson's Reds won the World Series in 1975 and 1976. Anderson left the Reds in 1978. Anderson became manager of the Detroit Tigers in 1979. Anderson's Tigers won the World Series in 1984. Anderson became the first manager to win championships in both the American and National leagues. Anderson was inducted into the Baseball Hall of Fame in 2000. Anderson died in 2010.

TECHNO TIP

For more on sentence variety, search the Internet for these videos:

Sentence Variety Lecture Part 1

Sentence Variety Lecture Part 2

MyWritingLab™ Visit ***MyWritingLab.com*** and complete the exercises and activities in the **Varying Sentence Structure** topic area.

RUN THAT BY ME AGAIN

LEARNING GOAL

1. Identify primary sentence functions

LEARNING GOAL

2. Use four sentence types to vary sentence construction

- **Sentence variety is concerned with adjusting the way sentences are structured, which keeps work from being . . .** monotonous.
- **Advanced writers vary . . .** the length, format, and rhythm of their sentences.
- **Sentence types are classified according to . . .** their *function* and their *construction.*
- **Simple sentences have . . .** one independent clause and no subordinate clauses.
- **Compound sentences have . . .** at least two independent clauses and no subordinate clauses.
- **Complex sentences have . . .** one independent clause and one or more subordinate clauses.
- **Compound-complex sentences have . . .** at least two independent clauses and one or more subordinate clauses.
- **Complex sentences have . . .** subordinate clauses introduced by a subordinate conjunction or a relative pronoun.

- **A series of short sentences will sound as if . . .** your ideas lack depth.
- **Varying the length of sentences makes . . .** writing stronger and more interesting.
- **Writers can vary the length of their sentences by . . .** combining sentences to create a mixture of simple, compound, complex, and compound-complex sentences.
- **A series of sentences that begin in the same way should be . . .** reworded so that they become less monotonous.
- **Sentences that begin with *There are* or *There is* can often be . . .** reworded to become more concise and powerful.

LEARNING GOAL
❸ Vary sentence length

LEARNING GOAL
❹ Vary sentence beginnings

SENTENCE VARIETY LEARNING LOG MyWritingLab™

Complete this Exercise

MyWritingLab™

Answer the questions below to review your mastery of sentence variety.

1. Sentence types are classified in what two ways?

   Sentence types are classified according to their function and their construction.

2. What are the four types of sentence functions?

   The four types of sentence functions are declarative, interrogative, imperative, and exclamatory.

3. What are the four types of sentence constructions?

   The four types of sentence constructions are simple, compound, complex, and compound-complex.

4. What is a simple sentence?

   A simple sentence has one independent clause and no subordinate clauses.

5. What is a compound sentence?

   A compound sentence has at least two independent clauses and no subordinate clauses.

6. What is a complex sentence?

   A complex sentence has one independent clause and one or more subordinate clauses.

7. What is a compound-complex sentence?

   A compound-complex sentence has at least two independent clauses and one or more subordinate clauses.

8. What is one drawback to writing a series of short sentences?

   A series of short sentences will sound as if the material lacks depth.

9. What can you do to vary sentence length?

   Combine sentences to create a mixture of simple, compound, complex, and compound-complex sentences.

10. To vary sentence beginnings, what types of words, phrases, or clauses might you use?

You might begin a sentence with an adjective, an adverb, or a prepositional phrase; an adverb or a noun clause; or an infinitive phrase, a present participial phrase, a past participial phrase, or a gerund phrase.

11. How do you reword a sentence that begins with *There are* or *There is*?

Eliminate *There are* or *There is*; then find the subject and verb of the sentence and rearrange the remaining words.

CHAPTER 26

Misplaced and Dangling Modifiers

LEARNING GOALS

In this chapter, you'll learn and practice how to

1. Identify misplaced modifiers
2. Identify and correct one-word misplaced modifiers
3. Identify and correct misplaced phrases and clauses
4. Identify and correct dangling modifiers

GETTING THERE

- Misplaced modifiers are words or groups of words that are located in of the wrong place in a sentence.
- Dangling modifiers are words or groups of words that do not clearly relate to another word in a sentence.
- Misplaced and dangling modifiers create sentences that are unclear, mangled, or even humorous.

What Are Misplaced Modifiers?

LEARNING GOAL

1. Identify misplaced modifiers

Modifiers are words, phrases, or clauses that give information about other parts of a sentence. They add precision and interest, and without them writing would be bland.

Misplaced modifiers are words, phrases, or clauses that are located in the wrong place in a sentence. Because what they modify (describe) in the sentence is unclear, the meaning of the sentence becomes unclear, or mangled, or even humorous. To prevent this, writers should be careful to place modifiers close to the words the modifiers describe.

Consider, for example, the phrase *peering through the window* in the following sentence:

Peering through the window, Jorge saw the man.

Here Jorge is peering.

In this sentence, the phrase is in a different position and it has quite a different meaning:

Jorge saw the man peering through the window.

Now the man is peering.

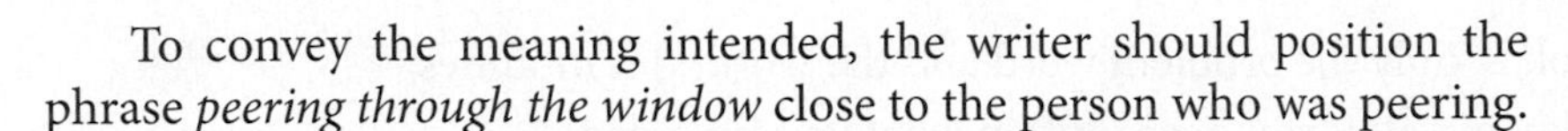

To convey the meaning intended, the writer should position the phrase *peering through the window* close to the person who was peering.

Misplaced Single-Word Modifiers

A single-word misplaced modifier

LEARNING GOAL

❷ Identify and correct one-word misplaced modifiers

Misplacing a single word sometimes creates a problem.

*Mike has **nearly read** all of the assigned articles.*

Mike hasn't nearly read the articles.

*After his race at the track meet, Chad **almost drank** a gallon of water.*

Chad didn't almost drink the water.

Place the problem word immediately before the word or words it modifies (describes)

In these examples, moving a single word makes the sentences correctly state what the writer intended.

*Mike has read **nearly all** of the assigned articles.*

*After his race at the track meet, Chad drank **almost a gallon** of water.*

These single words (sometimes called *limiting modifiers*) often create problems:

These words are also known as **simple modifiers**.

almost	hardly	only
even	just	scarcely
exactly	nearly	simply

If you're not sure that you have placed a word correctly

- look at only the problem word and the word(s) it modifies
- ask yourself if you have written what you intended

Eliza was concerned about whether she had placed the word *just* correctly in the following sentence, so she applied the fix.

Just I learned the vocabulary.

Do I mean that I was the only one who learned the vocabulary?

I just learned the vocabulary.

Did I mean that I recently learned the vocabulary?

I learned just the vocabulary.

Because Eliza meant that the vocabulary was the only thing she learned, this is the way she should word her sentence.

The Worst Offender

The word that most often creates a problem is *only*. Consider these three sentences:

Only Carol studied in the lounge in Sheffer Hall.

Carol was the only person who studied in the lounge in Sheffer Hall.

Carol only studied in the lounge in Sheffer Hall.

The only thing Carol did in the lounge in Sheffer Hall was study.

Carol studied only in the lounge in Sheffer Hall.

The only place Carol studied was in the lounge in Sheffer Hall.

Carol studied in the only lounge in Sheffer Hall.

Sheffer Hall has only one lounge, and that's where Carol studied.

TECHNO TIP

For a quiz about limiting modifiers, search the Internet for this site: Grammatically Correct: Misplaced Limiting Modifiers

Moving the word *only* changes the meaning of each sentence, so positioning it correctly is vital.

Another Single-Word Problem

Incorrect placement of the adverb *not* is another frequent problem. The following sentences convey different meanings when the word *not* is moved:

Because of the flood, not all the students were in class on Tuesday.

Some students were in class.

Because of the flood, all the students were not in class on Tuesday.

No students were in class.

TICKET to WRITE

26.1 Correcting Misplaced Words

Directions: Read each question and the two sentences that follow it; then indicate which of the two sentences is correct.

__b__ 1. If you mean some chairs are uncomfortable, which is correct?

a. All the chairs in the library are not comfortable.

b. Not all the chairs in the library are comfortable.

__a__ 2. If you mean you passed some classes, which is correct?

a. I passed almost every dance class I enrolled in.

b. I almost passed every dance class I enrolled in.

__a__ 3. If you mean you didn't have a crop other than tomatoes, which is correct?

a. I grew only tomatoes.

b. I only grew tomatoes.

__b__ 4. If you mean the quarterback didn't wave to anyone except the halfback, which is correct?

a. The quarterback just waved to the halfback.

b. The quarterback waved just to the halfback.

__b__ 5. If you mean that David has done most of the driving, which is correct?

a. David has nearly driven all the time we've been on the road.

b. David has driven nearly all the time we've been on the road.

__b__ 6. If you mean you saw many strangers in the lounge, which is correct?

a. I hardly knew anyone in the student lounge.

b. I knew hardly anyone in the student lounge.

(continued)

__a__ 7. If you mean the reason Victor went to class was to get Jessica's notes, which is correct?

a. On Tuesday, Victor went to class only to get Jessica's notes.

b. On Tuesday, only Victor went to class to get Jessica's notes.

__b__ 8. If you mean you learned the life story of most people, which is correct?

a. In the first meeting of our study group, I almost learned everyone's life story.

b. In the first meeting of our study group, I learned the life story of almost everyone.

__b__ 9. If you mean you were almost late, which is correct?

a. Because of taking the detour, we made it to class scarcely on time.

b. Because of taking the detour, we scarcely made it to class on time.

__a__ 10. If you mean you're not sure of the moment you understood the concept, which is correct?

a. I can't say exactly when the math concept clicked and I understood it.

b. I can't exactly say when the math concept clicked and I understood it.

Misplaced Phrases and Clauses

LEARNING GOAL

3 Identify and correct misplaced phrases and clauses

THE PROBLEM

A misplaced phrase or clause

If they are in the wrong place in a sentence, phrases and clauses can also cause problems.

Misplaced phrase

Cathy and David said the professor was riding a bike wearing a Scottish kilt.

The bike wore a Scottish kilt? **(What an odd bike.)**

Misplaced clause

Tony and Margaret discussed the nightlife they'd enjoy at their destination while flying.

Their destination was flying? **(Destinations can't fly.)**

THE FIX

Place the modifying phrase or clause as close as possible to what it modifies (describes)

Corrected phrase

Cathy and David said the professor wearing a Scottish kilt was riding a bike.

The professor wore a Scottish kilt. (That makes sense.)

Corrected clause

While flying, Tony and Margaret discussed the nightlife they'd enjoy at their destination.

Tony and Margaret are flying. (That makes sense.)

TECHNO TIP

For quizzes about misplaced modifiers, search the Internet for these sites:

Misplaced Modifiers: Exercise 1

Misplaced Modifiers: Exercise 2

26.2 Correcting Misplaced Phrases and Clauses

Directions: Rewrite the sentences below to eliminate misplaced phrases and clauses. Answers may vary.

1. When following a low-carb diet, rapid weight loss is experienced by many people.

   When following a low-carb diet, many people experience rapid weight loss.

2. Martha drove to the grocery store after her husband left for class to get some milk and eggs.

   After her husband left for class, Martha drove to the grocery store to get some milk and eggs.

(continued)

3. I found a twenty-dollar bill looking at the ground while crossing campus.
   While crossing campus and looking at the ground, I found a twenty-dollar bill.
4. Several students watched the building on campus that was being demolished.
   Several students watched the building that was being demolished on campus.
5. Discovered in 2010, archaeologists call the skeleton "Big Man."
   Archaeologists call the skeleton, which was discovered in 2010, "Big Man."
6. I gave the soda to my friend that I got out of the vending machine.
   I gave the soda that I got out of the vending machine to my friend.
7. After giving birth to three puppies, Jeannie was proud of her dog Lulu.
   Jeannie was proud of her dog Lulu after Lulu gave birth to three puppies.
8. Having studied for three hours, the questions on the test seemed easy to me.
   The questions on the test seemed easy to me, having studied for three hours.
9. Nick saw an accident driving to class.
   While driving to class, Nick saw an accident.
10. Receiving antibiotics, my symptoms improved almost immediately.
    My symptoms improved almost immediately after I received antibiotics.

What Are Dangling Modifiers?

LEARNING GOAL
4 Identify and correct dangling modifiers

Still another problem comes in the form of a **dangling modifier**, a word or group of words that does not clearly relate to another word in a sentence. Often the word the modifier should describe is missing from the sentence, so rewording the sentence becomes necessary.

Heads Up!
Dangling modifiers often come in the beginning of sentences.

THE PROBLEM

A dangling modifier

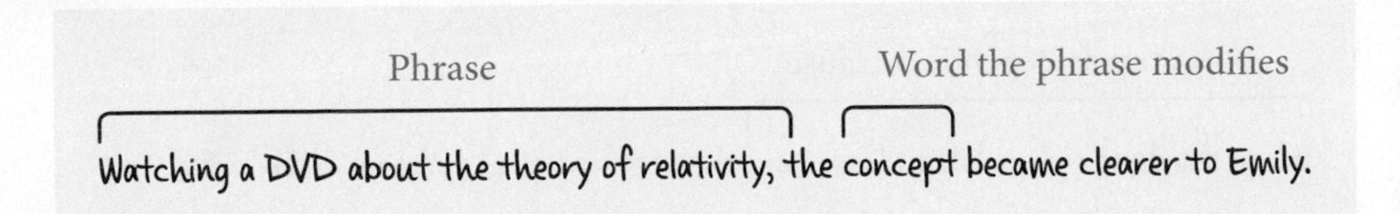

The way the sentence is worded, a concept watched a DVD. Because that makes no sense, the phrase *watching a DVD about the theory of relativity* is a dangling modifier, and the sentence should be reworded.

Phrase — Word the phrase modifies

Watching a DVD about the theory of relativity, **Emily found the concept clearer.**

Change either the modifier or the rest of the sentence

Fix #1: Leave the modifier and rewrite the rest of the sentence. Determine the word to which the modifier should refer, and rewrite the sentence so this word is the subject of the main clause of the sentence.

Fix #2: Rewrite the modifier so that it becomes a dependent clause (with a subject and a verb). Look at the following example:

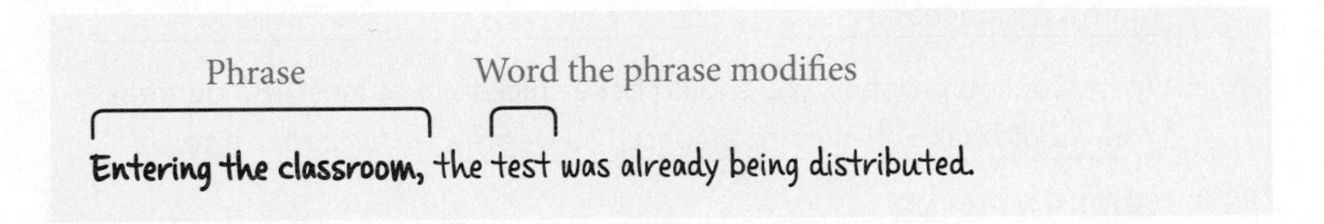

With this wording, a test was entering a classroom—clearly not what the writer intended. By rewording, the writer repairs the dangling modifier.

Using Fix #1: The writer could revise the sentence in the following way, making the word being modified (Marlene) the subject of the main clause of the sentence:

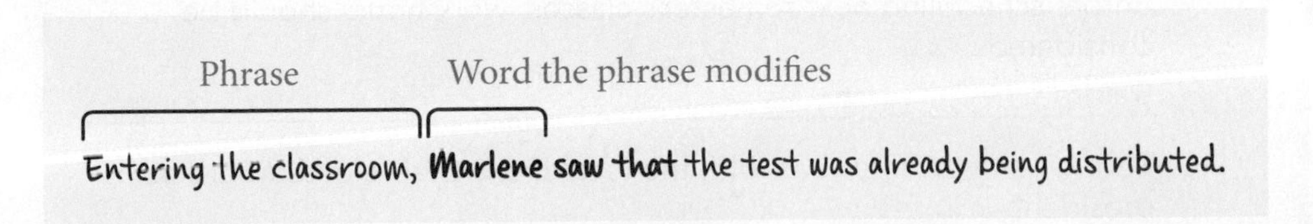

Using Fix #2: The writer could revise the sentence by rewriting the modifier as a dependent clause with a subject and a verb.

TECHNO TIP

For a quiz about dangling modifiers, search the Internet for this site:
Dangling Modifier Exercise

Rewritten as dependent clause

When Marlene entered the classroom, the test was already being distributed.

26.3 Correcting Dangling Modifiers

Directions: Rewrite the sentences below to eliminate dangling modifiers.
Answers may vary.

1. When in elementary school, phoning someone of the opposite sex was taboo.
   When we were in elementary school, phoning someone of the opposite sex was taboo.
2. Having arrived late to the movie, help was needed to find my friends.
   Having arrived late to the movie, I needed help to find my friends.
3. To improve Katherine's grade, the test was retaken.
   Katherine retook the test to improve her grade.
4. Driving with the top down, the essay flew away.
   As Jaci was driving with the top down, her essay flew away.
5. Before taking a test, reading the test material is necessary.
   Before taking a test, students should realize that reading the test material is necessary.
6. To be cooked properly, you should bake chicken to at least 180 degrees.
   You should bake chicken to at least 180 degrees in order for it to be cooked properly.
7. While watching, the professor tripped and fell and the class gasped.
   The class gasped when they watched the professor trip and fall.
8. After giving blood, the doctor told me to relax.
   The doctor told me to relax after I gave blood.
9. Before scheduling next semester's classes, work hours should be considered.
   Before scheduling next semester's classes, consider your work hours.
10. Though only twenty years old, the local Lions Club elected Barb as its president.
    The local Lions Club elected Barb as its president, even though she is only twenty years old.

TECHNO TIP

For more on misplaced or dangling modifiers search the Internet for these videos:
Grammar & Punctuation: What Is a Dangling Modifier?
Dangling Modifiers

MyWritingLab™ Visit *MyWritingLab.com* and complete the exercises and activities in the **Modifiers** and **Misplaced and Dangling Modifiers** topic areas.

RUN THAT BY ME AGAIN

- **Modifiers are words, phrases, or clauses that give . . .** information about other parts of a sentence.
- **Misplaced modifiers are words, phrases, or clauses that are . . .** located in the wrong place in a sentence.
- **Misplaced modifiers are problems because with them . . .** the meaning of the sentence is unclear, or mangled, or even humorous.
- **To prevent misplaced modifiers, writers should be careful to . . .** place modifiers close to the words the modifiers describe.
- **To repair single-word misplaced modifiers, writers should . . .** place the problem word immediately before the word or words it modifies.
- **The word that creates a problem most often is . . .** *only*.
- **To repair misplaced phrases or clauses, writers should . . .** place the modifying phrase or clause as close as possible to what it modifies.
- **A dangling modifier is a word or group of words that . . .** does not clearly relate to another word in a sentence.
- **To repair dangling modifiers, writers should . . .** change either the modifier or the rest of the sentence.

LEARNING GOAL
❶ Identify misplaced modifiers

LEARNING GOAL
❷ Identify and correct one-word misplaced modifiers

LEARNING GOAL
❸ Identify and correct misplaced phrases and clauses

LEARNING GOAL
❹ Identify and correct dangling modifiers

MISPLACED OR DANGLING MODIFIERS LEARNING LOG MyWritingLab™

Complete this Exercise

MyWritingLab™

Answer the questions below to review your mastery of misplaced or dangling modifiers.

1. **What are misplaced modifiers?**
   Misplaced modifiers are words, phrases, or clauses that are located in the wrong place in a sentence.
2. **Because misplaced modifiers do not modify (describe) the proper part of a sentence, what happens?**
   Because misplaced modifiers do not modify (describe) the proper part of a sentence, the meaning of the sentence is unclear, or mangled, or even humorous.
3. **What should you do to prevent a problem with misplaced modifiers?**
   To prevent a problem with misplaced modifiers, you should be careful to place modifiers close to the words the modifiers describe.
4. **What should you do to correct a single-word misplaced modifier?**
   To correct a single-word misplaced modifier, you should place the problem word immediately before the word or words it modifies.
5. **What word most often creates a problem in misplaced modifiers?**
   The word that most often creates a problem is *only*.
6. **How can you repair a problem with a misplaced modifying phrase or clause?**
   To repair a misplaced modifying phrase or clause, you can place the phrase or clause as close as possible to what it modifies.
7. **What is a dangling modifier?**
   A dangling modifier is a word or group of words that does not clearly relate to another word in a sentence.
8. **How can you repair a dangling modifier?**
   You can repair a dangling modifier by changing either the modifier or the rest of the sentence.

CHAPTER 27

Word Choice

LEARNING GOALS

In this chapter, you'll learn and practice how to

1. Recognize and eliminate wordiness
2. Recognize and eliminate vague language
3. Recognize and eliminate clichés

GETTING THERE

- Wordiness bloats writing and weakens the points a writer is striving to make.
- Vagueness prevents writers from presenting a clear picture of what they want to say.
- Clichés prevent writing from being original.

Avoid Wordiness

When you write well, you present your points concisely, without extra words that are unnecessary.

LEARNING GOAL

1. Recognize and eliminate wordiness

Redundancy (You've Said It Before; Don't Say It Again)

One example of wordiness is **redundancy**, the excessive or unnecessary repetition of an idea. Redundant writing not only bores readers, but it also weakens what a writer is saying. Look at these sentences:

If you want two fifty-cent synonyms for *redundancy*, try *tautology* and *pleonasm*.

Redundant No one disputes the true fact that the sun rises in the east.

Redundant The two suspects met at precisely 10:00 a.m. in the morning.

Redundant Does the bookstore have an ATM machine?

The sentences above all have some extra weight and need to go on a "word diet"; each contains an example of *redundancy*. Look at them again to determine why they are redundant.

Redundant No one disputes the true fact that the sun rises in the east.

By definition, a fact must be true, so *true fact* is redundant. If you eliminate the word *true*, the sentence is no longer redundant.

Correct	No one disputes the fact that the sun rises in the east.
Redundant	The two suspects met at precisely 10:00 a.m. in the morning.

Both *a.m.* and *in the morning* mean the same thing, so the writer should correct the redundancy by using only one of them.

Correct	The two suspects met at precisely 10:00 a.m.
Redundant	Does the bookstore have an ATM machine?

Heads Up!
Redundant quotes by famous people include these:
"I never make predictions, especially about the future."
—Samuel Goldwyn
"It's déjà vu all over again."
—Yogi Berra

The acronym *ATM* stands for *Automated Teller Machine*, so this question is asking about an *Automated Teller Machine machine*. The writer should use only *ATM*.

Correct	Does the bookstore have an ATM?

Below is a list of redundant phrases. These phrases are so commonly heard and used that writers often fail to notice their pointless repetition (their redundancy). Examine the entries and determine which phrases you have used. Note that you can remove the redundancy by eliminating the word or words in parentheses.

Redundant Phrases

(absolutely) certain	(clearly) evident	end (result)
(already) exist	combine (together)	evolve (over time)
(alternative) choice	completely (surround)	(exact) same
(anonymous) stranger	connect (together)	(favorable) approval
blend (together)	could (possibly)	filled (to capacity)
blue (in color)	depreciate (in value)	(final) result
(careful) scrutiny	descend (down)	first (and foremost)
cash (money)	(desirable) benefits	(first) began
chili con carne (with meat)	each (and every)	follow (after)
	earlier (in time)	(foreign) imports
circle (around)	eliminate (altogether)	(free) gift
classify (into groups)	emergency (situation)	fuse (together)

(future) plans
gather (together)
(general) consensus
HIV (virus)
(illustrated) drawing
(important) essentials
join (together)
(joint) collaboration
lag (behind)
large (in size)
lift (up)
may (possibly)
meet (together)
mix (together)
more (and more)
(necessary) requirement
never (before)
none (at all)
nostalgia (for the past)
(now) pending
(old) custom
(ongoing) evolution
over (and over)
(overused) cliché
(past) experience
period (of time)
(personal) friend
pick (and choose)
PIN (number)
plan (ahead)
point (in time)
(possibly) may
postpone (until later)
protest (against)
pursue (after)
RAM (memory)
refer (back)
repeat (again)
(rough) estimate
scrutinize (in detail)
share (together)
small (in size)
soft (to the touch)
start (off)
(still) remains
(sum) total
surrounded (on all sides)
tall (in height)
ten (in number)
(tiny) bit
total (number)
(ultimate) goal
(unexpected) emergency
unite (together)
UPC (code)
(usual) custom
(valuable) asset
visible (to the eye)
warn (in advance)
while (at the same time)
write (down)

TICKET to WRITE

27.1 Eliminating Redundant Words or Phrases

Directions: Cross out unneeded words or phrases in the sentences below to eliminate redundancy. Answers will vary.

1. We started out poorly, but now we've developed a mutual respect ~~for each other~~.
2. With enough ~~advance~~ planning, I can finish my reading for English class and still have time to get to the ball game.
3. The emergency ~~situation~~ called for the class to go to the basement until sirens stopped sounding.
4. While our class was assembled ~~together~~, the instructor gave assignments for tomorrow.
5. Lexie tried to pick ~~and choose~~ from the freebies, but she ended up taking them all.

(continued)

6. The package was too big ~~in size~~ to fit in the overhead bin.
7. My family lives in ~~close~~ proximity to three parks.
8. I wish that last grade would just disappear ~~from sight~~.
9. The loss from the fire was ~~estimated at~~ about $500,000.
10. Superstars are fewer ~~in number~~ today than ten years ago.
11. When the arena was full ~~to capacity~~, the fire marshal closed the doors.
12. The DNA presented by the prosecutor was ~~an~~ identical ~~match~~ to that of the defendant.
13. When Brian's contact lens dropped, he had to kneel ~~down~~ on the floor to find it.
14. Every time I feel like relaxing rather than studying, I look ahead ~~to the future~~ and tell myself that I want an associate's degree.
15. Work is done manually ~~by hand~~ by many people in Third World countries.
16. Claudia Jennings declared that she wanted to be reelected ~~for another term~~.
17. Cancer is a deadly ~~killer~~ in every country in the world.
18. "Raise ~~up~~ your hand if you agree that the college needs lower tuition!" the candidate for student government shouted.
19. After Sherina and Jon separated ~~apart~~ for a year, they finally decided to divorce.
20. The mole that my dermatologist was concerned with was round ~~in shape~~.

Wordy Phrases You Can Omit

"[U]se plain, simple language, short words and brief sentences. That is the way to write English . . . . Stick to it; don't let fluff and flowers and verbosity creep in."
—Mark Twain

Just as redundancies bloat your writing, so do **wordy phrases**, which can cause your readers to miss your point entirely. Eliminating these phrases makes your writing more direct and gives it greater impact.

Wordy	All things considered, the news stories today were pessimistic.
Wordy	Because of the fact that the computers are down, the lab isn't open.
Wordy	The east parking lot has been closed for the purpose of repaving.

Each sentence becomes stronger when the writer eliminates the wordy phrases:

Stronger	~~All things considered~~, the news stories today were pessimistic.
Stronger	Because ~~of the fact that~~ the computers are down, the lab isn't open.
Stronger	The east parking lot has been closed for ~~the purpose of~~ repaving.

Below is a list of wordy phrases that are commonly heard and used. Examine the entries in the list and determine which phrases you have used without realizing that they are wordy. Note the word or phrase you can substitute to eliminate the wordy phrase.

"He can compress the most words into the smallest ideas better than any man I ever met."
—Abraham Lincoln, describing a wordy lawyer

Wordy phrase . . .	Substitute	Wordy phrase . . .	Substitute
ahead of schedule	early	an estimated	about
as a means of	to	as long as	if, since
a small number of	a few	as well as	and, also
at the present time	now	because of the fact that	because
by the name of	named	call an end to	end
came to an agreement	agreed	come to an end	close
concerning the matter of	about	despite the fact that	although
due to the fact that	because	extend an invitation	invite
for the purpose of	to, for	for the reason that	because
give rise to	cause	has the ability to	can
if that is not the case	if not	if that is the case	if so
in an effort to	to	in a timely manner	promptly
in back of	behind	in connection with	about
in excess of	more than	in favor of	for
in light of the fact that	because	in possession of	has, have, had
in proximity to	near	in regard to	about
in the amount of	for	in the neighborhood of	about
in view of the fact	since	it would appear that	apparently
made a statement	said	not in a position to	cannot
on a daily basis	daily	on most occasions	usually
on the part of	by, for	over the course of	during
present time	now	previous (prior) to	before
provided that	if	reach a conclusion	end
relating to	about	some of the	some
sufficient number of	enough	the majority of	most
until such time as	until	with regard to	about

TECHNO TIP

For more on wordiness, search the Internet for these videos:

Avoiding Wordiness video

SAT Writing and ACT English: Recognizing Wordiness

For online quizzes on eliminating wordiness, search the Internet for these sites:

Exercises in Writing Concise Sentences

Exercises in Eliminating Wordiness

Rewriting Bloated Sentences

27.2 Eliminating Wordy Phrases

Directions: Revise the sentences below to eliminate wordy phrases.
Answers will vary.

1. I have no need of the expanded cable package ~~at the present time~~. [now.]
2. Jon and his wife will both be sorry when football season ~~comes to an end~~. [ends.]
3. [Although] ~~Despite the fact that~~ the warning light came on in his car, Dave kept driving.
4. Our college's literary magazine ~~goes by the name of~~ [is named] *Community College Connections*.
5. "If you fail to give me your assignments ~~in a timely manner~~ [on time], I'll deduct points," the professor said.
6. [Most] ~~The majority~~ of the people on the cruise ship became ill.
7. [Because] ~~In light of the fact that~~ the new vampire movie will open today, the theater will open at midnight.
8. Even after her yard sale, Julie ~~was in possession of~~ [had] more than ten pairs of shoes.
9. Until the semester is over, Madison ~~is not in a position to~~ [cannot] volunteer for extra hours at work.
10. Campus news is e-mailed to all students ~~on a daily basis~~ [daily].

Avoid Vague Language

LEARNING GOAL
❷ Recognize and eliminate vague language

When writers use vague words and phrases, they do not give their readers a clear picture of what they mean. Readers should not have to guess what they think writers are saying, so writers should be as specific as possible.

Read this example of a sentence that is written vaguely:

Vague	In my new job, I hope for a good salary.

Because readers have no idea what the writer thinks is a good salary, the sentence is vague. By changing the word *good* and replacing it with a specific detail, the writer gives a clearer picture of what he or she hopes to earn.

Explicit	In my new job, I hope for a salary of at least $50,000 a year.

Here is another example of vague writing:

Vague	Carlos and Kristen said the movie they saw last night was awful.

What made the movie awful? Were the actors terrible? Was the plot boring? The writer should state the reason behind the poor evaluation, as in the revised sentence below:

Explicit	Carlos and Kristen said the actors in the movie they saw last night were so miscast they were unbelievable.

The often-used but vague phrase *a large number* is a problem in this sentence:

Vague	A large number of citizens voted against the amendment.

Readers have no idea what the writer thinks is *a large number.* In the revised sentence, the writer gives both a number and a percentage, so readers now know how many citizens were opposed to the amendment.

Explicit	More than fifty thousand citizens (78% of those who cast a ballot) voted against the amendment.

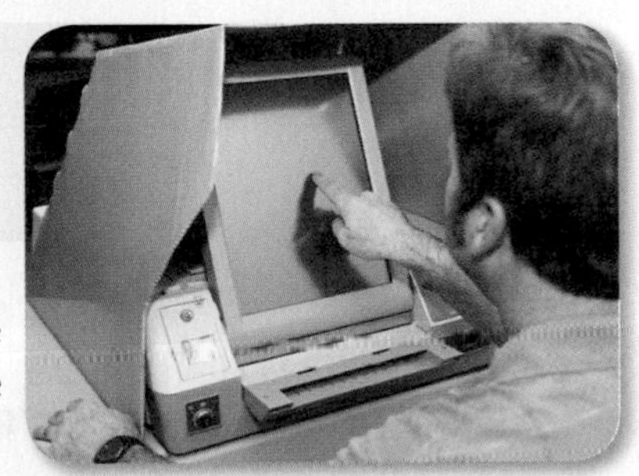

Below is a list of commonly used vague words and phrases. If you use one of these in your draft, revise by substituting more specific or precise words or phrases.

Vague Words and Phrases			
about	certain extent	many	slightly
a few	considerable amount	most	somehow
a large number of	extremely	nearly	something
a little	fairly	pretty	somewhat
a lot	good	process	sort of
approximately	great	quite	stuff
at least	hardly any	quite a few	thing
awful	just	rather	too
better	kind of	really	very
big	little	several	well

27.3 Eliminating Vague Language

Directions: Revise the sentences below to eliminate vague words or phrases. Answers will vary.

1. Because the room was extremely hot, class was moved to another building.
2. Jose put three things in his backpack.
3. The products sold were sort of defective.
4. Approximately twenty people signed up for the workshop.
5. The Career Center charges students for taking part in a few events.
6. Slightly over ten people will be at the dinner tonight.
7. I can tell you many reasons not to buy from that grocery.
8. Hardly any of the people who applied were rejected for employment in the new factory.
9. That restaurant has good food.
10. Shateria said that, to a certain extent, she agreed with Jo's argument about texting.

Avoid Using Clichés

LEARNING GOAL

❸ Recognize and eliminate clichés

Stale writing, like stale food, is unappealing because it has no flavor. One way to keep your writing from being stale is to avoid using **clichés**—expressions that are so overused they have lost their impact.

Writers, readers, and speakers are surrounded by clichés. Read the beginnings of sentences below and see if you can finish each one.

I could see the writing . . .

Try to think outside . . .

That was as easy as taking candy . . .

Better late . . .

When life gives you lemons . . .

Clichés are also known as *stock phrases* and *pat expressions*.

Because these phrases are so commonplace, you probably had no trouble completing them. These are clichés, and they convey no precise or original information.

To help ensure your writing doesn't bore your readers, look for any clichés you have used and reword them. Suppose you are revising your discovery draft and you see a sentence like this:

Alexandria has been there for me through thick and thin.

You realize that the sentence has two clichés: *been there for me* and *through thick and thin*. To revise the clichés, ask yourself what you mean by each phrase; then reword your sentence using your alternatives. You might have reworded with these alternatives:

Cliché	Alternative
been there for me	been a dependable friend
through thick and thin	in peaceful and in challenging times

Your revised sentence would now read this way, which is much more original and thought-provoking:

Alexandria has been a dependable friend in peaceful and in challenging times.

Now the sentence is not boring, and readers get a more precise picture of your relationship with Alexandria.

TECHNO TIP

For more on eliminating clichés, search the Internet for these sites:

Clichés: Avoid Them Like the Plague

Avoiding Trite Language and Overused Expressions

For more on eliminating clichés, search the Internet for these videos:

Dogging the Words Writing Tip 23 Editing Clichés: Take 2

27.4 Eliminating Clichés

Directions: Revise each sentence below to eliminate clichés. Answers will vary.

1. I knew I had to do my homework, but finding the time was easier said than done.

(continued)

2. The factory shutting down was a fate worse than death for the small town.
3. The week before finals is always the calm before the storm.
4. As the young couple danced, Pat and I could tell they were head over heels in love.
5. Even after his death, Michael Jackson seems larger than life.
6. The summer I worked on the lobster boat, breaks for sleep seemed few and far between.
7. After six months of back pain, Erin was going to bite the bullet and consult her doctor.
8. Shanté tried to act as cool as a cucumber after the officer pulled her over for speeding.
9. The committee took a step in the right direction when Sharon suggested creating an in-store policy addressing customer complaints.
10. Down by eighteen points at the half, the team knew that facing Coach Thompson would be a fate worse than death.

MyWritingLab™ Visit *MyWritingLab.com* and complete the exercises and activities in the **Standard and Non-Standard English** topic area.

RUN THAT BY ME AGAIN

LEARNING GOAL
❶ Recognize and eliminate wordiness

- **Redundancy is . . .** the excessive or unnecessary repetition of an idea.
- **Some redundant phrases are so commonly heard that . . .** writers often fail to notice the pointless repetition in them.
- **Wordy phrases can cause readers to . . .** miss your point entirely.
- **Eliminating wordy phrases makes . . .** writing more direct and gives it greater impact.

LEARNING GOAL
❷ Recognize and eliminate vague language

- **Vague words and phrases often prevent writers from . . .** giving readers a clear picture of what the writers mean.
- **Readers should not have to guess . . .** what they think writers are saying, so writers should be as specific as possible.

LEARNING GOAL
❸ Recognize and eliminate clichés

- **Clichés are . . .** expressions that are so overused they have lost their impact.
- **Clichés do not convey . . .** precision or originality.

WORD CHOICE LEARNING LOG MyWritingLab™

Answer the questions below to review your mastery of word choice. Answers will vary.

MyWritingLab™

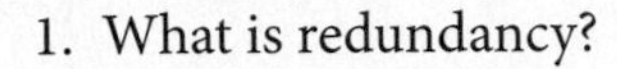

1. What is redundancy?

   Redundancy is the excessive or unnecessary repetition of an idea.

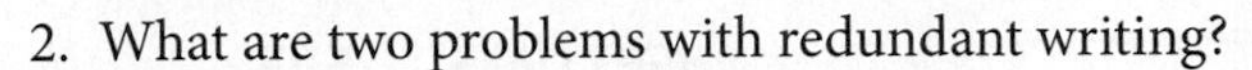

2. What are two problems with redundant writing?

   Redundant writing (1) bores readers and (2) weakens what the writer is saying.

3. What can wordy phrases cause readers to miss?

   Wordy phrases can cause readers to miss the writer's point.

4. What two benefits come from eliminating wordy phrases?

   Eliminating wordy phrases (1) makes writing more direct and (2) gives writing greater impact.

5. What occurs when writers use vague words and phrases?

   When writers use vague words and phrases, they do not give their readers a clear picture of what they mean.

6. What are clichés?

   Clichés are expressions that are so overused they have lost their impact.

CHAPTER 28 Figurative Language

LEARNING GOALS

In this chapter, you'll learn and practice how to

1. Differentiate between literal and figurative language
2. Use figurative language to enhance your writing

GETTING THERE

- Figurative language is often used to present ideas in a fresh and artistic manner.
- Identifying figurative language and learning how to use it is important in college writing.

What Is Figurative Language?

LEARNING GOAL

1. Differentiate between literal and figurative language

As you learned in Chapter 6, language is either *literal* or *figurative*. The difference is in interpretation of the words. **Literal language** is like the expression "what you see is what you get"; it has no room for interpretation. **Figurative language**—found in prose, poetry, and everyday speech—uses language in a unique way. Paragraph and essay writers occasionally use figurative language. When used correctly, the freshness and vitality of figurative language helps writing become clearer, deeper, more effective, or more interesting.

Suppose you're working outside in the middle of a heat wave and you say, "I'd give a dollar for a bottle of cold water." In that case, you're speaking literally; you mean exactly what you said. If, however, you said, "I'd give a million dollars for a bottle of cold water," you're speaking figuratively. Everyone understands your exaggeration is just that: an overstatement of the truth. You'll learn later that conscious exaggeration is a type of figurative language called *hyperbole*.

Use Figurative Language

LEARNING GOAL

2. Use figurative language to enhance your writing

Academic writing is concerned with facts, so it lends itself more to literal writing than figurative writing. All writers, however, use figurative language to some extent. For example, *simile*, *metaphor*, and *imagery* are effective in descriptive writing.

Figurative Term	Definition	Example
Alliteration (uh-lit-er-A-shun)	the repetition of beginning consonant sounds in neighboring words	A *gaggle of geese*
Allusion (al-LOU-zhun)	a reference to someone or something a reader should know	It's rained so long I think I'll have to *build an ark.*
Anachronism (a-NACK-ro-nism)	placement of someone or something in the wrong time period	The caveman began *typing on his computer.*
Aphorism (AF-or-ism)	a brief statement of opinion or truth	"Lost time is never found again." —Ben Franklin
Epithet (EP-i-thet)	a term used to describe a person, place, or thing	Ivan *the Terrible*

For more about alliteration, search the Internet for this video: Extreme Alliteration

28.1 Figurative Language, Part 1

Alliteration Allusion Anachronism Aphorism Epithet

Directions: Identify which type of figurative language is used in each sentence below.

1. When the armistice was signed to end World War II, the first e-mail General Eisenhower received was from President Truman.
   anachronism
2. Josh is finally going to visit Paris, the City of Light. epithet
3. Susie and Cindy sat side by side. alliteration
4. "I find that the harder I work, the more luck I seem to have." —Thomas Jefferson aphorism
5. "The guy in the third row is a real Don Juan," Sally whispered to Sherida. allusion
6. George Washington's troops kept many provisions warm by stashing them in a red, white, and blue cooler. anachronism
7. Three chunks of cheddar cheese looked inviting. alliteration
8. Frank Sinatra was known as "The Chairman of the Board." epithet
9. "When good men do nothing, evil triumphs." —Edmund Burke
   aphorism

(continued)

10. Rob felt like a Scrooge when he didn't contribute to the flower fund.
allusion

TICKET to WRITE

28.2 Figurative Language, Part 1—Your Turn

Directions: Write a sentence using each type of figurative language listed below. When you finish, compare your sentences with others in your class to see if your examples are easy to identify.

Example: Lovely Lucy lazily licked her lollipop. (alliteration)

Alliteration **Allusion** **Anachronism** **Aphorism** **Epithet**

Figurative Term	Definition	Example
Euphemism (U-fem-ism)	a word or phrase used to replace a word or phrase considered harsh or offensive	My dog *passed away.*
Hyperbole (high-PER-bo-lee)	a statement that deliberately exaggerates for emphasis or heightened effect	I'm so hungry I could eat *a horse.*
Idiom (ID-ee-um)	an expression with a meaning different from the literal meaning of its individual words	Randy *let the cat out of the bag* about the surprise party.
Imagery (IMM-ij-ree)	the use of words to appeal to one of the five senses	a *bright red* cardinal; the *sweet tang* of an orange; the *bumpy texture* of the sheet
Irony (I-run-ee)	the use of words to express something opposite to their literal meaning	When Brad came to class in a three-piece suit, Mick said, "Is this casual Friday?"

For more on hyperboles, search the Internet for this video:
Hyperbole and Understatement—A Short Digital Piece

For lists of idioms, search the Internet for these sites:
Idioms and Phrases
The Free Dictionary Idiom Site

For online quizzes about idioms, search the Internet for these sites:
Self-Study Idiom Quizzes
Tracking Creativity: In Step with Idioms

28.3 Figurative Language, Part 2

Euphemism **Hyperbole** **Idiom** **Imagery** **Irony**

Directions: Identify which type of figurative language is used in each sentence below.

1. My head is killing me! hyperbole
2. The politician said he had misspoken about his prior claims of being in the military. euphemism
3. The aroma of the freshly brewed coffee hit me as soon as I opened the door. imagery
4. Greg tried to get Candace to spill the beans about his present, but she wouldn't do it. idiom
5. Looking at the mud hut, the missionary said, "This palace will be my home for a year." irony
6. Isn't *bathroom tissue* another way of saying *toilet paper*? euphemism
7. Our real estate agent advised us to bake fresh cinnamon bread just before our open house. imagery
8. "Go ahead and toot your horn about your promotion," Norma's husband urged. idiom
9. "I see you're studying hard for your test," my wife said when she saw my book closed and noticed that I was catching up on e-mail. irony
10. Aurora's dog is so old that he put his paw print on the Declaration of Independence. hyperbole

28.4 Figurative Language, Part 2—Your Turn

Directions: Write a sentence using each type of figurative language listed below. When you finish, compare your sentences with others in your class to see if your examples are easy to identify.

Euphemism **Hyperbole** **Idiom** **Imagery** **Irony**

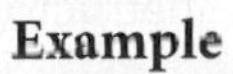

Figurative Term	Definition	Example
Metaphor (MET-a-for)	a comparison between two unlike things, without using *like* or *as*	It's an oven outside today!
Onomatopoeia (Ah-nuh-MAHT-a-PE-uh)	a word that sounds like its meaning	the *buzz* from the bees hear the crowd *murmur*

(continued)

For more on onomatopoeia, search the Internet for this video: Onomatopoeia

For more on oxymorons, search the Internet for this site: Some Funny Oxymorons

For more on oxymorons, search the Internet for this video: Oxymoron Song—Schoolhouse Rock

Oxymoron (OX-ee-MORE-on)	contradictory terms that appear side by side but still make sense	"Parting is such *sweet sorrow*." —William Shakespeare
Paradox (PARE-a-dox)	a statement that seems to contradict itself	"It was the best of times, it was the worst of times." —Charles Dickens
Personification (per-SAHN-i-fi-CAY-shun)	giving qualities of a person to something not human	The wind screamed at me to return to the house.

28.5 Figurative Language, Part 3

Metaphor Onomatopoeia Oxymoron Paradox Personification

Directions: Identify which type of figurative language is used in each sentence below.

1. "Hope is a good breakfast, but it is a poor dinner." —Francis Bacon metaphor
2. "What a pity that youth must be wasted on the young." —George Bernard Shaw paradox
3. The sizzle of the steak on the grill was the first sign of summer. onomatopoeia
4. Your GPA speaks well of you. personification
5. "The rain came down in long knitting needles." —Enid Bagnold, *National Velvet* metaphor
6. When the ball swished through the net, Vera and the other fans went wild. onomatopoeia
7. Buying six or a half dozen is the same difference. oxymoron
8. In *The Great Gatsby*, Jordan Baker said, "I like large parties. They're so intimate." paradox
9. When our instructor asked for an answer to his question about our homework, the silence was deafening. oxymoron
10. Winter limped in with velvety flakes of soft snow. personification

28.6 Figurative Language, Part 3—Your Turn

Directions: Write a sentence that illustrates each type of figurative language listed below. When you finish, compare your sentences with others in your class to see if your examples are easy to identify.

Metaphor **Onomatopoeia** **Oxymoron** **Paradox** **Personification**

Figurative Term	Definition	Example
Proverb (PRAH-verb)	a short, memorable saying expressing an idea many believe to be true	"Fools rush in where angels fear to tread." —Alexander Pope
Rhetorical Question (reh TORE-ik-cal)	a question asked only for effect with no answer expected	When will I ever have a perfect day?
Simile (SIM-i-lee)	a comparison between two unlike things, using *like* or *as*	"Life is rather like a tin of sardines: we're all . . . looking for the key." —Alan Bennett
Symbol (SIM-bol)	something that stands for or represents something else	The American flag is a symbol of freedom.
Understatement	a sentence that deliberately makes something important seem trivial in order to achieve a larger effect	We'd had rain for three days, but the weather reporter said, "It's a little damp out there."

For more on understatements, search the Internet for this video: Hyperbole and Understatement—A Short Digital Piece

28.7 Figurative Language, Part 4

Proverb **Rhetorical Question** **Simile** **Symbol** **Understatement**

Directions: Identify which type of figurative language is used in each sentence below.

1. Are you nuts? rhetorical question
2. Oprah Winfrey has earned a little money over the years. understatement
3. "A house without books is like a room without windows." —Horace Mann simile

(continued)

4. Don't cry over spilt milk. proverb
5. I'm pretty pleased with the 99% I earned on my communications exam. understatement
6. When will the world realize the futility of war? rhetorical question
7. The wedding ring stands for love. symbol
8. "I wandered lonely as a cloud." —William Wordsworth simile
9. A friend in need is a friend indeed. proverb
10. Two roads in the poem "The Road Not Taken" represent paths in life. symbol

28.8 Figurative Language, Part 4—Your Turn

Directions: Write a sentence that illustrates each type of figurative language listed below. When you finish, compare your sentences with others in your class to see if your examples are easy to identify.

Proverb **Rhetorical Question** **Simile** **Symbol** **Understatement**

MyWritingLab™ Visit *MyWritingLab.com* and complete the exercises and activities in the **Standard and Non-Standard English** topic area.

RUN THAT BY ME AGAIN

LEARNING GOAL
1 Differentiate between literal and figurative language

LEARNING GOAL
2 Use figurative language to enhance your writing

- **Literal language has . . .** no room for interpretation.
- **Figurative language uses language in a unique way to present . . .** a clearer, more vivid, more precise, or more concise picture.
- **Academic writing lends itself more to . . .** literal writing than figurative writing.
- **Simile, metaphor, and imagery are . . .** effective in descriptive writing.
- **A simile is . . .** a comparison of two unlike things, using *like* or *as*.
- **A metaphor is . . .** a comparison between two unlike things, without using *like* or as.
- **Imagery is . . .** the use of words to appeal to one of the five senses.
- **Other types of figurative language include . . .** alliteration, allusion, anachronism, aphorism, epithet, euphemism, hyperbole, idiom, irony, onomatopoeia, oxymoron, paradox, personification, proverb, rhetorical question, symbol, and understatement.

FIGURATIVE LANGUAGE LEARNING LOG MyWritingLab™

Answer the questions below to review your mastery of figurative language.

1. Where do you find the difference between literal and figurative language?
   The difference between literal and figurative language is in interpretation of the words.
2. How can figurative language help essay writers?
   When used correctly, the freshness and vitality of figurative language helps writing become clearer, deeper, more effective, or more interesting.
3. "I'd give a million dollars for a bottle of cold water" is an example of what type of figurative language?
   "I'd give a million dollars for a bottle of cold water" is an example of *hyperbole*.
4. What is alliteration?
   Alliteration is the repetition of beginning consonant sounds in neighboring words.
5. What is irony?
   Irony is the use of words to express something opposite to their literal meaning.
6. What is personification?
   Personification is giving qualities of a person to something not human.
7. What is a symbol?
   A symbol is something that stands for or represents something else.
8. "That vampire novel is a new classic" is an example of what type of figurative language?
   "That vampire novel is a new classic" is an oxymoron.
9. What is onomatopoeia?
   Onomatopoeia is the use of a word that sounds like its meaning.
10. "The kids were champing at the bit for spring break because we were going to the beach" is an example of what type of figurative language?
    "The kids were champing at the bit for spring break because we were going to the beach" is an idiom.

Easily Confused Words and Phrases; Spelling, Abbreviations, and Numbers

LEARNING GOALS

In this chapter, you'll learn and practice how to

1. Correct mistakes in homophones and other easily confused words
2. Correct mistakes in easily confused phrases
3. Learn five main spelling rules
4. Correctly use abbreviations and numbers in academic writing

GETTING THERE

- Homophones—words that sound alike but are spelled differently—often present problems for writers.
- In addition to homophones, writers also have difficulty with other words and phrases.
- Some words are considered to be nonstandard; writers should not use them in academic writing.
- Spelling mistakes can be lessened by applying five main rules.
- Academic writing allows abbreviations for some words and phrases.
- Academic writing requires that some numbers be written as figures and some be written as words.

What Are Homophones?

LEARNING GOAL

1. Correct mistakes in homophones and other easily confused words

Some words are easily confused because they are **homophones**, words that sound alike but are spelled differently and have different meanings. For instance, the words *air* (meaning "atmosphere") and *heir* (meaning "inheritor") are homophones, as are *eight* (meaning "the number between seven and nine") and *ate* (meaning "consumed"). Even though these word pairs are pronounced the same, their meanings are different.

The word **homophone** comes from combining the Greek *homos*, meaning *same*, and *phone*, meaning *sound.*

Because of their identity in sound, some homophones are easily confused. In speech, these words pose no problems, but in writing they are often mixed up.

Homophones: The Top Six

hear, here Use your *ear* to **hear**. The opposite of ***there*** is ***here***.

I'm **here** to **hear** the guest speaker in our astrology class.

To watch a video about *here* and *hear*, search the Internet for Here & Hear - Commonly Confused Words in English

its, it's **It's** means only **it is** (think of the apostrophe as standing in for the missing *i* in *is*). **Its** means only **belonging to it**.

It's frustrating to wait for the lab computer to finish **its** download.

there, their, they're *Here* is the opposite of *there*. If you want **they are**, use **they're** (think of the apostrophe as standing in for the missing *a* in *are*). ***Their*** means **belonging to *them***.

Students over **there** are raising **their** grades because **they're** doing extra credit.

To watch a video about *they're*, *there*, and *their*, search the Internet for Confused Words: They're, There, and Their

to, too, two If you mean **additional**, use the one with the **additional *o*** (***too***). **Two** comes after **one**; **to** means **in the direction of**.

The instructor gave **two** students permission **to** go on a field trip and attend the lecture, **too**.

To watch a video about *to*, *too*, and *two*, search the Internet for To, Two, Too - Commonly Confused Words in English

whose, who's **Whose** means **belonging to whom**; **who's** means **who is** (think of the apostrophe as standing in for the missing *i* in *is*).

Who's going to decide **whose** contribution to the project was the best?

To watch a video about *who's* and *whose*, search the Internet for Confused Words: Who's and Whose video

your, you're The **opposite of *our*** is *your*. If you mean **you are**, use **you're** (think of the apostrophe as standing in for the missing *a* in *are*).

If **you're** going to the subway, take **your** tokens.

29.1 Homophones and Other Easily Confused Words and Phrases, Part 1

Directions: Underline the correct word in each sentence.

1. Are you on your way (to, too, two) the cafeteria?
2. (Whose, Who's) going to offer the first presentation in class?
3. My paper seemed to take on a life of (its, it's) own.
4. All my roommates said (there, their, they're) going to the game.
5. If (your, you're) going to be late, I'll give you my notes after class.
6. Will you be going home for break (to, too, two)?
7. (Its, It's) clear to me that I need to get to the tutoring center.
8. If Frank is still (hear, here) after class, we're going to the library together.
9. "Please turn (your, you're) cell phones off immediately," Professor Albritten said.
10. Group A worked on (there, their, they're) peer reviews.

Other Common Problem Homophones

a lot, alot, allot	**Alot** (spelled without a space) is not a word. If you mean **a whole lot**, use **two whole words**: **a lot**. If you want to ***allo*cate** something, use ***allo*t**. This weekend, I'll **allot a lot** of my time to writing my paper.
aid, aide	**Aid** = **help**; an **aid*e*** = a helper, a **support*er*** or an **advis*er***. The **aid** from the instructor's **aide** was very helpful.
all ready, already	**All ready** means **all is** (or **are**) **ready**; **already** has to do with **what happened in the past**. Our study group **already** said they're **all ready** to crack the books.
alright, all right	**Alright** (spelled without a space and with one *l*) is not a word. You'd never say something is **aleft** or **alwrong**, would you? **All right** is correct only as **two words**. Is it **all right** if I ride to class with you?

all together, altogether

If you mean **simultaneously** or **all at once**, use **all together**. If you can substitute **entirely** or **wholly** in your sentence, use **altogether** (the word that's **entirely and wholly one word**).

"The class is **altogether** wrong about groups working **all together**; everyone is to work separately," Professor Florence announced.

aloud, allowed

If you mean **speaking out *loud***, use **a*loud***. When you mean **permission**, use **allowed**.

Professor Rosenbloom said we're not **allowed** to speak **aloud** during his presentation.

cite, sight, site

If you **cite**, you mention a **particular source**. **Sight** is your **vision** or a **scene**. A **site** is a **location**.

The picture on the Internet **site** I **cite** in my paper is a **sight** to behold.

coarse, course

If something is **coarse**, it's **rough**; if your throat is ***coarse***, your voice is ***hoarse***. A **course** can be a **route**, a **class**, or part of the phrase "**of course**."

Of **course**, we expect outdoor running in this phys ed **course** would lead us over some **coarse** terrain.

complement, compliment

When one thing ***completes*** another thing, it ***complements*** it. A ***compliment*** is a form of **praise** (***I*** like to receive a **compliment**).

Your extra credit will **complement** what you've already done in class, and you're sure to get a **compliment** on your hard work.

ensure, insure

If you **ensure** something, you **make certain of it**; **insure** refers to **a contract** that pays in the case of loss of life or property.

We will **ensure** that we **insure** our car by sending in the payment.

everyday, every day

If you mean **routinely** or **daily** (an **everyday** habit), use **everyday**; if you mean **every single day**, use **every day** (studying **every day**).

Mnemonic Alert! If you mean **every single day (single words)**, use **every day**.

The **everyday** low prices in the cafeteria assure students get a bargain **every day**.

led, lead *Led* means **head*ed* or guid*ed***; in all other cases, use **lead**.

Tuesday, Lori **led** the talk regarding **lead**; today, Larry will **lead** it.

pore, pour When you **read something thoroughly**, you **pore** over it. When you ***pour*, liquid goes *out*** of a container.

After I **pored** over my books for an hour, I **poured** myself a glass of iced tea as a reward.

principle, principal A **princip*le*** is a **ru*le*** or **valu*e***. **Principal** means **major** or **head**; it also means **money** earning interest in a bank.

Our speaker, a local school **principal**, believes in the **principle** that every student should read by age eight.

role, roll Your **role** is your **position** or **part** (as in a play); a **roll** is a type of **bread**; to **roll** is to **turn**.

In my **role** as family cook, I often bake a big pumpkin **roll**.

29.2 Homophones and Other Easily Confused Words and Phrases, Part 2

Directions: Underline the correct word in each sentence.

1. The article I read on the Internet will (complement, compliment) the article in the text.
2. Our sociology class read about the correspondence (principle, principal), which speculates a close relationship between social standing and the educational system.
3. Before Barry left biology class, he had to (pore, pour) out the liquid he had used.
4. The financial (aid, aide) Darrell received meant that he could stay another semester.
5. Jacob decided to do some extra credit to (ensure, insure) he kept his high grade.

6. When Kerri discovered that she had (a lot, alot, allot) of mistakes in her paper, she was mortified.
7. (Everyday, Every day), Luke reviews his notes for at least an hour.
8. Carmin's decision to drop a class (lead, led) to having more time to concentrate on other studies.
9. Ellie felt (all right, alright) on Friday, in spite of her bout with the flu on Thursday.
10. Graham is not (all together, altogether) surprised that his roommate is leaving.
11. Mike wasn't sure what his (role, roll) as leader of the peer review group would be.
12. During the (coarse, course), Helen felt she came to know her professor well.
13. Franklin has (all ready, already) written his paper and turned it in.
14. Why aren't we (aloud, allowed) to use our calculators?
15. May we (cite, sight, site) the same book three times?

Other Easily Confused Words

In addition to homophones, writers are often confused by other English words. Sometimes, reading a list of troublesome words—along with the difference between the words—helps writers recognize the specific words that they find tricky. Other times, writers find that using mnemonics helps identify words they misuse and then decide the correct form of the word they want.

a, an	The difference is in the sound of the next word. Use **a** before a word that begins with a **consonant sound** (a book); use **an** before a word that begins with a **vowel sound** (**an** earring). "It's **an** honor to be in **a** history class you offer," Ben told Professor Maltby.

Look at this example: It's **an** honor to meet **a** hero. The *h* in *honor* isn't pronounced, so you write ***an** honor*; the *h* in *hero* is pronounced, so you write ***a** hero*.

accept, except	**Accept** can mean **believe, receive, undertake**, and **consent**; ***except*** means ***excluding***. The Science Society will **accept** any recycled glass items to recycle **except** broken bottles.

Mnemonic Alert! To ***advise***, be **wise**; to get ***advice***, be **nice**.

advise, advice	**Advise** is **what you do**; **advice** is **what you give**. My **adviser** offered **advice** to help me get the right courses.

To watch a video about *advice* and *advise*, search the Internet for Confused Words - Advice & Advise

To watch a video about *affect* and *effect*, search the Internet for Confused Words - Effect & Affect.

affect, effect	If you mean **change** or **shape**, you want ***a**ffect*. If you want **result** or **appearance**, you want ***e**ffect*. The **effect** of Joann's getting a low grade on this test will not **affect** her overall grade in the class.
allusion, illusion	An **allusion** is a **reference**; an ***i**llusion* is a **m*i*staken *i*dea**, a **false *i*mpression**. When my roommate said she hoped I'd be back by midnight, I asked if she was under the **illusion** I was Cinderella; she didn't understand the **allusion**.
bad, badly	If you're writing about **the way you feel**, use **bad**. If you're writing about **how you reacted**, **performed**, or **did something**, use **badly**. After taking the test cold, Dora felt **bad** she scored so **badly**.
beside, besides	**Beside** means **by the side of**. The one **that has the additional *s*** (**beside*s***) means **in addition to**. **Besides** his roommates, Andrew wanted his brother **beside** him at his bachelor party.
between, among	If **two people or groups** divide something, use **between**; if **more than two people or groups** divide something, use **among**. After class, the pencils were divided **between** the men and women; the paper was divided **among** Allison, Jonathan, and Kwame.
breath, breathe	You **take** a **breath**; when you **inhal*e*** and **exhal*e***, you **breath*e***. During the worst of the winter semester, I found it hard to **breathe** every time I took a **breath** outside.

Mnemonic Alert! For a ***group***, use **among**; for ***two*** use **between**.

can, may If you *can* do something, you are *capable* of doing it. If you *may* do it, you have **permission** to do it.

You **can** turn in a paper with many misspelled words, but you **may** not.

29.3 Homophones and Other Easily Confused Words and Phrases, Part 3

Directions: Underline the correct word in each sentence.

1. Quentin's decision to change majors will (affect, effect) his class schedule for next semester.
2. (Beside, Besides) the paper we have to revise, we also need to read an article.
3. Let's divide the review questions (among, between) the four of us in our study group.
4. Did you have to wear (a, an) uniform at your high school?
5. (Can, May) we turn our papers in a day late?
6. When Professor Payne gives (advice, advise), students tend to listen.
7. The YouTube video of the war took Thom's (breath, breathe) away.
8. Sarah felt so (bad, badly) that all she wanted to eat was a bowl of soup.
9. Nell hasn't had trouble with any papers (accept, except) the latest one.
10. Rene realized that she'd been under the (allusion, illusion) that she wouldn't have to study.

capitol, capital A **capitol** is a **building** in which a legislature meets. Use **capital** with any other meaning (**assets**, **center**, **main city**).

Our field trip to the **capitol** took lots of our **capital**.

Mnemonic Alert! A **capitol** is a building that usually has a **dome**.

conscience, conscious Your **conscience** directs your **principles**; if you're **conscious**, you're **awake** and **aware**.

During the chat about plagiarism, Kent said his **conscience** wouldn't allow him to copy material and say he had written it; Karen, who was taking a nap, wasn't **conscious** of the discussion.

The noun and verb forms of **desert** are **heteronyms**: words with the same spelling but a different pronunciation.

Mnemonic Alert! A **dessert** is ***so sweet***.

desert, dessert	As a noun a **desert** is a **dry, arid region** or (usually used in plural) a **reward or punishment someone deserves** (*just deserts*). The verb that means **to leave** is also **desert**. After-dinner **food** is a **dessert**. After our geography class looked at pictures of an arid **desert**, we all chose a cold milk shake as a **dessert**.
device, devise	A **device** is a **machine** or an **apparatus**; to **devise** means to **invent** or **plan** something. If you **dev*ise***, you must be w*ise*. Does this n*ice* **dev*ice*** work on *ice*?
elicit, illicit	If you **elicit** something, you **extract** it or **bring it out**; something ***ill*icit** is ***ille*gal**. The student was a victim of **illicit** identity theft; someone was able to **elicit** her credit card information.
emigrate, immigrate	When you **emigrate**, you ***ex*it** a country; when you **immigrate**, you go ***in*to** a country. A man on our floor **emigrated** from Somalia; he **immigrated** to the United States two years ago.
farther, further	If you're writing about **distance** (like the ***ar*ea you can travel**), you want **f*ar*ther**. If you mean **additional** or **to promote**, you want **further**. After walking five miles in phys ed class, I couldn't go any **farther**; to add **further** points to my grade, I'd wait for the weekend.
fewer, less	When describing **plural** words, use **fewer**; when describing ***singular*** words, use **le*ss***. The peanut butter my roommate likes has **fewer** calories and **less** fat than the brand I usually buy.

figuratively, literally **Figuratively** means in a **symbolic** or **metaphoric** way; **literally** means **just as described.**

When Patsy came out of speech class and said she had cold feet, I thought she meant the phrase **figuratively** and was nervous about class. However, she meant the phrase **literally** because the temperature in the room was in the fifties.

good, well **Good** means **high-quality** or **fine**; if you mean **healthy, competently**, or **correctly**, use **well.**

Since I did so **well** in all my classes, my GPA should be **good.**

29.4 Homophones and Other Easily Confused Words and Phrases, Part 4

Directions: Underline the correct word in each sentence.

1. In our culinary class next week, we will concentrate on a favorite (desert, dessert).
2. The culinary class cooked the casserole so (good, well) that the instructor praised each student.
3. While I was tasting the pie, I wasn't (conscience, conscious) that I was smacking my lips.
4. Today's casserole has (fewer, less) ingredients than the one we cooked Tuesday.
5. Our professor will show us how to use the convection oven, a new (device, devise) that's recently been installed in the classroom.
6. When we cook, the smells that drift from the classroom (elicit, illicit) good comments from people passing by.
7. To get in our instructor's good graces, we want to butter her up, (figuratively, literally) speaking.
8. Our class has been asked to serve at a function in the dining hall at the state (capital, capitol).
9. When Raul decided to (emigrate, immigrate) to our country, he packed several cookbooks.
10. Before we added (farther, further) ingredients, we checked the recipe.

graduate, graduate from	An educational institution, like a college, will **graduate** you; you will **graduate from** an educational institution. The year after I **graduated from** high school, my school **graduated** only one hundred students.

hanged, hung	**People** are **hanged**; **art** is **hung**. The spurs of the **hanged** man were **hung** on the wall.

Hopefully is often misused to mean "I hope."

hopefully, I hope	**Hopefully** means **with confidence** or **with anticipation**. **I hope** means that you **wish** for something. Peter waited **hopefully** for an announcement that classes would be cancelled due to snow. **I hope** he's wrong about the weather.

imply, infer	A ***speaker implies***; a ***listener infers***. Joey thought the instructor had **implied** the test would be postponed, but Micah did not **infer** that.

in, into	If you mean something **with*in*** something else, use ***in***; if you mean going from the outside ***to*** the ***in*side**, use ***into***. Go **into** the building and look **in** Room 388 to find the lost book.

The most frequent problem comes with people incorrectly writing or saying "I laid down for a nap" or "We laid out in the sun." Both of these should be "lay."

lay, lie	Problems come in verb forms. **Lay** means **to put or place**; today you **lay** your cup down; yesterday you **laid** your cup down; in the past, you **have laid** your cup down. **Lie** means **rest or recline**; today you **lie** down to nap; yesterday you **lay** down to nap; in the past, you **have lain** down to nap. As Marcus **lay** in bed, he wondered where he had **laid** his notes.

loose, lose Loose (rhymes with **goose**) means **not tight**. **Lose** is the opposite of **find**.

Going across our windy campus, I'll **lose** my hat if it's too **loose**.

Mnemonic Alert! Think of the phrase "loosey-goosey." Someone who's "loosey-goosey" isn't "uptight."

pacific, specific *Pacific* means *peaceful*; **specific** means **precise** or **exact**. (When capitalized, *Pacific* only refers to the ocean.)

To be **specific**, the **pacific** view I note in my paper is from mountains.

passed, past Passed is always a verb (passed a test, passed a ball); **past** is most often used as an **adjective** (as when it means **historical** or **beyond**) or a noun meaning **the preceding time**.

In the **past**, both Ed and Ruth Ann easily **passed** all their classes.

persecute, prosecute People who **persecute** are ones who **oppress** or **maltreat**; people who **prosecute** are ones who **bring legal action**.

Our sociology class read about a case in which a state **prosecuted** students who **persecuted** a roommate.

29.5 Homophones and Other Easily Confused Words and Phrases, Part 5

Directions: Underline the correct word or phrase in each sentence.

1. From what Professor Kim said, did you (imply, infer) we'll watch *The Graduate* in film class?
2. The documentary we watched dealt with Nazis who (persecuted, prosecuted) others during World War II.
3. Surely you didn't (loose, lose) the notes we took in class today!
4. (Hopefully, I hope) the film class will take a field trip to a local cinema.
5. As soon as our class walked (in, into) the theater, the lights went dark.
6. We were amazed when our professor said we could (lay, lie) on the floor to watch the film.
7. Each student needs to note the (pacific, specific) film he or she will review.
8. When you (graduate, graduate from) this institution, how many hours will you have?

(continued)

9. In the movie our class watched, several "bad guys" were (hanged, hung) by a Western posse.
10. When we (passed, past) the midpoint of the semester, our professor brought in popcorn.

precede, proceed Something that will **precede** another will **come first**; if you **proceed**, you **keep on**.

Because of our schedules, John will **precede** me in arriving in class; when I get there, we will **proceed** to review our homework.

When you're writing these words, check their endings; that's where problems occur.

quiet, quite **Quiet** means **calm**, **silence**, or **silent**; **quite** means **very** or **rather**.

I'm **quite** sure the library is a **quiet** place to study.

In academic writing, don't use *real* as a substitute for *very*.

real, really **Real** means **factual** or **valid**; **really** means **in truth** or **actually**.

When Tom **really** examined his study habits, the **real** importance of reviewing instead of cramming hit him.

set, sit When you **put** or **place** something, you **set** it. When you're **in an upright position** (like in a **chair**), you **sit**.

Set your calculators on your desks before you **sit** down.

slow, slowly If you're **describing someone or something**, use **slow**. The mistake is almost always made with **slowly**, the adverb form. If you're **using the word after** *walk*, *go*, *drive*, or any adverb, adjective, or other verb, use **slowly**.

After Marvin worked at a **slow** pace on his math questions, he **slowly** walked out of class, wondering about his score.

suppose, supposed to If you **suppose**, you **assume** or **believe**. If you mean **expected to** or **designed to,** make sure to put the *-d* at the end of **supposed**.

	You're **supposed** to write a short paragraph for homework. I **suppose** you've begun your discovery draft.
than, then	If you're showing a **comparison**, use **than**. If you mean **next** or **therefore** or **at that time**, use **then**. On the first two tests, Bryan scored higher **than** I did; **then** he began to slough off and my grades beat his.
that, which	With **restrictive clauses** (ones that **don't need commas**), use **that**. With **nonrestrictive clauses**, which **need commas**, use **which**. My night classes, **which** meet after 6:00 p.m., are the ones **that** I have found the most interesting.
try and, try to	If you mean **aim to** or **attempt to**, then you want **try to**. I said I would **try to** e-mail yesterday's notes to the absent student. I did **try, and** the e-mail bounced back.
use to, used to	If you mean **employ for the purposes of**, you want **use to**; if you mean **formerly** or **in the past**, you want **used to** (often misspelled without the *-d*). In my last class, I **used to** like to listen to the excuses other students would **use to** get extensions on their deadlines.
weather, whether	**Weather** deals with the **climate**. **Whether** can mean **which**, **whichever**, or **if it is true that**. The **weather** is so odd I can't decide **whether** to take an umbrella.

In speaking, we often say "try and" when we mean "try to."

Mnemonic Alert! Heat is a form of weather.

who, which, that	Use **who** only with **people**; use **which** with **places** or **things**. Some instructors require **that** be used only with places or things; some also allow **that** be used with people. My math professor, **who** gives quizzes every week, said **that** the latest quiz, **which** she had just graded, was my best to date.

woman, women	One ***woman***, two or more ***women***. One **woman** in class asked if she could bring two other **women** as guests.

29.6 Homophones and Other Easily Confused Words and Phrases, Part 6

Directions: Underline the correct word or phrase in each sentence.

1. The class was so (quiet, quite) that I wondered if everyone had decided to take a nap.
2. (Weather, Whether) we review together or separately is up to you.
3. You're more (than, then) welcome to see my notes.
4. Isn't that (woman, women) the one who is in our sociology class?
5. "Please (set, sit) wherever you feel comfortable," said Professor Kasenow.
6. The (real, really) reason for my delay was that the bus ran late.
7. Many students (who, which, that) are enrolled in math class are smart enough to get tutoring.
8. Good ideas for time management, (that, which) are presented later in the unit, will help me.
9. I'll (try and, try to) put some of those ideas for time management into place immediately.
10. Gerry is (suppose, supposed to) meet me after class.
11. We'll drive (slow, slowly) through campus to see if we can spot where I dropped my book.
12. Natalie will (precede, proceed) Julia in giving her presentation.
13. I (use to, used to) cram before tests, but now I review almost every night.

TECHNO TIP

To learn about fifty words that are commonly confused, search the Internet for this video:
Eng 50 Commonly Confused Words

CAREFUL! Because we often slur words, sometimes we don't realize certain words are not spelled the way we pronounce them. Below are some frequent problem words and phrases.

gonna, gotta, wanna	Instead, use **going to**, **got to**, and **want to**. If you're **going to** the bookstore, I **want to** ask you to see if we have **got to** get a particular text for our criminology class.

may of, might of, must of, should of, would of, could of	Instead, use **may have, might have, must have, should have, would have, could have.** Our professor **must have** thought we **could have** written this paper without any further instructions; she **should have** known that we **might have** needed additional help.

Other often-heard words and phrases are considered to be **nonstandard**. Because of this, you should not use them in academic writing.

Instead of Using	Use	Instead of Using	Use
ain't	isn't, aren't, am not	nother	other, another
anyways	anyway	nowhere (if used with another negative)	anywhere
can't hardly	can hardly		
can't scarcely	can scarcely		
everywheres	everywhere	off of	from
hisself	himself	theirselves	themselves
irregardless	regardless		

29.7 Homophones and Other Easily Confused Words and Phrases, Part 7

To watch a video about *irregardless*, search the Internet for Irregardless Video

Directions: Underline the correct word or phrase in each sentence.

1. We'll take the field trip, (regardless, irregardless) of the weather.
2. (Anyway, Anyways), I'll see you after class.
3. Is it my imagination, or do you think Ernie hasn't been (hisself, himself) lately?
4. If Ellen doesn't (wanna, want to) join our peer review group, I guess that's her business.

(continued)

5. Did Jessica get this information (from, off of) the Internet?
6. Walking across campus in the blizzard, I (can't scarcely, can scarcely) see the sidewalk.
7. Whether to write a third revision is a whole (other, nother) decision.
8. Mandy (can't hardly, can hardly) stop grinning after seeing her math grade.
9. Peer Review Group B said they were turning in their revisions (theirselves, themselves) and then heading to the Student Center.
10. If you missed the announcement, you (must of, must have) been late in getting to class.
11. After class, Daniel looked (everywhere, everywheres) for Janet but couldn't find her.
12. Everyone in class had brought (themselves, theirself) a cup of coffee.
13. "I can't find a pen (nowhere, anywhere)!" Barb shouted.
14. "We (ain't, aren't) having our test today, are we?" Rob asked anxiously.

Easily Confused Phrases

LEARNING GOAL

❷ Correct mistakes in easily confused phrases

In addition to words, a number of phrases are often confused. The confusion sometimes comes as a result of mishearing part of a phrase (as in hearing someone say *all of a sudden*, but thinking the person said *all of the sudden*). Other times the confusion comes because the correct phrase contains a homophone. After hearing the phrase, a writer substitutes an incorrect spelling and botches the phrase (as in hearing *en route*, but thinking the person said *in root*).

Reading a list of problematic phrases alongside their correct forms helps writers recognize these close-but-no-cigar mistakes and isolate the particular parts of the phrases that are incorrect. Below are thirty-two phrases that are often misused.

The Correct Phrase	What's Often Incorrectly Written	The Correct Phrase	What's Often Incorrectly Written
all of a sudden	all of **the** sudden	**en route** to a party	**in root** to a party
Alzheimer's disease	**old-timer's** disease	for all **intents and purposes**	for all **intensive** purposes
amusing **anecdotes**	amusing **antidotes**	free **rein**	free **reign**
beck and call	**beckon** call	**guerrilla warfare**	**gorilla war fair**
biding your time	**biting** your time	going off on a **tangent**	going off on a **tandem**
by accident	**on** accident	he's just a **gofer**	he's just a **gopher**
can't **fathom** it	can't **phantom** it	I **couldn't** care less	I **could** care less
chest of drawers	**chester** drawers	I'd just **as soon**	I'd just **assume**
dog-eat-dog world	**doggy-dog** world		

The Correct Phrase	What's Often Incorrectly Written	The Correct Phrase	What's Often Incorrectly Written
mind your ***p's*** and ***q's***	mind your **peas** and ***cues***	**supposedly** doing something	**supposably** doing something
moot point	**mute** point	take it for **granted**	take it for **granite**
nip it in the **bud**	nip it in the **butt**	tongue **in** cheek	tongue **and** cheek
old **wives'** tale	old **wise** tale	tough **row** to hoe	tough **road** to hoe
out of **whack**	out of **wack**	**uncharted** territory	**unchartered** territory
quashed an appeal	**squashed** an appeal	up and **at 'em**	up and **adam**
right-of-way	right **away**	**whet** my appetite	**wet** my appetite
spitting image	**spit and** image		

29.8 Homophones and Other Easily Confused Words and Phrases, Part 8

Directions: Underline the correct word or phrase in each sentence.

1. Just (by accident, on accident) Travis discovered that his class had been cancelled.
2. Our class is (supposably doing something, supposedly doing something) after the final.
3. We have (free rein, free reign) on the topic of our first paper.
4. The professor's (tongue and cheek, tongue in cheek) comment drew a few laughs.
5. If you'll let me copy your notes, I'll be at your (beckon call, beck and call) to study with you.
6. (I'd just as soon, I'd just assume) get started on my paper now and relax later.
7. Our guest speaker seemed to be (going off on a tandem, going off on a tangent).
8. Beth decided to (take it for granted, take it for granite) that her peer review group liked her essay.
9. At 6:00 a.m., my roommate said it was time to get (up and adam, up and at 'em)!
10. In this (doggy-dog world, dog-eat-dog world), I like seeing a random act of kindness.
11. All the veterans in class have had a (tough row to hoe, tough road to hoe) in their careers.
12. Carol said she finished her discovery draft in an hour, but I (can't phantom it, can't fathom it).
13. The engine of my car sounded (out of wack, out of whack) yesterday.

(continued)

14. In the case we're studying in criminal justice class, the judge (squashed an appeal, quashed an appeal).
15. "You know that idea you had about partying rather than studying? Let's (nip it in the bud, nip it in the butt)," my roommate said.
16. (All of a sudden, All of the sudden) the lights went out in the room!
17. That (chest of drawers, chester drawers) in my apartment doesn't hold much.
18. My idea of how to word my thesis statement came to me (right-of-way, right away).
19. In our agriculture class, Professor Winstead related an (old wise tale, old wives' tale) about when to plant tomatoes.
20. Are you just (biting your time, biding your time) until the semester is over?
21. Pictures of Amy's sister revealed she was the (spit and image, spitting image) of Amy.
22. While they were (in root to a party, en route to a party), three members of the freshman class were involved in a wreck.
23. In Mark's essay on the last year of the Vietnam War, he read a great deal about (guerrilla warfare, gorilla war fair).
24. I just turned in my final paper, so (for all intensive purposes, for all intents and purposes), I've finished the class!
25. The class I want to take at 9:00 has already filled, so it's a (mute point, moot point) as to whether or not the book for it has sold out in the bookstore.
26. (I could care less, I couldn't care less) about what happens on most reality television shows.
27. Professor Buckley entertains class with many (amusing anecdotes, amusing antidotes).
28. This English class will (whet my appetite, wet my appetite) for the next one, I'm sure.
29. Alan is working at a downtown law firm, but for now (he's just a gofer, he's just a gopher) until he decides if he wants to pursue law as a career.
30. (Mind your *p*'s and *q*'s, Mind your peas and cues) when you have that interview!
31. When Brad finishes all his classes, he hopes to work in a nursing home that specializes in patients with (old-timer's disease, Alzheimer's disease).
32. Both Bob and Lorraine felt that they were entering (uncharted territory, unchartered territory) when they enrolled in college.

TECHNO TIP

For interactive quizzes on easily confused words, search the Internet for these sites:

The Notorious Confusables

The Notorious Confusables: Second Quiz

The Notorious Confusables: Third Quiz

The Notorious Confusables: Fourth Quiz

Why Is Spelling Difficult?

Why is correct spelling such a problem? If all English words were spelled phonetically (as they sound), spelling them correctly would not be nearly as tricky as it is. Part of the explanation to the problem of correct spelling lies in the age of the language. English originated over fifteen hundred years ago, when people spoke what is now called Old English. Over time, pronunciations, spellings, and even meanings of words have changed considerably. These Old English words have each changed in spelling and pronunciation, but not meaning:

LEARNING GOAL
❸ Learn five main spelling rules

Old English	**Modern English**
seafon	seven
mancynn	mankind
pund	pound
andswaru	answer
feor	far
geat	gate

To view the Lord's Prayer in Old English, Middle English, Early Modern English, and Late Modern English, search the Internet for Words In English: A Brief History.

To view the poem Beowulf in both Old English and Modern English, search the Internet for Beowulf: Old English and Modern English.

Although the spelling of each of these words is different from what it was long ago, you see the similarity in the original words and their modern equivalents. Other words have changed much more dramatically. For instance, you probably won't recognize these Old English words: *beadorinc* (*warrior*), *derian* (*harm*), *grymetan* (*roar*), *geteld* (*tent*), and *trymian* (*encourage*).

Another cause for strange or irregular spelling in English came when the printing press was invented in the 1400s. Before that time, many people were illiterate; they had no need for spelling rules because they could not read or write. With the invention of the printing press, however, books were more readily available and many common people learned how to read. Some early printers, however, settled on spellings that seem peculiar.

English has become a melting pot for words from other languages, and this has resulted in still more spelling oddities. Many words that originated elsewhere have been adopted into English but have retained the spelling they had in their original language.

Word Adopted into English	**Native Language**	**Word Adopted into English**	**Native Language**
raccoon	Algonquian	yacht	Dutch
sauna	Finnish	parachute	French
blitz	German	ukulele	Hawaiian
goulash	Hungarian	geyser	Icelandic
umbrella	Italian	karate	Japanese
guitar	Spanish	boulder	Swedish

How You Can Improve Your Spelling

See the eText chapter "Vocabulary" for more ideas about using print or online dictionaries.

You can rely on the spell-check feature of your computer to a certain degree, but you definitely should have a print or an online dictionary handy. Your computer won't know, for instance, if you should have used the word *there* when you typed *their* or *they're*. Always proofread your paper after you have sent it through spell-check.

See the eText chapter "Mnemonics" for more ideas about mnemonics.

When you have words that create spelling problems for you, make a list of them and then create mnemonics to help remember how to spell them correctly. For instance, if you question yourself about the spelling of the word that means "buddy" (is it *friend* or *freind*?), look the word up, find the correct spelling, and then think of a way to remember it. You might devise something like "a fri<u>end</u> to the <u>end</u>."

TECHNO TIP

Word games or puzzles can also improve your spelling. For crossword puzzles to solve, search the Internet for these sites:

Free Online Crossword Puzzles

Free Crossword Puzzles Online Daily

To play hangman online, search the Internet for this site:

Hangman Games Online

For online word search puzzles, search the Internet for this site:

Online Word Search Games

Five Handy Spelling Rules

Rule 1: Adding a Prefix

Only one spelling rule has no exceptions: Adding a prefix to a word does not alter its spelling.

Prefix	Root Word	New Word
mis-	understand	misunderstand
un-	known	unknown
pre-	pay	prepay
semi-	circle	semicircle

Rule 2: *I* Before *E*

You may have learned this spelling mnemonic in elementary school:

I before e, except after c, or when sounded as ay, as in neighbor or weigh.

That rule often works, especially with this condition added to the first part of it:

Use i before e (except after c) if the sound made by ie is a long e (as in niece).

Ironically, one of the most commonly misspelled words is *misspell*. Correctly spelling this word should be easy, though, because it just adds a prefix (*mis-*) to a root word (*spell*).

I before *e*,	this often works when the sound made by the *ie* vowels is a long *e* (as in *see*)	believe, chief, piece
Except after *c*,	this often works	deceive, receipt
Or when sounded as *ay*, as in *neighbor* or *weigh*.	this often works	beige, vein, eight, their

Here are some common exceptions to the *i* before *e* rule:

ancient	efficient	forfeit	neither	seizure
caffeine	either	geiger	poltergeist	sheik
codeine	Fahrenheit	heifer	protein	sleight
conscience	feisty	height	reveille	sovereign
conscientious	financier	kaleidoscope	seismograph	stein
counterfeit	foreign	leisure	seize	weird

TECHNO TIP

For an online spelling quiz concerning "ie" and "ei" words, search the Internet for this site:

Hyper Grammar: "ie" or "ei" Quiz

For more on spelling rules, search the Internet for these videos:

Ie/ei Spelling Rules Video

Spelling Rule: I Before E Video

TICKET to WRITE

29.9 Spelling (*I* Before *E*)

Directions: Underline the correctly spelled word in the sentences below.

1. "This paper is really trying my (patience, pateince)!" screamed Autumn.
2. Online identity theft is a form of (deciet, deceit).
3. "I'd like you to meet my (friend, freind) Blake," Brittney said to Antonio.
4. "It's better to give than to (recieve, receive)" is a well-known adage.
5. We're studying the (grief, greif) process in my course on death and dying.

Rule 3: Adding a Suffix to a Word That Ends in a Silent -*E*

When adding a suffix to a word ending in a silent -*e*

- **drop the -*e* if the suffix begins with a vowel**

wage + -ing = waging   refuse + -al = refusal

- **keep the -*e* if the suffix begins with a consonant**

hope + -ful = hopeful   large + -ly = largely

Some common exceptions:

argue + -ment = argument   intervene + -tion = intervention
due + -ly = duly   true + -ly = truly

TECHNO TIP

For an online quiz about the silent "e," search the Internet for this site:

Review: Final Silent "e"

29.10 Spelling (Adding a Suffix to a Word That Ends in a Silent -*E*)

Directions: Underline the correctly spelled word in the sentences below.

1. Russell thanked Lacey for being a (caring, careing) coworker.
2. Ramon was (carful, careful) when he let his boss know about his mistake at work.
3. Jamie was worried that he had (scarcely, scarcly) enough money to cover the bill.
4. The (scarecity, scarcity) of available computers proved to be a problem.
5. Both Troy and Daisy remarked that the lecture was (engageing, engaging).

Rule 4: Adding a Suffix to a Word Ending in -*Y*

When adding a suffix to a word ending in -*y*

- **change the *y* to an *i* when the -*y* is preceded by a consonant**

happy + -est = happiest   crazy + -er = crazier

This does not apply if the suffix begins with an i, as in:

carry + -ing = carrying

- **keep the -*y* when it is preceded by a vowel**

pay + -s = pays   obey + -ing = obeying

Some common exceptions:

shy + -ness = shyness   wry + -ly = wryly   day + -ly = daily

TECHNO TIP

For an online quiz about the final "y" before a suffix, search the Internet for this site:

Review: Final "y" Before a Suffix

29.11 Spelling (Adding a Suffix to a Word Ending in -Y)

Directions: Underline the correctly spelled word in the sentences below.

1. I would estimate the woman to be in her (fortys, forties).
2. The (fortyish, fortiish) man was in front of Hayden in line.
3. Victor and Miranda were invited to the same (partys, parties) over the weekend.
4. That new little restaurant is located between (alleys, alleies) downtown.
5. When Marcus (marrys, marries) Monica, they will still have one semester of classes.

Rule 5: Words with Double Consonants

When adding a suffix beginning with a vowel, double the final consonant of the root word if

- **the consonant ends a stressed syllable of the root word or the word is only one syllable and**
- **the consonant is preceded by only one vowel**

brag + -ed = bragged   hot + -er = hotter

"Take care that you never spell a word wrong. Before you write a word, always consider how it is spelled, and, if you do not remember, turn to a dictionary."
— Thomas Jefferson

Words ending in *-w*, *-x*, *-v*, or *-y* do not double the consonants.
Some common exceptions:

saw + -ed = sawed   fix + -es = fixes   profit + -ed = profited

TECHNO TIP

For an online quiz about double consonants, search the Internet for this title: Review: Double Consonants

29.12 Spelling (Words with Double Consonants)

Directions: Underline the correctly spelled word in the sentences below.

1. It never (occurred, occured) to me that I might have thrown my cell phone out with the trash.
2. Was Professor Strawn (referring, refering) to last night's television lineup?
3. (Stopping, Stoping) smoking was one of the hardest experiences of Hector's life.

(continued)

4. Omar and Kylie agreed that the concert had been (delighttful, delightful).
5. Tabitha (bagged, baged) the eggs separately from the canned goods.

Correctly Use Abbreviations and Numbers

LEARNING GOAL

4 Correctly use abbreviations and numbers in academic writing

Abbreviations

Abbreviations are shortened forms of words or phrases.

In academic writing, **abbreviate**

• a courtesy title before a name	Mr., Mrs., Ms.
• part of a person's name	Martin Luther King, Jr.
• references to time of day	a.m.
• organizations commonly known by their initials	UN
• commonly used Latin abbreviations	etc.

In academic writing, **spell out**

• days	~~Wed.~~ Wednesday	Using abbreviations in these instances is not acceptable in academic writing.
• months	~~Feb.~~ February	
• street names	~~Blvd.~~ Boulevard	
• state names	~~FL~~ Florida	
• courses	~~Eng~~ English	
• sections in a book	~~Pt. 3~~ Part 3	
• units of measurement	~~two ft.~~ two feet	

29.13 Abbreviations in Academic Writing

Directions: In the following sentences, underline the form that is correct in academic writing.

1. Fernando's (<u>psychology</u>, psych) class was cancelled.
2. Was Summer's grandfather Hector Sanford, (Senior, <u>Sr.</u>)?
3. Bring any equipment you'll need—(exempli gratia, <u>e.g.</u>), extra socks, flashlight, and batteries.
4. The first part of (<u>November</u>, Nov.) was wet and dreary.
5. How long did Hassan take to travel across (<u>Texas</u>, TX)?

Numbers

In academic writing, **spell out**

- **Numbers that can be expressed in one or two words**

five hundred (two words)	**501** (more than two words)
thirty-three (two words)	**3,456** (more than two words)
fifty-two dollars (two words)	**$152** (more than two words)

CAREFUL! Be consistent. If one number in a sentence requires a figure, change all the numbers to figures.

> 400
> Of the ~~four hundred~~ people attending, 356 were from the same town.

- **Figures that begin a sentence**

> Seventy-five percent
> ~~75%~~ of the nursing students passed their state exam.

- **Fractions**

> one-fourth
> Only ~~1/4~~ of the students took advantage of the review session.

In academic writing, **use figures** with

• days and years, but not months	October 16, 1947
• times	7:30 p.m.
• street numbers	436 Prince Avenue
• exact amounts of money	$6.98
• scores	21-14
• parts of a written work (book, play)	Chapter 13

Roman Numerals

Figures that are used predominantly today are "Arabic numbers" or "Arabic numerals." Another way of numbering is with Roman numerals, which are now used mainly in

• publication dates of books	published MMXIII
• introductory page numbers of books	page ix
• volume and chapter numbers of books	Chapter VIII
• some forms of outlining	II. Reasons for Identity Theft
• games	Super Bowl IX
• historic events	World War II
• inscriptions on buildings	MCMVIII
• names of kings, queens, and popes	Queen Elizabeth II
• children whose names are identical to their parents	John Michael Hernandez III

CAREFUL! In academic writing, use Roman numerals if they—not Arabic figures—are commonly used in material you are writing about.

VIII

King Henry ~~the Eighth~~ had six wives.

Roman Numerals			
Arabic Number	**Units**	**Tens**	**Hundreds**
1	I	X	C
2	II	XX	CC
3	III	XXX	CCC
4	IV	XL	CD
5	V	L	D
6	VI	LX	DC
7	VII	LXX	DCC
8	VIII	LXXX	DCCC
9	IX	XC	CM

In Roman numerals, 1000 = M.

TECHNO TIP

To learn more about Roman numerals, search the Internet for these videos:

Roman Numerals 1–25 (1–XXV)

Roman Numerals 101 Video

29.14 Numbers in Academic Writing

Directions: In the following sentences, underline the form that is correct in academic writing.

1. All (24, twenty-four) people in the campsite felt the earthquake.
2. (3, Three) kittens at the shelter looked as if they wanted to go home with Vaughn and Leann.
3. The package was delivered to (12, Twelve) North Ingram Street.
4. (½, One-half) of a point was all Davin needed to get a higher grade.
5. Of the (10,000, ten thousand) fans at the game, only (200, two hundred) pulled for the other side.

MyWritingLab™ Visit ***MyWritingLab.com*** and complete the exercises and activities in the **Easily Confused Words, Spelling, and Abbreviations and Numbers** topic areas.

RUN THAT BY ME AGAIN

LEARNING GOAL
❶ Correct mistakes in homophones and other easily confused words

- **Homophones are easily confused because they . . .** sound alike but are spelled differently and have different meanings.
- **Reading a list of troublesome words—along with the difference between the words—helps writers . . .** recognize the specific words they find tricky.
- **Other times, writers identify words they misuse and decide the correct form of the word they want by . . .** using mnemonics.
- **Because of the way we slur words, sometimes we don't realize . . .** certain words are not spelled the way we pronounce them.
- **Words like *hisself*, *everywheres*, and *ain't* should not be . . .** used in academic writing.

LEARNING GOAL
❷ Correct mistakes in easily confused phrases

- **Confusion with phrases sometimes comes as a result of . . .** mishearing part of a phrase or because the phrase contains a homophone.

LEARNING GOAL
❸ Learn five main spelling rules

- **Early printers sometimes . . .** settled on particular spellings that seem peculiar.
- **Many words that originated elsewhere have been adopted into English but . . .** have retained the spelling they had in their original language.
- **You can rely on the spell-check feature of your computer to a certain degree, but . . .** you should have a print or an online dictionary handy.
- **When you have words that create spelling problems for you, make a list of them and then . . .** create mnemonics to help you remember how to spell them correctly.
- **Adding a prefix to a word . . .** does not alter its spelling.
- **A common spelling mnemonic is . . .** *i* before *e*, except after *c*, or when sounded as *ay*, as in *neighbor* or *weigh*.
- **Exceptions to the *i* before *e* rule include . . .** *caffeine*, *efficient*, *foreign*, *seize*, *either*, *conscience*, and *height*.
- **When adding a suffix to a word ending in a silent *-e* . . .** drop the *-e* if the suffix begins with a vowel; keep the *-e* if the suffix begins with a consonant.
- **When adding a suffix to a word ending in *-y* . . .** change the *y* to an *i* when the *-y* is preceded by a consonant; keep the *-y* when it is preceded by a vowel.
- **When adding a suffix beginning with a vowel, double the final consonant of the root word if . . .** the consonant ends a stressed syllable of the root word or the word is only one syllable and the consonant is preceded by only one vowel.

LEARNING GOAL
❹ Correctly use abbreviations and numbers in academic writing

- **Abbreviations are . . .** shortened forms of words or phrases.
- **In academic writing, abbreviate . . .** a courtesy title before a name, part of a person's name, references to time of day, organizations commonly known by their initials, and commonly used Latin abbreviations.
- **In academic writing, spell out . . .** days, months, street names, state names, courses, sections in a book, and units of measurement.

EASILY CONFUSED WORDS AND PHRASES; SPELLING, ABBREVIATIONS, AND NUMBERS LEARNING LOG MyWritingLab™

Answer the questions below to review your mastery of easily confused words and phrases.

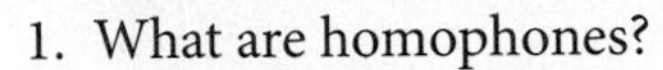

1. What are homophones?

   Homophones are words that sound alike but are spelled differently and have different meanings.

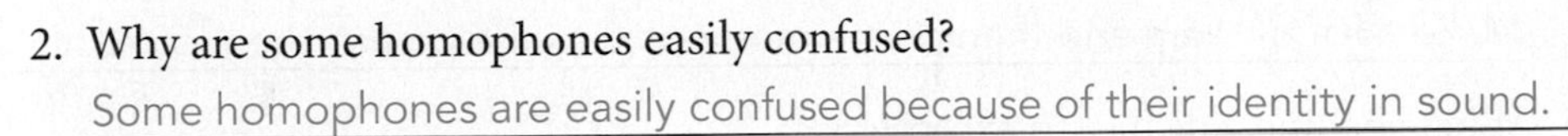

2. Why are some homophones easily confused?

   Some homophones are easily confused because of their identity in sound.

3. Sometimes, reading a list of troublesome words—along with the difference between the words—helps writers do what?

   Sometimes, reading a list of troublesome words—along with the difference between the words—helps writers recognize the specific words that they find tricky.

4. What do some writers use to help identify words they misuse and decide the correct form of the word they want?

   Some writers use mnemonics to help identify words they misuse and decide the correct form of the word they want.

5. Why do we sometimes not realize that words are not spelled the way we pronounce them?

   Because of the way we often slur words, sometimes we don't realize certain words are not spelled the way we pronounce them.

6. Why should words and phrases like *ain't*, *can't hardly*, and *irregardless* not be used in academic writing?

   Those words are considered nonstandard.

7. What can reading a list of problematic phrases alongside their correct forms help writers do?

   Reading a list of problematic phrases alongside their correct forms helps writers recognize their mistakes and isolate the particular parts of the phrases that are incorrect.

8. How long ago does the English language have its roots?

   English has its roots over fifteen hundred years ago.

9. What happened in the 1400s that changed the spelling of English words?

   When the printing press was invented in the 1400s, some early printers settled on particular spellings that seem peculiar.

10. What is a third cause of spelling oddities?
English has become a melting pot for words from other languages. (or) Many words that originated elsewhere have been adopted into English but have retained the spelling they had in their original language.

11. After you have sent your paper through spell-check, what should you do?
Always proofread your paper again.

12. What should you do when you have words that create spelling problems for you?
Make a list of them and then create mnemonics to help spell them correctly.

13. What spelling rule has no exceptions?
Adding a prefix to a word does not alter its spelling.

14. The "*i* before *e*" rule often works with what condition?
Use *i* before *e* (except after *c*) if the sound made by *ie* is a long *e* (as in *hygiene* or *niece*).

15. Name five common exceptions to the "*i* before *e*" rule.
Answers will vary.

16. When adding a suffix to a word ending in a silent -*e*, drop the -*e* under what condition?
Drop the -*e* if the suffix begins with a vowel.

17. When adding a suffix to a word ending in a silent -*e*, keep the -*e* under what condition?
Keep the -*e* if the suffix begins with a consonant.

18. When adding a suffix to a word ending in -*y*, change the -*y* under what condition?
Change the -*y* to an *i* when the -*y* is preceded by a consonant.

19. When adding a suffix to a word ending in -*y*, keep the -*y* under what condition?
Keep the -*y* when it is preceded by a vowel.

20. When adding a suffix beginning with a vowel, double the final consonant of the root word under what two conditions?
Double the final consonant of the root word if the consonant ends a stressed syllable of the root word or the word is only one syllable and the consonant is preceded by only one vowel.

21. Name three types of abbreviations that are acceptable in academic writing.
Answers will vary.

22. In academic writing, should numbers that can be expressed in one or two words be spelled out or written as figures?

    Numbers that can be expressed in one or two words should be spelled out.

23. In academic writing, should figures that begin a sentence be spelled out or written as figures?

    Figures that begin a sentence should be spelled out.

24. In academic writing, should fractions be spelled out or written as figures?

    Fractions should be spelled out.

25. In academic writing, should days and years be spelled out or written as figures?

    Days and years should be written as figures.

26. In academic writing, should parts of a written work be spelled out or written as figures?

    Parts of a written work should be written as figures.

27. In academic writing, you should use Roman numerals under what condition?

    Use Roman numerals if they—not Arabic figures—are commonly used in material you are writing about.

Punctuation Marks

LEARNING GOALS

In this chapter, you'll learn and practice how to

1. Use correct end punctuation
2. Use quotation marks to indicate quotes, short titles, and specific terms
3. Use apostrophes to indicate contractions and possession
4. Use commas correctly to ensure your writing is clear and precise
5. Use colons to introduce lists, explanations, and long quotations
6. Use semicolons to join related ideas
7. Use hyphens, dashes, parentheses, italics, underlining, and ellipsis points correctly

GETTING THERE

- Punctuation marks are a set of symbols you use to communicate to readers.
- Using correct punctuation marks reduces the chances of your sentences being misread or misinterpreted.

Final Punctuation

LEARNING GOAL

1. Use correct end punctuation

In English, all sentences end in a period, a question mark, or an exclamation mark. Each end mark sends a particular signal to your reader.

Periods

By far the most commonly used final punctuation mark—especially in academic writing—is the period.

1. Use a **period** to signal the end of a sentence that states a fact or an opinion or that makes a request.

Alaska is the largest state. (fact)

Mammoth Cave National Park is the most impressive national park. (opinion)

Please reserve a room at the Commonwealth Convention Center. (request)

CAREFUL! One common mistake of novice writers is to use a question mark at the end of a sentence that states a fact but includes a question in it, as in this one:

The instructor asked if anyone had a calculator.

Taken as a whole, the sentence states a fact (the instructor asked something), so the end mark should be a period.

2. Use a period in abbreviations, like **St.**, **Ms.**, **e.g.**, and **Gov.**

CAREFUL! In a sentence that ends in an abbreviation, use only one period at the end.

Only one period ↓

The Franks invaded Gaul in 257 A.D.

Heads Up!
When typing, don't put a space between the last letter of a sentence and the final punctuation mark.

TECHNO TIP!
For more on periods, search the Internet for this site:
Punctuation Periods Commnet

Heads Up!
Periods in URLs and e-mail addresses are called "dots." If you see a URL or an e-mail address at the end of a sentence, remember that the period at the end of the sentence is the final punctuation and not a dot in the URL or e-mail address.

Question Marks

Question marks signal uncertainty, either in what is asked or what is noted.

1. **With Direct and Tag Questions:** Use question marks at the end of direct questions

Are we supposed to complete Ticket to Write 30.1?

and sentences that end in questions.

We're supposed to complete Ticket to Write 30.1, aren't we?

A question at the end of a sentence is sometimes called a *tag question*; because of what is tagged on, the sentence as a whole becomes a question rather than a statement.

2. **To Note Other Uncertainty:** Use question marks to designate that something is uncertain. Often, this uncertainty denotes a date, as in this sentence:

Mark Allen (1754?–1825) built this home.

TECHNO TIP
For more on question marks, search the Internet for these sites:
Punctuation Quotation Marks Commnet
Punctuation Quotation Marks Rules

> **TECHNO TIP**
>
> For more on exclamation marks, search the Internet for these these sites:
>
> Punctuation Exclamation Marks Commnet
>
> English Club—Exclamation Mark

Exclamation Marks

Use **exclamation marks** (exclamation points) to convey strong feelings.

> I had no idea my brother could get a leave and come home for my party!

CAREFUL! In academic writing, however strongly you feel about a subject, do not use exclamation marks unless you are quoting material.

30.1 Using Periods, Question Marks, and Exclamation Marks

Directions: Working alone or in groups, use the rules about periods, question marks, and exclamation marks to identify which of the following sentences are correctly punctuated. Write *C* for correct or *I* for incorrect. Correct the sentences you marked *I*.

__I__ 1. Cece wondered if she knew the way to get to Paul's house?

__C__ 2. You won't believe what just happened!

__I__ 3. Rebecca Wells is one of your favorite authors, isn't she. ?

__I__ 4. Wasn't the telephone invented by Alexander Graham Bell. ?

__I__ 5. I can now address my uncle as Donald Bralty, M.D.. D.

__C__ 6. Is anyone using this computer?

__C__ 7. Please read Chapter 21 for homework for the next class.

__C__ 8. Let's party!

__I__ 9. A notable date in Roman history is the Battle of Adrianople in 378 A.D.. D.

__I__ 10. You said I could borrow your notes, didn't you. ?

Quotation Marks

Quotation marks alert readers to words a person said or wrote, to titles of short works, and to slang and technical terms.

LEARNING GOAL
❷ Use quotation marks to indicate quotes, short titles, and specific terms

Direct Quotations

In academic writing, use quotation marks to show readers

- words you or others spoke (this is called a *direct quotation*)
- words you copy verbatim from someone else

In a direct quotation, use quotation marks around the words you or someone else spoke. The words, however, must be exact and in the same order the person spoke them. If you change the order or the wording, you have an *indirect quotation*, and you may not use quotation marks.

This direct quotation is punctuated correctly because the words inside the quotation marks are Henry's exact words, in exactly the order he said them.

Patrick Henry said, "Give me liberty or give me death."

This *indirect quotation* is punctuated correctly because Henry's words are not reported exactly, and they don't come in the exact order he said them.

Patrick Henry said to give him liberty or give him death.

CAREFUL! A common mistake is using quotation marks with indirect quotations that contain words like *said* or *asked*. For example, suppose Donna asked Jim to hand her the newspaper. When writing that request, students sometimes mistakenly punctuate the sentence this way:

Donna asked, "If Jim would hand her the newspaper."

Because the words inside the quotation marks are not the exact words Donna spoke, in exactly the order she spoke them, the sentence is punctuated incorrectly. As it is written, the quotation marks and the comma should be omitted, and the word *If* should be in lowercase, like this:

Donna asked if Jim would hand her the newspaper.

Quotation Marks with Dialogue

When quoting dialogue, show readers a change in speakers by starting a new paragraph every time a speaker changes.

Rick was surprised to see Rhoda. "Hey. What's up?" he asked.

"Hi, old friend." Different speaker, new paragraph

"It's been a long time." Different speaker, new paragraph

"Yes, and I came with a surprise." Different speaker, new paragraph

"What's that?" Different speaker, new paragraph

"I found your wallet on the sidewalk." Different speaker, new paragraph

Quotation Marks with One Paragraph from a Single Source

If you're quoting a paragraph that has multiple sentences from the same person or text, put opening quotes at the beginning of the first sentence and closing quotes only at the end of the last sentence, not at the beginning and end of each sentence.

Alan continues to speak, so no new quotation marks

Alan said, "After I study for an hour, I'll eat a snack. Then I plan to take a short nap. When I wake up, I'll hit the books again. I want to be ready for the test." ← Closing quotation marks show Alan has finished speaking

Quotation Marks with Multiple Paragraphs from a Single Source

Heads Up!
If you're required to use a certain style, such as MLA or APA, you'll find there are additional requirements for quoting and for quotation marks. See Chapter 15.

If you're quoting more than one paragraph from the same person or the same text, place opening quotation marks at the beginning of each paragraph and closing quotation marks only at the end of the final paragraph. This lets your readers know that the same person or text is being quoted throughout.

Rupert said, "When my family arrives at the beach, the first thing we do is unpack the car and get our swimming gear out of our suitcases. We all usually have a quick dip in the ocean as our first priority.

Rupert continues to speak, so no closing quotation marks

Opening quotation marks mean Rupert continues in a new paragraph

"If we arrive in the middle of the night, though, we unpack and go on to bed, knowing that when we wake up the ocean will be there to greet us."

Closing quotation marks show Rupert has finished speaking

Quotation Marks with Titles

Use quotation marks around titles of short works such as the following:

Short stories	"Calling Me Home"	Short poems	"Trees"
Magazine articles	"Invest Now!"	Essays	"A Modest Proposal"
Chapters of books	"The Allure of Coffee"	Songs	"Written in Reverse"
Newspaper articles	"Mayor Backs Tax Hike"		

Quotation Marks with Slang and Technical Terms

Use quotation marks around slang, technical terms, or words you're using outside of their normal or literal usage.

Heads Up!
In academic writing, slang is not usually acceptable, except in dialogue.

Granddad looked perplexed when I said he was "da bomb."

As a beginning sailor, Taylor didn't know the meaning of "ready about."

Quotation Marks and Other Punctuation

1. **Periods and commas:** Place periods and commas *inside* closing quotation marks; place colons and semicolons *outside* closing quotation marks.

Dorene said, "I won't spend money on that movie," but Harry enjoyed it.

Dorene said, "I won't spend money on that movie"; however, Harry enjoyed it.

While periods always go inside closing quotation marks, the two other end marks (question marks and exclamation marks) go either inside or outside closing quotation marks, depending on what you are quoting.

2. **Questions marks:** If *the words you are quoting contain a question*, put the question mark *inside* your closing quotation marks.

> The student government president asked Stacy, "Why don't you run for secretary?"

Because the words that the president spoke are a question, the question mark goes inside your closing quotation marks.

If *the sentence as a whole makes a question*—not the words you're quoting—put the question mark *outside* your closing quotation marks.

> Did the student government president say, "I hope Stacy runs for secretary"?

The words the president said ("I hope Stacy runs for secretary") don't form a question; the sentence as a whole does. That's why the question mark goes outside the closing quotation marks.

3. **Exclamation marks:** The rules above also apply to exclamation marks. If *the words you're quoting are exclaimed*, put the exclamation mark *inside* your closing quotation marks.

> Cheryl said to Gerald, "I can't believe I have three tests in one day!"

However, if *the sentence as a whole forms the exclamation*, put the exclamation mark *outside* your closing quotation marks.

> Franklin actually said, "I'll pay for your lunch"!

CAREFUL! If you have a construction in which both your quoted material *and* your whole sentence are questions or exclamations, use only one end mark, and place it inside your closing quotation marks.

> Did you say, "What time did you call?"

Heads Up!
You'll notice different punctuation used in books published in Britain, Canada, and other English-speaking countries.

4. **Other quotation marks:** If you have a quotation within a quotation, use single quotes to show readers where the quotation or title begins and ends.

Single quotes inside double quotes

> "I liked the short story 'Telephone Line' so much I want to read its sequel," Anna said.

TICKET to WRITE

30.2 Using Quotation Marks

Directions: Work alone or in groups and use the rules about quotation marks to mark the sentences below. Mark *C* if correct or *I* if incorrect. Correct the sentences you marked *I*.

___I___ 1. Jeremy said, "We need to get gas before we leave for the trip", but I didn't have money.

___C___ 2. After I read the article "Blues Festival Heats Up" in the paper, I decided to attend.

___I___ 3. Jimmy said "He liked tea better than coffee." [corrected: he]

___I___ 4. Maria's boss asked, "Can you work overtime tonight"? [corrected: ?"]

___I___ 5. When my niece said that the couple walking down the street was cupcaking, I was almost afraid to ask what that meant. [corrected: "cupcaking,"]

___I___ 6. "Langston Hughes's poem "Life Is Fine" really moved me," Keith said. [corrected: 'Life Is Fine']

___I___ 7. After Shawn read Poe's short story The Fall of the House of Usher, he had trouble getting to sleep. [corrected: "The Fall of the House of Usher,"]

___C___ 8. When Deidre called, she said, "Please help me with this paper."

___I___ 9. For both German and British forces, Lili Marlene was perhaps the most popular song of WWII. [corrected: "Lili Marlene"]

___I___ 10. Robin shouted, "Stand with your feet apart." "Hold one dumbbell in each hand." "Curl the dumbbells toward your shoulders." "Push the dumbbells overhead." "Rotate your palms away from your body." "Hold this for one second." "Lower the dumbbells."

TECHNO TIP

For more on quotation marks, search the Internet for the site "How to Use Quotation Marks" and for these videos:

English Grammar & Punctuation: How to Use Quotation Marks in Articles

English Grammar & Punctuation: How to Use Quotation Marks in Dialog

English Grammar & Punctuation: How to Use a Comma Before Quotation Marks

Apostrophes

An **apostrophe** is used in a word that is a contraction; the apostrophe indicates that at least one letter has been omitted. In the following list of contractions, the first four are commonly misused because they are *homophones*, words that are pronounced the same way but have different meanings.

LEARNING GOAL

❸ Use apostrophes to indicate contractions and possession

The two words . . .	use an apostrophe to form the contraction . . .	commonly confused with . . .	
it is	it's	its	possessive pronoun (shows ownership)
they are	they're	there	an adverb meaning "that location"
		their	a possessive pronoun (describes a noun)
who is, who has	who's	whose	a possessive pronoun (shows ownership)
you are	you're	your	a possessive pronoun (shows ownership)
he will	he'll		
we have	we've		
I had (or I would)	I'd		
could have	could've		

Odd Contractions

While most commonly used contractions include a verb, here are some that do not:

The word(s) . . .	use an apostrophe to form the contraction . . .	The word(s) . . .	use an apostrophe to form the contraction . . .
of the clock	o'clock	you all	y'all
madam	ma'am	old	ol'
never-do-well	ne'er-do-well	and	'n' (rock 'n' roll)
cat-of-nine-tails	cat-o'-nine-tails		
jack-of-the-lantern	jack-o'-lantern		

CAREFUL! Some instructors do not allow contractions in academic papers, except in quotations or in words (like some in the list of odd contractions above) that have come into common usage as contractions.

Ownership or Possession

Apostrophes also indicate possession or ownership.

CAREFUL! A frequent mistake is to use an apostrophe to show a plural. If you are writing

President Washington made three points in his speech.

the word *points* is simply a plural. It does not own or possess anything, and it is not a contraction. No apostrophe is needed. Here are some simple rules for using apostrophes to show ownership or possession:

Rule 1: Look at the word that owns or possesses something. If that word doesn't end in *-s*, form the possessive by adding *-'s*.

The word that possesses something

the tail of the cat (a cat possesses a tail)

Because *cat* doesn't end in *-s*, you show possession by adding *-'s*.

The 's shows the cat possesses the tail

the cat's tail

Rule 2: Look at the word that is owning or possessing. If that word is plural and ends in *-s*, form the possessive by adding an apostrophe after the final *-s*.

The word that possesses something

the tails of two cats (two cats possess tails)

Because *cats* is plural and ends in *-s*, show possession by adding an apostrophe.

Because two cats possess tails, use s'

the two cats' tails

Rule 3: Look at the word that owns or possesses. If that word is singular and ends in *-s*, you may form the possessive in one of two ways:

- by adding an apostrophe after the final *-s*
- by adding *'s* after the final *-s*.

Heads Up!
Check to see if your instructor has a preference for one of these forms.

The word that possesses something

the hat of Charles Sommers (one person owns a hat)

Because *Sommers* is singular and ends in *-s*, show possession by adding an apostrophe after the final *-s* or by adding *'s* after the final *-s*.

A singular person, with a name ending in -s,

use s' → Charles Sommers' hat

or 's → Charles Sommers's hat

CAREFUL! In situations in which pronunciation would be difficult, such as *Moses* or *Achilles*, use the *-s'* option only.

Rule 4: If you have more than one person, place, or thing owning or possessing something, you need to show readers whether the ownership or possession is joint or individual. You indicate this by the number of apostrophes you use.

If the people, places, or things own or possess something together, put the appropriate apostrophe after only the final name. Suppose you're borrowing books from your friends Ike and Betty. If they own the books together, punctuate your phrase like this:

Apostrophe after only the final name, to indicate joint ownership

Ike and Betty's books

But if your friends own books separately, punctuate like this:

Apostrophes after both names, to indicate individual ownership

Ike's and Betty's books

TECHNO TIP

For more on apostrophes, search the Internet for the sites Punctuation Apostrophe Commnet and OWL: Apostrophe and for these videos: Grammar & Punctuation: How to Use Apostrophe Worksheets English Grammar & Punctuation: How to Use Apostrophes

Denoting Plurals

Use an apostrophe to form a plural only when omitting the apostrophe might result in misreading the material (a rare occurrence). For instance, to form the plural of a lowercase letter, add *'s* to the letter.

Without the apostrophes, the words would read "is" and "ts."

I'll dot my i's and cross my t's before handing in my paper.

TICKET to WRITE

30.3 Using Apostrophes

Directions: Use the rules about apostrophes to mark the sentences below. Write C if correct or *I* if incorrect. Correct the sentences you marked *I*.

I 1. The six remaining ~~contestant's~~ contestants seemed to be bickering with each other.

C 2. Who would get ownership of Steve and Ashley's dog was the hardest problem to solve in their breakup.

C 3. Kelly was confused because the two books' titles were the same.

I 4. Three fishing ~~hooks'~~ hooks were stuck in my hat.

I 5. Is that a new shirt ~~your~~ you're wearing?

I 6. The ~~Philbin's~~ Philbins visit to the ~~McKendree's~~ McKendrees was cut short by the storm.

C 7. Two cords were tangled together.

C 8. Jack Jones was a popular singer in the 1960s.

C 9. Lou Ann's and Michelle's children played together in the park.

I 10. ~~Its~~ It's my party, and ~~Ill~~ I'll have another piece of cake.

Commas

Commas are used inside a sentence to help readers understand a sentence's exact meaning.

LEARNING GOAL

4. Use commas correctly to ensure your writing is clear and precise

Commas in a Series

Use commas to separate a series of items.

Heads Up!
Some instructors prefer a comma between the last two items in a series; some prefer it be eliminated. Check to see if your instructor has a preference.

Please pass the salt, pepper, and mustard.

Use commas to separate two or more adjectives that modify a noun.

That loud, grating sound nearly drove me crazy.

Commas are the most frequently used punctuation mark.

CAREFUL! If you use *and*, *or*, or *nor* to connect all the items in the series, eliminate the commas:

Please pass the salt and pepper and mustard.

Please pass the salt or pepper or mustard.

The picnic basket contained neither salt nor pepper nor mustard.

Commas with a Compound Sentence

Heads Up!
Remember the *boysfan* mnemonic: *but, or, yet, so, for, and, nor.*

Use a comma to join two or more independent clauses (thoughts that could stand alone as sentences) connected by *but*, *or*, *yet*, *so*, *for*, *and*, or *nor*.

Independent clause | Independent clause

That loud, grating sound nearly drove me crazy, but suddenly it stopped.

Independent clause | Independent clause

Marcy drove six hours, and she arrived at the beach just in time for the party.

Commas with Quoted Material

If a sentence you're quoting is interrupted by words such as *he said* or *she replied*, insert one comma before the first closing quotation mark and another before the second beginning quotation mark.

Comma before closing quotation mark

"For the next class," Prof. Thompson said, "read and annotate Chapters 3–5."

Comma before beginning quotation mark

Commas with Clauses, Participial Phrases, Appositives, and Unnecessary Words

1. **Commas with Clauses:** Use commas to set apart a nonrestrictive clause (a group of words with a subject and verb that can be removed from a sentence without changing the meaning of the sentence).

   Josh's art class, which meets on Tuesday nights, is the one he likes the most.

   If you remove

   which meets on Tuesday nights

   the meaning isn't changed; you still have the gist of the sentence. Because that's the case, put commas around that clause.

2. **Commas with Participial Phrases:** Use commas with a participial phrase (a participle plus any words that complete its thought).

   If the participial phrase comes at the beginning of the sentence, separate it from the noun it modifies with a comma.

   Participial phrase

   Jumping for joy, Lori shared the news of her grade with her family.

   If the participial phrase comes in the middle or end of the sentence, separate it from the rest of the sentence with commas if it is not necessary for the meaning of the sentence.

   Participial phrase not necessary for meaning

   Cindy Reid, looking frantically through the crowd, finally spotted her cousin.

   Participial phrase necessary for meaning

   The woman asking about you is across the hall.

Participial phrase not necessary for meaning

The bird returned to its nest, singing softly.

Participial phrase necessary for meaning

Maura watched the bird flying through the air.

3. **Commas with Appositives:** An appositive is a noun or noun phrase that gives more information about another noun in a sentence.

Noun Noun phrase that's an appositive

Madison, my beautiful goddaughter, has just become a middle-schooler.

Use commas to set off appositives from the rest of the sentence. If the appositive comes at the end of the sentence, use only one comma. If the appositive comes in the middle of the sentence, use two commas.

Noun Noun phrase that's an appositive

Grandmother was sad when As the World Turns, her favorite soap opera, was canceled.

Noun Noun phrase that's an appositive

Grant ordered his favorite breakfast, a Western omelet.

4. **Commas with Unnecessary Words:** Use a comma after an introductory word or phrase that is not essential for the meaning of the sentence.

Now, I told you not to do that!

Since you asked, I'll let you in on a little secret.

Oh, that's okay.

Use a comma before a nonessential word or phrase that comes at the end of a sentence.

Want to grab a quick bite to eat in the cafeteria, Chris?

Call back when you can, please.

If unnecessary words or phrases interrupt a sentence, use commas on both sides of them.

Drop by between six and seven, Matt, and John will help with the biology homework.

The movie I saw last night, like the one I watched last week, was a real dud.

Use a comma after an introductory participle, gerund, infinitive, or verbal phrase.

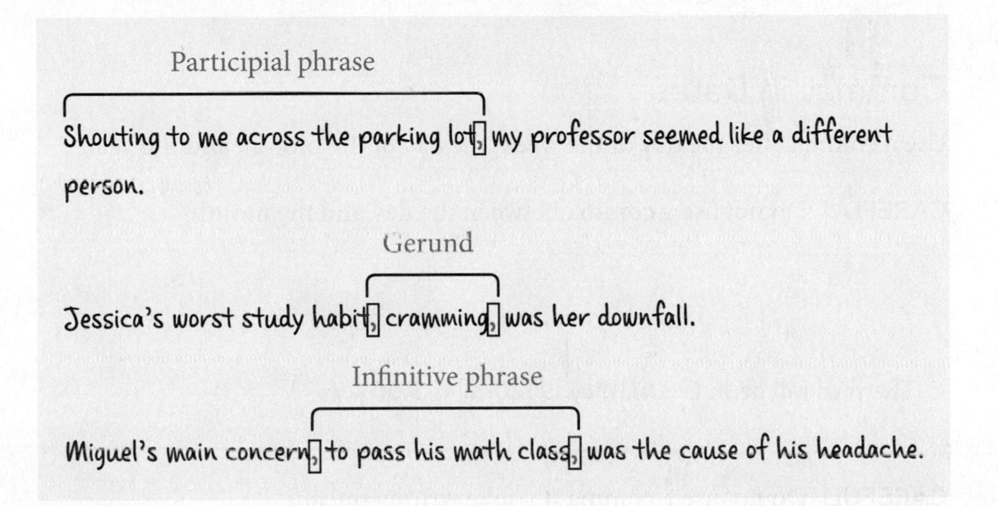

Commas in Addresses

When writing an address in the body of your text, put one comma between the person's last name and the beginning of the street address, another comma after the street address, and another between the city and the state. Do not put a comma between the state and the zip code.

For an address on separate lines, use a comma only between the city and state.

Margaret Eberhart
335 Wayville Ct.
Conrad, CO 80948

If you write a city and state in text, enclose the state in commas.

The festival will be held in Bloomington, Indiana, in early June.

CAREFUL! If you write a city and country in text, enclose the country in commas.

A visit to Inverness, Scotland, is on my wish list.

Commas in Dates

Use a comma after the day of the week, the day of the month, and the year.

CAREFUL! Do not use a comma between the day and the month.

No comma

The final will be on Friday, May 15, 2013, at 3:00 p.m.

CAREFUL! Do not use a comma if you're writing only

- the day and the month
- the month and the year

No comma

The final will be on May 15.

No comma

The November 2011 meeting of the Science Club was held in Baronow Hall.

Commas in Letters

Use a comma after

- the salutation of all friendly letters (not business letters)

Dear George and Lucy,

- the closing of all letters

Sincerely, With lots of love,

Commas with Titles or Degrees

Use commas around a person's title or degree.

Jon Moore, M.D., will be our speaker.

Commas with Long Numerals

In numerals with four digits or more, use a comma after every three digits from the right.

The drive from my house to my dorm is exactly 1,065 miles.

Heads Up!
Don't use commas in years, zip codes, phone numbers, page numbers, serial numbers, or house numbers.

TECHNO TIP

For more on commas, search the Internet for these sites:

Punctuation Commas

Commnet

English Composition Comma Quiz

TICKET to WRITE

30.4 Using Commas

Directions: Working alone or in groups, use the rules about commas to mark the sentences below. Mark *C* if correct or *I* if incorrect. Correct the sentences you marked *I*.

I 1. Enzymes, proteins catalyzing chemical reactions, work by lowering the activation energy for a reaction.

I 2. "I think that success," Jennifer stated, "comes when you achieve something you have long desired."

I 3. Some on-site campus programs, such as Kids at College, can help parents who need daycare facilities.

I 4. The hot, spicy pizza was just what Neal needed to relax.

C 5. The boy running toward me is my oldest child.

C 6. Three things I'll never eat are eels, reindeer steak, and crickets.

I 7. Abraham Lincoln stated, "Honor is better than honors."

(continued)

___I___ 8. Pat's new baby arrived on Monday, February 12, 2012.

___C___ 9. Our outing included a trip to Animal Land, a local petting zoo.

___I___ 10. You will call me back promptly, won't you?

___C___ 11. I'm giving up my favorite time-waster, playing Wii, until after finals.

___C___ 12. Deborah's next project, to clean out her closet, will take all day Saturday.

___I___ 13. Overall, Marty is happy with the classes he's taking.

___I___ 14. Tony's address is 5644 Comflent Dr., Scherrville, Ohio 46899.

___I___ 15. We'll meet for the family reunion in Phoenix, Arizona, in the middle of June.

___C___ 16. The spring 2011 semester proved to be the hardest for me.

___I___ 17. Wen Chang, M.D., will speak to my biology class on Thursday, January 15.

___C___ 18. Searching for support, Karen found help in unexpected places.

___I___ 19. At least 14,987 people died in the war.

___I___ 20. Keep the car you own for three extra years, and you can save a considerable amount of money.

Colons

LEARNING GOAL

5 Use colons to introduce lists, explanations, and long quotations

Colons appear at the end of complete thoughts and introduce lists, explain material, or precede lengthy quotations. Additionally, colons are used with business letters, time, titles, and certain citation formats.

Introducing a List

Use a colon to let readers know a list will immediately follow.

In the trunk, I stored the following: a blanket, a first-aid kit, soda, and crackers.

CAREFUL! If the list is the object of a verb or preposition, omit the colon.

No colon

In the trunk, I stored a blanket, a first-aid kit, and a six-pack of soda.

Explaining Material

Use a colon to explain or elaborate on preceding material in a sentence.

We booked this singer for three reasons: he has a haunting voice, he was available when we needed him, and we could afford his fee.

Preceding a Lengthy Quotation

In academic writing, a lengthy quotation is usually preceded by a colon.

Heads Up! Styles such as MLA and APA have particular requirements for quotations. See Chapter 15.

President John F. Kennedy concluded his inaugural address this way:

> With a good conscience our only sure reward, with history the final judge of our deeds, let us go forth to lead the land we love, asking His blessing and His help, but knowing that here on earth God's work must truly be our own.

Introducing a Business Letter

Use a colon after the greeting of a business letter.

Dear Sir or Madam:

Separating Hour and Minute

Use a colon to separate the hour and minute.

Class begins at 9:15.

Dividing a Title and Subtitle

Use a colon to separate the title and subtitle of a book.

We read from Body Language: Understanding Yourself.

Separating Chapter and Verse of the Bible

Use a colon to separate a chapter and verse of the Bible.

Fr. Hammerstein's favorite Biblical passage is Psalm 16:8.

Citing Volume and Page Number of a Magazine

Use a colon to separate the volume and page number of a magazine.

Part of my research came from an article in American Journey 134:33.

Separating City and Publisher in a Bibliographical Entry

In a bibliographical entry, use a colon to separate the city and publisher.

Liles, Lorraine. Women of High School. New York: Sisterhood Press, 1967.

TECHNO TIP

For more on colons, search the Internet for the sites

Punctuation Colon Commnet and Punctuation Made Simple: Guide to the Colon

and for the following videos:

The Colons That Bind (Grammar Film)

Grammar & Punctuation: When to Use a Colon

Semicolons

LEARNING GOAL

6 Use semicolons to join related ideas

Semicolons link closely related ideas and are used in place of joining words, commas, or periods.

Joining Two Independent Clauses

Use a semicolon to join two independent clauses that are not joined by a *boysfan* word (*but, or, yet, so, for, and*, or *nor*).

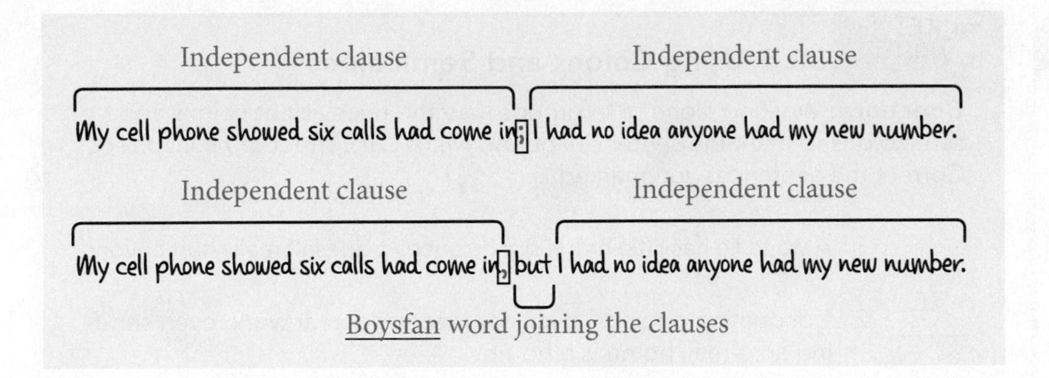

Use a semicolon to join independent thoughts only if the thoughts are related to each other.

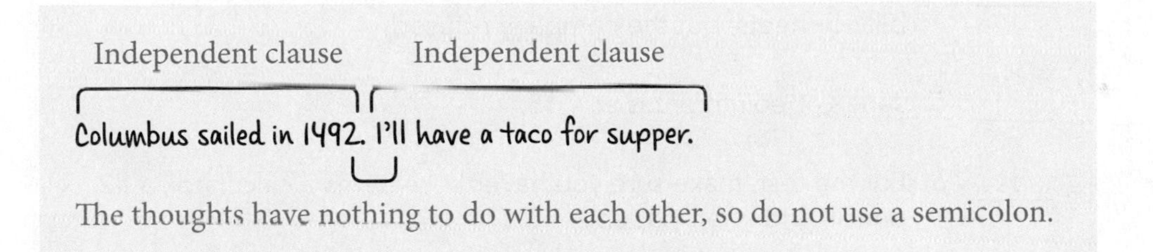

With Transitional Words or Phrases

When you join two independent clauses with a transitional word or phrase, use a semicolon *before* the transitional word or phrase and a comma *after* it.

> I'd like to go out to eat; however, I'm short on funds until Saturday.
>
> I went out to eat anyway; as a result, my credit card is maxed out.

> "I have grown fond of semicolons in recent years. . . . It's almost always a greater pleasure to come across a semicolon than a period."
>
> —Lewis Thomas, M.D.

Instead of a Comma

To clarify material for readers, use a semicolon when a comma might otherwise be used.

> My sister is considering colleges in Tallahassee, Florida, Springfield, Illinois, Denver, Dallas, and Sacramento.

To make this sentence read more clearly, use a semicolon to show which cities go with states and which are separate.

> My sister is considering colleges in Tallahassee, Florida; Springfield, Illinois; Denver, Dallas, and Sacramento.

TECHNO TIP

For more on semicolons, search the Internet for the site Our Friend, The Semicolon and Punctuation Made Simple: Guide to the Semicolon and for these videos:

English Grammar & Punctuation: Do You Use a Semicolon Before or After Conjunctive Adverbs?

Grammar & Punctuation: When to Use a Semicolon

Basic Punctuation: How & When to Use Semicolons

TICKET to WRITE

30.5 Using Colons and Semicolons

Directions: Working alone or in groups, use the rules about colons and semicolons to mark the sentences below. Mark *C* if correct or *I* if incorrect. Correct the sentences you marked *I*.

;

I 1. A white lie has little moral significance, it doesn't really hurt anyone.

C 2. Conducting personal business is forbidden at work; even sending an e-mail home is a no-no.

C 3. To Whom It May Concern:

,

I 4. Members of Congress called for the company to produce its billing sheets; but the company refused.

:

I 5. Jane's appointment is at 4,15.

:

I 6. For the test, make sure you have these items a calculator, a #2 pencil, and scrap paper.

1:1

I 7. Paul opened the Bible and began reading at Genesis 11.

C 8. For my definition paper, I read an article in *Chinese Monthly* 78:35.

C 9. David McCall stated that 78% of students would return next semester; we had hoped the percentage would be higher.

C 10. I made over $10,000 last year; but that's not what I took home.

. M

I 11. Two panda bears frolicked in the zoo; my computer suddenly crashed.

;

I 12. I'd like to have a big bowl of spaghetti, on the other hand, I'd like to lose some weight.

C 13. *Sense of Morals: Doing What's Right* is required reading for the course.

:

I 14. The recent storm had devastating effects three were killed, and hundreds were injured.

,

I 15. Dear Uncle Lewis:

Use Other Punctuation Correctly and Effectively

Other common forms of punctuation—*hyphens, dashes, parentheses, italics, underlining,* and *ellipses*—used correctly help make your writing more accurate and easier to understand.

LEARNING GOAL
7 Use hyphens, dashes, parentheses, italics, underlining, and ellipses correctly.

Hyphens

Hyphens are odd ducks because they can be used to both divide and to join words.

Dividing Words

The most common use of hyphens is to divide words at the ends of lines.

Heads Up!
When dividing words between lines, put the hyphen at the end of the first line, not at the beginning of the second line.

CAREFUL! The division is allowable only

- between syllables (hyphenating a one-syllable word is not allowed)
- if the word is not a proper noun or proper adjective
- if the word is not an acronym (a word formed from the initial letters of the several words in a name, like NATO), contraction, or numeral
- if you can position at least two letters plus the hyphen on the first line and three letters on the second line
- if both parts of the divided word will appear on the same page

Joining Numerals

Use a hyphen to show a continuation of dates or page numbers.

The Persian Gulf War lasted from August 1990-February 1991.

Read pages 45-92 for next week.

Joining Written Numbers

Use a hyphen when compound numbers from *twenty-one* to *ninety-nine* are written as words rather than numerals. Also, use a hyphen in all fractions written as adjectives (*one-third, three-fourths*).

Joining Compound Adjectives

Two or more adjectives joined together to form a single thought or image are *compound adjectives.* When these adjectives come before the noun they modify, use a hyphen between them.

Noun the adjectives modify

The ear-splitting scream was heard for blocks.

Noun the adjectives modify

The new big-box store drove two small businesses away.

"[H]yphens are used for many purposes, apart from the very useful one of making people's names sound grander."
—Stephen Murray-Smith

Joining Prefixes and Suffixes to Root Words

Hyphenate any prefix before a capitalized word, and hyphenate all prefixes that are capitalized.

anti-American sentiment   A-frame house

Hyphens are almost always used with the prefixes *ex*-, *self*-, and *all*- and with the suffix *-elect*.

ex-athlete   self-confidence   all-district   president-elect

Clarifying Meaning

On rare occasions, hyphenate a word so its meaning will be clear. The sentence

My friend Dana has reformed herself. (she's gotten rid of bad habits)

has a different meaning from

My friend Dana has re-formed herself. (her physical appearance is different)

CAREFUL! Other words that may need to be hyphenated to clarify meaning are *recreation*, *recollect*, and *resign*.

TECHNO TIP

For more on hyphens, search the Internet for these sites
Punctuation Hyphen Commnet and OWL Hyphen Use and for this video:
Grammar & Punctuation: When to Use a Hyphen in a Sentence

Heads Up!
With some word processors, typing two hyphens turns the characters into a dash.

Dashes

Use a **dash** to show a sudden change in thought or tone.

"Let's start with the discussion of—now, where did I put my glasses?" our instructor said.

Use a dash to emphasize or clarify material previously stated in a sentence.

The journey the pioneers began would take them over harsh terrain—a type of terrain they had never seen.

When I arrived at the Tutoring Center, three people from my math class—Jorge, Frannie, and Carl—were already there.

30.6 Using Hyphens and Dashes

Directions: Work alone or in groups and use the rules about hyphens and dashes to mark the sentences below. Mark *C* if correct or *I* if incorrect. Correct the sentences you marked *I*.

three-fourths
__I__ 1. I think ~~threefourths~~ of the population would vote for a tax reduction.

__C__ 2. What five-year-old boy or girl can compete in Little League?

7-July
__I__ 3. The convention will take place from July ~~7 July~~ 11.

__C__ 4. At age fifty-six, Carol's father decided to join her in college.

self-respect
__I__ 5. Vincent is trying hard to increase his children's ~~selfrespect~~

a—is
__I__ 6. "Do you want to grab ~~a is~~ that a spider crawling on the floor?" John asked.

—
__I__ 7. Censorship bans many activities ^ reading books, watching movies, accessing Internet sites.

Pro-African
__I__ 8. ~~ProAfrican~~ sentiment eventually became dominant.

re-collect
__I__ 9. The Math Club collected aluminum cans last year and will ~~recollect~~ them in two weeks.

small-appliance
__I__ 10. The ~~small appliance~~ industry is suffering in the recession.

TECHNO TIP

For more on dashes, search the Internet for these sites:

Punctuation Commnet: The Handy Informal Dash

Punctuation Made Simple: The Dash

Parentheses

Parentheses alert readers that you're giving additional information that isn't essential to the meaning of the sentence.

Elizabeth Blackwell (1821–1910) became the first female American doctor.

Ruth's surprise party (which wasn't really a surprise) lasted four hours.

Senator Rick Crowdy (R-Ky) declared he would vote for the amendment.

TECHNO TIP

For more on parentheses, search the Internet for the site

Punctuation Parentheses Commnet and for these videos

English Grammar & Punctuation: How to Use Parentheses

Basic Punctuation: How to Use Parentheses

Italics and Underlining

In punctuation, italics and underlining are the same. If you're writing by hand, use underlining. If you're word processing, use italics.

Punctuating Long Titles

Italicize or underline the titles of long works, such as books.

The Financial Lives of the Poets by Jess Walter

or

The Financial Lives of the Poets by Jess Walter

CAREFUL! Titles of sacred books are not italicized or underlined.

My religious studies class read from the Bible, the Koran, and the Torah.

Other titles that should be italicized or underlined are these:

Book-length poems	*Leaves of Grass*	Dramas	*Rent*
Operas	*Don Giovanni*	Movies	*Captain America*
Television programs	*American Idol*	Works of art	*Mona Lisa*
CDs and albums	*Rebirth*	Magazines	*Ebony*
Newspapers	*The Longwood Times*	Aircraft	*Hindenburg*
Spacecraft	*Voyager*	Trains	*Lincoln Service*
Ships	U.S.S. *Constitution* (**CAREFUL!** Don't italicize the U.S.S.)		
Pamphlets	*Answers to Financial Aid Questions*		

Italicizing or Underlining for Emphasis

Use italics or underlining to show readers emphasis on a particular word or phrase. When you write

I have to go now.

readers see the importance of the word *now*.

In Special Situations

If you're using a word or phrase outside its normal context, or if you're using a foreign word or phrase, italicize it or underline it. This provides clarity for your readers.

In my paper, I mixed up *affect* and *effect*.

Samuel Clemens used the nom de plume Mark Twain.

TECHNO TIP

For more on italics and underlining, search the Internet for the sites Using Italics and Underlining: Commnet and About Grammar and Composition: Italics and for this video: English Grammar & Punctuation: Do You Underline the Name of a Television Show?

Ellipsis Points

An **ellipsis** is made up of three spaced periods (called points) and is used to show that words are missing or when a deliberate pause takes place.

Noting Omitted Material

Ellipsis points let readers know that material from a quoted source has been omitted. When using ellipsis points in place of part of a quotation, be sure the omission does not change the meaning of the quotation.

For a paper you're writing, suppose you want to quote part of this paragraph from Jesse Jackson's 1984 speech at the Democratic National Convention:

> Our present formula for peace in the Middle East is inadequate. It will not work. There are 22 nations in the Middle East. Our nation must be able to talk and act and influence all of them. We must build upon Camp David, and measure human rights by one yard stick. In that region we have too many interests and too few friends.

If you see that you need to use only the first and last sentences of the paragraph to highlight the point of your paper, you would quote the portions of the speech this way:

Our present formula for peace in the Middle East is inadequate. . . . In that region we have too many interests and too few friends.

The period of the first quoted sentence is followed by three periods (the ellipsis points).

Noting Deliberate Pauses

To note a deliberate pause in a person's speech or thoughts, use ellipsis points.

When the alarm clock went off, Laurie thought, "I'll get up in five seconds . . . or five minutes . . . ten minutes sounds better . . . maybe I'll turn over for a little more shut-eye."

TECHNO TIP

For more on ellipsis points, search the Internet for the site Punctuation Ellipsis Commnet and for this video: English Punctuation & Grammar: When to Use an Ellipsis

30.7 Using Parentheses, Italics (or Underlining), and Ellipses

Directions: Working alone or in groups, use the rules about parentheses, italics (or underlining), and ellipses to mark the sentences below. Mark *C* if correct or *I* if incorrect. Correct the sentences you marked *I*.

(1905–1984)

I 1. Lillian Hellman~~, 1905–1984,~~ wrote *The Little Foxes*.

C 2. South Vietnam's Ngo Dinh Diem (who came to power in 1955) visited the United States in 1957.

I 3. Carla was interested in the article "Clueless Cousins" in the magazine *Your Family Tree* ~~"Your Family Tree."~~

C 4. The whole class is going to see the campus production of the play *Romeo and Juliet*.

I 5. Everything Marty purchased in Paris—new shoes, new clothes—was *au courant.* ~~"au courant."~~

C 6. Lincoln said, "Four score and seven years ago, our forefathers brought forth on this continent a new nation . . . dedicated to the proposition that all men are created equal."

Harry Potter and the Philosopher's Stone

I 7. J. K. Rowling finished writing ~~"Harry Potter and the Philosopher's Stone"~~ on a manual typewriter.

C 8. Two of Columbus's ships were the *Nina* and the *Pinta*.

Nomen est omen

I 9. Romans had the expression ~~Nomen est omen~~, meaning that your name was your destiny.

"Everything About Italics"

I 10. ~~*Everything About Italics*~~, the third chapter in the book *Grammar Is Good*, was interesting.

MyWritingLab™ Visit *MyWritingLab.com* and complete the exercises and activities in the **Final Punctuation, Commas, Quotation Marks, Apostrophes, Semicolons, Colons, Dashes, and Parentheses** topic areas.

RUN THAT BY ME AGAIN

- **Use a period . . .** to signal the end of a sentence that states a fact or an opinion or that makes a request, or use a period in an abbreviation.
- **Question marks signal . . .** uncertainty.
- **Exclamation marks convey . . .** strong feelings.
- **Quotation marks alert readers to . . .** words a person said or wrote, to titles of short works, and to slang and technical terms.
- **Apostrophes are used . . .** in words that are contractions, to indicate possession or ownership, or to form a plural if omitting the apostrophe would result in misreading material.
- **Commas are used . . .** to separate a series of items; to join two or more independent clauses connected by *but*, *or*, *yet*, *so*, *for*, *and*, or *nor*; to indicate an interruption in quoted material; to denote an appositive; in most addresses; in dates; in parts of letters; with titles or degrees; and with long numerals.
- **Commas are used inside a sentence in many ways to help readers . . .** understand a sentence's exact meaning.
- **Colons are used at the end of complete thoughts to . . .** introduce lists, explain material, and precede lengthy quotations as well as in business letters, time, titles, and citations.
- **Semicolons link** closely related ideas and are used in place of joining words, commas, or periods.
- **Hyphens can be used to . . .** both divide and to join words.
- **Dashes show . . .** a sudden change in thought or tone.
- **Parentheses alert readers to . . .** additional information that is not essential to the meaning of the sentence.
- **Use italics or underlining . . .** in titles of long works, to show readers emphasis on a particular word or phrase in a sentence, and to indicate that a word or phrase is foreign or is being used outside its normal context.
- **Ellipsis points let readers know that . . .** material from a quoted source has been omitted, but the omitted part does not change the meaning of the sentence.

LEARNING GOAL
❶ Use correct end punctuation

LEARNING GOAL
❷ Use quotation marks to indicate quotes, short titles, and specific terms

LEARNING GOAL
❸ Use apostrophes to indicate contractions and possession

LEARNING GOAL
❹ Use commas correctly to ensure your writing is clear and precise

LEARNING GOAL
❺ Use colons to introduce lists, explanations, and long quotations

LEARNING GOAL
❻ Use semicolons to join related ideas

LEARNING GOAL
❼ Use hyphens, dashes, parentheses, italics, underlining, and ellipses correctly

PUNCTUATION MARKS LEARNING LOG MyWritingLab™

MyWritingLab™

Answer the questions below to review your mastery of punctuation marks.

1. What three types of punctuation signal the end of a sentence?
   The end of a sentence is signaled by a period, question mark, or exclamation mark.
2. When do you use quotation marks in a direct quotation?
   Use quotation marks around the words you or someone else spoke if you are quoting the words exactly and in the same order that the person spoke them.
3. When quoting dialogue, how do you show readers a change in speakers?
   Show readers a change in speakers by starting a new paragraph every time a speaker changes.
4. With what types of works do you use quotation marks?
   Use quotation marks around titles of short works such as short stories, short poems, articles in magazines or newspapers, essays, chapters of books, or songs.
5. In relation to closing quotation marks, where do periods, commas, colons, and semicolons go?
   Periods and commas go inside closing quotation marks; colons and semicolons go outside closing quotation marks.
6. In a contraction, what does an apostrophe indicate?
   An apostrophe indicates that at least one letter has been omitted.
7. If a word that owns or possesses something does not end in *-s*, how do you form its possessive?
   Form its possessive by adding *'s*.
8. If a plural word that owns or possesses something ends in *-s*, how do you form its possessive?
   Form its possessive by adding an apostrophe after the final *-s*.
9. If a singular word that owns or possesses something ends in *-s*, how do you form its possessive?
   Form the possessive by adding an apostrophe after the final *-s* or by adding *'s* after the final *-s*.

10. **When do you use commas to separate a series of items?**
Use commas to separate a series of more than two items not joined by *and, or,* or *nor.*

11. **When do you use commas to join independent clauses?**
Use commas to join two or more independent clauses connected by *but, or, yet, so, for, and,* or *nor.*

12. **Where is a comma placed in an address in a text?**
Put one comma between the person's last name and the beginning of the street address, another comma after the street address, and another between the city and the state.

13. **Where are commas placed in dates that provide the week, day, and year?**
Use a comma after the day of the week, the day of the month, and the year.

14. **Where are commas used in letters?**
Use a comma after the salutation of all friendly letters and the closing of all letters.

15. **When are colons used?**
Colons are used at the end of complete thoughts to introduce lists, explain material, or precede lengthy quotations; they are also used with business letters, time, titles, and certain citation formats.

16. **When are semicolons used to join two independent clauses?**
Semicolons join two independent clauses not joined by a *boysfan* word (*but, or, yet, so, for, and,* or *nor*).

17. **When used at the end of a line, what does a hyphen indicate?**
When used at the end of a line, a hyphen indicates a word has been divided into syllables.

18. **When is a hyphen used in numbers?**
Use a hyphen when compound numbers from *twenty-one* to *ninety-nine* are written as words rather than numerals or in any fraction written as an adjective (*one-third, three-fourths*).

19. How are parentheses used?

    Parentheses alert readers that the material inside them is not essential to the meaning of the sentence.

20. With what types of works do you use italics or underlining?

    Italicize or underline the titles of long works, such as books, book-length poems, operas, television programs, CDs and albums, newspapers, spacecraft, ships, and pamphlets.

21. What do ellipsis points show?

    Ellipsis points show that words are missing or that a deliberate pause takes place.

Reading Tips and Additional Readings

As you read and studied the various chapters in the first four parts of *Ticket to Write*, you examined a number of essays. Part 5 gives you ten additional essays to help you build upon what you have learned and to make critical connections in reading and writing. It also includes a chapter of reading tips that will help you to think critically about what you read, use skimming and scanning to find information, ask questions, annotate, and summarize main ideas and key details in reading assignments.

"Today a reader, tomorrow a leader."
– Margaret Fuller

CHAPTER 31 Reading Tips

LEARNING GOALS

In this chapter, you'll learn and practice how to

1. Read critically
2. Prepare to read critically
3. Use specific skills while reading critically
4. Use specific skills after reading critically

GETTING THERE

- Critical reading is usually more demanding than casual reading.
- Skimming and scanning are two skills used in both casual and critical reading.
- Using background knowledge, reporter's questions, annotation skills, vocabulary skills, and questioning skills are all important in critical reading.
- Skills to use after reading include summarizing, taking on a different persona, collaborating, and seeking help from others.

What Is Critical Reading?

LEARNING GOAL

1. Read critically

Obviously, as a college student, you know how to read. What you might not realize, however, is that the kind of reading you'll most use in college—*critical reading*—differs from casual reading, the kind of reading you do for pleasure. **Critical reading** (sometimes called *active reading*, *academic reading*, or *detailed reading*) is characterized by careful, precise appraisal and judgment of what you read.

When you read for pleasure, you may be escaping to a world of vampires or wizards, catching up on the latest sports news, or investigating some interesting trivia. In your college classes, though, you probably won't encounter those topics.

A recent report by the National Endowment for the Arts found that reading for pleasure also has academic advantages. Students who read for pleasure score better in both reading and writing tests than do nonreaders.

Critical reading is generally more challenging than casual reading. In critical reading, you'll be much more active because, to get the most out of your reading (to learn it rather than to be entertained by it), you should annotate and reread material. These skills help you better comprehend the material as you read. They also help after you read, when you may be

- quizzed on the material
- asked to evaluate or apply what you have learned from it
- asked to write a reaction to or summary of it
- instructed to connect it with additional reading.

"Reading is to the mind what exercise is to the body."
—Joseph Addison

When you read critically, you actively connect with the text. The reading level or the concepts of much college material may be challenging but will contribute to your intellectual growth. You can become a better critical reader by taking several steps before you begin reading, while you read, and when you finish reading.

Before You Begin Reading

LEARNING GOAL
❷ Prepare to read critically

Several pre-reading strategies help you gain a deeper connection with the material you're assigned to read.

Create a Positive Reading Environment

Take note of your environment. Before you read, check the following:

Lighting	You need sufficient lighting, and you need lighting that is without glare.
Ventilation	You need enough air so you don't get groggy, and you need the temperature to be cool enough so you don't want to take a nap.
Background distractions	Some students study best with no background noise; others study best with soft background noise (like lyrics-free music).

Get Acquainted with Your Textbook

Before you jump into a textbook, spend a little time getting acquainted with it and the way it's organized. Identify the author, the publisher, and the date it was published. If the text includes an introduction, foreword, or preface, read that section to get a clearer idea of its structure and purpose.

Next examine the table of contents, which outlines the organization of the text. Note the division and titles of the

- main parts (units, numbered parts, etc.)
- chapters or sections
- headings and subheadings

Then look at one chapter and note any study aids, like

- lists of learning goals or objectives at the beginnings of chapters
- pre-reading questions or other introductory material
- marginal notes
- diagrams, charts, photographs, boxed information, or other graphic material
- vocabulary or other bolded material

Usually, all chapters in a text have the same format.

- reviews or summaries at the ends of chapters
- questions, exercises, or other activities within or at the ends of chapters or sections
- suggested supplemental readings for deeper understanding of chapters

Finally, turn to the material in the back of the book, right before the index. Note if the book includes any supplementary material like

- a bibliography
- a glossary
- appendixes
- additional readings
- maps, charts, or other graphic material

31.1 Becoming Acquainted with a Textbook

Directions: Answer the following questions about *Ticket to Write.*

1. With what does Part 1 deal? The Writing Process
2. With what does Part 2 deal? Types of Paragraphs and Essays
3. With what does Part 3 deal? Writing Situations
4. With what does Part 4 deal? Grammar and Mechanics
5. With what does Part 5 deal? Reading Tips and Additional Readings
6. With what does Part 6 deal? Study Skills
7. Sample essays are included in Chapters 6–14. Choose an essay in one of those chapters. Where can you find definitions of vocabulary from that essay? Definitions can be found next to the selection.
8. All chapters provide a section of review questions. What is that section called? Learning Log
9. Where do you find the learning goals for each chapter? Learning goals are listed at the beginning and within each chapter.
10. Exercises within chapters have what general title? Ticket to Write

Get Acquainted with Shorter Material

Before you begin to read an article or other shorter reading material, look

- to see if the selection begins with an *abstract* (a *summary* of important points of the selection)
- to see how the article is divided (with sections? headings? subheadings?)
- at diagrams, charts, photographs, or other graphic material
- for vocabulary or other bolded material
- for a review, summary, or questions at the end

Activate Your Background Knowledge

Studies show that your reading comprehension is more extensive if you take a short time before you begin reading to review what you already know about the topic. This is called **activating your background knowledge**. You may know quite a lot about the topic or you may know very little, but taking the time to raise your awareness helps when you begin reading.

"Books were my pass to personal freedom. I learned to read at age three, and soon discovered there was a whole world to conquer that went beyond our farm in Mississippi."
—Oprah Winfrey

Look at the topic and ask yourself

- Do I have any experience with the topic?
- Have I read or heard anything else about the topic?
- Have I read anything else by the author?

Students often find they can easily activate their background knowledge by composing some reporter's questions (*Who? What? When? Where? Why?* and *How?*) about the topic and then answering their questions.

The professional essay "Mick, Mom, and Me" centers on the author's experience attending a Rolling Stones concert with her mother. Before reading the essay, students could activate their background knowledge by composing reporter's questions like these:

Whom . . . would I like to see in concert if I could go with one of my parents or children??

What . . . might I gain by going to a concert with one of my parents or children?

When . . . would be a good time to see a concert?

Where . . . would be the closest place to see this kind of concert?

Why . . . would my parent or child like to go to this concert?

How . . . would I surprise my parent or child with tickets to a concert?

TICKET to WRITE

31.2 Activating Background Knowledge

Directions: Compose three reporter's questions for each topic below and write your answers. When you have finished, share your background knowledge with members of your writing group. Compare your questions and answers to those in your group to learn other ways of composing reporter's questions. Answers will vary.

1. Outcasts from society
2. City life
3. Problems with vehicles
4. Unemployment
5. Current fashion

"In the case of good books, the point is not to see how many of them you can get through, but rather how many can get through to you."
—Mortimer Adler

Determine Your Purpose for Reading

Just as your purpose is important in writing, it's also important in reading. In critical reading, *why* you read (your *purpose*) determines *how* you read. Sometimes you don't need to read material in depth; in those cases, you can use either *skimming* or *scanning*. At other times, you should study material thoroughly, so you use *critical reading*.

Skimming

Skimming is a kind of surface reading you use when you don't need to read every word in a text. Instead, you read quickly and casually. Use skimming to get an overview of content and decide if the text suits your needs.

- **Getting an overview of content** If, for example, in a history class you're assigned to read a chapter about World War II, skimming the chapter will help you learn its format and increase your comprehension when you return to the material for more careful reading.
- **Deciding if the text suits your needs** Sometimes all you need to see is if certain material fits your needs. For instance, if you're writing a paper about the results of a recent election, you have many available resources. To see which ones might be useful, you can skim the material.

How to Skim

Don't spend a great deal of time when you're skimming. Just read key information like

- the title
- the introduction
- headings and subheadings (these state major points)
- main ideas (these are often found in the first sentence of body paragraphs)
- tables, graphs, charts, or other visuals
- bolded or italicized words and phrases
- transitional words like *therefore, in addition*, *because*, *resulting in*, and *since*

Scanning

In **scanning**, you look for particular information, like a name or date, so you skip over material that's not pertinent. In casual reading, you use scanning skills when you find a score on a sports page, a word in a dictionary, or a number in a phone book. You scan through entries until you find what you need.

Scanning is helpful in critical reading as well as casual reading. Suppose for history class you need to check your textbook for the exact date in 1945 when the war with Japan ended. In that case, you scan the chapter until you see the year 1945; then you read the whole sentence to be sure it contains the information you needed.

How to Scan

Like skimming, scanning is quick reading. To scan effectively,

- decide what information you're looking for
- condense the information into a shortened form (a key word or phrase)
- consider variations on the way the information might be presented (e.g., if you're looking for a date, you might find it in written or in numerical form)
- move your eyes over your text until you find what you think might be the right information

TECHNO TIP

For more on determining purpose for reading, search the Internet for this site:
About Academic Reading: What's Your Purpose?

Think of skimming as "getting the gist" of material.

TECHNO TIP

For activities on skimming, search the Internet for these sites:
RMIT Study & Learning Centre: Reading Skills: Activity 2
UVIC Skimming Exercise
Reading Skills for Academic Study: Efficient Reading Skills

Scanning involves even less time than skimming.

TECHNO TIP

For activities on scanning, search the Internet for these sites:
UVIC Scanning Exercise
UNE-About Academic Reading: Exercise 3

To determine whether to skim or scan, search the Internet for this site:
UNE-About Academic Reading: Exercise 3

- read the accompanying text (the rest of the sentence, or more if necessary) to see if it contains the information you're seeking

While You're Reading

LEARNING GOAL
3 Use specific skills while reading critically

Critical reading involves more than running your eyes across the page. You gain the most from reading when you become actively involved with the material.

Connect with the Material

Begin critical reading by looking at the title of the selection, which sometimes gives a clue to the topic, the direction of the topic, or the slant of the author. For instance, a psychology text chapter titled "Helping Kids with Behavioral Challenges" leaves little doubt about what the chapter will address. Similarly, an article titled "The Myth of Body Language" alerts readers to the author's view on the subject.

Heads Up!
You may or may not agree with an author's viewpoint.

Next, again *activate your background knowledge*. What do you already know about the topic? Employ reporter's questions (*Who?*, *What?*, *When?*, *Where?*, *Why?*, and *How?*) to think of personal experience, of past reading or viewing related to the topic, or of material you have read by the same author.

If you must answer study or review questions after you read the selection, read those questions *before* you begin reading the selection. When you find a possible answer to a question, make a note beside it. You may not find all the answers the first time you read, but you may save time by finding some.

Also, think about your purpose for reading. If you need to find only a certain detail or answer a specific question, scan the material. However, if you need to read the entire selection and evaluate it, interpret its message, or apply what you learn from it, begin by skimming it all to get a content overview. Then apply critical reading skills like annotating, defining unfamiliar vocabulary, and questioning.

Annotate

Don't hesitate to mark in your text.

Annotating is important in critical reading. When you **annotate**, you interact with your text, marking it with notes or symbols as you read. Annotation helps you understand material on a deeper level and also remember it longer, both of which are key in critical reading.

Important material to annotate includes

- the thesis statement
- the main ideas
- vocabulary terms you already know
- vocabulary terms you don't know
- questions you have about any material
- personal comments you have as you read the piece

Devise any annotation method that suits you, but use the system consistently. Here are some annotation symbols you may choose to use.

What to Annotate	Suggested Symbols
the thesis statement of a selection	mark TS (thesis statement) in margin
the main ideas of a selection	mark MI (main idea) in margin
vocabulary you already know	underline or circle
vocabulary you don't know	mark with ?
important examples	mark ex (example)
questions about confusing material	mark with ?
ideas with which you strongly agree or disagree	mark with !
answers to study questions	mark with SQ (study question)
ideas that relate to others in the selection	mark with connecting arrows in margin, or note on separate paper
personal comments about what you read	jot in margin or on separate piece of paper

While you read, keep your class notes handy. If you see connections between your reading and your notes, annotate those as well. Devise your own label of annotation, or use this one:

On your notes, write C w/ p. 345, text (Connect with p. 345 in text)

In your reading, write C w/ notes, Feb. 13 (Connect with notes on Feb. 13)

31.3 Annotating a Selection

Directions: Select an article or use one provided by your instructor. Annotate the article, using the symbols above or others that suit your style. After you have finished, share your annotations with those in your writing group. Compare what you annotated to the annotations of other members of your group, looking for areas that you or they may have missed. Then compare the symbols you used to annotate, looking for symbols that might better suit your style. Answers will vary.

Increase Your Vocabulary

As you read critically, you will encounter vocabulary words important for understanding the subject. Some books include definitions of new or difficult words. These definitions may be found

- in the margins of the books or selections
- at the beginning or end of a chapter or selection
- in a glossary in a separate section of a book

Other college reading material may have no vocabulary defined for you. Part of your responsibility as a student is to learn the meanings of any unfamiliar words.

Sometimes you can determine meanings of words from their **context** (the way they are used in the sentence). The meanings of the words in blue in the examples below are probably obvious when you read the whole sentence:

Open this tome to page 145 and read the next chapter.
Let's just loll around the Student Center until our next class.
I began the arduous task of sorting through my deceased aunt's belongings.

If you're still confused after looking up a word, speak with your instructor.

From their context, you could probably determine that a *tome* is a book, to *loll* is to relax, and an *arduous* task is one that's difficult.

When you come upon a word or phrase that is unclear and you are not sure about its meaning from its context, use a dictionary or other reference material to establish the meaning. Many words have multiple definitions, so first determine the meaning that fits the way the word is used in the sentence. Write that definition in the margin of your text, next to the word or phrase. Then reread the sentence, substituting the dictionary's definition. That should clarify the meaning of the sentence.

TECHNO TIP

For more on vocabulary studies, search the Internet for these sites:
Guide to Vocabulary—Academic Support
English-Word Information

Ask Questions

To be sure you understand what you are reading, stop and question yourself about the material as you read. If the material seems easy, stop every few paragraphs.

If the material is difficult, stop after every paragraph. Ask yourself questions like these:

- What are the main ideas of the section?
- In my words, what did the author say in this section?
- How does this section fit with preceding material?
- How does this section fit with other material from class?
- How does this section fit with other material I have read?
- With what points do I agree or disagree?

"Books had instant replay long before televised sports."
—Bert Williams

After you finish reading a section, turn the title or heading of the section into a question. Then answer the question. Refer back to the section if you're stumped. If you feel you didn't answer the question sufficiently (that is, in the way your instructor would expect), reread the section as many times as necessary.

Section Heading	Turned into a Question
The Role of an IT Specialist	What is the role of an IT specialist?
Manhattan's Native Americans Before Hudson's Arrival	Who were the Native Americans living on Manhattan before Hudson arrived?
Sportswriters in the News	Why are certain sportswriters in the news?

31.4 Creating Questions to Increase Comprehension

Directions: Read the section headings below and create two possible questions about the content of each. Answers will vary.

1. Methods of Voter Registration
2. Secrets to Living Past 100
3. Cause of Change in Social Media
4. Decreasing Global Warming
5. Italian Renaissance Artists

Read and Reread

In critical reading, a single reading is usually not sufficient to master the content. In Chapter 3, you learned that one key to good writing is revising, or rewriting, your material. The same applies to critical reading: one key to effective reading is rereading and revising your understanding of the material.

When they encounter a difficult passage, some students find that reading aloud, emphasizing different words with successive readings, aids their comprehension.

After You Read

LEARNING GOAL

4 Use specific skills after reading critically

When you have finished reading, ask yourself

- Do I fully understand what I've read?
- Can I answer any study or review questions connected with the selection?

If you answer "no" to either of these questions, go back and reread. If you answer "yes" to both questions, use the skills below to increase your retention and understanding.

Summarizing

Summarizing is an effective tool in reinforcing both retention and comprehension. The key to **summarizing** is to use your words to reduce large reading selections down to their main ideas, key details, and important words and phrases. Unlike a summary for a writing class assignment, in a summary for a reading assignment you can include any personal observations or reactions to the text.

Heads Up!
See Chapter 15 for more detailed information on summarizing in writing.

Some students prefer to summarize orally, some in a written manner. Others find that, with especially difficult material, summarizing both aloud and on paper or on screen helps them. To summarize a part of a selection:

Oral Summary	Written Summary
Using your words, explain the material aloud; use a recording device to document your summary.	Using your words, write the material, either in longhand or by word processing.

Take on a Different Persona

After you finish reading, think of yourself as the instructor. Ask yourself these questions:

- What are the most important parts of this selection?
- How do these parts interrelate with other material from class or from other assignments?
- If I were writing a test on this material or thinking of topics I'd address in class, what questions would I include?
- How would I answer those questions?

Collaborate in a Study Group

Collaborate with others in a study group to relate and compare your analyses of what you read. Compare your impressions, opinions, and interpretation of the material. You may find this additional opportunity to summarize orally helps both your comprehension and your retention.

If you composed possible test, quiz, or discussion questions related to the readings, present them to your study group. Take part in answering possible questions others bring. If some reading passages are still unclear to you, use this opportunity to ask others for their interpretation of the troublesome sections.

Get Help from the Professionals

Sometimes, in spite of your best efforts, parts of a chapter or article may remain unclear and you need additional help. Don't be embarrassed about this, and don't be shy about taking action. Make an appointment with your instructor or with a tutor. Bring your reading selection and point out the troublesome areas.

TECHNO TIP

For more on reading skills for college students, search the Internet for these videos:
Reading Strategies
How to Use Your Textbook (For Something Other than a Doorstop)

MyWritingLab™ Visit *MyWritingLab.com* and complete the exercises and activities in the **Study Skills Module** topic areas.

RUN THAT BY ME AGAIN

LEARNING GOAL
❶ Read critically

- **Critical reading is characterized by . . .** careful, precise appraisal and judgment.
- **Critical reading is generally more challenging than . . .** casual reading.
- **Before you begin reading, make sure your environment is . . .** helpful to reading.

LEARNING GOAL
❷ Prepare to read critically

- **Before you begin reading, examine . . .** the layout and features of your textbook.
- **Before you begin reading, activate . . .** your background knowledge of the topic you'll be reading about.
- **Before you begin reading, determine . . .** your purpose for reading.

LEARNING GOAL
❸ Use specific skills while reading critically

- **Skim to get . . .** an overview of content or to decide if the text suits your needs.
- **Scan to look for . . .** particular information.
- **While you're reading, again activate . . .** your prior knowledge.
- **When you annotate, you . . .** mark your text with notes or symbols as you read.
- **Part of your responsibility as a student is to learn the meanings of . . .** unfamiliar words.
- **As you read, stop and question yourself about . . .** the material to be sure you understand what you are reading.

LEARNING GOAL
❹ Use specific skills after reading critically

- **When summarizing what you read, you . . .** use your words to reduce large reading selections down to their main ideas, key details, and important words and phrases.
- **After you read, think of yourself as . . .** the instructor.
- **Collaborate with others in a study group to . . .** relate and compare your analyses of what you read.
- **If all else fails, do not hesitate to . . .** make an appointment with your instructor or with a tutor.

READING TIPS LEARNING LOG MyWritingLab™

MyWritingLab™

Answer the questions below to review your mastery of reading tips.

1. To read critically, what should you do?
   You should annotate and reread material.
2. In the room where you read, what do you need in lighting?
   You need sufficient lighting, and you need lighting that is without glare.
3. To get a clearer idea of a textbook's structure and purpose, what should you read?
   You should read any introduction, foreword, or preface in a textbook.
4. What type of supplemental material should you note in the back of a book?
   You should note whether a book contains a glossary, appendixes, a bibliography, additional readings, or maps, charts, or other graphic material.
5. What is an abstract?
   An abstract is a summary of the important points of an article or other short reading.
6. To activate your background knowledge of a subject, what three questions should you ask yourself?
   Ask yourself if you have any experience with the topic, if you have read or heard anything else about the topic, and if you have read anything else by the author.
7. You can activate your background knowledge by composing what type of questions?
   You can activate your background knowledge by composing reporter's questions.
8. What is skimming?
   Skimming is a kind of surface reading you use when you don't need to read every word in a text.
9. For what two reasons should you skim?
   Skim to get an overview of the content or to decide if a text suits your needs.
10. What type of information do you look for when you are scanning?
    You look for specific information.

11. **Why should you begin critical reading by looking at the title of the selection?**

    The title sometimes gives a clue to the topic, the direction of the topic, or the slant of the author.

12. **In critical reading, what should you do after you look at the title of the selection?**

    You should activate your background knowledge.

13. **What do you do when you annotate?**

    You interact with your text, marking it with notes or symbols as you read.

14. **What should you do when you come upon a word or phrase that is unclear and you can't be sure about its meaning from its context?**

    Use a dictionary or other reference material to establish the meaning of the word the way it's used in the sentence; write the definition in the margin of the text, next to the word or phrase; then reread the sentence, substituting the dictionary's definition.

15. **How often should you stop to question yourself about material you're reading?**

    If the material is difficult, stop after every paragraph.

16. **What is the key to summarizing?**

    The key to summarizing is to use your words to reduce large reading selections into their main ideas, key details, and important words and phrases.

17. **After you finish reading, you should ask questions as if you were whom?**

    After you finish reading, ask questions as if you were the instructor.

18. **How can collaboration in a study group help your reading?**

    You can compare impressions, opinions, and interpretation of the material; you can compare possible questions; you can get additional help interpreting difficult material.

19. **If, after several readings, you're still uncertain about a passage, what should you do?**

    Make an appointment with your instructor or with a tutor.

Additional Readings

My Home Is New Orleans

by Mike Miller

Mike Miller was a New Orleans resident in 2005 when Hurricane Katrina struck. After four months, Miller returned to New Orleans, the place he calls home. This essay aired on National Public Radio's Morning Edition, *August 28, 2006, the one-year anniversary of Katrina.*

1 I believe in attachment to place. I believe that watermarks fade, tears dry, and lives mend.

2 A year after the flood, the nation is remembering Hurricane **Katrina**. And some of us, whether labeled "**displaced**," "evacuated," or "back home," will wonder if we still believe. We will wonder—sitting on our porches, in our bar rooms, and in our **gutted homes**—if we still *should* believe.

3 When I left New Orleans, I found myself, like thousands of displaced Gulf Coast residents, living on the generosity of others. People opened their homes to me. In some ways, life was easier. I'd almost forgotten how tough it is to live in New Orleans. In Chicago I was offered jobs that pay three times more than anything I could make in New Orleans. I thought about moving: Seattle, Anchorage, New York, Key West, Tucson, and everywhere in between. But looking at a map spread on a table I already knew. My home is New Orleans . . . still.

4 I moved back into an apartment uptown in the **Twelfth Ward**—on the third floor this time. I'm a little **paranoid** about flooding. But now I can really hear the **foghorns** of the ships on the river.

5 Life in New Orleans is hard nowadays. I work for the **Louisiana Family Recovery Corps**, and the mental health scene is not good: Depression is **rampant**. Suicides and substance abuse have been on the rise since Katrina.

6 I'm also back bartending and, mixed in with the grief, I can feel the pulse still there. We live the best we can. It's like this street musician in the **Quarter** who always says, "Man, we're just trying to get back to abnormal!"

7 I believe the soul of this place cannot be easily destroyed by wind and rain. I believe the music here will live and people will continue to dance. I believe in "Darlin'" and "Baby." I believe in "Where 'yat?'"

Katrina a destructive hurricane that landed in New Orleans on August 28, 2005

displaced those forced to move or leave

gutted homes homes stripped of all belongings

Twelfth Ward a section of New Orleans; the lower, densely populated part of it flooded in 2005

paranoid suspicious, distrustful

foghorns deep, loud horns on ships, used for sounding warnings

Louisiana Family Recovery Corps an organization created to help residents recover from the effects of Hurricane Katrina

rampant extensive, widespread

Quarter a reference to the French Quarter, a specific section of New Orleans

and "**Makin' groceries**." I believe in neighborhoods where **Mardi Gras Indians** sew beaded costumes, kids practice trumpet in the street, and recipes for okra can provide conversation for an entire afternoon.

8 My family asked me why I wanted to return to New Orleans. "Why do you want to live somewhere where garbage is piled up, rents have doubled, there are no jobs, and houses are filled with black mold? Is it safe? Is it healthy?" They ask if New Orleans is still worth it. I don't have an answer to satisfy them; I can't really even give myself an answer. I keep hearing Louis Armstrong saying, "Man, if ya gotta ask, you'll never know."

9 I'm just 26, my clothes can all fit in a backpack; I've got a graduate degree in social work and a 65-pound bulldog. I could move anywhere at all, but I believe in this place. I believe I belong here. As hard as it is to live in New Orleans now, it's even harder to imagine living anywhere else.

makin' groceries an idiomatic phrase, unique to New Orleans, meaning "to go to the supermarket or corner store for groceries"

Mardi Gras Indians African-Americans, dressed in elaborate Native American costumes, who march on their own parade route during the Carnival season

A CLOSER LOOK

1. What is the author's thesis statement?

   Answers may vary. Possible answers: (1) "I believe in attachment to place." (2) "I could move anywhere at all, but I believe in this place." (3) "My home is New Orleans."

2. What question(s) does the thesis statement answer?

   Why do you want to live somewhere where garbage is piled up, rents have doubled, there are no jobs, and houses are filled with black mold? Is it safe? Is it healthy? Is New Orleans still worth it?

3. What supporting ideas does Miller offer to back up his thesis statement?

   Possible answers: He returned to New Orleans after the flood. He states he "could move anywhere at all," but he chose New Orleans.

4. What could readers in a general audience learn from this essay?

   Possible answers: Readers could learn how someone can be connected to a certain place.

5. How does Miller use elements of the descriptive essay to accomplish his thesis?

   Answers may vary. Miller uses sensory detail to describe New Orleans, illustrating how well he knows the city and how much he appreciates the sights and sounds of New Orleans.

MAKING CRITICAL CONNECTIONS

1. Miller's essay originally aired as a radio essay on August 28, 2006, one year after Hurricane Katrina devastated much of New Orleans. Miller contends that life in New Orleans after Katrina came with difficulties and fears and, although he states he could have moved anywhere, he returned. If you were in Miller's situation and had to leave a place you call home after a natural disaster such as a hurricane, tornado, or flood, would you return to the place or not? Discuss your reasons for your answer with those in your writing group.
2. Miller's personal essay was submitted to the public radio series *This I Believe* and was broadcast on NPR. Review the *This I Believe* essay guidelines at **thisibelieve.org/** and write your own "I believe" essay. The essay should tell a story and reveal the occasion when you adopted a specific belief.
3. Miller refers to unique ingredients of New Orleans's culture: its people and their music, food, language, dialect, activities, celebrations, and respect for their city and way of life. Write an essay about some place you know well—perhaps where you live—and describe how its culture defines that specific place. Use at least three culturally significant topics, such as specific groups of people, music, food, language, activities, and celebrations.

Stepfather's Day

by Rick Bragg

When he was in his mid-forties, author and journalist Rick Bragg became a stepfather to Jake, who was then eleven. As Jake finishes high school, Bragg reflects on the influence he has had on his stepson in the years since their bond began.

(1) I think the boy might turn out all right. Next year he goes to college, far from me and any bad habits I have left to teach.

(2) His name is Jake. When he was 11, I taught him how to cheat at cards. "You didn't teach me," he said. "I just caught you." I taught him how to throw a punch at 12. He drew a peace sign on his shirt.

3 I taught him how to shoot a jump shot at 13, and throw an elbow. He quit the team, and joined the drama club. One day he was a small forward. The next he was **Bassanio** in ***The Merchant of Venice***.

4 At 15, I taught him the words Hank Williams wrote. He took up the guitar, and played John Lennon. I tried, after his second or third girlfriend, to tell him that pretty women come and go, and variety—when you are young—makes an old man rich in memories. He picked love songs on his **Epiphone**.

5 Now, it is too late to improve him anymore. But, in a way, that is what I have done. It is as if the boy studied me as he grew, and decided I would be the **template** for what he would not be. How odd, to be so proud to fail.

6 I guess a man deserves that, when he plays in the fields of the Lord, living life the way he has damned well pleased, and decides, at 46, that being a father is the one thing he never tried. Maybe that is why there is no such thing as Stepfather's Day. My own heart has always broken on the third Sunday in June. I was, almost all my life, a fatherless boy. On Father's Day I call my mother, to thank her for shouldering the weight alone.

7 The boy, I figured, might be a remedy. I got him when he was 10. He came in the package with his mom, like an extra biscuit or that ninth piece of chicken. I immediately began teaching him bad habits because I did not have any good ones. I bought him a .22 rifle and a go-kart that would run with traffic on the interstate. Every time his mother, Dianne—whom we shall hence refer to as "The Warden" —was out of town, we ate pancakes at **IHOP**, or chicken in a box.

8 "Life is an adventure," I told him. "Have some." He had his own definition of adventure. Like me, he wants to see the world. I tell him about Africa, about camel trains on the far horizon, and voodoo priests in the slums of Haiti. I lived to chase stories. He craves adventure, too, but says he might pursue it through the Peace Corps.

9 He is not, of course, perfect. He forgets to take the garbage out until our home smells like a South Georgia chicken house. But he has a fine heart, a fine mind. More than anything, he has a peace in him. He picks his guitar some slow evenings on Mobile Bay, and sings to the gulls, and the snowbirds in their high black socks. I do not have the patience to watch a whole sunset. I end up staring down into the murky water, to the fish I should have caught.

10 I search his face for a sign of me, but I am missing. There is, maybe one thing. I never had the **gall**, the hypocrisy, to lecture him on being good. The only **high ground** I ever tried to claim was this: In a world that grows more selfish every day, where people use politics and even religion to build higher walls between the lucky and unlucky, he should

Bassanio a character in Shakespeare's *The Merchant of Venice*

The Merchant of Venice a tragic comedy written by William Shakespeare in the 1590s

Epiphone a type of guitar

template a pattern used as a guide

IHOP the restaurant chain International House of Pancakes

gall boldness

high ground moral superiority

refuse, and see the value in the lives of people who work hard for a living but never had that much luck.

(11) I would like to believe he heard what I said, but the truth is it was in him all the time. He routinely gives his allowance to charities at school. He even gave his sneakers, and came home barefoot. Once, when a basketball coach had forgotten a boy at the far end of the bench, Jake walked over to him—like a man—and reminded him the boy had not played. That took more courage than it takes to punch someone in the nose. So he leaves soon, **shy of ruination**.

(12) Almost. When Jake turned 16, The Warden insisted he get a safe, slow, boxy car, as ugly as possible. When my boy leaves, he will leave in a Mustang.

shy of without
ruination destruction

A CLOSER LOOK

1. This selection has an implied thesis statement. In your words, what is the thesis statement?

   Answers may vary. Possible answer: Bragg is proud of the young man his stepson has become.

2. What does Bragg mean when he writes, "My own heart has always broken on the third Sunday in June"?

   The third Sunday in June is Father's Day. For almost all of his life, Bragg was "a fatherless boy," so readers know that Father's Day was no cause for celebration when he was younger.

3. Bragg's colorful language is one of his specialties. In this essay, Bragg uses two similes (comparisons of unlike persons, places, or things, using *like* or *as*) that involve chickens. Cite these similes.

   He [Jake] came in the package with his mom, like an extra biscuit or that ninth piece of chicken. He [Jake] forgets to take the garbage out until our home smells like a South Georgia chicken house.

4. What does Bragg mean by "How odd, to be so proud to fail"?

   Answers may vary. Possible answer: Bragg cited instances of when he had tried to steer Jake in one direction, but his stepson chose another. In other words, Jake chose to be his own person. Bragg called this a "failure" of which he (Bragg) was proud.

5. Bragg says he told Jake to "see the value . . . of people who work hard . . . but never had that much luck." What supporting details does Bragg provide to show that Jake had this truth "in him all the time"?

Jake donated his allowance to charities; Jake gave away his sneakers (and came home barefoot); Jake reminded a coach to play a boy who had not yet been in a basketball game.

MAKING CRITICAL CONNECTIONS

1. The title of this article is "Stepfather's Day," but nowhere does Bragg advocate the creation of such a holiday. Would you support the establishment of a national Stepfather's Day and Stepmother's Day? Why or why not? Discuss reasons for your answer with those in your writing group.
2. Bragg mentions a number of "skills" he taught Jake (like how to throw a punch). Jake responded to these by doing something opposite (like drawing a peace sign on his shirt). Write about a time you were told to do something but did the opposite. Give details for the reasons behind your action.
3. Bragg's love for Jake is evident, even though Jake is not his biological son. Think of someone to whom you are close, someone who is not a biological relative. What are the circumstances that brought the two of you together? What has that person done for you, and what have you done for that person? In your writing group, discuss this relationship and compare it with the experiences of your group members.

My Journey from Scribbling to Art in Only 60 Years

by Bob Allen

In this article, author Bob Allen recounts his early negative experience with one art form and his lifelong involvement with a different art form.

1 I got off to a bad start with art. My third-grade teacher humiliated me in front of the entire class because my drawing of a burning house wasn't very good. Actually, it was terrible: a purple square with a triangle on top for the house, and orange scribbles for flames. But that's no reason to be mean to an 8-year-old.

2 I never was good at drawing, although in the late 1980s I took a drawing class that focused on **left-brain/right-brain** drawing. If I could get into my right brain, I could draw something recognizable, but it was difficult to turn off my left brain. Some people just weren't cut out to draw.

left-brain/right-brain based on the theory that one side of an individual's brain is dominant; those with left-brain dominance are verbal, analytical, and rational; those with right-brain dominance are nonverbal and intuitive

(3) I found another route to art. I took Air Force **ROTC** in college, figuring that being an officer in the Air Force was better than getting drafted into the Army. My college degree was in biology, leading to a high score in the science and technology sections of the officer qualification test.

(4) That led to an **initial** assignment to electronics school, followed by an assignment to the **Office of Special Investigations**. I went into criminal and **counterintelligence** investigations. One skill I learned was photography.

(5) The emphasis was on photography in difficult situations—low light, extreme distance, and not being seen taking pictures. Shooting under those conditions usually resulted in poorly exposed film, so we used darkroom tricks and techniques to produce usable prints. Not beautiful photos, but recognizable pictures of people.

(6) A few years later, I was stationed in the **Middle East** and traveled extensively throughout the region, to Greece, Africa, and India. I bought myself a nice camera (for 1969), a **Pentax Spotmatic**, and accumulated an assortment of lenses and gadgets.

(7) Often I used my camera on operations. Traveling to exotic places, I always found time to go sightseeing. I started to notice that some pictures were pleasing while others just **documented** that I had been there. Studying those pictures, I could see characteristics that made the good ones good and the boring ones boring.

(8) I retired from the Air Force and lost my access to a photo lab. I still tried to take more interesting pictures, but felt I'd reached a **plateau** in quality. Then, about 10 years ago, I got a digital camera. And with my computer, I now had a color photo lab—a lab that didn't need a darkroom or a lot of water and chemicals. Best of all, there was an "un-do" button: I could try something to improve a photo, but if it didn't work, I could go back to the original version with the click of my mouse.

(9) After retiring from my post-military job as a science writer and **technical editor** at Pacific Northwest National Laboratory, I got involved with Allied Arts as their publicity chair. A few months after I started, a **juried show** was announced. By now, I was feeling good about my photography. I picked out a picture taken on an island off the coast of Norway, enlarged it, and did a bit of computer enhancement. I submitted the digital file to the show, and it was accepted. I didn't win a prize, but just being selected (only about a third of the entered works were selected) made me feel like my artistic talent had been **validated**. Take that, you mean third-grade teacher!

ROTC Reserve Officer Training Corps, a college-based program that trains students to become officers in various military branches

initial beginning

Office of Special Investigations a U.S. Air Force agency that investigates terrorist threats and espionage

counterintelligence efforts made to prevent a nation's enemies from gathering intelligence about the nation

Middle East the area that includes western Asia and northern Africa

Pentax Spotmatic a 35-mm camera manufactured between 1964 and 1976

documented proved, supported

plateau a state of little or no change

technical editor a person who reviews and identifies errors in technical text (e.g., computer hardware and software, engineering, and the aerospace industry)

juried show an art show that accepts only entries that have been judged to be of high value

validated confirmed

10 I now have a new camera with **higher resolution** and expanded capabilities. After six months, I'm still learning new features and how to use those features. And getting better pictures.

11 Maybe someday I'll sell one.

higher resolution the capability of giving greater details of images

From *The Entertainer*, February 2011, p. 16. Reprinted by permission of *The Entertainer Newspaper*, Kennewick, Wash.

A CLOSER LOOK

1. The thesis statement in this work is implied. In your words, what is the author's thesis statement?

   Possible answer: Even after being shamed at an early age, a person can later prove his or her talent.

2. What background information does the author provide for his "bad start in art"?

   He relates the story of being humiliated by his third-grade teacher for the picture he had drawn of a burning house.

3. Why does the author include details about his military background?

   Possible answer: The author wants the audience to know why he was trained in photography, the "difficult situations" in which he used his photography skills, and how his time in the military led him to "exotic places" that he could photograph.

4. As the title suggests, this article spans several decades. What transitional words or phrases does the author use to indicate a transition in time?

   Transitional words indicating time include *later, still, then, now, after, a few years later,* and *about ten years ago.*

5. The author uses a simple conclusion. What is the effect of that one-sentence conclusion?

   Possible answer: The author's one-sentence conclusion emphasizes to the reader that he is still interested in his art form and has hope for his future work in it.

MAKING CRITICAL CONNECTIONS

1. Allen recounts a time when he was humiliated. Think of a time you felt you were wronged by a teacher, an employer, or anyone else in authority. In your

writing group, compare and contrast your circumstances with those of other group members.

2. Like Allen, write a narrative essay about an occasion in which you felt you were wronged. In your essay, describe what happened and discuss any insight you gained or change you experienced.
3. Allen initially learned photography through his job, but now photography is his hobby. Write an essay about a hobby you would like to pursue if you were retired. Cite reasons for your interest in this hobby and how you would follow it if you had time.

Ah, to Return to the Halcyon Days of Academe

by Craig Wilson

After attending parties for a number of college-bound freshmen, USA Today *columnist Craig Wilson, at age sixty, looks back at his college days and offers a lighthearted proposal.*

1 The old **cliché** is true, as old clichés often are. **Youth is wasted on the young**. I'll take it one step further. College is wasted on the young, too.

2 I've been attending a round of farewell parties for college-bound freshmen, all sons and daughters of friends. I look at these kids and want to shake them. Do you have a *clue* what lies ahead of you? I want to ask. A *clue*? They don't, of course. They're bright, not wise.

3 I will admit I didn't have a clue, either, when I was heading off to Syracuse. I look back on my college years now—what I remember of them, anyway—and realize I could have done so much more. I could have gone to class, for instance.

4 It was indeed **pearls before swine**. Just think about it. Someone—usually willing parents—gives us four years to read books in an **idyllic** setting, to hang out with our friends until late into the night, to explore any new world we want to explore, often abroad. They also throw in long holiday vacations—sometimes a month—and summers off. Nice work if you can get it. And what did we do? We slept until noon. I know because I did.

5 So I'm going to make a proposal. Perhaps Congress could put it on its **docket** when it returns from its summer recess. (The similarities between Congress and college are not lost on me.) I propose every 60-year-old in America gets sent back to college. Maybe not for four years. I'd take one. We'd appreciate the second chance, perhaps even learn something this time around.

6 The details have to be worked out, of course. Employers might like to know where we've gone, but imagine how much brighter we'll be when we return. The cost of such a venture also will have to be covered, but

halcyon untroubled, peaceful

academe college or university

cliché an overused expression

youth is wasted on the young an expression meaning that those who are young do not appreciate the advantages of their situation

pearls before swine an expression that means that items of quality are offered to those who are not refined enough to appreciate them

idyllic peaceful, relaxing

docket official calendar

surely Congress can find the money somewhere. It seems to find money for everything else. We wouldn't even ask for a clothing allowance.

7 We still have our **Bass Weejuns**. Room and board is all we ask. Not that we don't have a few requirements. First, my dog will have to come with me. I also have to live in a single. I have no desire to train a new roommate at 60. It was hard enough at 18. And I prefer not to eat in the dining hall. Some kind of dine-around card could work, along with a laundry service that picks up and delivers. Hanger, please. Light starch. I also take my coffee black. Thanks.

8 In return, I promise to call home every Sunday night. **Collect**, of course. Just for old time's sake. Can you even do that anymore?

Bass Weejuns slip-on shoes (loafers) popular with college students in the 1960s

collect a type of telephone call that allows the person receiving the call to have the option of paying for the call

A CLOSER LOOK

1. What is the thesis statement of this article?

   "College is wasted on the young."

2. What does Wilson offer as the reason he began thinking about the proposal he suggests?

   Answers will vary. Possible answer: The parties for today's college-bound freshmen led him to reminisce about his days as a student.

3. Wilson admits he "didn't have a clue" about what lay ahead of him when he began college. What two mistakes does Wilson admit he made when he was a student?

   He didn't attend class and he slept until noon.

4. What is the proposal that Wilson suggests?

   Wilson suggests that every sixty-year-old in America be sent back to college for a year.

5. Besides room and board, name two of the requirements that Wilson says will need to be fulfilled for this proposal.

   Answers will vary. Possible answers: Dogs will be allowed; a single room will be available; a dine-around card will be available; laundry service that picks up and delivers clothes on hangers will be available; black coffee will be available.

MAKING CRITICAL CONNECTIONS

1. Wilson says that his years in college were "pearls before swine." Think of a time when you were given an opportunity you didn't appreciate. In your writing group, compare your situation with those of your peers.
2. Like Wilson, look at those younger than you and write an essay about an opportunity they do not appreciate. Include details about what they should do differently in order to value that particular opportunity.
3. Wilson says that the people in his age group "still have our Bass Weejuns." Write an essay about some mode of fashion currently in style for college students. Explain its popularity and cite reasons why you think it will or will not remain popular.

Boxers, Briefs, and Books

by John Grisham

Acclaimed author John Grisham writes about his life before he became a well-known novelist.

1 I wasn't always a lawyer or a novelist, and I've had my share of hard, dead-end jobs. I earned my first steady paycheck watering rose bushes at a nursery for a dollar an hour. I was in my early teens, but the man who owned the nursery saw potential, and he promoted me to his fence crew. For $1.50 an hour, I labored like a grown man as we laid mile after mile of chain-link fence. There was no future in this, and I shall never mention it again in writing.

2 Then, during the summer of my 16th year, I found a job with a plumbing contractor. I crawled under houses, into the cramped darkness, with a shovel, to somehow find the buried pipes, to dig until I found the problem, then crawl back out and report what I had found. I vowed to get a desk job. I've never drawn inspiration from that miserable work, and I shall never mention it again in writing, either.

3 But a desk wasn't in my immediate future. My father worked with heavy construction equipment, and through a friend of a friend of his, I got a job the next summer on a highway asphalt crew. This was July, when Mississippi is like a sauna. Add another 100 degrees for the fresh asphalt. I got a break when the operator of a Caterpillar bulldozer was fired; shown the finer points of handling this rather large machine, I contemplated a future in the cab, tons of growling machinery at my command, with the power to plow over anything. Then the operator was back, sober, **repentant**. I returned to the asphalt crew.

repentant feeling or showing regret for misconduct

4 I was 17 years old that summer, and I learned a lot, most of which cannot be repeated in polite company. One Friday night I accompanied my new friends on the asphalt crew to a honky-tonk to celebrate the end of a hard week. When a fight broke out and I heard gunfire, I ran to the restroom, locked the door and crawled out a window. I stayed in the woods for an hour while the police hauled away rednecks. As I hitchhiked home, I realized I was not cut out for construction and got serious about college.

5 My career sputtered along until retail caught my attention; it was indoors, clean and air-conditioned. I applied for a job at a Sears store in a mall. The only opening was in men's underwear. It was humiliating. I tried to quit, but I was given a raise. Evidently, the position was difficult to fill. I asked to be transferred to toys, then to appliances. My bosses said no and gave me another raise.

6 I became **abrupt** with customers. Sears has the nicest customers in the world, but I didn't care. I was rude and **surly** and I was occasionally watched by spies hired by the company to pose as shoppers. One asked to try on a pair of boxers. I said no, that it was obvious they were much too small for his rather ample rear end. I handed him an extra-large pair. I got written up. I asked for lawn care. They said no, but this time they didn't offer me a raise. I finally quit.

7 Halfway through college, and still drifting, I decided to become a high-powered **tax lawyer**. The plan was sailing along until I took my first course in tax law. I was stunned by its complexity and **lunacy**, and I barely passed the course.

8 Around the same time, I was involved in mock-trial classes. I enjoyed the courtroom. A new plan was hatched. I would return to my hometown, hang out my shingle and become a hotshot **trial lawyer**. Tax law was discarded overnight.

9 This was 1981; at the time there was no **public-defender** system in my county. I volunteered for all the **indigent** work I could get. It was the fastest way to trial, and I learned quickly.

10 When my law office started to struggle for lack of well-paying work—indigent cases are far from **lucrative**—I decided to go into yet another low-paying career: in 1983, I was elected to a House seat in the Mississippi State Legislature. The salary was $8,000, which was more than I made during my first year as a lawyer. Each year from January through March I was at the State Capitol in Jackson, wasting serious time, but also listening to great storytellers. I took a lot of notes, not knowing why but feeling that, someday, those tales would come in handy.

11 Like most small-town lawyers, I dreamed of the big case, and in 1984 it finally arrived. But this time, the case wasn't mine. As usual, I was loitering around the courtroom, pretending to be busy. But what I was

abrupt hasty, quick

surly impolite, bad-tempered

tax lawyer an attorney who specializes in cases regarding government levies on economic transactions

lunacy insanity, senselessness

trial lawyer an attorney who specializes in defending clients

public-defender an attorney assigned by a court to defend people who cannot afford an attorney

indigent poor; not having enough money to meet one's needs

lucrative profitable, well-paying

really doing was watching a trial involving a young girl who had been beaten and raped. Her testimony was **gut-wrenching**, graphic, heart-breaking, and **riveting**. Every juror was crying. I remember staring at the defendant and wishing I had a gun. And like that, a story was born.

12 Writing was not a childhood dream of mine. I do not recall longing to write as a student. I wasn't sure how to start. Over the following weeks I refined my plot outline and **fleshed out** my characters. One night I wrote "Chapter One" at the top of the first page of a legal pad; the novel, "A Time to Kill," was finished three years later.

13 The book didn't sell, and I stuck with my day job, defending criminals, preparing wills and deeds and contracts. Still, something about writing made me spend large hours of my free time at my desk.

14 I had never worked so hard in my life, nor imagined that writing could be such an effort. It was more difficult than laying asphalt, and at times more frustrating than selling underwear. But it paid off. Eventually, I was able to leave the law and quit politics. Writing's still the most difficult job I've ever had—but it's worth it.

gut-wrenching sickening, difficult to tolerate

riveting engrossing, fascinating

fleshed out completed; added details and traits to make full

A CLOSER LOOK

1. What is Grisham's implied thesis statement in this essay?

   Although he has had many jobs, writing is the most rewarding.

2. Who would make good audience members for this essay? Why these people in particular?

   Possible answer: One audience may be young people who are starting out with their first jobs. Grisham lets them know they should keep working toward pursuing the vocation they find rewarding.

3. Throughout the essay Grisham traces his work history. List the jobs he held prior to attending law school.

   Grisham was a nursery worker, fence crew member, plumbing laborer, asphalt crew member, bulldozer operator, and retail salesman.

4. What primary perception pushed Grisham to either leave or want to leave most of the jobs he held?

   Possible answer: He left each job because the job was not fulfilling.

5. In the last line, Grisham writes, "Writing's still the most difficult job I've ever had—but it's worth it." Explain what he means by this.

   Possible answer: He has had many jobs that were boring, physically difficult, mentally challenging, or time-consuming, yet none were as difficult as writing. He finds fulfillment in the challenges of writing—something he couldn't find anywhere else.

MAKING CRITICAL CONNECTIONS

1. Grisham lists the jobs he had over the years before discovering writing as a rewarding vocation. He briefly states why these jobs prior to writing novels were unpleasant. Think of a job you have had that you found unpleasant and knew you did not want to have on a permanent basis. In your writing group, compare and contrast your unpleasant job experiences with those of other members in your writing group.
2. Write a short essay similar to Grisham's in which you discuss work you have done—full-time, part-time, volunteer, community service-related—that you know you would not like to do for the rest of your life. In your essay, describe the work you did, discuss why you would not want that job on a full-time basis, and explain what you learned from the experience.
3. The last paragraph of Grisham's essay concludes with the statement, "Writing's still the most difficult job I've ever had—but it's worth it." Write an essay about a job that you would like to have. Discuss the reasons you would like to have this job and explain why you believe it would be rewarding. Research this job by going online to the *Occupational Outlook Handbook* (**www.bls.gov/oco/**) and looking up information about this job or field.

Popular Culture: Shaping and Reflecting Who We Are

by Billy Wilson

Popular culture influences all kinds of art, music, literature, beliefs, and values not only in America but in other countries as well.

(1) Have you ever tried to escape popular culture—to give up Oprah and not rush out to read the books she recommends; to turn off your

favorite DJ or talk radio station on the way to work? Have you ever refused to engage in small talk about the next *Survivor* castaway or about who killed JonBénet? Can you resist the tailgate party at the big game on Saturday? Can you boycott the homecoming dance, where you could dress in elegance, or the **masquerade** party, where you could be anything from an Osmond to an Osbourne? If so, you are fighting popular culture—and you have probably already lost the battle!

2 Few among us ever really escape popular culture—those who don't know that wrestlers and action heroes can be governors; who don't care what Nicole Kidman wears to the Oscars; who are not turned on by an XKE or a 350 Z; who can't tell hip-hop from punk rock; who haven't a Clue who killed Colonel Mustard in the library or how many degrees Kevin Bacon is separated from Queen Latifah; who don't know Calvin Klein from Calvin Coolidge or Shaquille O'Neal from Ally McBeal; who are convinced that Siberian tigers don't live in Las Vegas and don't realize that **Neverland** is just outside of Santa Barbara! The few who escape the popular culture phenomena may live simpler lives—but they probably have a lot less fun. Their lives might even be "D-U-L-L-dull" as Barney Fife would say (or was it Gomer Pyle?).

3 Dr. Ray Browne, the father of the academic study of popular culture, says that popular culture includes "all aspects of our daily lives that are not narrowly academic and are free from the elitist standards which dominate the fine arts." Dr. Joan Fedor, a member of **Phi Theta Kappa**'s Honors Committee, adds that popular culture often incorporates "the boldest sights and sounds of society." Dr. Fedor's definition emphasizes not only the word "bold," but also the word "society," stressing that popular culture becomes part of our shared beliefs and values.

4 Popular culture is often bold because of those responsible for it. Some **purveyors** of popular culture, such as Bob Marley or Ellen DeGeneres, are often restless, with causes that cry out for recognition. Others, like James Dean, are merely rebels—perhaps even "rebels without a cause." Stars like Madonna, television's Fonzie, Cher, and Britney Spears might also fall into this category. Some, like Janis Joplin, Prince, and Michael Jackson simply march (or dance) to the beat of a different drum. Whatever drives them to create the "bold sights and sounds," they make a name for themselves in popular culture, and also in the history books. Historians say that Elvis Presley was at the forefront of the "youth movement" in America and that Bob Dylan helped shape America's conscience during the Civil Rights Movement. These same historians argue that certain pop culture events were "**epochal**," in that they helped to define the times or change attitudes and values. One

masquerade a party of guests wearing costumes and masks

Neverland Michael Jackson's California estate

Phi Theta Kappa an honor society that recognizes the academic achievement of two-year college students

purveyors providers

epochal highly significant, especially marking the beginning of a new era

such event was the raunchy and defiant **Woodstock**; another was Elvis's first appearance on *The Ed Sullivan Show*, which some see as a symbol of **the establishment**'s **acquiescence** to rebellious youth and its music.

5 But popular culture is not always rebellious, confrontational, or bold. It can be a purring pussycat as well as a roaring tiger—unifying as well as **divisive**. Consider those two characters from the Andy Griffith television show: Barney, the **wiry caricature** of police incompetence, and Gomer, the good-hearted. Why do we know them? Because, at any given time, we can find on our 80 channels a rerun of the exploits of Andy, Opie, Barney, Gomer, Aunt Bee, and Floyd the Barber. Boomers and children alike find comfort in the eternal Mayberry, just as we find comfort in the ageless Mickey Mouse and Yuletide celebrations accompanied by Bing Crosby's "White Christmas." Generations find common ground in the popular culture of today as well. When high school and college students flock to see *The Lord of the Rings* trilogy, their teachers are often occupying seats in the same theater. Parents are standing in the same lines as their sons and daughters for an autograph from Michael Jordan, Venus and Serena Williams, or Julia Roberts. Adults are discussing Harry Potter books in book clubs just as their pre-teens are reading them. And today's youth can look forward to sharing this popular culture with their children, because, like Mayberry and Mickey, Michael Jordan and Harry Potter may be "keepers" for generations to come.

6 Americans of all generations have been proud to share our popular culture with people of other nations. Mayberry, Mickey, and "White Christmas" have traveled the globe. America has not only created popular classics with an international following, but has given birth to entirely new classifications of popular culture. These include motion pictures, two of the world's most popular sports—baseball and basketball—and three kinds of music—jazz, rock and roll, and country. The stars of these American genres are known and welcomed throughout the world. Michael Jordan soars in the remote villages of the Third World. Elvis Presley may be better known abroad than any of the world's political figures, literary giants, or philosophers. Blue jeans, cowboy boots, and Garth Brooks are almost as popular in some European cities as in Texas. New Orleans jazz and Memphis blues have found large international markets. In many ways, American pop rules the world, but America gets a wonderful exchange for its exports. Just as Jordan is an icon abroad, Norway's Sonja Henie and Brazil's Pelé have been heroes in the United States. Ichiro, the Japanese baseball star, and Yao Ming, the Chinese basketball star, have made smooth transitions between cultures and find themselves heroes in both.

Woodstock a large rock music festival held in 1969 at Bethel in New York state

the establishment people holding most of the power and influence in government or society

acquiescence acceptance without protest

divisive causing great, unfriendly disagreement in a group of people

wiry lean, stringy, thin in appearance

caricature a cartoon-like character

(7) Popular culture is unique in its power of universal appeal. Unlike most other cultural phenomena, it appeals to individuals and groups that are markedly different from each other. As a result, it is a great **leveler of classes**. Historians and **anthropologists** have shown that even elitists, who typically resist popular culture when it is new, eventually embrace it in the same ways and venues as people in the middle and lower classes. As popular culture has expanded, it has encouraged and affirmed the participation of **ethnic minorities** and others whose voices are often suppressed. By appealing to people of different ethnic groups, different socioeconomic status, and different political ideologies, popular culture connects and unites people who might otherwise be mutually **antagonistic**.

(8) A close look at popular culture reveals a phenomenon that is difficult to define. It is irresistible, annoying, marvelously entertaining, frequently out of control, and ultimately inevitable for all but those who hide under a rock. Popular culture can be a vehicle for change, a generational and international unifier, a leveler of classes, and a tool for recognizing and understanding even the least among us. Popular culture boldly manifests itself in anger and rebellion, but also provides comfort through shared experiences. Some expressions of popular culture establish deep roots, while others put up a bold front, die quickly, and leave no roots for future growth. We barely remember such popular products as the pet rock or one-hit wonders like Tiny Tim, yet Elvis may live forever. In the end, most popular culture falls somewhere between the **evanescent** and the eternal. It may not last forever, but it leaves an impression and produces a flame that lights the way for others, whose own sights and sounds will both reflect and perhaps also shape our society.

leveler of classes something that erases economic or social inequalities for all people

anthropologists people who study the origins of human beings and their worldwide cultures

ethnic minorities groups of people who share common traits and customs but who differ racially or politically from a larger group of which they are a part

antagonistic hostile, unfriendly

evanescent vanishing like vapor

A CLOSER LOOK

1. An *allusion* is an indirect reference to a person, place, or thing readers should already know. In paragraph 2, the author writes of people "who don't know that wrestlers and action heroes can be governors." To whom is Wilson alluding in that quotation?

   Wrestler Jesse Ventura was governor of Minnesota; actor Arnold Schwarzenegger was governor of California.

2. In paragraph 4, Wilson says that popular culture "is often bold because of those responsible for it." As support for this, he lists three categories of

"purveyors of popular culture." Name the categories and the famous people he lists for each category.

Wilson lists Bob Marley and Ellen DeGeneres as members of the "restless" category; James Dean, Madonna, Fonzie, Cher, and Britney Spears as "rebels without a cause"; and Janis Joplin, Prince, and Michael Jackson as those "who march (or dance) to the beat of a different drum."

3. In paragraph 5, Wilson states that "[g]enerations find common ground in the popular culture." What examples does he offer to support this statement?

   Wilson offers the characters from the Andy Griffith television show (Barney, Gomer, Andy, Opie, Aunt Bee, Floyd), Mickey Mouse, Bing Crosby, *The Lord of the Rings*, Michael Jordan, Venus and Serena Williams, and Julia Roberts.

4. Cite three examples of support Wilson offers for the topic sentence of paragraph 6, "Americans of all generations have been proud to share our popular culture with people of other nations."

   Answers will vary. Possible answers: Wilson offers Mayberry, Mickey, "White Christmas," motion pictures, baseball, basketball, jazz, rock and roll, country, Michael Jordan, Elvis Presley, blue jeans, cowboy boots, Garth Brooks, New Orleans jazz, Memphis blues.

5. In paragraph 7, Wilson says that popular culture "is a great leveler of classes." Cite one example of support he offers for this.

   Answers will vary. Possible answers: Elitists embrace popular culture in the same ways and venues as people in the middle and lower classes; popular culture has encouraged and affirmed the participation of ethnic minorities and others whose voices are often suppressed; and popular culture connects and unites people who might otherwise be mutually antagonistic.

MAKING CRITICAL CONNECTIONS

1. Wilson mentions "such popular products as the pet rock or one-hit wonders like Tiny Tim." In your writing group, discuss another person or a fad that was once popular but is no longer considered trendy, talented, or fashionable.
2. Paragraph 6 offers a number of aspects of American popular culture that "have traveled the globe." Write an essay about one of those aspects, detailing what you think is its appeal.
3. Wilson begins by asking if the reader has ever "tried to escape popular culture." Write an essay about one aspect of popular culture that you would like to escape, detailing why you think it does not merit its popularity.

The Death of My Father

by Steve Martin

In this essay, comedian and actor Steve Martin discusses what he has learned from moments in his father's life and from his father's last days.

1 In his death, my father, Glenn Vernon Martin, did something he could not do in life. He brought our family together.

2 After he died at age 83, many of his friends told me how much they loved him—how generous he was, how outgoing, how funny, how caring. I was surprised at these descriptions. During my teenage years, there was little said to me that was not criticism. I remember him as angry. But now, ten years after his death, I recall events that seem to contradict my memory of him. When I was 16, he handed down to me the family's 1957 Chevy. Neither one of us knew at the time that it was the coolest car anyone my age could have. When I was in the third grade he proudly accompanied me to the school tumbling contest where I won first prize. One day, while I was in single digits, he suggested we play catch in the front yard. This offer to spend time together was so **anomalous** that I didn't quite understand what I was supposed to do.

3 When I graduated from high school, my father offered to buy me a tuxedo. I refused because my father always shunned gifts. I felt with my refusal, that somehow in a **convoluted**, perverse logic, I was being a good son. I wish now that I could have let him buy me a tuxedo, let him be a dad.

4 My father sold real estate but he wanted to be in show business. I was probably five years old when I saw him in a bit part at the **Callboard Theater** on **Melrose Place** in Hollywood. He came on in the second act and served a drink. The theater existed until a few years ago and is now finally **defunct** and, I believe, a lamp shop.

5 My father's attitude toward my show business accomplishments was critical. After my first appearance on ***Saturday Night Live*** in 1976, he wrote a bad review of me in the newsletter of the Newport Board of Realtors where he was president. Later, he related this news to me slightly shamefaced, and said that after it appeared, his best friend came into his office holding the paper, placed it on his desk, and shook his head sternly, indicating a wordless "no."

6 In the early '80s, a close friend of mine, whose own father was killed walking across a street and whose mother committed suicide on Mother's Day, said that if I had anything to work out with my parents,

anomalous unusual, inconsistent

convoluted difficult, hard to understand

Callboard Theater a once-popular theater for up-and-coming actors in Hollywood

Melrose Place a section of Melrose Avenue in West Hollywood, California, that is home to antique shops, boutiques, and salons

defunct no longer in existence

Saturday Night Live a weekly television comedy sketch and variety show that began in 1975

I should do it now, because one day that opportunity would be over. When I heard this remark, I had no idea that I would ever want to work anything out with them, that, in fact, there was anything to work out at all. But it stewed in my brain for years, and soon I decided to try and get to know my parents. I took them to lunch every Sunday I could, and would **goad** them into talking.

7 It was our routine that after I drove them home from our lunches, my mother and father, now in their 80s, would walk me to the car. I would kiss my mother on the cheek and my father and I would wave or awkwardly say goodbye. But this time we hugged each other and he whispered, "I love you," with a voice barely audible. This would be the first time these words were ever spoken between us. I returned the phrase with the same awkward, broken delivery.

8 As my father ailed, he grew more irritable. He made unreasonable demands, such as waking his 24-hour help and insisting that they take him for drives at three a.m., as it was the only way he could relax. He also became **heartrendingly** emotional. He could be in the middle of a story and begin to laugh, which would provoke sudden tears, making him unable to continue.

9 In his early 80s, my father's health declined further and he became bedridden. There must be an instinct about when the end is near, as we all found ourselves gathered at my parents' home in Orange County, California. I walked into the house they had lived in for 35 years and my weeping sister said, "He's saying goodbye to everyone."

10 A **hospice** nurse said to me, "This is when it all happens." I didn't know what she meant, but soon I did.

11 I walked into the bedroom where he lay, his mind alert but his body failing. He said, almost **buoyantly**, "I'm ready now." I understood that his intensifying rage of the last few years had been against death and now his resistance was **abating**. I stood at the end of the bed and we looked into each other's eyes for a long, unbroken time. At last he said, "You did everything I wanted to do."

12 I said the truth: "I did it for you."

13 Looking back, I'm sure that we both had different interpretations of what I meant.

14 I sat on the edge of the bed and another silence fell over us. Then he said, "I wish I could cry, I wish I could cry."

15 At first, I took this as a comment on his condition but am forever thankful that I pushed on. "What do you want to cry about?" I finally said.

16 "For all the love I received and couldn't return."

17 He had kept this secret, his desire to love his family, from me and from my mother his whole life. It was as though an early misstep had

goad provoke, stimulate

heartrendingly causing extreme sadness or grief in others

hospice a type of medical care that focuses on making terminally ill patients emotionally and physically comfortable

buoyantly cheerfully

abating gradually lessening, fading

kept us forever **out of stride**. Now, two days from his death, our pace was **aligning** and we were able to speak.

out of stride not in pace together, having unmatched steps

aligning bringing into line, cooperating

morbidity unhealthy gloominess or sadness

tangible capable of being understood, real, touchable

(18) My father's death has a thousand endings. I continue to absorb its messages and meanings. He stripped death of its spooky **morbidity** and made it **tangible** and passionate. He prepared me in some way for my own death. He showed me the responsibility of the living to the dying. But the most enduring thought was expressed by my sister, Melinda. She told me she had learned something from all this. I asked her what it was. She said, "Nobody should have to die alone."

A CLOSER LOOK

1. In paragraphs 2 and 3, Martin briefly mentions a few interactions between his father and himself. What do these incidents show about their relationship?

   Answers will vary. Possible answers: These incidents illustrate their father-son relationship when Martin was a child and show what Martin sees as the basis for their relationship as adults.

2. Overall, Martin describes his father as having a critical, irritable, and distant personality, but in paragraph 11 Martin describes his father as speaking "buoyantly." Why has his father's personality changed so drastically?

   His has accepted that he is dying and will soon be free of his illness.

3. In paragraph 11, Martin's elderly father says to him, "You did everything I wanted to do." Why is this comment on Steve Martin's career so important to the writer?

   Answers will vary. Possible answers: Martin's father had always been critical of his acting and writing career. Now, his father had finally offered him praise.

4. Why does Martin keep starting new paragraphs in paragraphs 11–16?

   He starts a new paragraph each time the speaker changes.

5. Martin concludes his essay stating that he has learned much from his father's death. List three points he makes that he has taken from his father's death.

   Possible answers: He came to have a new understanding about what people who are dying go through; he came to realize that those close to someone who is dying have a responsibility to make that person comfortable; he learned that no one should die alone.

MAKING CRITICAL CONNECTIONS

1. Martin begins his essay by stating that his father did something through his death that he couldn't do while he was living: "he brought our family together." Discuss in your group how Martin's father accomplished this.
2. In the conclusion, Martin states that he is still gaining "messages and meanings" from his father's death. Fastwrite a list of significant events in your life. Then, choose one and write an essay about that event, discussing the message or meaning you have gained from that event.
3. Martin describes what he and his family experienced as his father's health declined. He writes of the things he said and did to make his father more comfortable in his father's final years. Conduct research on hospice care and write a short essay on why it might be important to make someone comfortable when facing death. Review these Web sites or others:

   Mahogany Hospice

   American Cancer Society: Hospice Care

Can't We Talk?

by Deborah Tannen

Professor Deborah Tannen frequently writes and speaks about the varied ways relationships are affected by everyday language. Below is an excerpt condensed from her bestselling book You Just Don't Understand: Women and Men in Conversation.

(1) A married couple was in a car when the wife turned to her husband and asked, "Would you like to stop for a drink?"

(2) "No, thanks," he answered truthfully. So they didn't stop.

(3) The result? The wife—who had indeed wanted to stop—became annoyed because she felt her preference had not been considered. The husband, seeing his wife was angry, became frustrated. *Why didn't she just say what she wanted?*

4 Unfortunately, he failed to see that his wife was asking the question not to get an instant decision, but to begin a negotiation. And the woman didn't realize that when her husband said no, he was just expressing his preference, not making a ruling. When a man and woman interpret the same interchange in such conflicting ways, it's no wonder they can find themselves leveling angry charges of selfishness and **obstinacy** at each other.

5 As a specialist in **linguistics**, I have studied how the conversational styles of men and women differ. We cannot, of course, lump "all men" or "all women" into fixed categories—individuals vary greatly. But research shows that the seemingly senseless misunderstandings that haunt our relationships can at least in part be explained by the different conversational rules by which men and women often play.

6 Whenever I write or speak about this subject, people tell me how relieved they are to learn that what they had previously **ascribed** to personal failings, is, in fact, very common. Learning about the different (though equally valid) conversational frequencies men and women are tuned to can help banish blame and help us truly talk to one another. Here are some of the most common areas of conflict.

7 **Status vs. Support.** Men grow up in a world in which a conversation is often a contest—either to achieve the upper hand or to prevent other people from pushing them around. For many women, however, talking is typically a way to exchange confirmation and support. I saw this firsthand when my husband and I had jobs in different cities. When people made comments like "That must be rough" and "How do you stand it?" I accepted their sympathy.

8 But my husband would react with irritation. Our situation had advantages, he would explain. As **academics**, we had long weekends and vacations together.

9 Everything he said was true, but I didn't understand why he chose to say it. He told me that he felt some of the comments implied: "Yours is not a real marriage. I am superior to you because my wife and I have avoided your misfortune." It had not occurred to me there might be an element of **one-upmanship**, though I recognized it when it was pointed out.

10 I now see that my husband was simply approaching the world as many men do: as a place where people try to achieve and maintain status. I, on the other hand, was approaching the world as many women do: as a network of connections, in which people seek consensus.

11 **Independence vs. Intimacy.** Since women often think in terms of closeness and support, they struggle to preserve intimacy. Men, concerned with status, tend to focus on establishing independence. These

obstinacy stubbornness
linguistics the scientific study of language
ascribed credited to
academics employees (usually professors) of a college or university
one-upmanship the art of outdoing or showing up someone

traits can lead women and men to starkly different views of the same situation.

12 When Josh's old high-school friend called him at work to say he'd be in town, Josh invited him to stay for the weekend. That evening he told Linda.

13 Linda was upset. How could Josh make these plans without discussing them with her beforehand? She would never do that to him. "Why don't you tell your friend you have to check with your wife?" she asked.

14 Josh replied, "I can't say I have to ask my wife for permission!"

15 To Josh, checking with his wife would mean he was not free to act on his own. It would make him feel like a child or an **underling**. But Linda actually enjoys telling someone, "I have to check with Josh." It makes her feel good to show that her life is entwined with her husband's.

16 **Advice vs. Understanding.** Eve had a **benign** lump removed from her breast. When she confided to her husband, Mark, that she was distressed because the stitches changed the contour of her breast, he answered, "You can always have plastic surgery."

17 This comment bothered her. "I'm sorry you don't like the way it looks," she protested. "But I'm not having any more surgery!"

18 Mark was hurt and puzzled. "I don't care about a scar," he replied. "It doesn't bother me at all."

19 "Then why are you telling me to have plastic surgery?" she asked.

20 "Because *you* were upset about the way it looks."

21 Eve felt **like a heel**. Mark had been wonderfully supportive throughout her surgery. How could she snap at him now?

22 The problem stemmed from a difference in approach. To many men, a complaint is a challenge to come up with a solution. Mark thought he was reassuring Eve by telling her there was something she could *do* about her scar. But often women are looking for emotional support, not solutions.

23 **Information vs. Feelings.** A cartoon I once saw shows a husband opening a newspaper and asking his wife, "Is there anything you'd like to say before I start reading?" We know there isn't—but that as soon as the man begins reading, his wife will think of something.

24 The cartoon is funny because people recognize their own experience in it. What's not funny is that many women are hurt when men don't talk to them at home, and many men are frustrated when they disappoint their partners without knowing why.

25 Rebecca, who is happily married, told me this is a source of dissatisfaction with her husband, Stuart. When she tells him what she is

underling inferior person
benign of no danger to health
like a heel an idiom meaning "thought he or she was dishonorable"

thinking, he listens silently. When she asks him what is on his mind, he says, "Nothing."

26 All Rebecca's life she has had practice in verbalizing her feelings with friends and relatives. To her, this shows involvement and caring. But to Stuart, like most men, talk is information. All his life he has had practice in keeping his innermost thoughts to himself.

27 Yet many such men **hold center stage** in a social setting, telling jokes and stories. They use conversation to claim attention and to entertain. Women can wind up hurt that their husbands tell relative strangers things they have not told them.

28 To avoid this kind of misunderstanding, both men and women can make adjustments. A woman may observe a man's desire to read the paper, for example, without seeing it is a rejection. And a man can understand a woman's desire to talk without feeling it is an intrusion.

29 **Orders vs. Proposals.** Diana often begins statements with "Let's." She might say, "Let's park over there" or "Let's clean up before lunch." This makes Nathan angry. He has deciphered Diana's "Let's" as a command. Like most men, he resists being told what to do. But to Diana, she is making suggestions, not demands. Like most women, she wants to avoid confrontation and formulates requests as proposals rather than orders. Her style of talking *is* a way of getting others to do what she wants—but by winning agreement first.

30 With certain men, like Nathan, this tactic backfires. If they perceive someone is trying to get them to do something indirectly, they feel manipulated and respond more resentfully than they would to a straightforward request.

31 **Conflict vs. Compromise.** In trying to prevent fights, some women refuse to openly oppose the will of others. But at times it's far more effective for a woman to assert herself, even at the risk of conflict.

32 Dora was frustrated by a series of used automobiles she drove. It was she who commuted to work, but her husband, Hank, who chose the cars. Hank always went for automobiles that were "interesting," but in continual need of repair. After Dora was nearly killed when her brakes failed, they were in the market for yet another car.

33 Dora wanted to buy a late-model sedan from a friend. Hank fixed his sights on a 15-year-old sports car. Previously, she would have **acceded** to his wishes. But this time Dora bought the boring but dependable car and **steeled** herself for Hank's anger. To her amazement, he spoke not a word of **remonstrance**. When she later told him what she had expected, he scoffed at her fears and said she should have done what she wanted from the start if she felt that strongly about it.

hold center stage an idiom meaning "become the center of attention"

acceded given in, agreed

steeled braced, steadied

remonstrance protest, objection, opposition

34 As Dora discovered, a little conflict won't kill you. At the same time, men who habitually oppose others can adjust their style to opt for less confrontation.

35 When we don't see style differences for what they are, we sometimes draw unfair conclusions ("You're illogical," "You're self-centered," "You don't care about me"). But once we grasp the two characteristic approaches, we stand a better chance of preventing disagreements from spiraling out of control. Learning the other's ways of talking is a leap across the communication gap between men and women, and a giant step toward genuine understanding.

A CLOSER LOOK

1. What is the thesis statement of this essay?

   Answers will vary. Possible answers: (1) "But . . . the seemingly senseless misunderstandings that haunt our relationships can at least in part be explained by the different conversational rules by which men and women often play." (2) "When a man and woman interpret the same interchange in such conflicting ways, it's no wonder they can find themselves leveling angry charges of selfishness and obstinacy at each other."

2. What detail does the author include so readers will accept her authority on the subject?

   She lets readers know that, as a specialist in linguistics, she has studied how the conversational styles of men and women differ.

3. In each section explaining a common area of conflict, Tannen offers an anecdote involving two or more people. What do these anecdotes provide?

   Answers will vary. Possible answer: The anecdotes illustrate Tannen's points and give a more subjective view to the explanation of the areas of conflict.

4. Because Tannen presents two different sides to areas of conflict, she uses a number of transition words and phrases that show contrast. Cite three of these transition words.

   Answers will vary. Possible answers: *however, on the other hand, since, like, yet, but, when*

5. In her conclusion, what does Tannen offer as advantages of understanding how the opposite sex speaks?

   Advantages are (1) a leap across the communication gap between men and women, and (2) a giant step towards genuine understanding.

MAKING CRITICAL CONNECTIONS

1. Tannen says, "Men grow up in a world in which a conversation is often a contest, either to achieve the upper hand or to prevent other people from pushing them around. For many women, however, talking is typically often a way to exchange confirmation and support." In your writing group, discuss whether you agree with this statement or not. Cite reasons for your opinion.
2. Tannen writes of the "seemingly senseless misunderstandings that haunt our relationships." Write about a time when you had what felt like a senseless misunderstanding with a person of the opposite sex. Include both details and consequences of the misunderstanding.
3. Tannen offers anecdotes and general information about each of the six separate areas of conflict she describes. Choose one of the areas and write another anecdote that is appropriate for the topic.

The Pitfalls of Linking Doctors' Pay to Performance

by Sandeep Jauhar, M.D.

Dr. Sandeep Jauhar writes about medicine and cardiology for both the New York Times *and the* New England Journal of Medicine. *He is also the director of the Heart Failure Program at Long Island Jewish Medical Center.*

1 Not long ago, a colleague asked me for help in treating a patient with **congestive heart failure** who had just been transferred from another hospital.

congestive heart failure a common form of heart failure resulting in a patient retaining fluids

2 When I looked over the medical chart, I noticed that the patient, in his early 60s, was receiving an **intravenous** antibiotic every day. No one seemed to know why. Apparently it had been started in the emergency room at the other hospital because doctors there thought he might have pneumonia.

3 But he did not appear to have pneumonia or any other infection. He had no fever. His white blood cell count was normal, and he wasn't coughing up sputum. His chest X-ray did show a vague marking, but that was probably just fluid in the lungs from heart failure.

4 I ordered the antibiotic stopped—but not in time to prevent the patient from developing a severe diarrheal infection called **C. difficile colitis**, often caused by antibiotics. He became dehydrated. His temperature spiked to alarming levels. His white blood cell count almost tripled. In the end, with different antibiotics, the infection was brought under control, but not before the patient had spent almost two weeks in the hospital.

5 The case illustrates a problem all too common in hospitals today: patients receiving antibiotics without solid evidence of an infection. And part of the blame lies with a program meant to improve patient care.

6 The program is called pay for performance, P4P for short. Employers and insurers, including Medicare, have started about 100 such initiatives across the country. The general intent is to reward doctors for providing better care.

7 For example, doctors receive bonuses if they prescribe **ACE inhibitor** drugs to patients with congestive heart failure. Hospitals get bonuses if they administer antibiotics to pneumonia patients in a timely manner.

8 On the surface, this seems like a good idea: reward doctors and hospitals for quality, not just quantity. But even as it gains momentum, the initiative may be having **untoward** consequences.

9 To get an **inkling** of the potential problems, one simply has to look at another quality-improvement program: surgical report cards. In the early 1990s, report cards were issued on surgeons performing coronary bypasses. The idea was to improve the quality of **cardiac** surgery by pointing out deficiencies in hospitals and surgeons; those who did not measure up would be forced to improve.

10 But studies showed a very different result. A 2003 report by researchers at Northwestern and Stanford demonstrated there was a significant amount of "**cherry-picking**" of patients in states with mandatory report cards. In a survey in New York State, 63 percent of cardiac surgeons acknowledged that because of report cards, they were

intravenous injected into a vein and absorbed by the bloodstream

C. difficile colitis the common name of a bacterial infection of the intestines

ACE inhibitor a drug that makes the heart's work easier by blocking chemicals that constrict capillaries

untoward troublesome or unexpected

inkling a vague understanding or notion

cardiac related to the heart

cherry-picking an idiom meaning "selectively choosing only the best to benefit someone personally"

accepting only relatively healthy patients for heart bypass surgery. Fifty-nine percent of **cardiologists** said it had become harder to find a surgeon to operate on their most severely ill patients.

11 Whenever you try to legislate professional behavior, there are bound to be unintended consequences. With surgical report cards, surgeons' numbers improved not only because of better performance but also because dying patients were not getting the operations they needed. Pay for performance is likely to have similar **repercussions**.

12 Consider the requirement from Medicare that antibiotics be administered to a pneumonia patient within six hours of arriving at the hospital. The trouble is that doctors often cannot diagnose pneumonia that quickly. You have to talk to and examine a patient and wait for blood tests, chest X-rays, and so on.

13 Under P4P, there is pressure to treat even when the diagnosis isn't firm, as was the case with my patient with heart failure. So more and more antibiotics are being used in emergency rooms today, despite all-too-evident dangers like antibiotic-resistant bacteria and antibiotic-associated infections.

14 I recently spoke with Dr. Charles Stimler, a senior health care quality consultant, about this problem. "We're in a difficult situation," he said. "We're introducing these things without thinking, without looking at the consequences. Doctors who wrote care guidelines never expected them to become performance measures."

15 And the guidelines could have a chilling effect. "What about hospitals that stray from the guidelines in an effort to do even better?" Dr. Stimler asked. "Should they be punished for trying to innovate? Will they have to take a hit financially until performance measures catch up with current research?"

16 The incentives for physicians raise problems too. Doctors are now being encouraged to voluntarily report to Medicare on 16 quality indicators, including prescribing aspirin and **beta blocker** drugs to patients who have suffered heart attacks and strict cholesterol and blood pressure control for diabetics. Those who perform well receive cash bonuses.

17 But what to do about complex patients with multiple medical problems? Forty-eight percent of Medicare beneficiaries over 65 have at least three chronic conditions. Twenty-one percent have five or more. P4P quality measures are focused on acute illness. It isn't at all clear that they should be applied to elderly patients with multiple disorders who may have trouble keeping track of their medications.

cardiologists doctors who treat people with heart problems

repercussions consequences or resulting effects

beta blocker a drug that slows heart rate, lowers blood pressure, and protects from future heart attacks

18 With P4P doling out bonuses, many doctors have expressed concern that they will feel pressured to prescribe "mandated" drugs, even to elderly patients who may not benefit, and to cherry-pick patients who can comply with pay-for-performance measures.

19 And which doctor should be held responsible for meeting the quality guidelines? On average, Medicare patients see two primary-care physicians in any given year, and five specialists working in four practices. Care is widely **dispersed**, so it is difficult to assign responsibility to one doctor. If a doctor assumes responsibility for only a minority of her patients, then there is little financial incentive to participate in P4P. If she assumes too much responsibility, she may be unfairly blamed for any **lapses** in quality.

20 Nor is it clear that pay for performance will actually result in better care, because it may end up benefiting mainly those physicians who already meet the guidelines. If they can collect bonuses by maintaining the **status quo**, what is the incentive to improve?

21 Doctors have seldom been rewarded for excellence, at least not in any tangible way. In medical school, there were tests, board exams, and **lab practicals**, but once you go into clinical practice, these traditional measures fall away. At first glance, pay for performance would seem to remedy this problem. But first its deep flaws must be addressed before patient care is compromised in unexpected ways.

dispersed spread or distributed
lapses slight errors or slipups
status quo a Latin phrase meaning "the way things are"
lab practicals occasions for medical students to work in hospitals and medical test centers

A CLOSER LOOK

1. Jauhar introduces the subject of the essay in paragraphs 5 and 6 but does not offer his thesis until much later in the essay. What is Jauhar's thesis?

   Possible answer: "At first glance, pay for performance would seem to remedy this problem. But first its deep flaws must be addressed before patient care is compromised in unexpected ways."

2. Jauhar compares P4P to the surgical report cards program of the early 1990s. What problem did the surgical report cards produce that Jauhar fears P4P could produce?

   Possible answer: Doctors "cherry-picked" only the relatively healthy patients because the doctors knew those patients would more than likely respond well to treatment.

3. In paragraphs 17 and 20, Jauhar asks rhetorical questions, which are questions asked merely for effect (the author does not expect us to answer him). What is the effect of the question in each paragraph?

   Possible answer: The question at the beginning of paragraph 17 introduces a problem the P4P program does not address. The question at the end of paragraph 20 shows the reader that doctors would have no incentive to improve under the P4P program.

4. How does the title of the essay connect to the anecdote (in paragraphs 1–4) concerning a patient with congestive heart failure?

   Possible answer: The anecdote illustrates one of the pitfalls of the P4P program: patients receiving certain drugs before doctors have enough evidence the drugs are needed.

MAKING CRITICAL CONNECTIONS

1. Jauhar is a medical doctor, so he is familiar with the quality incentive programs in the medical profession. Instead of quietly benefitting from the P4P program's financial rewards, Jauhar writes of the program's flaws. How is writing about the program's flaws beneficial to Jauhar? Discuss this with those in your writing group.
2. This is a persuasive essay in which Jauhar suggests the P4P program not be supported until its flaws have been addressed. Although you are probably not in the medical profession, you can discuss the program with your physicians and even government representatives. Write a letter to your local or regional hospital's director, sharing your position on the P4P program.
3. Similar to Jauhar, write a persuasive essay in which you make your readers aware of a problem with an existing program or common practice at your college or workplace. Following Jauhar's example, illustrate the problem with detailed examples, and discuss the causes of the problem and the specifics that need to be addressed.

Designer Dogs: Fabulous Fact or Fiasco?

by Therese Backowski

When Therese Backowski was a freshman in Professor Beth Franz's composition class at North Central State College in Mansfield, Ohio, she wrote the following essay. She then submitted it to Pearson's Writing Rewards Student Essay Contest. In 2011, this essay won first place.

On the inside back cover of this text, you can find details about submitting your essay in this year's contest.

1 In 1988, Wally Conran, manager of the Royal Institute of the Blind in Australia, received a letter from a lady whose husband had severe allergies. The man needed a guide dog, one that didn't shed hair and **dander**. Conran had what he conceived to be a brilliant idea. He crossbred two purebred dogs: a friendly Labrador Retriever and a smart, non-shedding Standard Poodle. He **dubbed** his creation the Labradoodle. Conran intended the resulting puppies to be **hypoallergenic**, **tractable**, and **devoid** of any of the genetic defects that the parent breeds were **prone** to harbor. The puppies were intended to be **amenable** and of superior intelligence. Thus, begun by one man, was born the designer-dog phenomenon. The popularity of these mutts is astounding. Jennifer Aniston is just one of the high-profile movie stars who just "loves" her designer dog. Tiger Woods owns two. Even President Obama considered getting one when he was about to fulfill his promise of a puppy to his daughters. While some people like the idea of designer dogs, I and others see the dishonesty and inhumane treatment involved in this trend, and I think it should stop.

2 Some people like the idea of designer dogs. Like many of the misguided, Ginger Danto, author of "The Year of the Designer Dog," thinks designer dogs are wonderful. Her bias is obviously something that began by talking to the wrong people. When referring to a purebred dog, she says, "the people's choice, according to pet stores, puppy breeders and happy owners everywhere, is more along the lines of the Labradoodle." I think her choice of survey sources is questionable. Puppy breeders love the dogs for one huge reason: they sell them. Pet stores love the dogs for the same reason: profit. People love them for the same reason we all love our pets, at least the ones we keep. The dogs give us unconditional love, no matter how awfully we treat them or how strange our hair looks. Danto's **giddy** fascination with the **hybrid** is obvious by her word choices. She claims the dogs have a soft, hypoallergenic coat that comes in a variety of types and colors. She describes the hair colors in words that should be used to describe foods or home décor, like *apricot cream*, *gold*, *chalk*, and *café*. These are words used

dander small scales from skin
dubbed named
hypoallergenic having little likelihood of causing an allergic reaction
tractable easily led, docile
devoid without
prone likely (to)
amenable agreeable
giddy frivolous, silly
hybrid mixture, cross

to make people grab their credit cards to buy objects, not ones used to describe dog hair colors. Danto advises us to forget about purchasing trendy clothing or designer purses. All we need to be noticed and envied as we walk down the street is a designer dog. That mentality is what has been the strongest force in the popularity of these dogs.

3 These dogs don't live up to the claims made about them. Sellers claim these furry creatures are much better buys than mutts or purebreds. Designer dogs are obscenely expensive. Generally, they cost upward of five hundred dollars—quite a lot of money for a mutt. The hybrids do not display predictable behavior characteristics, since the genetic mix can go in any direction. If they don't have an active **congenital** abnormality, they are often carriers of recessive genes that cause those disorders. The cost to produce them is **negligible** if the breeder chooses to cut corners concerning the care of the parent dogs. Designer dogs don't have to be registered, although there are several phone registries that will take a dog's parental history and record it for a fee. Designer dogs are touted as non-shedding, easily trained, fashionable, and free of genetic problems. However, in reality, they are not an improvement over Lassie.

4 Worst of all, this designer dog trend leads to massive abuse and inhumane treatment of these dogs. In an interview in the *Australian*, Conran said, "I wish I could turn the clock back . . . I **rue** the day that I made that decision." Just as Frankenstein turned on his creator and made him sorry for his efforts, so has the Labradoodle. The Labradoodle and his cousins, the Goldendoodle, Maltipo, Mastidane, and myriad other ridiculously named crossbred dogs have suffered **incalculable** pain and **exploitation** because of the nature of man to make a buck at any cost. During the process of grinding out puppies for mass consumption, heartrending case after case of neglect has been inflicted on the parents of these dogs. *Newsweek* author Suzanne Smalley point out that designer dogs are nothing more than a cash crop for **unscrupulous** puppy mill breeders. In many cases, the dogs aren't even treated with the same consideration as animals bred for slaughter.

5 In conclusion, the designer dog phenomenon is just another way to exploit animals. Wally Conran wishes he had never conceived the plan to cross that first Standard Poodle and Labrador Retriever. Conran had no clue that it would lead to the exploitation of not only the Labradoodle, but of all the other ridiculously named crosses that originated from his idea. He is convinced that his experiment resulted in the breeding of thousands of unwanted dogs that are ending up rejected and, if lucky, in shelters. I agree with him that designer dogs just aren't what they

congenital inherited
negligible insignificant, small
rue regret
incalculable countless
exploitation selfish use, mistreatment
unscrupulous dishonest, corrupt

are represented to be, cute name or not. A dog is a dog, and it comes with needs. Expensive or not, a dog is not a decoration, a watch, or a purse. It is a living, breathing creature that requires care and consideration. While some people like the idea of designer dogs, I and others see the dishonesty and inhumane treatment involved in this trend. I think it should stop now.

A CLOSER LOOK

1. Backowski's title uses two examples of alliteration (the repetition of beginning consonant sounds in neighboring words). Why might Backowski have chosen to use alliteration in her title?

   Answers will vary.

2. What is Backowski's thesis statement?

   "While some people like the idea of designer dogs, I and others see the dishonesty and inhumane treatment involved in this trend, and I think it should stop."

3. How could Backowski have reworded the thesis statement to be in third person rather than first person?

   Answers will vary. Possible answer: While some people like the idea of designer dogs, others see that the trend should be stopped because of the dishonesty and inhumane treatment involved.

4. Identify the topic sentences in body paragraphs 2, 3, and 4.

   The topic sentence for paragraph 2 is "Some people like the idea of designer dogs." The topic sentence for paragraph 3 is "These dogs don't live up to the claims made about them." The topic sentence for paragraph 4 is "Worst of all, this designer dog trend leads to massive abuse and inhumane treatment of these dogs."

5. Backowski presents her arguments against designer dogs in emphatic order. What transitional phrase does she use to show the strongest of her arguments?

   The author begins her last argument with the transitional phrase "Worst of all."

MAKING CRITICAL CONNECTIONS

1. Backowski writes, "All we need to be noticed and envied as we walk down the street is a designer dog." In your writing group, discuss whether someone walking down the street with a designer dog would be noticed or envied. Then write an essay in which you reject or defend the quotation.
2. Throughout paragraph 2, Backowski uses words that show partiality (e.g., *misguided, bias, wrong, questionable*). In your writing group, discuss whether these words help or hinder the author's argument. Next, devise alternative wording for the sentences that contain these words. Then compare your revised sentences to the originals and determine which have a greater impact.
3. Take the opposing viewpoint and write an essay that defends designer dogs. Conduct your own research, or use information from these Web sites:

   In Defense of "Designer Dogs"
   "Designer Mutts" Not Likely To Cause New Health Concerns
   Are Designer Dog Trends Bad For Dogs?

Credits

Text

Page 95: Interview used by permission of Penny Pennington; **Page 120:** Interview used by permission of Joey Goebel; **Page 150:** Interview used by permission of Erik Peterson; **Page 177:** Interview used by permission of Marvin Bartlett; **Page 184:** Zosima A. Pickens, "Choosing a College Major." Pearson Writing Rewards Student Essay Contest, 2010. Copyright © 2010 by Pearson Education, Inc.; **Page 205:** Interview used by permission of Michael Minerva, Jr.; **Page 214:** Judy Brady, "I Want a Wife," *Ms. Magazine,* December 31, 1971. Copyright 1970 by Judy Brady. Reprinted by permission of the author; **Page 235:** Interview used by permission of Earl Brandon and Chip Stauffer; **Page 267:** Interview used by permission of Renee L. LaPlume; **Page 296:** Interview used by permission of Judy Carrico; **Page 329:** Interview used by permission of Breck Norment. **Page 403:** © EnchantedLearning.com. Reprinted with permission; **Page 699:** "Ac-cent-tchu-ate the Positive," Lyrics by Johnny Mercer. Music by Harold Arlen. © 1944 (Renewed) Harwin Music Co. All Rights Reserved. Reprinted by permission of Hal Leonard Corporation.

Photo

Page 1: ©Photosani/Fotolia; **Page 5:** ra2 studio/Shutterstock.com; **Page 8:** marekuliasz/Shutterstock.com; **Page 16:** Tom Grundy/Shutterstock.com; **Page 19:** Library of Congress; **Page 22:** ©Alexander Raths/Fotolia; **Page 24:** Svetlana Larina/Shutterstock.com; **Page 29:** ©auremar/Fotolia; **Page 33:** Bridge and Tunnel Club; **Page 41:** ©Judith Collins/Alamy; **Page 49:** almagami/Shutterstock.com; **Page 53:** ©Corbis Flirt/Alamy; **Page 57:** Dec Hogan/Shutterstock.com; **Page 57:** Hannamariah/Shutterstock.com; **Page 64:** Pixsooz/Shutterstock.com; **Page 68:** ekler/Shutterstock.com; **Page 76:** ©olly/Fotolia; **Page 81:** wavebreakmedia ltd/Shutterstock.com; **Page 87:** Dmitriy Shironosov/Shutterstock.com; **Page 91:** ©V. Yakobchuk/Fotolia; **Page 93:** ©Erik Isakson/agefotostock; **Page 93:** John Prior/Cutcaster Images; **Page 93:** AP Photo/Capitol Records Nashville; **Page 93:** AP Photo/Island Def Jam Music Group; **Page 95:** Courtesy of Penny Pennington; **Page 95:** samotrebizan/Shutterstock.com; **Page 97:** ©Vicki Beaver/Alamy; **Page 104:** REUTERS/Leon Neal; **Page 108:** ©20th Century Fox Film Corp. All rights reserved, Courtesy: Everett Collection; **Page 118:** ©mediablitzimages (uk) Limited/Alamy; **Page 118:** ©Image Source/Alamy; **Page 118:** ©image100/Alamy; **Page 118:** ©Scott Griessel/Fotolia; **Page 120:** Joey Goebel; **Page 123:** ©Corbis Bridge/Alamy; **Page 147:** ©diego cervo/Fotolia; **Page 147:** kwest/Shutterstock.com; **Page 147:** Jami Garrison/Shutterstock.com; **Page 147:** ©Ted Horowitz/Alamy; **Page 150:** Courtesy of Erik Peterson; **Page 156:** iofoto/Shutterstock.com; **Page 162:** ©imagebroker.net/SuperStock; **Page 174:** manzrussali/Shutterstock.com; **Page 174:** ©Wendy White/Alamy; **Page 174:** AP Photo/LM Otero; **Page 174:** AP Photo/ Matt Houston; **Page 177:** Courtesy of Marvin Bartlett; **Page 180:** ©Paul Debois/age fotostock; **Page 183:** ©Robert Kneschke/Fotolia; **Page 189:** ©Mark Burnett/Alamy; **Page 191:** ©Sam Dao/Alamy; **Page 191:** Andre Jenny Stock Connection Worldwide/Newscom; **Page 203:** ©Dmytro Konstantynov/Fotolia; **Page 203:** ©UpperCut Images/Alamy; **Page 203:** ©Megan Q Daniels/age fotostock; **Page 203:** Morgan DDL/Shutterstock.com; **Page 205:** Courtesy of Michael Minerva; **Page 213:** ©CORBIS/age fotostock; **Page 218:** Wendy Kaveney Photography/Shutterstock.com; **Page 222:** Monkey Business Images/Shutterstock.com; **Page 230:** ©culture-images GmbH/Alamy; **Page 230:** Daniel Ochoa |FOTOCHOA|/Shutterstock.com; **Page 230:** ©Michael Chamberlin/Fotolia; **Page 230:** AP Photo/Adam Butler; **Page 231:** Fatseyeva/Shutterstock.com; **Page 233:** Yuri Arcurs/Shutterstock.com; **Page 235:** Courtesy of Chip Stauffer; **Page 235:** Courtesy of Earl Brandon; **Page 235:** Photo Researchers/Getty Images; **Page 236:** ©Lisa F. Young/Fotolia; **Page 248:** Edyta Pawlowska/Shutterstock.com; **Page 249:** withGod/Shutterstock.com; **Page 264:** irisdesign/Shutterstock.com; **Page 264:** Iurii Osadchi/Shutterstock.com; **Page 264:** ©Mark Richardson/Alamy; **Page 264:** Stacie Stauff Smith Photography/Shutterstock.com; **Page 267:** Courtesy of Renee LaPlume; **Page 269:** Vitalliy/Shutterstock.com; **Page 270:** Supri Suharjoto/Shutterstock.com; **Page 276:** ©TP/Alamy; **Page 292:**

AP Photo/Tony Dejak; **Page 292:** AP Photo/Erie Times-News, Christopher Millette; **Page 292:** ©i love images/Alamy; **Page 292:** ©Ariel Skelley/age fotostock; **Page 296:** Courtesy of Judy Carrico; **Page 296:** ©Minerva Studio/Fotolia; **Page 296:** ©Dmitry Lobanov/Fotolia; **Page 299:** ARENA Creative/Shutterstock.com; **Page 306:** ©Eric Carr/Alamy; **Page 312:** ©Marmaduke St. John/Alamy; **Page 326:** ©Biletskiy Evgeniy/Fotolia; **Page 326:** J.D.S./Shutterstock.com; **Page 326:** Juice Images/age fotostock; **Page 326:** ©lightpoet/Fotolia; **Page 327:** Rob Wilson/Shutterstock.com; **Page 329:** Courtesy of Breck Norment; **Page 331:** Alistair Scott/Shutterstock.com; **Page 336:** Martin Meyer/age fotostock; **Page 346:** Monkey Business Images/Shutterstock.com; **Page 359:** ©goodluz/Fotolia; **Page 361:** Fox/AdMedia/Newscom; **Page 361:** Charles Harris/AdMedia/Newscom; **Page 361:** martan/Shutterstock.com; **Page 361:** AP Photo/Reed Saxon; **Page 365:** Visions of America/SuperStock; **Page 378:** Monkey Business/Fotolia; **Page 378:** United States Army; **Page 378:** stavklem/Shutterstock.com; **Page 381:** Library of Congress; **Page 385:** Susan Law Cain/Shutterstock.com; **Page 385:** almagami/Fotolia; **Page 385:** Kzenon/Shutterstock.com; **Page 385:** ©vege/Fotolia; **Page 385:** ©librakv/Fotolia; **Page 395:** ©PhotoEdit/Alamy; **Page 395:** ©NetPhotos/Alamy; **Page 395:** lznogood/Shutterstock.com; **Page 395:** Denis Vrublevski/Shutterstock.com; **Page 402:** PhotoEdit/Alamy; **Page 402:** Robert Neumann/Shutterstock.com; **Page 402:** Stephen Oliver/Dorling Kindersley; **Page 402:** Jeff Greenberg/Alamy; **Page 415:** Gloria Brown; **Page 421:** Urbano Delvalle/Getty Images; **Page 425:** Pixsooz/Shutterstock.com; **Page 428:** JinYoung Lee/Shutterstock.com; **Page 429:** ©razihusin/Fotolia; **Page 434:** ©chasingmoments/Fotolia; **Page 443:** ©Deklofenak/Fotolia; **Page 445:** AP Photo/Chris O'Meara; **Page 448:** ©Lucidio Studios/ age fotostock; **Page 451:** Jason Stitt/Shutterstock.com; **Page 453:** ©elnariz/Fotolia; **Page 460:** AlexKalashnikov/Shutterstock.com; **Page 466:** Pavel V Mukhin/Shutterstock.com; **Page 473:** ©Andre/Fotolia; **Page 475:** Josh Resnick/Shutterstock.com; **Page 488:** Tomas Abad/agefotostock; **Page 497:** ©Sharpshot/Fotolia; **Page 510:** Kamira/Shutterstock.com; **Page 514:** Paul Matthew Photography/Shutterstock.com; **Page 525:** ©JUNIORS BILDARCHIV/age fotostock; **Page 533:** ©WavebreakmediaMicro/Fotolia; **Page 548:** ©TheSupe87/Fotolia; **Page 553:** ©Anton Ishchenko/Fotolia; **Page 560:** Vibrant Image Studio/Shutterstock.com; **Page 565:** Lisa F. Young/Shutterstock.com; **Page 570:** dean bertoncelj/Shutterstock.com; **Page 570:** maggee/Shutterstock.com; **Page 578:** ©lznogood/Fotolia; **Page 578:** ©WavebreakMediaMicro/Fotolia; **Page 585:** ©JJAVA/Fotolia; **Page 589:** ©Evgenia Smirnova/Fotolia; **Page 597:** ©Everett Collection Inc/Alamy; **Page 603:** ©AlShadsky/Fotolia; **Page 611:** AP Photo /National Park Service, Gary Berdeaux; **Page 622:** ©charlesknox/Fotolia; **Page 629:** AP Photo/File; **Page 643:** Diego Cervo/Shutterstock.com; **Page 648:** bikeriderlondon/Shutterstock.com; **Page 658:** ©plus69free/Fotolia; **Page 665:** Stocktrek Images, Inc./Alamy; **Page 670:** ©vejasdrugelis/Fotolia; **Page 677:** ©JcJg Photography/Fotolia; **Page 683:** ©silver-john/Fotolia; **Page 693:** ©Corbis/Glow Images; **Page 695:** ©Anatoly Vartanov/Fotolia; **Page 697:** Malcolm McGregor/Dorling Kindersely; **Page 701:** ©North Wind Picture Archives/Alamy; **Page 702:** Subbotina Anna/Fotolia; **Page 702:** Aaron Amat/Fotolia; **Page 702:** Vladimir Voronin/Fotolia; **Page 750:** Orange Line Media/Shutterstock.com; **Page 753:** Cheryl Casey/Shutterstock.com; **Page 755:** artjazz/Shutterstock.com; **Page 760:** Orange Line Media/Shutterstock.com; **Page 763:** ©VIPDesign/Fotolia; **Page 775:** Arcady/Shutterstock.com

Index

H

I

O

P

T

U